A Final Story

A Final Story

Science, Myth, and Beginnings

NASSER ZAKARIYA

The University of Chicago Press

CHICAGO AND LONDON

The University of Chicago Press, Chicago 60637
The University of Chicago Press, Ltd., London
© 2017 by The University of Chicago
Published 2017
Printed in the United States of America

26 25 24 23 22 21 20 19 18 17   1 2 3 4 5

ISBN-13: 978-0-226-47612-4 (cloth)
ISBN-13: 978-0-226-50073-7 (e-book)
DOI: 10.7208/chicago/9780226500737.001.0001

Published with the support of the Susan E. Abrams Fund.

Library of Congress Cataloging-in-Publication Data

Names: Zakariya, Nasser, author.
Title: A final story : science, myth, and beginnings / Nasser Zakariya.
Description: Chicago ; London : The University of Chicago Press, 2017. |
Includes bibliographical references and index.
Identifiers: LCCN 2017015353 | ISBN 9780226476124 (cloth : alk. paper) |
ISBN 9780226500737 (e-book)
Subjects: LCSH: Science—Historiography. | Science—Philosophy.
Classification: LCC Q125 .Z353 2017 | DDC 507.2/2—dc23
LC record available at https://lccn.loc.gov/2017015353

♾ This paper meets the requirements of ANSI/NISO Z39.48-1992
(Permanence of Paper).

To Rachel Zakariya, who took a different view

Contents

Introduction

Relating a natural history of the world is a canonical and iconic scientific practice. In the present, its scientific and scientifically invested chroniclers promote a narrative in a mythic register in which they adopt a self-avowedly epic form. The epic relates a history of the universe from its earliest conceivable moments up until and often past the emergence of humanity, performing what the narrators perceive as humanity's own evolving task of determining the breadth and truth of that story. This historical narrative bears different names indicating different emphases, including "the scientific epic," "the evolutionary epic," "cosmic evolution," "the new cosmic myth," "the new Genesis," or simply "the history of the universe."[1] The historical analysis of the variants of this story, defined through the coevolution of matter and life, including the acts of its authorship and its anticipations of the future of the living and the physical world, is the central subject of this work.

The terms of this historical narrative as an object of analysis declare the tensions running through its many exemplars. The locution "history of the universe" implies the related and still more pedigreed "universal history." And the invocation of "natural history" suggests that any historical analysis of a historicizing narrative cannot be confined to histories alone. "Natural history" still retains museological connotations, suggesting how much its own history as a category was also invested in apparently nonnarrative representations of the world and in the institutions and practices that formed the context of such representations.

Given the unqualified embrace of the totality that the epic of a universal history details, the analysis of the history of such narratives therefore requires taking account of other modes in which such a totality might be embraced, other natural historical attempts to synthesize the results of what were once often referred to as the "special sciences." From the varied contributions that the different scientific disciplines might make to capturing that totality, these attempts stage the construction of a whole of knowledge. As a result, the historical analysis here must treat different, overlapping modes of synthesis. These modes are themselves suggestive of how scientists and other historical actors concede disunity (sometimes explicitly) while attempting to work toward unification. Their efforts are a reminder that the historiographic analysis of the diversity of the sciences does not come to an end with the lesson that there may be no one scientific method that the natural sciences practice in common. Such analysis also requires attending to how historical actors attempt to synthesize disparate disciplines, to answer fears of disunity or promote hopes of a future unity.

GENRES OF SYNTHESIS

A central argument pursued here is that despite apparently divergent paths, human/political universal history and the natural history of the universe share recurrent narrative structures and topics, as well as recurrent ambitions and central assumptions, even while their idioms and modes of proof or compulsion may differ vastly. I will call the orienting frameworks structuring these totalizing ambitions "genres of synthesis." These are synthesizing modes calling upon a constellation of representational devices including narrative forms, structures of formal and more informal argument, and visual or visually evocative schemes of survey and coordination. Once the basic vocabulary of modern scientific concern with a totalized account of the universe was established in the nineteenth century, these genres have been employed as different strategies for advancing and constituting an organized and unifying synthesis of scientific and broader knowledge. The strategies at play in the construction of synthetic frames have posited ways in which the apparently balkanizing trends in the sciences could still be regarded as enfolded into one whole project, even as the articulation of that project reinflected the understanding of the individual sciences it molded together.

Four genres of synthesis have largely shaped attempts at achieving a final, natural scientific synthetic frame: the scalar, the historical, the foundational, and the fabulaic. Scalar syntheses construct accounts via a diagramming of space and/or time, often involving the expansion and contraction of spatial scales, or scales of both space and time. Historical syntheses deploy some version of a timeline or thicker historical narrative. Foundational syntheses attempt to reduce (via presumed or implied mathematical and logical consequence) their accounts to the existence of atemporal universal laws. And fabulaic syntheses often rely on quasi-fictional narratives that employ storytelling devices such as a journey taken through different domains. These representations are discernable to an individual understanding in a crisp enough fashion that they can supply coordinates for specific individual research within the disciplines and in relation to each other. More broadly, they position a conception of the human within the genre, in the role that this emplacement may dictate. In professional, generalist, or popular representations, they thus establish understandings of the categories of the scientist and of science.

They consequently orient all knowledge-producing scientific work as such, which, in the abstract, always undertakes the same task: to sharpen the contours of the very representation upon which the given genre itself is founded. For the cases explored here, the tasks they set include attempting to cover all the scales, from the microcosmic to the macrocosmic, from instant to eon; to tell the full universal history; to drill down to the enduring foundations of things or to construct a scaffold on which they can be said to rest; or to come to the end of the road or quest, passing every station in between. Generically, the scalar is thus often linked to a spectrum, map, or outline; the historical is linked to a timeline, a chronological spine revealed in the growth of scientific knowledge in epic terms; the foundationalist is linked to postulations or claims to the discovery of first principles and the potential ladder of deductive argument extending from them; and the fabulaic is related to and told along the path of a science adventure story.

In addition, more recently, interdisciplinary and mathematical/statistical analyses promote still other, related candidate syntheses, themselves promoted as sciences of knowledge visualization, potentially establishing new orientations for knowledge-making and new modes for addressing a potential totality. These can appeal to cartographic and organic imaginaries (maps, landscapes, trees of knowledge) in generally descriptive modes attempting to capture the active and evolving links between scholars. But these totalizing visualizations can be prescriptive, as well, in attempting to predict trends and promote research innovations. A proliferation of potential visualizations emerges, for example, on the basis of algorithms appealing to bibliographic data, structured through mathematically developed metaphors of foraging or searching. These may result in new synthesizing genres to a degree independent in character from those that precede them and, perhaps, in the celebration of their plasticity and adaptability to use, offering new orienting and meaning-making frames.[2]

In this regard, the synthesis of a body of knowledge tends itself to be rendered as the multidimensional object to be discovered (what scientists are up to, captured by varying measures emphasizing different units of analysis) according to what are regarded as the best principles to determine active research pathways and potential sites of meaning. What such contemporary representations or form-producing algorithms point to in their apparent flexibility is not only the extent to which the entirety of active scientific practices cannot easily be reconciled to any synthetic form, but also the extent to which any such forms must to varying degrees be dynamic and fluid if they are to sustain a connection to the ongoing research, purposes, and desires of scientific practitioners. There is a tension in synthesizing genres, then, between motioning toward the finalization of their forms, as toward a final theory or story, on the one hand, and the project of an open-ended inquiry into ultimate truths, on the other. The notion of finalizability can be anathema to more freely evolving and proliferating representations, the referents for which might prove to be more plastic and therefore less conforming than prior genres of synthesis already being employed. When plasticity is celebrated over and above the retention of a specific form, as may be the case with knowledge visualization, making meaning can tend to center on the promise of innovation as such. Addressing a potential totality of knowledge, this science in particular concentrates more on the organization of a given network through the pragmatics of connecting to information and knowledge understood as useful to some potential searcher. This pragmatism can in turn stipulate that the value of information or knowledge be continually redetermined by the actual or potential use of an actual or potential user.

Even in such a case, treating abstract principles of use and innovation as fundamental can constrain consequent representations of knowledge as a function of the conventional choices available for operationalizing these principles (as, for example, in measures of use/usefulness based on the circulation or impact of papers and journals). And these visualizing algorithms can extend to representations of the using and innovating human subject as information hunter-gatherer set to the task of rendering that hunting-gathering ever more comprehensive and efficient, pointing to a notion of finalization at work in and beyond the operationalization of these attributes. At the same time, structuring metaphors, such as trees and landscapes, may constrain representational algorithms in such a way as to produce visualizations no more plastic than all the varied modulations of histories or maps of the world, which such visualizations also may to a degree rely on or reproduce.

A vision of a scientific enterprise/practice as a whole is at play in how, more generally, the genres of synthesis link elements they posit as pivotal to different disciplines. The question of organization therefore is central: the genres conceptualize and have increasingly informed organizational principles and policies for science. The generic representation of the scientific product thus has entailed the representation of scientific organization. But then what constitutes scientific synthesis is itself at stake through that organization, articulating the relationship between disciplines. In the nineteenth century, a central issue of scientific historical debate was the primacy of physical versus biological reasoning, and contested pivotal elements were historical events—a conflict between arguments of physical cosmology and human evolution. In decades thereafter, as we will see, synthesizers posited several different ways of unifying these once historically conflicting sciences to the satisfaction of scientific expert opinion. And as we will also see, these genres influenced resolutions of how research projects should be designed, advocated for, defended, and funded. In providing frameworks for the whole of knowledge, these syntheses established putative objectives for that knowledge. In narrow terms, the syntheses could articulate specific targets: some specific scientific research that should determine this or that chapter in the history of the world, or unexplored region of spacetime or energy frontier.

In short, then, genres of synthesis structure an approach to the production of knowledge. But they also provide a kind of grammar of ignorance, delimiting what it means not to know, shaping what progress or its lack amounts to and further indicating what might be in principle unknowable: those parts of the representation that can never be resolved or will forever be open to further resolution.

None of these frameworks has existed in purity, however. Scalar mappings of the universe assume the operation of fundamental laws at one or a multitude of scales, with central sites and often central historical events providing salient markers, while the narratological principles of historical accounts are expressed through the unfolding of the consequences of such laws. Foundational accounts privilege certain ultimate phenomena and laws—ultimate domains of experience revealed historically in space and time—even if this privileging is understood as the universe's doing. When those experiential foundations are indexed to different scales, so that changes in scale are changes in fundamentals, different scales become different worlds. In turn, fables recounting travels between worlds generally require not only calculations to chart that travel, but the election of fundamental parameters deciding how to reckon that a domain is a world. Given their interweaving, these genres of synthesis can be understood as principles active within myriad instances of universalizing accounts, rather than as pure syntheses in themselves. Nevertheless, one or another is often the ruling figure, with the logics of other genres enabling and coordinated within that figure. A synthesis might for example be housed and structured within a history, the individual chapters of which relate an ongoing emergence at different scales, or the plot and causal structure of which may conform or appeal to fundamental laws.

The various topics characteristic of these genres of synthesis also were often implicit in each genre and implicitly run together by both historical actors and analysts. The earliest, the greatest, the most distant, the most exotic, the most profound—these were contrasted to the present, the mundane, the near, the familiar, the superficial. But these topical correlations also could be broken. In some of the narratives that resulted, humanity was represented as being of cosmic consequence in its ability to destroy its own and potentially all life; in others, such self-destruction would instead represent nothing more than the disappearance of creatures of no cosmic significance whatever. The correlations could break down still further when the representation threatened to dismiss the importance of an entire discipline, as foundational accounts might relegate to secondary status the importance of anything but the study of the most energetic physics.[3]

Importantly, in all these schemata for achieving an account of or revealing an underlying unity in all phenomena, the genres are essentially aspirational. Such syntheses may promise a unity to come, but it is a unity that may or may not be inherent in knowledge itself or in those who author it. It is a unity that may turn out never to exist. In deference to the tension between the potential unity and increasing fragmentation of scientific knowledge and of science as a practice, these synthetic efforts have functioned as a reminder of the acknowledged absence of unity within the sciences.

To treat accounts of universal history in terms of genres of synthesis thus is in effect simply to name the commitments a number of existing works have in common, works that are perceived to be similar in some respects. Scientific epics or natural historical texts (or, more recently, films and other mixed media) establish a canonical set—a scientific canon of texts or materials—the abstracted features of which amount to a genre. In this sense, many of these synthesizing works were to become or are becoming "classics," simultaneously outdated and celebrated by research carrying out the very goals they had enshrined.

Alternatively, these evolving architectonic frames involve a number of related, consonant investigations or dimensions of analysis that are re-encodings of each other. By representing the effort of scientists and other cultural actors trying to enact, express, and/or represent a meaningful totality, genres of synthesis invoke varied modes for attempting to address totality as such—a totality of nature as given or nature as it becomes known—suggesting how such modes stand in relation to each other. These modes therefore constitute a set of synonyms for the same effort, each taking as central or as most salient a different aspect of scientific materials.

The first of these saliencies or synonyms is a history of a narrative *as* synthesis, the development of a story that is itself a contemporary history of the world. As we will see in the first chapters of what follows, in the modes of natural history posited in the nineteenth century by John Herschel, William Whewell, or Mary Somerville, universal history could not provide a synthetic frame. But across the nineteenth century, attempts to claim a unity for the sciences and to provide a synthetic account of the nature of the universe became expressly narratological, set alongside and struck through with other synthetic principles and possibilities. These attempts helped frame the terminological and narrative elements of the genres of synthesis that were to come.

Intimately related to the historical trajectory of ideas and images involved in the developing genres of syntheses since then are the contestations that also reveal to what extent the synthesizing project was itself a historiographic and more broadly cultural enterprise, an extended debate over whether to address scientific universal history and how such a history should be drafted—whether to construct a synthesis as a whole of knowledge and/or simply to invest faith in a unity that is to be revealed or that transcends human understanding.

The question of what it means for scientists to carry out and write a history of a totality becomes itself a historiographic concern. Natural history, in its later historical variations, required deciding or discovering (depending on philosophical viewpoint) what might be pivotal universal-historical events and pivotal structures needing elaboration and explanation, what would cause that history to unfold and even allow the universe to be seen as given to a history subject to unfolding. Alternatively, other modes of synthesis, such as the scalar, might establish a spatiotemporal frame and a chain of being housed in that frame. Whatever the genre, the possible and/or implicit limits of such representations have tended to be overshadowed by the project's totalizing ambitions—ambitions that at the same time must treat as so many unregarded ellipses the majority of events or scales, objects, implications of fundamental laws, or minor stations on the road to scientific insight into the nature of things.

For the question of universal history, the first and last elements might instead be the most important to understand, the impossible alpha and omega implied by any presentation of a narrative based on the unity of science. Or alternatively, a missing chain in a sequence of what are regarded as formative events might be crucial to the historical account. Here the ratio between the known and unknown, the representation and its completion, is an epic ratio in which a few events structure an immensity of time.[4] The many architects and draftspeople of varied synthesizing and synthetic genres have attempted to summarize all natural history and to determine specific elements they affirm as pivotal, as with the place of the emergence of life in universal history. Universal historical authors have worked in concert and parallel to stabilize that history by schematizing it through narrative elements it refines and reiterates: its own formative moments (the sequence of its pivotal events) and its shaping principles (how its story unfolds and resolves).

EPIC, SCIENCE, AND MYTH

As explanatory models, the relations between such attempts at universal synthesis and myths thus become evident. Functionally, the stories that science has told accounting for beginnings and endings, for the nature of everywhere and everywhen, can be seen as scientific myths. New accounts replacing old ones were being put forward, not stories in a vacuum. Earlier generations had embraced myths as a way to root explanations of what is in the heavens, how old the world is, where humanity—or some part of it—came from, what produced human difference. But part of what became only more distinctive in the period from the mid-nineteenth century and into the earlier part of the twentieth in relation to the natural historical or synthetic work was that there was no simple, certain, or secure cosmological myth, scientific or otherwise, in which to root explication—not in the US and European contexts, at the very least. The multiplicity exhibited by the various genres of synthesis and announced with increasing self-consciousness was often interpreted as deprivation, the absence of a collective myth, a unifying truth, rather than as plenitude, the celebration of many diverse accounts, the acceptance of the multiplicity of truths. New stories were being argued for, while old ones—often understood as well-defined religious accounts—failed to carry the conviction that might have made the composition of new accounts seem foolish.

The frameworks articulated in the different genres of synthesis mattered on this level because they raised the question of whether it is the business of science, or of the tribe of scientists, to supply society with myths, and as we will see, the supporters of a historical synthesis answered this question with increasing affirmation. Their stories articulated positions as to what falls within the province of science, what could belong only to faith, and what could be rendered into a naturalistic story, what could count as a new secular, rational account, and what was smuggled into such accounts in obscurity. Put another way, the question was how comprehensive an epic history science could produce, and increasingly not whether it should or could seek comprehensiveness. And here, the presumption was that the answers that science can provide to questions about the origins, ends, and dynamics of the universe are all-important, that there is something for science to be presumptuous about: its ability to answer questions previously answered in essentially mythic terms, whether metaphysical, spiritual, or religious.

If science is seen as the prime figure of institutionalized, experientially determined rationality, and myth as a prescientific explanation of things in the form of a story, long since untethered from direct, publicly producible experience, then a scientific myth presents itself as an enlightened and reasonable tale, a self-interrogated superstition, the rational submission of reason to the need for meaning. Judged by the central sense of its individual terms, the notion of a scientific myth is oxymoronic, or nearly so. But in the multitude of concepts that the term could house, these are only the most immediate tensions. More salient tensions and cultural resonances lay within the inversion of the terms: "mythic science."

Mythic science is an already tamed, if multivalent phrase, the adjectival form of myth bearing little suggestion of "objectively false," but rather the sense of "epically scaled" or "famously successful." Both of these latter senses are active in the circulation and reception of these natural histories. But scientific epic is an incongruity in a different way. Like myths, epic tales recount struggles above experience. Alternatively put, they split experience into what is true for their heroic kings and what is true for the subjects who celebrate and suffer the hero.[5] Science apostrophized as an ambition, by contrast, is alien to such a rupture. The reality to which this scientific ambition still directs itself, regardless of all the half-heard or misheard critiques voiced by critical disciplines, is a reality it posits as shared by all, whether or not that ultimate experience can be made intelligible only by or to a few.

These contraries within terms mixing science and myth depend on committed distinctions or at least conceptual purifications, science held up on the ground of reason that can penetrate as far as it can latch itself to an actual or even potential experience, uprooting any unnecessary assumption, false presumption, conventional belief—any nonscience. As a conceptual matter, therefore, science is represented as grounding itself. Whatever the breadth of the sources stitched into it, however much its knowledge might re-encode conventions, political resolutions, projected social orderings, or historical contingencies great and small—not generally understood themselves as entangled in the practice or knowledge of science—scientific results are represented as self-justifying.

But even according to the logic of these representations of science, the divisions between science and myth or between science and epic can stand firm only in certain frameworks. For Theodor Adorno and Max Horkheimer, myth and the Enlightenment remained tightly and problematically intertwined, despite the belief that the inductive sciences were based on the rejection of traditional, mythic authority. "Just as myths already entail enlightenment, with every step enlightenment entangles itself more deeply in mythology. Receiving all its subject matter from myths, in order to destroy them, it falls as judge under the spell of myth."[6] Likewise, epic and myth diverge and are bound together. For these critics, the epic was demonstrably linked to "classifying reason," Odysseus's journey anticipating a scientific calculating mastery of nature achieved through estrangement from nature—an epic as pregnant with a future science and technology as was the modernity Francis Bacon found in the printing press, gunpowder, and the nautical compass.[7] Successful journeying entails empirical knowledge and at times a self-renouncing, (partial) mastery of nature, the latter itself once a popular formula for the project of science itself. The metaphor of the road to knowledge was present in antique thought and remains a controlling metaphor of knowledge acquisition. The absence of a royal road to geometry revealed the presence of a rougher, longer road, closer to epic travel. The epic understood in terms of quest, from a standpoint taking account of the history and pervasiveness of that metaphor alone, seems to slide closer to science to the degree that science is structured by metaphors of quest. Examined in terms of the history of the coevolutionary account of humanity and the universe, this sliding becomes still more precipitous.

Regardless of common usage in distinct discourses, myth is often less distant from science than the denotations of the individual terms and their cultural connotations suggest. The terms are repeatedly juxtaposed in a way that does not imagine either dialectic or dissonance.[8] The expansive present at times elects as the aim of the totality of scientific knowledge the bridging of the distance between the scientific representation of the universe, on the one side, and the resonance, promise, and subject matters of myth, on the other. In this line of thought, "origin" is a middle term between science and myth, science seeking the ultimate origins (of matter, of life, of thought) once apparently imagined as disclosed by older mythologies. Myth can establish a discourse representing what, within the mythic representation, are the ultimate origins of everything, origins pregnant with ultimate ends. These mythic origins need not narrate a beyond-themselves; myth voices its truth out of a silence that it does not narrate or approach. Just as science provides its own ground, myth likewise can remain silent over the possibility of a further origin, over another ground in which to root out a further license to relate its account. Once science, as celebrated by prominent expositors, explicitly includes in its manifest ambitions the discovery of such an ultimate ground for everything, a ground that itself can be expressed narratively and that does not admit an "outside" of its own story, the designation of myth is to hand, and indeed, it fired the imagination of a plurality of scientists and commentators over the past several decades.[9]

But myth and epic each bear heavy conceptual and cultural loads separate from science. Myths shoulder histories, persons, and practices foreign to even the most expansive formulation of the task of science. The concepts of scientific myth or evolutionary tales are by these lights generic monsters that themselves bear prodigies. Odysseus becomes Sisyphus, the hours of Gilgamesh's dark run are prolonged: the ritual practices and finalizations posited by nonscientific cosmologies or secured by the homecoming of the king are not as clearly won by a scientific history looking with less certainty to the future. For some practitioners, the central reward of a scientific history is simply the possibility of determining more of it.

If the relevant concepts are rendered crisply enough, "scientific myth" can be included in the wider category of "scientific narrative" or "scientific story." These more general terms are suffused with their own, more dispersed tensions, depending on how close or far apart "truth" and "story" are held for consideration. Narrative is no more foreign to truth, and no less biased by individual or collective perspective, than is history. Like myth, "story" in reference to science is used both to advocate scientific history and to dismiss it. A history so extensive may, as in some exemplars, dizzyingly project a historicization of time, exciting in some of its producers and consumers awe for what science has been able to accomplish or for the vision it projects. For others, that same extensiveness immediately negates itself, revealing how far beyond the reach of empirical license or conceptual integrity such a history must apparently extend and provoking a diagnosis of scientific "storytelling" or nonsense.[10]

But a totalizing or universalizing history is not the only conceptual framework providing such an inclusive shelter for the varied products and producers of knowledge. Myth and epic as forms have themselves been theorized in terms of their aggregative or totalizing character. That character is also found in synthesizing genres or frameworks often not overtly presented in narrative terms. For example, the mapping of the world or of the knowable or surveyable universe engages a similarly totalizing ambition. A second, more fantastic example is the possibility of visiting every world, with or without a map, as a mode to capture everything, an idea itself constituting a fable with an indefinite number of ellipses. Mathematics or mathematical logic holds out the hope of a still more abbreviated form of synthesis, in which some ultimate predicate or law is valid without qualification and out of which the domains of knowledge and experience can be derived by actual or in-principle mathematical and logical manipulations.

These different modes of synthesis all have shared histories, foremost because, as we have noted, they rely on each other. But in relation specifically to the history of the universe, treating these genres together also serves to illuminate how narrative itself can function as a synthetic form—how story can achieve or constitute what scientific theory promises. To make this latter claim, however, requires positing a slippery distinction between synthesizing or generic activity (efforts to actualize such syntheses) and dogmatic belief that the unity promised by these syntheses is already active, without any need for extended proof or elaboration. A belief in a foundational law and its mathematical consequences might be dogmatic, a matter of belief in no derisive sense. But such a dogmatic belief can be distinguished from any scientific practice actively attempting to establish a more thorough, foundationalist frame, theorizing and organizing scientific practice according to that conception. Similarly, the idea that the universe as a whole has a history can be a dogmatic assertion, one that has at times been rejected as, for example, an anthropomorphic conceit or as hypostatizing the totalizing, possibly fictive idea of "world" at play within it. By contrast, presenting, promoting, and working to establish the aim of the sciences by way of authoring a historical account reveals and turns on the synthetic character of narrative itself. The history and functions of totalizing scientific narratives underscore differences between belief in unity and synthesizing practice.

BETWEEN STORY AND THEORY:
THREE EXAMPLES

To clarify and concretize these arguments and distinctions, it is necessary to examine at length and in depth the extended development of contemporary genres of synthesis, their composers and composition, their refinement, distribution, promised finality and form, the dilemmas at stake within them, the organizational structures their architects promoted through these genres and that promoted, shaped, and refined them in turn. That is the task of the many pages that will follow. However, a particularly crisp, accessible moment of roughly simultaneous synthetic effort in the late 1970s offers a preface to this more extended history of synthetic genres, underscoring the issues broached here insofar as many of the terms of analysis are explicitly invoked.

At a moment of summation in his first generalist, popularist text, *The First Three Minutes: A Modern View of the Origin of the Universe* (1977), the physicist Steven Weinberg stated: "We are now prepared to follow the course of cosmic evolution through its first three minutes."[11] Having presented his reader with the relevant scientific background, Weinberg steps back in a kind of dramatic suspension before delivering the central matter of the work, a fluid exposition of the first moments in the history of the universe as the temperature of the universe fell quickly with its expansion. It is not an even, straightforward chronological presentation: "Events move much more swiftly at first than later, so it would not be useful to show pictures spaced at equal time intervals, like an ordinary movie. Instead, I will adjust the speed of our film to the falling temperature of the universe, stopping the camera to take a picture each time that the temperature drops by a factor of about three."[12] Weinberg details the first six frames, from a period when the universe was a hot "cosmic soup" of elementary particles to three and three-quarter minutes later, when the basic ingredients of matter formed. At the end of those minutes, Weinberg observes: "The universe will go on expanding and cooling, but not much of interest will occur for 700,000 years."[13] After this, conditions will change in such a way as to allow matter to begin to form stars and galaxies, but even this is of less interest to Weinberg, since "the ingredients with which the stars would begin their life would be just those prepared in the first three minutes."[14] Concluding in an autobiographical register that includes all of humanity, Weinberg dryly observes that "after another 10,000 million years or so, living beings will begin to reconstruct this story."[15]

Weinberg is still more precise with his timeline, specifying more exactly the moment when the authorship of that scientific story became possible.

According to him, during most of the history of modern science, "there simply has not existed an adequate observational and theoretical foundation on which to build a history of the early universe."[16] But from the point of view of the middle 1970s, "now, in just the past decade, all this has changed," and advances in scientific theory and practice allow the scientific story to be told. As a story with "adequate observational and theoretical foundation," it has to Weinberg's mind greater merit than the origin myths that preceded it—Norse myths are his chief example—while looking to resolve the same questions: "What could be more interesting than the problem of Genesis?"[17]

In a similar spirit, in the same period, the sociobiologist E. O. Wilson found the scientific story to be one that had virtues beyond those of the accounts that preceded it: "Indeed, the origin of the universe in the big bang of fifteen billion years ago, as deduced by astronomers and physicists, is far more awesome than the first chapter of Genesis or the Ninevite epic of Gilgamesh. When the scientists project physical processes backward to that moment with the aid of mathematical models they are talking about everything—literally everything—and when they move forward in time to pulsars, supernovas, and the collision of black holes they probe distances and mysteries beyond the imaginings of earlier generations."[18] Wilson quotes the biblical Book of Job, where Job is chastised by God, asked to give an answer to whether he has "descended to the springs of the sea," "walked in the unfathomable deep," "comprehended the vast expanse of the world." "Come, tell me all this, if you know." To which Wilson in the place of Job responds, "Yes, we *do* know and we have told."[19] For Wilson, the fundamental theoretical approach of science, its material and unifying basis, or as he terms it, "scientific materialism," has provided and told all these answers, allowing him and the modern generation to answer the biblical God. And for Wilson, "the core of scientific materialism is the evolutionary epic," his term for humanity's story of the world and humanity's place in it.

Like Wilson, the astronomer Carl Sagan in his 1977 *Dragons of Eden: Speculations on the Evolution of Human Intelligence* also found a resonant figure in Job, invoking in an epigraph words in which Job bemoans his condition: "I am a brother to dragons, and a companion to owls." That epigraph, cueing it would seem the title of the text,[20] reflected Sagan's belief in the deep-seated place of myth in humanity's self-understanding, a vision in a somewhat different spirit, but consonant with Wilson's connection of myth and science. In sentiments on similar themes made in an address that same year and republished in 1979 in an essay entitled "The Golden Age of Planetary Exploration," Sagan lists what he saw as the many virtues of planetary exploration. The last dwells on the present moment in history and the histories that science could tell: "And it [exploration] permits us, for the first time in history, to approach with rigor, with a significant chance of finding out the true answers, questions on the origins and destinies of worlds, the beginnings and ends of life, and the possibility of other beings who live in the skies—questions basic to the human enterprise as thinking is, as natural as breathing."[21] For Sagan, "Interplanetary unmanned spacecraft of the modern generation extend the human presence to bizarre and exotic landscapes far stranger than any in myth or legend."[22]

This reference point of myth and legend involves not only the origins and fates of other worlds, but those of this world and the history of the universe as a whole. When Sagan comes elsewhere to modern cosmological theories and finds in them uncomfortable analogies and connections to the experience of human birth—analogies that he ultimately takes to be at the root of religion—he queries almost anxiously: "Are we such limited creatures that we are unable to construct a cosmology that differs significantly from one of the perinatal stages? Is our ability to know the universe hopelessly ensnared and enmired in the experiences of birth and infancy? Are we doomed to recapitulate our origins in a pretense of understanding the universe?" But his final question in this same series is aligned more with an optimism for an unconstrained capacity to know, an optimism that was the dominant key of his works: "Or might the emerging observational evidence gradually force us into accommodation with and an understanding of that vast and awesome universe in which we float, lost and brave and questing?"[23]

In the composition and reception of this history of the world, the word "story," its cognates and synonyms and the ambitions they express for the rendering scientific truths are pervasive.[24] For those advocating a new history or myth, those terms were generally not dismissive; they did not suggest that scientists and other synthesizing architects attempted to build a false structure. For these authors, the opposite was the case: the use of mythic categories and registers reflected a belief in and an ambition to relate a true story of the universe, a story that might translate, structure, capture, or serve as the theory that finally gives account of everything.[25]

MODELS AND OBSTACLES:
HISTORY AND EPISTEMOLOGY

In their primary representations, the scientific ways for addressing this narrative were neither remade from whole cloth nor simple iterations of past efforts. The language they employed in relation to natural science, the language and structures of story, of history, of myth, and of epic, clearly did not begin with the twentieth century or with the nineteenth. Generic models of scientific narrative existed well before modern scientific histories offered the possibility of concise, thoroughly naturalistic accounts. The cultural reception of these scientific histories would not only juxtapose them to sacred histories, but compare them with the scholarly and popular tradition of universal history.

Universal histories and their connection to sacred history itself were under ongoing extensive interrogation and critique. From earlier universal histories on, there was a pattern of fluctuation and diversity in natural scientific views of the history of the world—even of whether the world has a history and whether science could speak to it—varying from acceptance to denial and from contestation, to indifference, to celebration. For the larger part of the nineteenth and twentieth centuries, no one view commanded so much respect as to render its skeptical responses unscientific, even if there were dominant claims. Likewise, the advisability and true possibility of world/human/species history would become subject to extensive debate and critique.

At the time when the primary account here begins, in the second third of the nineteenth century, deliberations and debates over natural history established strong conceptual obstacles to framing a naturalistic synthesis of natural knowledge in the form of history. Natural history, understood in the sense of providing an intelligible order of the natural world, had included elements in that order—the pieces of its imagined museum—understood as unchanging. A stable solar system or the fixed stars did not betray or disclose their origin. Whereas the eighteenth-century French mathematician Pierre-Simon Laplace's physical astronomy was held to demonstrate the perpetuation of the planets in their orbits, his theory of the creation of the solar system was received not only with enthusiasm, but also with the "défiance," the distrust, with which he advanced it.[26] According to this theory, a great solar cloud contracted and at its fringes, as it condensed and cooled, it successively formed the planets. The theory thereafter was described as the "nebular hypothesis," a term coined by William Whewell to name one of in fact a collection of varied hypotheses dealing with the variety, dynamics, and potentialities of such celestial clouds or nebulae.[27] These theories generally argued for the historicity of entire stellar systems and, at times, of all the fine-grained chains of being they might support. But the ahistorical objects of the natural world demanded fields of knowledge that themselves could not be suborned to the field of history and to its repertoire of historical productions.

This departure point, when history could not easily establish a synthesis of natural knowledge, helps to identify the marks of the genre of more recent scientific universal history—its debts to and contrasts with scientific history of an earlier period, clarifying as well aspects of those past natural histories. The presumptions or explicit arguments that related and countered the physical, biological, and geological sciences as they themselves developed were steeped in and dwelled on the cultural dissonances and consonances discernible in new histories and syntheses.

That longstanding world or universal historical tradition in turn suggests that modern natural history drew from persistent narrative conventions, themselves preceding the nineteenth century. This persistence will be overstated if such older conventions and narrative models are imagined merely to reappear in the future—if, for example, new epic histories are themselves understood as merely iterations of sacred history. Much historiography basing itself on a widening arc distinguishing natural history and the world history of peoples from at least the seventeenth century on argues compellingly against any such simple iterative view of historical genre. From such a perspective, for centuries preceding the nineteenth century, the dominant framework for Western histories was being stretched almost out of recognition. Earlier voyages of discovery had revealed a natural geography and a distribution of cultures, peoples, and recorded histories that placed new and stringent demands on the old, scripturally sourced accounts. The newer modes of empiricist temper arising over the same centuries argued for new kinds of proof for historical claims.

Across this period, according to this genealogy, historical knowledge about the world appeared to fragment into different modes, depending on whether rocks and nations were read to confirm scripture or to threaten it. Differences in methods, in temporal and spatial scales, and in the subject matter of natural and civic or human histories became only more multiple and manifest. Human history as a whole, much less the history of states, organizations, populations, or individuals within it, occupied an increasingly small moment of chronological/mechanical time, a time itself posited as an ever smaller fraction of a more quickly aging world. Conversely, the history of the natural world could be excluded from human history. There remained some few overt connections between written history and the history of the universe—as when the annals of world history were plumbed for confirmation of the periodicity or nature of astral events—but the documentary emphasis and methodology of each kind of history only intensified differences attributable to paradigmatic historical and natural scientific methods: the reading of strata and skies versus the reading of texts.[28]

The general point of view of such a history is that the seventeenth century nursed a process of sacred historical revision in the production of a history of the world (as for example in the genre of "sacred histories of the Earth" or in relation to conceptions of and politics toward the "other" in the world) and that the eighteenth century saw the features of a newer universal natural history break more distinctly from earlier patterns (with the arguments of those such as Jean-Andre Deluc, Georges-Louis Leclerc, Comte de Buffon, Johann Friedrich Blumenbach, and many others).[29] The widening, aging, pluralistic world had placed pressures on the natural and scriptural history of the West, pressures that at the same time produced innovations, reinterpretations of sacred accounts, or recalibrations. These provided new models for scientific/naturalistic histories, which appear to be cast out of radically different molds from either sacred or (human) universal histories. That same extended period of time saw the further elaboration of all-embracing universal history as a scholarly genre, a genre itself drawing on the forms and dilemmas of sacred history and embracing the content of natural history. But even by the lights of this clear a genealogy, in those newer molds, the older templates still played a structuring role. Through to the turn of the nineteenth century, such reliances were often overt: Genesis might continue to provide an initial chapter of a world history, or, less conspicuously, scripture could sanction the imagined alliance of the character and history of peoples and places.[30]

There is much that is enlightening in the sharp lines of such a genealogy. It helps make clear how, by the early nineteenth century, a cultural ground had been prepared for the widening intellectual foment and subdivision in the years thereafter, giving the intellectual West a multiplicity of candidate cosmologies and so, at the same time, no cosmology. A virtual absence of learned consensus—the absence of an accepted cosmology—lasted for most of the nineteenth and twentieth centuries. There was dissensus between disciplines studying the natural world. Biology and geology at times required an earth that outlived the nourishment that seemed available to sustain it; observational astronomy found stars older than the heavens in which they were embedded. At the same time, the divisions between civil and natural history were stark, not only in content and methodology, but according to the lessons and character they were imagined to instruct and instill. Apostles of humanistic values collided with disciples of the varieties of natural scientific ethos and cultures, a collision ramifying through and constructed by a changing matrix of social class and of political and pedagogical movements. It was not until the latter part of the twentieth century that this period of foment appeared to come to some conclusion, when the chronologies of the material and organic world posited by the different branches of the natural sciences increasingly converged and synthetic accounts, finalizing theories and stories, became possible. This still-present moment, in the recent historical genre of "Anthropocene history," likewise has seen attempts to bridge humanistic and scientific historical narratives, promoted as a reconciliation establishing a species truth through a universalizing account.

The new contemporary "myth" or universal history, as narrated by scientific practitioners or historians, is often characterized as a late chapter in the unfolding of the Enlightenment. These histories emphasize that the origin of things is distant from the Word and Eden. The conclusion of the universal history they assemble is found in species extinction and cosmic death or in an unknown human potential or future human transformation, a conclusion immediately alien to, for example, an idea of reconciliation with divinity according to the dictates of providence. The overall account of the history of knowledge embedded in these modern universal histories instead has the flavor of a human epic. According to it, the falsehoods of the past are finally cast off, giving rise to the contemporary synthetic viewpoint, to a day when historically minded scientists and scientifically competent historians can relate the rise and future fall of humanity, of prehistoric and posthistoric cultures, and human nature itself. Such genealogies tend to posit their own telos to the evolution of knowledge: the composition of the history of the world is approaching the truth that the material history of the world is approaching no truth, no culmination at all.

The contrast between epistemological and material evolution and its implications can be best brought out and examined by looking at the steps in that genealogy. Scholars from a variety of disciplinary standpoints largely accept that the historical disciplines fragmented as natural and civil histories became more reliant on their different expert knowledges. The current scientific histories written by scientists, historians, and other cultural producers, such as film documentarians, see themselves as effecting reconciliation in the wake of that dissolution.

But any sense of current reconciliation raises the question of how different kinds of history have continued to rely on each other in deep ways, even across their apparent fragmentation, and how they have been understood as companions to each other. How does a progressive account of successively truer universal history potentially depart from *and* conform to important aspects of world/universal historical ambition and character? Likewise, how might contemporary universal histories continue to be invested in the sacred traditions they explicitly reject and from whose narrative content they have so long departed? How have histories of the universe and universal histories been brought closer together as a function of both repeated critiques of their universalist character and repeated attempts across the last century and more to stitch them together on the assumption that a reconciliation is necessary? The epic trajectory of knowledge embedded in the scientific epic tends to bracket the historical and conceptual connections between the modern categories of scientific and humanistic universal histories, connections often suggested by those past authors most often celebrated as forebears to modern cosmology and universal history.

The Enlightenment offers no simple picture or belated starting point for this trajectory. Recent historiography functions as a reminder that there is no simple alignment between secularity and Enlightenment thought, troubling as a result the picture of the divergence of varying historical modes—particularly between sacred and scientific/secular, but by extension, between sacred and universal/world-historical.[31] But even the central figures adduced within historiographies asserting increasing divergence present by the example of their efforts no simple picture of the distinctions between emergent categories of scientific and other histories.

In 1791, Kant's *Universal Natural History and Theory of the Heavens* was published, not long before Laplace's *System of the World* (1796), each putting forward varieties of "the nebular hypothesis." Kant's cosmology had been printed nearly forty years earlier in 1755 and well before his "Idea for a Universal History with a Cosmopolitan Purpose" in 1784—the same year as his famous essay "What Is Enlightenment?" and three years into his "critical period," inaugurated with the 1781 *Critique of Pure Reason*.[32] In 1755, the young Kant had rhapsodically enlarged the story of everything in *The Theory of the Heavens*, as it is often abbreviated. He theorized a "phoenix of nature," an unending creation taking place across an endless universe. Kant and Laplace addressed the origins of the solar system and (at different points in their works) the formations beyond it: Kant extended the Newtonian system in an extrapolation that Newton would not, at least at times, have held to be legitimate, reasoning with an idea of gravitation as active at all times and places and imagining the death of old worlds feeding the birth of new ones.[33]

Kant's universal natural historical reflections addressed and reconfigured a natural theological concern: What sustains the world? Without some form of sustenance, the world and the solar system, it seemed, would spend their force and halt. The concern as to what sustains the world system, a concern held more generally, persisted in different form in the natural histories of the following centuries. For those who found the possibility of universal death inconceivable, God provided the ultimate assurance of new energies to come. As a natural theological argument, this was close to circular, since the natural world does not in all respects give signs of its endless perpetuation.

Three decades later, his "Idea of Universal History" grounded future political hopes in the natural character of human interaction present from the state of nature. These different conceptions and treatments of universal histories, natural and human/political, suggest points of both divergence and connection between different synthetic histories, natural, universal, and, through his treatment of each individually, the sacred tradition. The divergence might be grounded in the disunity of Kant as a historical personality whose intellectual convictions change over time. But there are also the persistent themes and sources prompting the different works and theses circulating within a decade of each other that connect them through and beyond the biographical personality signing the same name to them. How did the matrix of these histories—all "universal" in different senses of that word—remain in dialogue, even as they adopted their own generic frames?[34]

In that generic and critical work, Kant provided an ongoing resource and target for considerations on how natural and universal history were integrated in the conceptual relations of the sciences, each to each. The sciences in broad terms could be distinguished and made possible by elucidating the grounds on which knowledge is possible—on a knowledge that structures experience, rather than being informed by it. In his view, to ground knowledge, and so to clarify what makes the sciences possible and their proper relations to each other, constrains the ambitions of both a natural history and a universal history.[35]

But Kant's own speculations on the heavens at times connected more immediately to other historical initiatives, just as his universal history also connected more immediately to the natural sciences. In *The Theory of the Heavens*, Kant tied his cosmological reflections to the massive universal historical project of the day, the multiauthored, multivolume *Universal History from the earliest Time to the Present, compiled from Original Authors, and illustrated with Maps, Cuts, Notes, Chronological and other Tables*,[36] looking to it as a justification for the propriety and sense of his natural historical labors: "I will therefore quote the Authors of the *Universal History*, when they say: 'However, we cannot but think the essay of the philosopher who endeavoured to account for the formation of the world in a certain time from a rude matter, by the sole continuation of a motion once impressed, and reduced to a few simple and general laws; or of others, *who since attempted the same, with more applause, from the original properties of matter, with which it was endued at its creation*, is far from being criminal or injurious to GOD, as some have imagined, that it is rather giving a more sublime idea of His infinite wisdom.'[37]

These universal histories had themselves been extensive, multigenerational, collective projects—and as in this quote, themselves relied on potential naturalistic cosmogonies and sacred sources. Roughly three-quarters of a century prior to Kant's effort, Bishop Bossuet, who had been tutor to the eldest son of Louis XIV, addressed the dauphin in a universal history rooted in the mosaic cosmogonic account and the biblical peoples, a chronicle up to the time of Charlemagne.[38] Voltaire built on and critiqued Bossuet, attempting to recount the histories and periods of nations that his predecessor ignored or did not reach.[39] At roughly the same time, the English Arabist George Sale, along with colleagues at Oxford and Cambridge, contributed to the production of a many-volume universal history; their joint work attempted to canvass the recorded histories of every nation, but was also grounded in a biblical cosmogony. It was the translation of this work by Siegmund Jacob Baumgarten that Kant apparently examined in 1755.[40] While these ambitious chronicles were being written, themselves across intergenerational, multilingual networks, natural historians and philosophers such as Buffon reworked the origin accounts in which the universal histories were rooted, ultimately suggesting more extended, naturalistic cosmogonies and anthropogenies.

Tying his efforts to these, Kant's early *Weltgeschichte* advocated a view of the world that saw in gravity a universal and timeless force that, in forces of attraction and repulsion (by more contemporary lights, a term for the more fictive centrifugal force), formed the galaxies out of matter distributed everywhere. He attempted to demonstrate his views by making sense of a set of observations concerning the orbits of the planets, their shared directions of revolution, the eccentricity of their nearly circular orbits, their masses, and so on, which in many ways were shared with the later, partly tentative reflections of Laplace. In aid of this demonstration, he argued for a view of the fixed stars as moving and for the hazy nebulae in the heavens as their own as yet unresolved and unknown star systems. He found in these hypotheses a way to unify contrary positions, the one holding with Newton that there was no way to imagine the present configuration of the solar system to have emerged without God's intervention, the other suggesting a mechanistic view of the world explaining the many coincidences of the system by appeal to a natural origin in time. He presented his speculations as tied to the strongest proofs of the existence of God.[41] If this was an antinomy of sorts that he posited and resolved, thirty years later, he would address antinomies of origin in starkly different fashion.

The older Kant provided an Enlightenment-era vision of the way in which such a scientific history, however compelling, might never amount to knowledge, confronting fears that long predated those of Carl Sagan that a possible anthropomorphism structures any cosmogony. In this later critical work, Kant believed himself to have proven that the ontological proof of God's existence is impossible. And on the way to that proof, he argued that a full history of the universe that would imagine itself to know the origins of everything also is impossible. The restriction against telling absolute origins, as he explicated it, is a constraint on the powers of reason itself.

These constraints were announced with the publication of his 1781 *Critique of Pure Reason*.[42] The tone of the exuberant *Theory of the Heavens* was gone. But however much the tone of his later works differed from his earlier cosmogony, they revealed abiding concerns dating from his first natural historical efforts. Regardless of how much the older and younger views and styles differed and evolved over time, it would be as much an oversimplification to treat the older Kant's work as simply contradicting that of his younger self as to treat the elder's work as a mere extension of the younger.[43]

The *First Critique* was the site of Kant's "Copernican Revolution," in which objects generally understood as external to the intuiting subject were reconfigured as appearances of that subject. Kant employed a formulation of truth as the agreement between concepts and objects of a possible experience. His emphasis on a possible experience was to determine those structures of the human subject that did not depend on any actual experience, but rather themselves determined the categories of experience. The revolution he posited was to treat those objects, and not just the concepts pertaining to those objects, as subjective products. This meant in turn forgoing a view of objects as things in themselves, and instead, strictly delimiting the human intuition of them to appearances that exist *for us*. No knowledge of things in themselves is possible according to this view—concepts apply only to *our* appearances, with "our" understood in the broadest sense of the cognizing subject. And thus Kant celebrates and grounds the extent of knowledge through its very chastising, pronouncing on its limits.

The historical consequences of that chastisement are perhaps at their clearest in the articulation of his first antinomy of reason. The antinomies are paired theses and antitheses, each by the lights of logic equally valid, even while positing opposing views. His first antinomy addresses the possibility of whether it can be known that the world has a beginning of time and is bounded in space. In the *Critique of Pure Reason*, Kant argues for the unknowability of an originary moment or the boundedness of the world.

"We should of ourselves," he argues, "desist from the demand that our questions be answered dogmatically, if from the start we understood that whatever the dogmatic answer might turn out to be it would only increase our ignorance, and cast us from one inconceivability into another, from one obscurity into another still greater, and perhaps even into contradictions."[44] In the case of cosmological claims, in particular, Kant notes that whenever the question is considered of whether the world does or does not have an origin, the questioner is thrown into an antinomy: "*First*, that *the world has no beginning*: it is then *too large* for our concept, which, consisting as it does in a successive regress, can never reach the whole eternity that has elapsed. Or suppose that *the world has a beginning*, it will then, in the necessary empirical regress, be *too small* for the concept of the understanding. For since the beginning still presupposes a time which precedes it, it is still not unconditioned; and the law of the empirical employment of the understanding therefore obliges us to look for a higher temporal condition; and the world [as limited in time] is therefore obviously too *small* for this law."[45]

There are many philosophical subtleties and historical shadings in the argument outside the scope of my treatment of it. However, elements of even a simplified reading bear drawing out, keeping in mind Kant's emphasis not only on truth understood as the agreement between concepts and objects of a possible experience, but also Kant's assertion that without knowledge of things in themselves, our concepts relate only to *our* appearances.

Finalization is an issue here, since whatever is apparently finalizable provokes the question of a further ground and whatever is not finalizable checks what can in principle be known. The cited passage suggests that it cannot be known in truth that the world has no beginning, since there is no appearance or object of human intuition that can stand for the infinite chain of cause and effect (the regress) that precedes that intuition. Our conceptions cannot capture the entirety of that regress in time—the world that would contain this infinity is too large. If it is instead supposed that there is a beginning, our concept of a "beginning" itself poses the question of the beginning or cause of the beginning. We place the beginning in time (for Kant, an "inner sense"), but to do so forces upon us the conception of a previous time. The world becomes too small for our concept, unable to contain its own beginning.

But the world is the crux, and this is of relevance philosophically and historically. Ultimately, Kant argues that both thesis and antithesis are based on an illusion, the idea of an unconditioned whole: "Since the world does not exist in itself, independently of the regressive series of my representations, it exists *in itself* neither as an *infinite* whole nor as a *finite* whole. It exists only in the empirical regress of the series of appearances and is not to be met with as something in itself. If, then, the series is always conditioned, and therefore can never be given as complete, the world is not an unconditioned whole, and does not exist as such a whole, either of infinite or of finite magnitude."[46] In other words, both arguments, that the world has a beginning and that the world does not have a beginning, presuppose the knowledge of the existence of the world as a thing in itself, something inaccessible to the human subject, in Kant's view.[47] The world may not be something of which "beginning" or "finitude" is predicated apart from the subject predicating them.

This critique of an inquiry into origins based in the notion of a world in itself that can be subordinated to human reason imposed limits on what the younger Kant had himself set out to do: capture a totality through a representation. The elder Kant did not argue that without the knowledge of origins knowledge is impossible. His argument was the opposite: in order to determine or establish a ground for knowledge, the grasping of the world at play in the search for origins must be perceived and dismissed as a distortive illusion. Thus, Kant provided an example of knowledge not grounded in originary conceptions. Origins, on this view of the grounding of science, are prohibitive of knowledge, rather than constitutive of it. His inquiry marked a historical and conceptual moment, providing a nontheological argument *against* a distortive possibility of ultimate origins while at the same time positing at least a naturalistic origin for the *natural* world of appearances. Both moments in the work of Kant and other thinkers continued to play a role in the debates and arguments over representations to come, at times in explicit dialogue with the younger or older Kant.

L'AMOUR PROPRE AND THE JUSTIFICATION OF NATURE

Contemplated as one grand whole, astronomy is the most beautiful monument of the human mind; the noblest record of its intelligence. Seduced by the illusions of the senses, and of self-love [*l'amour propre*], man considered himself, for a long time, as the centre of the motion of the celestial bodies, and his pride was justly punished by the vain terrors they inspired. The labour of many ages has at length withdrawn the veil which covered the system. Man appears, upon a small planet, almost imperceptible in the vast extent of the solar system, itself only an insensible point in the immensity of space. The sublime results to which this discovery has led, may console him for the limited place assigned him in the universe. . . . They have rendered important services to navigation and astronomy; but their great benefit has been the having dissipated the alarms occasioned by extraordinary celestial phenomena, and destroyed the errors springing from the ignorance of our true relation with nature; errors so much the more fatal, as social order can only rest in the basis of these relations. *Truth, Justice;* these are its immutable laws. Far from us be the dangerous maxim, that it is sometimes useful to mislead, to deceive, and enslave mankind, to insure their happiness. Cruel experience has at all times proved that with impunity, these sacred laws can never be infringed.[48]

These final words of Pierre-Simon Laplace's *System of the World* embraced the infinitude that nearly two hundred years earlier, in 1606, Kepler found and rejected in Giordano Bruno: "This very cognition carries with it I don't know what secret, hidden horror; indeed one finds oneself wandering in this immensity, to which are denied limits and center and therefore also all determinate places."[49] Alexander Koyré roots Kepler's rejection of the infinite in his Aristotelianism.[50] But if so, there is little to suggest that it was a matter of Laplace's *amour propre*. In the main, Kepler's objections did not indulge this "secret, hidden horror," but instead engaged what he took to be the absurdity of such views. Kepler's larger context was still one where however central the earth might have been, it was not treated as the happiest of seats ("the cosmic dustbin," as C. S. Lewis characterized the inherited antique view of earth as the "offscourings of creation"),[51] nor did medieval worldviews often flatter the metaphysical place of humanity. Humanity might be essential enough for the sacred history of redemption, but at the same time have no high place in the plenitude of being. Nevertheless, the vanity Laplace detected in humanity's past history resonated with concerns expressed in universal histories advanced by later astronomers, resounding in human histories embedded in cosmic evolutionary accounts.

Laplace likewise emphasized in this quote the service astronomy was to do for the active politics of humanity, once the social had established its proper relation to nature. His remarks had a particular charge, given their publication date, printed in the text as "L'An Quatre de la Republique Française." As Roger Hahn has documented, while working on the questions relating to the stability of the solar system, in 1790 Laplace wrote to Jean André Deluc in England in words resonant with his later, more famed conclusion to his

System of the World: "I congratulate you upon living under a regime that has for a long time been stable, and that seems one of the best results of the human intellect. It is about the same time at the end of the last century that the true foundations of the system of the world and the social system were laid down. We fought the first of these and finally adopted it; perhaps it will be the same for the social system."[52] Not for the last time would the search for cosmic stability be rooted in a period of political unrest and violence.

Such a relation, between the social and natural orders, had also been sketched by Kant. In his 1784 "Idea for a Universal History with a Cosmopolitan Purpose," he established a regulative vision of a future, a human universal and natural history, on the basis of a developing cosmopolitanism related to, but not reducible to, the political situation of the Holy Roman Empire during his life.[53] Launching nine propositions building up toward a hopeful future confederation of humanity, Kant noted that the reasoning power of humans itself suggests that "no history of them in conformity to a plan (as e.g. of bees or of beavers) appears to be possible." But at the same time, that reason apparently had failed to determine history: "One cannot resist feeling a certain indignation when one sees their doings and refrainings on the great stage of the world and finds that despite the wisdom appearing now and then in individual cases, everything in the large is woven together out of folly, childish vanity, often also out of childish malice and the rage to destruction."[54]

Kant likewise found further consternation at human *amour propre*, set against this sorry set of doings, so much so that "one does not know what concept to make of our species, with its smug imaginings about its excellences." The task he therefore set himself was to try to determine "an aim of nature in this nonsensical course of things human; from which aim a history in accordance with a determinate plan of nature might nevertheless be possible even of creatures who do not behave in accordance with their own plan."[55] In the brief compass of his essay, Kant limited himself to "finding a guideline for such a history," thereafter "to leave it to nature to produce the man who is in a position to compose that history accordingly." Nature had famously obliged in the past: "Thus it did produce a Kepler, who subjected the eccentric paths of the planets in an unexpected way to determinate laws, and a Newton, who explained these laws from a universal natural cause."[56]

Given his view of the unreasonable pageant presented by history and drawing in part on Montaigne, Kant produced this guideline for a path through human unreason—an "unsocial sociability" where a person strives in relation to others "whom he cannot *stand*," working toward the improvement of the social world despite its unreason, a collective, intergenerational movement toward a confederate destiny.[57] The maturation of the political state would of necessity be judged from the perspective of descendants who would fulfill what was only partly in growth in the present in which the universal histories were written—and that development asserted or queried the achievement of a certain conception of a just society. History and meaning were premised as tasks for humanity as a whole to assume in fulfilling itself by following out a plan that may be grounded in nothing other than the collective task of knowing—to realize a plan that may not exist outside itself, that may already be a self-sanctioned providence.

In such a way, in the consummation of its history according to plan, Kant attempted to find a "*justification* of nature—or better, of *providence*. . . . For what does it help to praise the splendor and wisdom of creation in the non-rational realm of nature, and to recommend it to our consideration, if that part of the great showplace of the highest wisdom that contains the end of all this—the history of humankind—is to remain a ceaseless objection against it, the prospect of which necessitates our turning our eyes away from it in disgust and, in despair of ever encountering a completed rational aim in it, to hope for the latter only in another world?"[58] The redemption of a natural, human, universal history would have to be found immanently, within natural human history itself, without reference to transcendent causes, and on the model not of sacred history alone, but of both sacred and natural history. An understanding of nature, including species nature, could ameliorate humanity's present condition (Laplace) and offer hope for humanity's future condition (Laplace and Kant). But that same nature had yet to supply the Kepler and Newton who would make a natural science of history.

These points were recurrent ones, stitched into works promulgating totalizing histories or, more generally, genres of synthesis thereafter: a notion of an end, perfection, plan, all according to an intrinsic human potentiality and nature while at the same time retaining a secular grammar.[59] This would come to be the justification of the later scientific epic and what is claimed for its spiritual dimensions. A final understanding of the meaning of the universe would be achieved as humanity fulfilled itself by heroically pursuing precisely that knowledge.

STABILITY AND CHANGE, DETERMINATION AND HISTORICAL TOTALITY

What was celebrated as Laplace's signal achievement among his many mathematical successes was a celestial mechanics attesting to the stability of the solar system in the face of the known perturbations of planetary movements—and in the face of the politically turbulent times in which he was crafting his theory.[60] Like Kant, Laplace saw himself in the Newtonian tradition, and unlike Kant, he was seen by many as having written the work that crowned and consummated that tradition.[61]

But that solar system was itself undergoing radical change in his lifetime, with detection of what would come to be known as the planet Uranus in 1781, along with the further scrutiny of nebulous matter in the skies—matter that sometimes could be clarified into stars by more penetrating telescopes and that sometimes appeared more likely to be other kinds of formations. William Herschel, a musician, conductor, composer, and astronomer, observed the new object yet to be named a planet and, as it became clearer that it might be in a closed, rather than parabolic orbit, Laplace presented an analysis of it as a planet in 1783.[62]

Like Kant, Laplace also speculated on the origins of the system whose stability he was establishing. For predecessors, he named only Buffon as having articulated an origin account insufficient in comparison with his own.[63] And the fact of such an origin, in Laplace's exposition, did not somehow undercut the fact of celestial stability.

In its maturation, if not its own inception, Laplace's account was indebted to the expansive astral nebular hypothesis of Herschel, with whom he had mediated connections.[64] Over the years, as Simon Schaffer has argued, Herschel categorized the different nebular clouds as a natural history itself—initially a kind of botanic, atemporal classification of the nebulae.[65] Laplace and others in turn connected and further historicized that natural history, regarding different forms of nebulae at different stages of maturation in the condensation of matter—extending ultimately to the primitive cloud or atmosphere predating the solar system.[66] As a result, in later editions of *The System of the World*, Laplace extended his nebular hypothesis to the emergence of the sun and tied it to Herschel's now historicized natural order of the nebulae.[67] His arguments for stability and condensation produced a narrative of a world that emerged as stable once it was assembled according to naturalistic principles. This was an account that could be and ultimately was taken to resonate with either natural theological or more aggressively secular grammars and with either revolutionary or more conservative politics.

But the nebular hypothesis was only one part of a set of resources for determining the history of the world. For Laplace himself, the future of the world might be dictated by a scientific or natural philosophical determinism, arguing that from the initial conditions of a physical system, the future development of such a system might be fully decided and understood. Thus, if such an origin/initial condition could be perceived, then according to such a physicalist view, the entire history of the universe would be knowable. Such a viewpoint therefore would suggest that if an original state of the world is both sensible and knowable, despite the conclusions of Kant in his critical modes, the history of the world thereafter was a given. Hahn has shown that in private reflections, Laplace extended such determinism to human will and, more publicly, applied his determinism-minded reflections on probabilities to governance—suggesting, however, that science, despite the rigid constraints of historical determinism, somehow might still help shape the predetermined course of history.[68] At the same time, the utopian possibilities of a science of history or politics apostrophized by Laplace and sketched by Kant would be emphatically declared in the next generation by Herschel's famous son John. The distance between human and universal history and between totalizing synthetic ambitions for a universal history and the status of human agency thus would contract further as—despite Kantian, natural scientific, and natural theological critiques—the discovery of the ultimate origins as the prize awaiting the end of the scientific quest became a rising, recurring refrain.

HUMAN NATURE AND THE POSSIBILITY OF A UNIVERSAL NATURAL HISTORY

Hovering over Laplacian determinism and cosmogony and Kantian universal history (natural and human) was the possibility of achieving a universal natural history, including humanity's place in it. This link between the natural and civic or human historical was an ongoing persistent topos in those conceptions of a synthetic history that saw themselves as focusing on the natural world and those focusing on the history of humanity's place and role in it.

With the turn of the eighteenth century, the category of universal history as such (whatever its focus) was brought to still greater self-consciousness with the articulation and spread of Hegelian history and philosophy. Hegel was among those gaining attention for positing universal histories in the civic sense and gaining infamy in discussions of the relations of science in at least the conceptual sense. Kant's writings were unsurprisingly insinuated in Hegel's, where for each, the hero of universal history, at least, was the will or reason. For Hegel, this was more emphatic and much more historically (narratively) deployed, placing the idea of reason or intellection or spirit as perhaps the chief natural principle, as what could determine and reform the categories of knowledge itself. Hegel's universal history of humanity could therefore be read in secular or nonsecular ways and helped define the meaning and prosecution of universal history in the civic sense of the term.

Regardless of the register in which it was read, that universal history functioned as the site of the articulation of spirit in the context of "world historical" peoples and was confluent with the development of all of nature itself.[69] Hegel saw his universal history as necessarily naturalistic, connecting one shared, if variegated, human history with natural totality.[70] Flatly put, it was the object of universal history to trace how matter comes to be conscious of itself through the increasing self-consciousness of spirit, reason, freedom—cognate and intimately related terms in Hegelian thought.[71]

Hegel then promoted a concrete synthesis of histories of "Oriental," "Greek," "Roman," and "German" worlds according to an arguably naturalistic logic that itself would meet with much more resistance and criticism by the nineteenth-century natural scientific synthesizers. At the same time, as early as the first years of the nineteenth century,[72] Hegel began putting forward a "philosophy of nature," arguing against much of the common language of a Newtonian worldview in a not straightforward language of its own. Employing the perspectives in his work on logic and spirit, he broadly related different branches of the sciences, primarily viewed by him as mathematics, physics, and physiology. In abstruse presentations, he distinguished them ultimately by how an idea of nature, or the Idea, presents itself, whether as external and universal (mathematics), as real and mutual (physics), or as living actuality (physiology).[73]

The consonance of aspects of Hegel's work and those of synthetic authors such as Whewell suggests how far certain universal historical understandings extended.[74] However, the obscurity of some of his own comments on Newtonian natural philosophy prompted frustration and even disgust in iconic nineteenth-century figures such as William Thomson, the later Lord Kelvin, and Hermann von Helmholtz.[75] Regardless of the speculative extent of Hegel's influence on natural science, of more importance here is the extent to which the successes of natural science and ambitions for its future possibilities fed the sense that a universal history encompassing and driven by human inquiry could be possible. Against the backdrop of an apparently irrational history that suggested the "purposelessness" of nature, Kant had argued that nevertheless, "one can regard the history of the human species in the large as the completion of a hidden plan of nature to bring about an inwardly and, to this end, also an externally perfect state constitution, as the only condition in which it can fully develop all its predispositions in humanity." From this perspective, the dismal nature of human historical experience "reveals a little; for this cycle appears to require so long a time to be completed that the little part of it which humanity has traversed with respect to this aim allows one to determine the shape of its path and the relation of the parts to the whole only as uncertainly as the course taken by our sun together with the entire host of its satellites in the great system of fixed stars can be determined from all the observations of the heavens made hitherto." And the new astronomical discoveries of the late eighteenth century could throw into greater uncertainty the principles guiding the careful computations of celestial paths determined by those such as Laplace. Nevertheless, regardless of such uncertainty, Kant argued that "yet from the general ground of the systematic constitution of the cosmic order and from the little one has observed, one is reliably able to determine enough to infer the actuality of such a cycle."[76] This placed the consummation of history in a potentially distant future, when more human or historical "cycles" might be revealed and where the most hopeful possibility is accelerating the realization of that returning cycle: "Nevertheless, in regard to the most distant epochs that our species is to encounter, it belongs to human nature not to be indifferent about them, if only they can be expected with certainty. This can happen all the less especially in our case, where it seems that we could, through our own rational contrivance, bring about faster such a joyful point in time for our posterity. For the sake of that, even the faint traces of its approach will be very important for us."[77]

Hegel, like Kant, emphasized the special role of humanity, not given as a natural object obeying a formulaically cyclical law: "It must be observed at the outset, that the phenomenon we investigate—Universal History—belongs to the realm of *Spirit*. The term '*World*,' includes both physical and psychical Nature. Physical Nature also plays its part in the World's History, and attention will have to be paid to the fundamental natural relations thus involved. But Spirit, and the course of its development, is our substantial object."[78] From this perspective, universal history set itself a different ambition than did the history of the universe: "Our task does not require us to contemplate Nature as a Rational System in itself—though in its own proper domain it proves itself such—but simply in its relation to *Spirit*."[79]

And yet at the same time, Hegel, like Kant, found hope for a plan in nature through Kepler's approach to the natural scientific: "Universal history . . . shows the development of the consciousness of Freedom on the part of Spirit, and of the consequent realization of that Freedom. This development implies a gradation—a series of increasingly adequate expressions or manifestations of Freedom."[80] That form freedom takes expresses itself differently at different periods: "In history this principle is idiosyncrasy of Spirit—peculiar National Genius. It is within the limitations of this idiosyncrasy that the spirit of the nation, concretely manifested, expresses every aspect of its consciousness and will—the whole cycle of its realization. Its religion, its polity, its ethics, its legislation, and even its science, art, and mechanical skill, all bear its stamp. These special peculiarities find their key in that common peculiarity—the particular principle that characterizes a people; as, on the other hand, in the facts which History presents in detail, that common characteristic principle may be detected."[81] This determination of "the peculiar genius of a people," according to Hegel, is to be "derived from experience, and historically proved." A certain training or stamp of mind is necessary for this task: "The investigator must be familiar *à priori* (if we like to call it so), with the whole circle of conceptions to which the principles in question belong—just as Keppler [*sic*] (to name the most illustrious example in this mode of philosophizing) must have been familiar *à priori* with ellipses, with cubes and squares, and with ideas of their relations, before he could discover, from the empirical data, those immortal 'Laws' of his, which are none other than forms of thought pertaining to those classes of conceptions."[82] As with an investigator confronting the laws of the universe without the proper training (as Hegel would himself be accused of doing), a person "unfamiliar with the science that embraces these abstract elementary conceptions, is as little capable—though he may have gazed on the firmament and the motions of the celestial bodies for a lifetime—of *understanding* those Laws, as of *discovering* them."[83]

Thus Hegel, like Kant, regardless of their different frames and solutions, established the problem of history on the model of the orderable heavens. And they each routed the development of history through the unreasonable behavior of reasoning individuals: "We assert then that nothing has been accomplished without interest on the part of the actors; and—if interest be called passion, inasmuch as the whole individuality, to the neglect of all other actual or possible interests and claims, is devoted to an object with every fibre of volition, concentrating all its desires and powers upon it—we may affirm absolutely that *nothing great in the World* has been accomplished without *passion*. Two elements, therefore, enter into the object of our investigation; the first the Idea, the second the complex of human passions; the one the warp, the other the woof of the vast arras-web of Universal History."[84] And likewise, each emphasized the awfulness of history, Hegel in his discussion of universal history in still starker terms than Kant: "When we look at this display of passions, and the consequences of their violence; the Unreason which is associated not only with them, but even (rather we might say *especially*) with *good* designs and righteous aims; when we see the evil, the vice, the ruin that has befallen the most flourishing kingdoms which the mind of man ever created; we can scarce avoid being filled with sorrow at this universal taint of corruption: and, since this decay is not the work of mere Nature, but of the Human Will—a moral embitterment—a revolt of the Good Spirit (if it have a place within us) may well be the result of our reflections."[85] The fate of exemplary lives and nations "forms a picture of most fearful aspect, and excites emotions of the profoundest and most hopeless sadness, counterbalanced by no consolatory result."[86] From this "mental torture" a retreat follows. There is also the challenge to the universal historian: "But even regarding History as the slaughter-bench at which the happiness of peoples, the wisdom of States, and the virtue of individuals have been victimized—the question involuntarily arises—to what principle, to what final aim these enormous sacrifices have been offered."[87]

The possibility of redeeming the horror of history by establishing the Kepler or Newton who will find in history reason or plan would remain part of universal history and the history of the universe thereafter. Laplace, for his part, would have been surprised if such an experiment in discerning this thread of reason would have been possible without natural science.

However much universal history was not a history of the universe, as Hegel emphasized, the history of the universe was increasingly posited as a universal history in senses promoted through the Kantian and Hegelian traditions. And however much the character of their and other synthetic historical work did not engage the same mathematical or observational methods of a Laplace or Herschel, the ambition of Kantian and Hegelian universal history was grounded as much in Keplerian and Newtonian orderings of the solar system as it was in a sacred, providential history.

UNIVERSAL HISTORY, NATURAL HISTORY, HUMAN HISTORY

The apparent divergence between natural history and earlier versions of universal history has at times been overstated, including by those authors articulating them. This apparent distance was only emphasized by identifying the Enlightenment with secularism.[88] As recent historiography and methodology in religious studies has undercut any simple picture of such secularization, aligning universal historical impulses with exclusively secular grammars becomes increasingly untenable.

Scientific histories of the natural world and humanity's place in it, like both universal histories and sacred histories, perhaps inevitably organize materials according to narrative structures of period and of event, largely in a linear progression or with a linear through-line. As such, the structure of narrative itself might impose progression and teleological finality.[89] But the later scientific histories that are so much the subject of this book also share with earlier attempts at universal histories the impulse to characterize humanity and the contemporary moment through a natural historical account—and recurrently to "descend" to the present and to moralize in the manner of what Hegel termed "pragmatic history." These modes of history—the natural historical and the human historical, the human universal histories and the natural universal histories—could chastise and encourage at once, often through the shared theme of the progress of history and its eventual outcome. The issue of what, if anything, has decided human orientation or species orientation across these histories is often explicit and always in play.[90]

That the history of the universe is presented as a species of universal history might sound a glib or fatuous formulation. How could a history of the universe not be a universal history? But the proximity of terms is in the first instance (as a conceptual matter) misleading, even if ultimately, as a historical matter, they converge. It might as easily be asked how a natural history cannot be a human history. The genealogies constructed by historians of science confronting the divergence of natural history and human history, or those of historians describing the apparent expulsion of universal history from conventional academic history, underscore the conceptual possibility of separation. And if natural history was understood through paradigmatic figures of the museum collection, cabinet, or ark, then it could appear to be distant from a narrative structure more typical of universal and especially sacred histories. Nevertheless, how far separate any of these synthetic frames were depended on how that divergence was gauged. Moreover, the combinatorics of terms are confusing and confused: natural history, world history, universal history, universal natural history, natural universal history, and so on. To argue too strongly for or against their interdependence might be an exercise in nomination.

The reconfiguration, dissolution, and postulation of cosmologies detail a history of candidate universal histories, the authors of which sought out a cultural status imagined to adhere to those of paradigmatic sacred histories of the past, if on the basis of different narrative content and in the context of a different culture. Compared with earlier sacred or world historical accounts, these scientific histories did not have an obviously commensurate variety of the narrative elements of character, event, voice, and thematics; the content they appeared to shape shared more obviously in synoptic natural histories. But however far from either human universal or religious sacred histories, the ambitions of and lessons drawn from scientific totalizing histories were by the same token close to their less (natural) scientific counterparts.

METHODOLOGY AND HISTORICAL SHAPE

A friend who is an academic and has had formal education in philosophy at all stages of his training told me of an exchange with a career philosopher in which he happened to make a casual remark about the relevance of truth, to which the career philosopher replied, "Truth? It's only because you're not a philosopher that you ask about that." In its turn away from claims such as those of the *Annales* school and in its long dwelling on case histories, is scientific historical analysis subject to a similar characterization—the more so given any tendency to demand that national borders and established periods be observed, periods of ever smaller duration, as the history in question approaches the present? Can historical analysis range over extended periods of time, finding ways to address studies of different durations, without at the same time lapsing into excesses of irresponsible generalization, or presumption, or bloated duration? How is it that by contrast, history, even critical history, might be more comfortable with the attitude toward space of comparative studies than with the attitude toward time of durational studies? If there is any truth to these characterizations, then some of what drives the intellectual claims aligning shorter periods with greater safety (whether epistemological or ethical) must itself be rooted in the disciplinary construction of history as a subject, in the modes of synthesis of the subfields of history themselves. In turn, some resistance to varying historiographic confrontations with time must be that which is provoked by trends that at their best encapsulate a wise, but forever limited, selection of possible ways of doing history, and that at their worst have little intellectual justification other than professional pragmatism.[91]

In attempting to embrace the merits of such historical conventions while at the same time sustaining the object of inquiry—of a particular mode of synthesis, among other such modes—my presentation is largely sequential, attempting to give due weight to periodization as well as to genealogy. I attempt to sustain a balance between continuity and discontinuity, innovation and repetition—an idea of balance that may itself be an oversimplification, depending on whether elements of the history (relating to ambitions, dilemmas, valuations, and so on) can be said to recur or whether what is new in a historical period may be a property more of the whole than of its parts. Genealogically, this work seeks to understand how an intellectual and cultural space was established in which scientists could function as the authors of a new universal history as one among related natural histories or genres of synthesis—as a myth taken seriously by a wide audience of both supporters and detractors.

As such, it reveals a persistent alternative tradition of critical history within the natural sciences in regard to universalization as such. It hazards a corrective of certain historical treatments of universal histories tending to mirror the poles of the myth's own reception: a dismissal on political, ethical, or epistemological grounds, on the one hand, or a more uncritical embrace, on the other. The accounts in each generation demonstrate sufficient diversity and criticality to merit more historical analytical responses. By peering into the past, the historicity of the current universal history is revealed and the fact that different elements of it may be subject to different temporalities. The "largest" contours of a naturalistic, developmental history might be of an antique vintage. On the other hand, an account that harmonizes different chronologies, different disciplines, in a largely linearizing story suggests a more recent tale.

To capture those temporalities does not in itself require a sequential history, but having experimented with different modes of organization, I have opted for a largely developmental account of an evolutionary universal history that, however much each chapter is invested in its historical moment, may share some of the problems and potential seductions of the natural and universal history that is the object of my account. To capture the historicity of the contemporary myth requires some comparison between elements of the past and the future, and this could be drawn still more selectively and sparsely than in my effort. Here, however, the method adopted instead is conventional: to trace a genealogy of scholars and scholarship together, to attempt to see one set of lines that slowly build up elements only belatedly celebrated as myth by more than a handful of practitioners who explicitly heralded such an account. Moreover, if, as the examples considered here repeatedly evince, (scientific) universal histories as a genre attempt to help establish and fix the conditions necessary for the determination of human meaning, or even more, attempt to characterize that meaning itself and actuate it politically and culturally—then however much my own work may be made of the same stuff as the synthetic ambitions and logics it traces, my own synthesis falls far short of being a universal history.

The historical pivots of the individual chapters relate to the changing meaning and valence of synthesis at work in natural and universal history. Across a limited set of historical actors, texts, and cultural documents, often in discussion and debate, entangled in their own political and cultural dilemmas, experiments in synthesis were conducted. These actors repeatedly confronted the question of what relation the natural sciences bore to each other, to other fields of knowledge, and to personal and collective orientations.[92] Repeatedly, prominent practitioners hazarded such connections, which themselves helped define the boundaries of the fields they hoped to bridge—many of whom therefore fielded attempted syntheses that often failed to establish the kind of consensus that captured not only an avant-garde, an elect, or a cultish following, but appealed to a mainstream, as well.

These accounts therefore demand further reconceptualization of the relationship between technical and lay work. The other face of the emergence of the sciences as professions is the emergence of the conception of "lay people" standing outside those professions. As the disciplinary landscape of the sciences grew more intricate and involuted as the professions proliferated or fragmented into subdisciplines, the other consequence of this historical process was the dissolution of the universal expert—or rather, it should have meant the dissolution of a category that was never realized, repeatedly affirmed as possible only in a less knowledgeable past, but also repeatedly celebrated. As such, as the scales and divisions of natural science were likewise perceived as growing, the idea of speaking to scientists as a whole (much less to a broader audience) itself became a more self-consciously mythic idea, regardless of whether a scientist conveyed a historical synthesis or any other overview of the sciences. The putative universal scientist would speak across all the different scientific professions, to tens of thousands with vastly different backgrounds and training, in a way that was found relevant to their work and with the expectation that the author would be heard. When speaking entirely within the idiom of their own expertise, actual authors or researchers could not expect to be understood by everyone who was a scientist by profession.[93]

But such synthetic activity in different periods and modes was itself carried out through a diversity of cultural and political valences in different national and cultural quarters. In one period, ethnic particularity might be celebrated by some scientists, universalism defended by others or by other cultural critics; in a later period, the majority of scientists might stand by the most thoroughgoing universalism of a cosmic tale, while critical philosophers could decry the more troubling and sinister histories embedded in universalist apologetics or positivist projects. These terms, particularity and universality, person and human, did not carry with them a fixed politics or a neutral epistemological field. And they advised different schemes for the conceptual and practical organization of the schools and bodies of knowledge, repeatedly leading to public addresses on the themes of what citizens and subjects need to know for their own welfare and the welfare of the state.

The work of the individual chapters that follow determined two structuring points for the work as a whole. First, the body of the work begins with the 1830s largely because of what the figures in the first chapter make clear: this period saw the theorization of natural history as such on grounds that largely preclude any *historical* synthesis. The natural historical theorization of the period precluded precisely the kind of synthetic experiments soon to emerge through the much-studied and contested syntheses of Robert Chambers, effected in part by the translations of Laplace's more expansive, Herschel-appropriated nebular hypothesis. It was also a period in which in the English context the term "scientist" was first hazarded, as a generalist category that to some might capture the apparent connections between the varied researchers of nature, but that to others was itself an admission that little of substance does connect them.

Here already is a dialectic between historical and ahistorical truth, between unity and diversity, unity and synthesis, between scientist and scientists, between scientist and nonscientist—whether the nonscientist is enfigured as classically trained patrician, as literary critic and humanist, or, more vaguely and problematically still, as non-Western non–truth-seeking subject. Attending to continuities and differences, to being a stranger and to being at home, is necessary in order to avoid so many methodological manipulations threatening to contort the analyst in such a way as to establish either that the world is created absolutely anew in every generation, or that the world never sees anything new. In so doing, this treatment also looks at which questions varied intermingling and uneven generations presume as perennial and seek to renew.

The second point requires departing to a degree from certain forceful inherited and intellectual conventions concerning disciplinary and national history, at the risk of appearing to revive older intellectual historical conventions or of appearing to relate an evolutionary account in some way uncomfortably close to the evolutionary accounts the history treats. Again, with regard to the latter, the explication of universal history given in this work, if by its own convenient lights, does not amount to a universal history or any of the genres of synthesis discussed. But to the extent that efforts at synthetic historical accounts in the present moment evoke an evolutionary synthesis, this historical and largely chronological account will evoke the same.

I do not believe conventions of time or nation are flouted here, nor do I imagine that the good sense of them is undermined by this work. At the same time, it impossible to accept that the conventions are themselves so universal that to violate them to some degree is somehow to render a historical analytical work more faulty or more outdated than other historical efforts. In attempting what must clearly be a fragmentary account of such synthetic work, too much obeisance to national boundaries or to established periods would prune some of the essential members of any given branch. It would also at times mistake the shifting terrain built out of the changing homes and developing networks housing these figures, or mistake the extent of some of the political dilemmas in which these figures were involved. Only the most obvious case of this is the migration of scholars out of continental Europe in the 1930s.

This work falls into three parts of four chapters each. The first part is centered initially on European scholarly networks in the nineteenth century, primarily English with strong connections to German and French intellectual discussions, and their deliberations on "Herschelian" natural histories—conditions for the possibility of genres of synthesis and totalizing visions in their own right. The second part focuses on twentieth-century networks increasingly based in the United States, networks that expanded with the emigration of scholars in the 1930s, but that include the earlier and enduring role of US scientists in their connection to primarily European-based research. These chapters trace the emergence of genres of synthesis along with the emergent theorization of science in relation to epic and myth. The final part, from the last third of the twentieth century, examines the fuller development and articulation of these genres, the scientific and institutional commitments they structured and reflected, the cultural products to which they gave shape and cultural positions they held. Each part examines universal historical and related syntheses, giving prominence to efforts to author an immanent historical account of the material world that simultaneously promoted a rationalizing vision of human history and the interrelations of knowledge—an account that is "immanent," rather than "secular," since an immanent account might still be religious, a religious account immanent, and neither tempting the idea of "secularization."

As will be seen repeatedly, many of these thinkers who addressed history and synthesis did not stop at the apparent borders of their disciplinary expertise and thus their apparent license.[94] This disciplinary reach and the diversity of training of those reaching out for other disciplinary, authoring hands combat any chance perception that the physical sciences were for much of their history free of the ambition for an expert, universal historical synthesis.

The demands of storytelling and synthesis shaped contemporary scientific research and the ways in which scientific authoring, through universal history, came to determine the present-day character of myth. Such demands entail examining different modes of contemporary natural history—its historical and ahistorical modes—how it inhabits culture, politics, and pedagogy across the proximate and partial genealogy addressed here. As endorsement, myth is largely understood by the synthesizing architects treated here as the most all-encompassing and truest of stories, a natural history that gives social and cultural orientation, names the essential things of the world, and offers a framework embracing diverse and potentially contradictory social values. The genealogy of the "scientific epic" and related synthesizing genres directs attention to the changing nature of expertise, popularization, and unification over the extended period during which it was established. These narratives were embedded in the often-fragmenting political and value-laden struggles of the scientific authors who researched, wrote, and filmed them, struggles that in turn drove and informed scientific results and goals. These efforts in media programming, in popular and pedagogical outreach, in conceiving unity and experimenting in synthesis, structured and advanced a contemporary myth of science, instituting different devices for domesticating the expanse of that history while positing a culturally charged and problematized species voice. As such, the prominent historical narratives and syntheses of science were and are forms of knowledge in themselves, revealing what science means in wider culture and focusing attention on the almost paradoxical nature of writing a myth where humanity is known to be the author.

PART I

CHAPTER ONE

Varieties of Natural History

The Whole of the Natural and the Known

Is this wondrous world of matter and of thought, of object and of subject, of blind force and of moral relation, a one indivisible and complete whole, or a mere fragmentary assemblage of parts . . . ?

JOHN HERSCHEL, Review of Alexander von Humboldt's *Kosmos*, 1848

Our classifications will come to be, as far as they can be so made, genealogies.

CHARLES DARWIN, *On the Origin of Species*, 1859

In 1830, William Herschel's son John (1792–1871) offered a telling definition of science: "Science is the knowledge of many, orderly and methodically digested and arranged, so as to become attainable by one."[1] The younger Herschel expanded on both parts of this definition—on the many and on the one. In the context of the physical sciences, the labor of the many addressed itself to the knowledge of the whole of nature. Herschel referred to that knowledge by the traditional category of "Natural History," and he proposed two characterizations of it, undigested and unarranged, on the one hand, digested and arranged, on the other. Undigested and unarranged, natural history was "a collection of facts and objects presented by nature, from the examination, analysis, and combination of which we acquire whatever knowledge we are capable of attaining both of the order of nature, and of the agents she employs for producing her ends, and from which, therefore, all sciences arise."[2] Natural history in this guise is knowledge concerned with the whole of nature, or as much as is humanly accessible, standing at the ready for future arrangement and use. It is all that is given in the natural world itself, prior to its reassembly into a structure of knowledge representing that whole in its totality. Seen in this aspect, natural history would be the starting point and ground for the natural sciences.

By contrast, natural history in its second sense was a goal. Digested and arranged, it was seen as an "assemblage of phenomena to be explained; of effects to be deduced from causes; and of materials prepared to our hands, for the application of our principles to useful purposes."[3] Natural history in the second instance, as an end, included any sufficient way to address and capture all potential knowledge through causal explanation.[4]

That arrangement of knowledge did not have to be presented as a narrative history, and for Herschel in 1830, such a choice would have seemed ill-considered. The controversial and sensational historical synthesis of knowledge that Robert Chambers was to offer in the following decade through his *Vestiges of the Natural History of Creation* was only one among the historical visions Herschel would reject. Natural history in its first sense, as undigested and unarranged knowledge of particulars, may itself have offered materials for a history, but natural history in its second sense, as knowledge of nature as a whole, was not essentially or even primarily historical. Instead, natural history *as an end* suggested a synthetic view of nature that might potentially offer a privileged knower an outlook on the whole of the natural world and on natural philosophy, as well, beyond any narrative of their development.

Herschel therefore regarded science in its essence as a matter of a *surveyable* synthesis, one that was not primarily historical. Scientific practice was to his mind unified through shared methods across a collective effort, but practitioners also strove toward a shared final structure. It was not for everyone to attain a sense of that ultimate goal. However it was conceived, that final structure was primarily to be adumbrated by the efforts of the few synthesizers who drew together different phenomena and eventually different branches of the physical sciences under increasingly general description. The "one" for Herschel was not only or primarily an abstract figure—a person who could follow all of science from an armchair—but was an architect of that future assemblage. The main subjects of this and the chapters that immediately follow are such candidate architects in the nineteenth century, their connections, cultural supports, and synthetic constructions.

At the time, Herschel was becoming as esteemed a figure as his father. His astronomical discoveries extended those of both his aunt Caroline and William, and the junior Herschel contributed to various domains of science and technology, from mathematics to photography.[5] His *Preliminary Discourse on Natural Philosophy*, in which he gave this exposition of natural history, would extend his reputation further. Alongside his voyage to the Cape of Good Hope in 1833 to carry out an observational survey of the Southern Hemisphere, the *Preliminary Discourse* would help to establish him as the picture of gentlemanly science.[6] In the context of an attack on the character of natural scientists as a group, Chambers targeted Herschel in particular, since he was "the very prince of modern philosophers."[7] However, Herschel was hardly alone at that moment in attempting to grapple with the question of what would constitute natural history in either of his senses, either as a growing body of knowledge of particulars or as a synthetic overview of nature as a totality. He and others thus confronted an implicit tension between science seen as the ongoing accumulation of knowledge of the world, a process that is open-ended, and science seen as offering a totalized view of the world as a whole, as the telos, outcome, and end of that process, linked to varying visions of underlying or potential unity. His friends Mary Somerville (1780-1872) in Chelsea and William Whewell (1794-1866) in Cambridge confronted the implications of this tension directly.[8]

By the early 1830s, Somerville, an unconventionally trained natural philosopher, had already completed her *Mechanism of the Heavens*, an authoritative exposition of Laplacian celestial mechanics. She followed this work three years later with an overview of the state of scientific research in her *On the Connexion of the Physical Sciences*. In it, she testified to the interdependence of the different branches of the sciences and their tendency to come together as a result of a widening application of mathematical analysis. Whewell, in his own self-formation as a gentleman of science and a Trinity don, also was widely canvassing different branches of the sciences himself. But in his laudatory review of Somerville's efforts, he saw fragmentation where she saw unification. To her, science was "daily extending its empire," expanding and coalescing at the same time.[9] To him, the fragmentation of science "was like that of a great empire falling to pieces." He lamented that "science, even mere physical science, loses all traces of unity."[10] But Whewell nevertheless saw in Somerville's work the portrayal of the "noble expanse" of scientific labor and results, a prospect on the totality of knowledge.[11] He dealt with the tension between particularity or process and end by distinguishing between unity and synthesis, a distinction that will be important to observe here. Whewell also used the opportunity afforded by the review to suggest, probably lightheartedly, that the word "scientist" could be used to name Herschel's many laborers.[12] Even given an initially humorous intent, the neologism asserted the fact of a shared undertaking.

Extending John Herschel's contemporaneous use of the term "natural history," with an emphasis on its second sense, this chapter and those following examine paradigmatic attempts to grasp the totality of the natural world in the middle of the nineteenth century and on, attempts troubled by and constituted through conflicts in conceptualizations of nature and of natural history difficult to reconcile. I use the term "natural history" therefore to apply to attempts on the part of popular and expert authors such as Somerville,

Herschel, and Whewell to address the whole of the natural world through the entirety of the natural sciences—together, ultimately, with the human sciences. These are efforts to find a framework to capture and address everything. They are constituted in part by investigations of the ways in which the sciences might be interconnected. Were the sciences part of one system of thought and practice? Would they be in the future? Were they predicated on an identifiable inherent unity, or would knowledge of that unity be deferred, to be established at some future point, as the nourished state of a finally arranged and digested natural history? Analysis of the natural historical efforts of such practitioners and their connection to those of others they admired, such as Alexander von Humboldt, or largely rebuked, such as Chambers, allows us to contextualize the ongoing task of writing universal history as one kind of synthetic program and the problematics involved in that effort.

HISTORICAL CONSCIOUSNESS, WRITERS, AND REFORMERS

The advent of "historical consciousness" at that time has received critical attention from a diversity of perspectives.[13] Focusing on the period from 1800 to 1830, the historian of science Richard Yeo observes that the merits of the claim to this novel historical consciousness rest on the twin facts of "new critical use of sources, especially primary, archival documents," and "a greater sensitivity to the social and cultural differences between various historical periods."[14] In reference to the specific question of science alone, Yeo notes that "these early decades of the nineteenth century also saw the emergence of new disciplines such as geology, archaeology, anthropology and the study of language, all having strong historical frameworks."[15] Historians have argued that the appearance of such a new historical consciousness was itself occasioned by still earlier encounters with different cultures, by a shifting geographical imagination and the possibility of new and different histories.[16]

In light of this historical consciousness, in the European and specifically English-speaking context in the middle third of the nineteenth century, it is striking that many of the natural philosophical commentators who reflected on the possibility of a universal history also applied themselves to questions concerning the nature of the interrelation of the different branches of knowledge. Such questions could complicate that project, because the conformity of all knowledge to history and more narrowly to a shared timeline was a hard constraint to satisfy, intrinsically placing demands on the topography of knowledge and natural history.

Scholars who focus on Victorian England also couple the new historical consciousness with an emphasis on the emergence of popular programs of education, the popular press, and popular literature at that time. Many see these ambitious pedagogical and universalizing efforts as tied both to voting reform and to a politically informed geographical and/or spatial imagination.[17] It was no accident, then, that in Britain and elsewhere, a community of scholars addressed as mutually implicated issues the interrelation of the sciences both theoretically and pedagogically, the possibility of natural history, and the origins of the world. Herschel, Somerville, and Whewell were prominent among them.[18]

Science writers such as these, along with the pedagogical reformers who published them, began to establish a self-conscious pattern of expert natural historical authorship that knowingly extended the philosophical and scientific traditions that preceded them. They helped initiate a genre of expert popularization adjacent to and defined in part by a contrast with popularization more widely and with pedagogy more narrowly. In doing so, they implicitly posited a kind of standing critique of the limits of educational systems and demonstrated the ways in which expertise itself is something of an elastic category. For Herschel and Whewell, the educational critique was linked to their own long-standing efforts at university reform, even while their individual efforts did not always share a consonant spirit.[19] For Somerville, the critique was produced by the necessity of achieving expertise and forging an audience from a position outside formal institutions. It was a critique made explicit as Somerville argued against educational systems and philosophies underestimating the intellectual capacity and potential of women.[20] But as we will see, they were not alone in this effort. Their publishers worked to establish alternative modes of education, while the work of synthesis in natural history was also carried out by more "amateur" writers. In stitching together fields of knowledge, however, some lack of expertise was inevitable. Even a celebrated "universalist" such as Whewell would fail to capture all knowledge expertly. And so in this way, as well, the possibility of realizing such a synthesis remained an issue.

HERSCHEL'S MANY AND ONE

An often-repeated observation in the relevant history and historiography is that John Herschel held a particularly special place among his more immediate colleagues and the broader intellectual community.[21] By contrast with his friends, Herschel's path to prominence and expertise was certainly better paved. Nevertheless, it was not self-evident that he would pursue the course that ultimately won him fame. He attended Eton in his youth, but was ultimately pulled from the school and tutored at home. At Cambridge, he followed his successes as senior wrangler, the top mathematics graduate, and as winner of the Smith Prize in mathematics and natural philosophy with an abortive foray into the law. It seems he was drawn to natural philosophy partly through an attempt to consummate the work of his aging father and aunt.[22] But unlike this more senior generation, the younger Herschel sharply distinguished the observational nebular work he took over from them from more cosmogonic conceptions.[23]

He published his popular and less technical *Preliminary Discourse* in 1830. The book appeared as the initial volume of the Cabinet of Natural Philosophy in the Cabinet Cyclopedia "conducted" (edited) by Dionysius Lardner. Lardner himself was something of a larger-than-life figure and, at the time, a lecturer and author of increasing popularity. The historical cabinet he edited included a universal history at the start of the 1830s, beginning with a discussion of the relationship between human and earth history.[24] Lardner also treated the nebular hypothesis in his own lectures on astronomy. But the project of the Cyclopedia as a whole was his signal effort. Prior to suffering a "fall" (from moving to Paris with a married woman) and relocating to the United States before ultimately settling in France, he attempted to advance and take advantage of "the March of the Mind"—the wave of new literacy in Victorian England that began to create a generalist audience for the new sciences.[25] Lardner emphasized the importance of such work both for artisanship and for liberal education. He joined Lord Henry Brougham of the Society for the Diffusion of Useful Knowledge (SDUK) in the belief that technology could not advance without scientific theory.[26] Lardner argued that knowledge of the sciences is crucial both to the moral and the intellectual welfare of the state.[27] As a result, he emphasized teaching theory and practice, in the universities and more broadly. For such a mission, John Herschel must have seemed an ideal contributor to the Natural Philosophy Cabinet.

The *Preliminary Discourse* was ultimately predicated on views not completely consonant with Lardner's, but Herschel worked to promote the signal importance of the natural sciences and to convey a sense of the physical sciences to a wide audience.[28] However, he certainly did not endorse the possibility that a thorough understanding of the sciences could be by and large accessible to a broad public. The three parts of the *Preliminary Discourse* dealt overall with the benefits, approaches, and place of the physical sciences. The first part addressed the mental and social advantages these sciences can offer. The second part treated their methods, providing an extended articulation of Baconian induction. The third and final part approached a theorization of the relations of the different branches of the physical sciences to each other and, more broadly, if more briefly, to other branches of knowledge.

Central to the first part was Herschel's definition of science, of its many laborers, and of the possibility that their work could be conceptually arranged by one. The definition was preceded by a moral defense of the sciences, emphasizing their favorable influence on practitioners and human society. It was in the second part, however, that he connected this problem of the one and the many to that of induction. He defined induction broadly as the process of moving via well-chosen classifications from individual objects to greater and greater levels of generality (with "general facts" pertaining to the groups, rather than to the individuals), ultimately arriving at laws and the "highest" axioms. On this view, induction is climbing the syllogistic ladder of a logical structure ordering objects in the world. This, in turn, he connected to two modes of induction, one involving many people comparing agreed-upon facts, the other involving the discernment of a principle of agreement in all objects of study (of individuals in a class). In Herschel's exposition, the former relies on "a division of labour," the latter on "individual penetration" requiring "a union of many branches of knowledge in one person."[29] Natural history should provide a prospect the whole of which was visible to a practitioner, even if initially only to the most gifted. A conception of scientific labor (and of a particular kind of scientific laborer) was here already intrinsic to the possibility of a conceptual synthesis and the potential unity that would flow from it.

Throughout the *Discourse*, however, Herschel asserted that natural science *already* formed a unity, standing as a paradigm for other disciplines of knowledge in its emphasis on its shared, exacting methods—a paradigm for those disciplines that were scientific or sought to be, such as politics, legislation, economy, and history. This was a claim that shared a moral vision with other conceptions of universal history: the bloody course of history must be redeemed by being reconciled to an idea of the progress of reason, realized in the emergence of a more perfect government.[30] The claim joined with other declarations of the exceptionalism of natural science precisely along these lines—that such science stands apart because it is singularly successful in reaching its aims. The place of natural science among the branches of knowledge was that it was free from any admixture of speculation and therefore stood as a model to every branch.

But in approaching that whole of natural philosophy unified through its methods, Herschel found it necessary to explain the relationship of its parts. In the final section of the book, he sketched the interconnection between the different disciplines of the physical sciences. He initiated his explanation of the relationship between the branches by first invoking the large difference between statics and dynamics and further dividing each category into the physical state in which their objects of inquiry were to be found: solid, liquid, or gas ("airs"). This division ultimately resulted in different branches such as hydrostatics or hydrodynamics or, drilling further into material structure, crystallography. But his was not a fully systematic delineation, and Herschel did not appear to intend it as such. Whatever the case, that more complete delineation would amount to a sketch of the totality of knowledge that the physical sciences attempted to assemble, and the ability to approach that systematization as an end was the unique success that natural science already exhibited, at least to Herschel's mind.

These divisions of knowledge carried into astronomy and geology. By the treatment of these branches, the *Preliminary Discourse* also suggested some of the reasons why historical narrative might not have presented itself as an obvious choice for a synthesis of knowledge, despite the era's historical self-consciousness. Herschel pointed to the power of geology over astronomy to produce knowledge of origins. Even more confidently, he asserted that physical astronomy was "confessedly incompetent to carry us back to the origin of our system, or to a period when its state was, in any great essential, different from what it is at present."[31] This statement was a quiet and subtle play of allusion. He both asserted his own position in potential contradistinction to his father's, whose nebular hypotheses could at least appear to lead to origins, and played with a famous phrase of James Hutton. Hutton, in the conclusion of his *Theory of the Earth,* had declared of the earth system that "we find no vestige of a beginning,—no prospect of an end."[32] Herschel in turn declared: "So far as the causes now in action go, and far as our calculations will enable us to estimate their effects, we are equally unable to perceive in the general phenomena of the planetary system either the evidence of a beginning, or the prospect of an end."[33] The statement seemed more a sly correction to and critique of Hutton than a testimony to him, for though neither astronomy nor geology spoke against the possibility of the earth having a beginning, it was only geology that gave positive proof of that origin.[34] Hutton's assertions were already regarded as outdated. What Herschel saw as distinguishing the geology of his day from that of what he called "a race of geologists now extinct" was the resistance to "wild theories of the formation of the globe from chaos," substituting instead theories based only on evidence of former times.

Yet this distinction between a historical earth and an ahistorical cosmos in certain ways leapt back over the figure of Hutton into a more distant past.[35] It is redolent of medieval cosmology, with its representation of the universe as changeable below the moon and unchangeable above it. Working to eliminate any such difference, in its own attempt at progress, Hutton's "race of geologists" had patterned its geology after Newton's cosmology.[36] So Hutton's advocate John Playfair in 1802 had connected Laplace's cosmology with Hutton's geology and concluded that "Dr Hutton's theory of the earth comes at last to connect itself with the researches of physical astronomy." To Playfair, this was a satisfying example of consilience, the linking of principles and evidence from different disciplines, though Whewell had yet to coin the term: "The conclusion to be drawn from this coincidence is to the credit of both sciences. When two travellers, who set out from points so distant as the mineralogist and the astronomer, and who follow routes so different, meet at the end of their journey, and agree in their report of the countries through which they have passed, it affords no slight presumption, that they have kept the right way, and that they relate what they have actually seen."[37]

But certainly by 1830, many didn't believe these travelers had arrived at the same point. With the Herschels, father and son, we have both the naming of cosmic beginnings and a distancing from pronouncements on these origins. The nebular hypothesis in the hands of the younger Herschel offered a way in which science partly defined itself in this period in contradistinction to what it could not offer and what therefore was *not* knowledge: the ultimate origins of things.[38]

But if he did not recount a foundation story for the physical universe, Herschel did present a foundation story for the emergence of different branches of knowledge. History did indeed organize the topography of the sciences, even if the content of knowledge was not always itself historical. Herschel's historical pattern went as follows: when scientists came to regard "ultimate facts" as "elementary," as not explicable in terms already provided by knowledge at hand, new branches emerged from old.[39] Establishing independent branches, Herschel argued, in turn opened up the possibility for still more expansive laws, revealing that analogies subsisting between branches were grounded in deeper principles. These laws then unveiled the formerly independent laws and phenomena as aspects of the same whole. Here, Herschel offered as evidence what eventually became a much-used example for this kind of delimited foundationalism: the analogies between electricity and magnetism uncovered the single phenomenon *and* the single branch of knowledge known as electromagnetism.

SOMERVILLE'S CONNECTIONS

The actual and potential connections between these branches of knowledge and the physical sciences as a whole were still more famously treated by John Herschel's friend Mary Somerville. But she first made her reputation by way of the mechanics of the solar system. As with the case of Herschel, outsized figures were involved in Somerville's introduction into popular writing. In 1827, Henry Brougham, who also was a founder of the *Edinburgh Review*, first solicited Somerville by a letter through her second husband, her cousin William Somerville, to write a treatment of Laplace's celestial mechanics for the Society for the Diffusion of Useful Knowledge, which had been instituted that year. Somerville later wrote of the output of the *Edinburgh Review* during that period of her life with admiration and was, it seems, flattered at the time by the commission. Brougham hoped Somerville would contribute to his efforts at what one commentator discerningly refers to as his "scientific 'induction' of the people"—Brougham's extensive educational initiatives addressing a rising middle class of which he was one of the most well-known and prolific members.[40]

Somerville's own course of study must have made her seem a natural choice for the effort Brougham had in mind. Socially well connected and celebrated, the Scottish, Whig-leaning Somerville was herself part autodidact, teaching herself classics, literature, mathematics, and the physical sciences from her teens on, despite family proscription. By her midthirties, she was living in Hanover Square in London in the society of varied illustrious figures, the more scientific among whom included Charles Babbage, William Herschel, William Buckland, Charles Lyell, William Hyde Wollaston, and Thomas Young.[41] Partly through travels abroad during the same period, she met and in some cases became more connected to scientists on the continent, such as François Arago, Georges Cuvier, and notably, if less firmly, Pierre-Simon Laplace. Through such contacts, she gained further expertise in the sciences, maintaining connections with the leading natural philosophers in Europe while conducting her own scientific research.[42] Indeed, the natural philosopher David Brewster urged her in press to devote more time to scientific research over and above her textual efforts, praising her work on the "magnetism of violet rays."[43]

Brougham wrote to William Somerville to enlist him in "the design I have formed against Mrs. Somerville." He had in mind for the SDUK "two subjects which I find it most difficult to see the chance of executing," one of which was an account of the Newton's *Principia* to be undertaken by an unnamed Cambridge source. Concerning the other subject, Brougham stressed, that "unless Mrs. Somerville will undertake—none else can, and it must be left undone, though about the most interesting of the whole, I mean an account of the Mécanique Céleste." With such treatises, he hoped to "carry ignorant readers into an understanding of the depths of science." And in the case of Laplace's treatise, he observed that "in England there are now not twenty people who know this great work, except by name; and not a hundred who know it even by name. My firm belief is that Mrs. Somerville could add two cyphers to each of those figures."[44] According to Somerville, Brougham visited Chelsea to urge her to undertake the project. She accepted the commission, but in the end, the book was published by John Murray, instead.

John Murray II had founded the ongoing *Quarterly Review* in 1809 and had in 1829 initiated the "Family Library" series, cheaply produced, copyrighted volumes, with the intent of producing a wide readership.[45] But Somerville wrote her exposition in a register that did not attempt to engage the wider readership Brougham and Murray were trying to construct and instruct. Somerville's 1831 *Mechanism of the Heavens* was mathematically demanding. Both Whewell and George Peacock wrote to her directly in 1832 to inform her that they would introduce it into Cambridge undergraduate studies.[46] She reported herself that the majority of the initial 750 copies were sold in Cambridge.

In *Mechanism*, she explicated and endorsed the expectation that the universe could maintain a perpetual stability, and in a manner similar to Herschel, she weighed the historicity of geology against the timelessness of astronomy. Overall, she constructed an atheistic Laplace friendly to natural theology[47] and amenable to the vocabulary of the English political arena.[48] But she subtly demurred with respect to scientific history while laying the ground for its articulation.

Her well-known "Preliminary Dissertation" to the work itself circulated as a popular treatise in the United States as early as 1832.[49] Somerville also distributed the "Dissertation" separately to acquaintances, among them Maria

Edgeworth, who in the same year found her own mind "so distended" by its portrayal of the "scientific sublime."[50] In the "Dissertation" and in work that followed it, Somerville interrogated stability through an examination of the various perturbations of planets from their axes: "Minute as these changes are, they might be supposed liable to accumulate in the course of ages sufficiently to derange the whole order of nature, to alter the relative positions of the planets, to put an end to the vicissitudes of the seasons, and to bring about collisions, which would involve our whole system, now so harmonious, in chaotic confusion."[51] Here, in looking to the "proof . . . that creation will be preserved from such a catastrophe," Somerville made a weighty concession: "nothing can be known from observation, since the existence of the human race has occupied but a point in duration, while these vicissitudes embrace myriads of ages."[52] She invited her readers to peer into an epistemological abyss by contrasting the human scales of empirical knowledge with the observational demands necessary to affirm permanent universal laws. But she immediately closed up this abyss with an appeal to the recent triumphs of celestial mathematics: the astronomical perturbations in question could be described by mathematical sinusoidal/cyclical functions and therefore as forever bound within certain limits.[53] She also stipulated in her work what would become a sticking point upsetting that demonstrated stability of the system: these arguments assumed that there is no "resisting medium," that space is a perfect vacuum.

Even given the emphasis on stability, Somerville recounted William Herschel's version of the nebular hypothesis, but in the case of the nebulae, as opposed to the perturbations, mathematics could not yet take over from observation. Giving pride of place in her account to the observational astronomy of Caroline and William Herschel,[54] she made short work of the actual statement of the hypothesis: "The nature and use of this matter scattered over the heavens in such a variety of forms [of cloudy/nebulous matter] is involved in the greatest obscurity. That it is a self-luminous, phosphorescent material substance, in a highly dilated or gaseous state, but gradually subsiding by the mutual gravitation of its particles into stars and sidereal systems, is the hypothesis which seems to be most generally received."[55] Here she also introduced a concession regarding observational evidence, averring that "the only way that any real knowledge on this mysterious subject can be obtained, is by the determination of the form, place, and present state of each individual nebula, and a comparison of these with future observations will show generations to come the changes that may now be going on in these rudiments of future systems." Generational time would have to do some more work to compete with cosmological scales. In the case of the nebulae, that generational time was a literal matter: Somerville observed that the younger Herschel had taken up the work of the older and that their common project was nearly complete.[56]

Moreover, she introduced here an inconspicuous asymmetry: the solar system might have had a determinable beginning in time, but might not have an end. That was because geology was historical; astronomy was not.[57] Somerville argued that despite not yet being able to refer the Earth to the same certain treatments as the heavens, it was in the study of the earth that data concerning the remotest ages lay: "The traces of extreme antiquity perpetually occurring to the geologist, give that information as to the origin of things which we in vain look for in the other parts of the universe." Those geological traces, Somerville argued, "date the beginning of time; since there is every reason to believe, that the formation of the earth was contemporaneous with that of the rest of the planets."[58] An earth with an origin contemporaneous with that of other planets would contradict Laplace's narrative, where the planets emerged successively as the nebulous solar material contracted. Somerville's statement, if read as a precise claim, thus was an implicit rejection of Laplace's nebular hypothesis in the course of the defense of his celestial mechanics.

But here, not atypically, expansive time itself was no obstacle to harmony between natural theology and natural historical speculation. Despite her qualified endorsement of Herschel's nebular hypothesis and her subtle rejection of Laplace's, she saw little reason for natural science not to confront the origin of things and even, perhaps paradoxically, the origins of time itself. Somerville concluded the "Preliminary Dissertation" with the moral that "creation is the work of Him with whom 'a thousand years are as one day, and one day as a thousand years.'"[59]

The relative weight of geology and astronomy and their relationship to each other was a subject for thinkers interested both in the interrelations of the branches of science and with origins.[60] The question of the relationship between the sciences was one that Somerville took up in force in her next serious endeavor, *On the Connexion of the Physical Sciences*, first published in 1834.[61] It was a work that ran into so many editions that the content of its later versions bore little resemblance to earlier ones.[62] In the very first words of the first edition of *Connexion*, Somerville claimed that "the progress of modern science, especially within the last five years, has been remarkable for a tendency to simplify the laws of nature, and to unite detached branches by general principles. In some cases identity has been proved where there appeared to be nothing in common, as in the electric and magnetic influences; in others, as that of light and heat, such analogies have been pointed out as to justify the expectation, that they will ultimately be referred to the same agent: and in all there exists such a bond of union, that proficiency cannot be attained in any one without a knowledge of others."[63] In this brief passage, the demonstration of which Somerville expanded on throughout *Connexion*, she indicated two stark ways of uniting the sciences: through reference to the same agent and through the imbrication or interpenetration of working scientific knowledge.

Just prior to her claims concerning the extension of the empire of mathematics, Somerville observed that the "connexion of heat with electrical phenomena, and the electricity of the atmosphere, together with all its energetic effects, its identity with magnetism and the phenomena of terrestrial polarity, can only be understood from the theories of these invisible agents, and are probably principal causes of chemical affinities."[64] By "agent," she meant the material actants lying behind and causing more evident phenomena. She juxtaposed this idea with mathematical analysis. To find a unified mathematical form connecting two fields of knowledge might have seemed nearly equivalent to connecting the fields by finding a common underlying agent. These, in turn, seemed to suggest and depend on the fact that several sciences must be known to practice any one.[65]

Ultimately, the work as a whole showed how one branch of the physical sciences (in application or theoretical exposition) was required in the understanding of another and how each naturally flowed into each. So studies of the "celestial bodies" might lead to the study of tides or to estimates of the velocity of light, each, in turn, leading to or connected with studies of geology, heat, or electricity, and so on. In the conclusion of the work, Somerville, like Herschel before her, more explicitly endorsed a view in which a form of mathematical analysis is tied to reducing the sciences to a fundamental material agency and to a fundamental set of axioms.

Though she retained the claims and stayed close to many of the phrases from her earlier more technical volume, she adopted a more strained rhetorical strategy in *Connexion*: the phrase from *Mechanism* that the "proof [of stability] is simple and compelling" was strengthened to the "proof is simple and conclusive," even as she made a greater concession that scientists by and large now accepted the fact that instead of being a perfect vacuum, space consisted of a resistive medium, "the ether." All the more intriguing is that in the 1834 edition and in editions thereafter, she did not alter this phrasing, continuing to endorse *as conclusive* the physico-theological argument for stability *and* conceding that a medium permeating the cosmos meant that decay was perhaps unavoidable.[66] For Somerville, then, the earth had a history, and the solar system might have an empirically inaccessible past, but likely had no future differentiable from the present; the disciplines were converging through mathematical analysis and through practice, but not through any narrative they could join in telling.

WHEWELL: EXPANSES AND INDUCTIONS

Another conspicuous point in this self-referring circle of commentary and learning was William Whewell, who, in reviewing the work of his friends, advanced his own views of their shared concerns: universal history, the interrelationship between the sciences providing and given by a representation of the totality of the world and of knowledge, and the tensions between historical and ahistorical disciplines and specific nebular hypotheses.

Whewell followed another path altogether to self-formation as a spokesperson of science. A master carpenter's son who was a talented student across a plurality of disciplines, Whewell eventually became an elder statesman and Victorian scientific idol.[67] His own scientific research included mineralogy, crystallography, and, fundamentally and most extensively, the study of tides, "tidology," as he coined the term (adapted from an earlier "tydology"). Tidology linked him to the Admiralty through the organization of globalizing tide experiments, in the course of which he attempted to gain a view of all scientific effort and its historical lines at once.[68] Among these treatments of shared philosophical and historical concerns, Whewell's might be considered both the most synthetic and the most critical, contrasting with Somerville's technically minded exposition and Herschel's more subtle, if starched tones.

Already in his 1831 *Quarterly Review* article on Herschel's *Preliminary Discourse*, Whewell expressed reservations about the general scholarly emphasis on Baconian induction and the multiplicity of its meanings, even while exempting Herschel from this criticism.[69] Whewell tellingly noted Herschel's omission of a step in the "inductive ascent" toward scientific truth: terminology as process, with which he, Whewell, as a fertile neologist, was very concerned.[70] In his strategic summary of Herschel, a sense of Whewell's own cast of thought was evident in his emphasis not only on the inductive, but the deductive in the characterization of science. Both the upward and downward movement along the syllogistic ladder were important, the latter of which Whewell claimed was too often neglected.[71]

Likewise, in his 1834 laudatory *Quarterly Review* response to Somerville's *Connexions*, Whewell took the opportunity to make a number of important declarations about the nature of science and of the popular scientific writing of the day.[72] The first point readdressed the problem of Herschel's one and the many: science itself was partly defined through the very tendency to branch out or to fracture. "The tendency of the sciences has long been an increasing proclivity to separation and dismemberment. Formerly, the 'learned' embraced in their wide grasp all the branches of the tree of knowledge. . . . But these days are past."[73] For Whewell, the separation divides philosophers and poets from science, but "the disintegration goes on, like that of a great empire falling to pieces; physical science itself is endlessly subdivided, and the subdivisions insulated." Whewell concluded poetically that science had lost any trace of unity, a lament in contrast to the prologue and conclusion of the very book he was reviewing.[74]

An indication of this dismemberment for Whewell was the fact that there was no term to designate "students of the knowledge of the material world collectively."[75] In the course of the previous meetings of the British Association for the Advancement of Science, "some ingenious gentleman proposed that, by analogy with *artist*, they might form *scientist*."[76] So the first published appearance of "scientist" was here, in Whewell's review of a generalist work of science by the generalist author Somerville. This signals a precondition of the term itself: that "scientist" is generalist. The coinage "scientist" intrinsically affirmed a unified identity across the different practitioners of the sciences, the crumbling of which Whewell also testified to and only partly resisted by the term's introduction.

The context of the word's appearance also affirmed the idea of popular treatments as privileged sites for expanding on the relations between the sciences. The strength of popularization, Whewell argued, was its ability to address difference. Precisely for that reason, popular science could give its readership a sense of the nature of scientific synthesis, if not unity: "There are two different ways in which Physical Science may be made popularly intelligible and interesting: by putting forward the things of which it treats, or their relations;—by dwelling on the substance of discoveries, or on their history and bearing;—by calling up definite images and trains of reasoning; or by taking these for granted, and telling what can be told in general terms concerning such matters."[77] Whewell saw Somerville as engaged in the latter task. And in every such respect—in treating the relations between the objects of science, in dwelling on the history and bearing of scientific discoveries, and in approaching science in general terms—Whewell was as much describing his own efforts. Through his own partial projection onto Somerville, he offered a curious defense of this second way of writing in a popular style.

So a seemingly dilated passage in the review began with the intriguing claim that "such general aspects of the processes with which science is concerned may be apprehended by those who comprehend very dimly and obscurely the nature of the processes themselves."[78] In a review of Somerville's attempt to canvas such general aspects, this may have appeared something of a swipe. But Whewell testified to her expert knowledge of the "leading branches" of the physical sciences.[79] The suggestion here might instead concern a concession to the limits of expertise, that no expert could take in the whole of science while having an intimate acquaintance with every subdiscipline. In this light, Whewell's claim was a concession to the impossibility of *universal* expertise—even before that fragmenting continued and perhaps even accelerated—coupled with an apology that the universal expert is unnecessary. And it was an argument paying a quiet compliment to the "omniscient" Whewell himself, who also sought a universalist understanding.[80]

In this same passage, Whewell directly turned to the claim that the "office of language is to produce a picture in the mind" evoking those "un-Pindaric" pictures where "we are struck by the profound thought and unity displayed in the colouring, while there is hardly a single object outlined with any tolerable fidelity and distinctness."[81] He conjured up one such image, against which to draw his conclusion: "The long-drawn vista, the level sunbeams, the shining ocean, spreading among ships and palaces, woods and mountains, may make the painting offer to the eye a noble expanse magnificently occupied; while, even in the foreground, we cannot distinguish whether it is a broken column or a sleeping shepherd which lies on the earth, and at a little distance we may mistake the flowing sleeve of a wood-nymph for an arm of the sea." The key to this mental picture is the sacrifice of detail for the broad contours: "In like manner, language may be so employed that it shall present to us science as an extensive and splendid prospect, in which we see the relative positions and bearings of many parts, though we do not trace any portion into exact detail—though we do not obtain from it precise notions of optical phenomena, or molecular actions."[82]

The object of this form of scientific popularization is the presentation of a "noble expanse." That vista can take a number of forms: the unity it broadcasts is in the manner of composition (for example, coloring), rather than in a specific principle uniting part and part (column or sleeve, shepherd or sea). There is a unity in the mere fact of the finished picture, having combined its elements into one, but that unity may not reveal any other trenchant, inherent connections between those elements, between the individual sciences—or by contrast any fundamental differences. Whewell almost suggests that the unity established depends on sacrificing any "tolerable fidelity and distinctness." The logic of the composition itself therefore also turns on the scale of composition, on the amount of detail and particularity effaced. To offer a synthesis of the sciences, then, promised at the very least this form of expansive, compositional, but qualified unity, based on a distant portrayal of the branches and an idealizing portrayal of the whole.[83]

Somerville's composition, to Whewell's mind, depended primarily on the historical discovery of general principles sometimes shared by different scientific disciplines.[84] But Whewell adds "that in the same way in which a kindred language proves the common stock and relationship of nations, the connexion of all the sciences which are treated of in the work now before us is indicated by the community of that mathematical language which they all employ."[85] It is unclear what he attributed to himself in this point and what to Somerville: whether ultimately he saw mathematics as demonstrating a past or future unity, an essential or strictly analogous unity, a unity enacted by a natural agent, or as established only by shared mathematical formulae.

Somerville, Herschel, and Whewell thus all saw themselves as engaged in what they regarded as a united project in more ways than the compositional, "un-Pindaric" form. Across the sciences there was "a bond of union," in Somerville's phrase, so strong that each of the sciences was necessary to the other. Analogies might indeed connect different branches of science more firmly, manifesting a logical, syllogistic connection through shared mathematical language. In some cases, the analogies could be taken further, to determine an underlying mathematical unity embracing the whole.[86] This mathematical unity is potentially grounded in the material unity of the world, such shared "agents" as Somerville points to in her preface and that Whewell reattests. But Whewell remained evenhanded. He emphasized as much the independence of branches as their suggestive analogies, lamenting the loss of unity even while propounding what unities there might be.

The same tension between process or particularity and totality in nature and natural history, and between difference and unity in the natural sciences, appears in Whewell's treatment of the issue of the stability or instability of the world as raised by the nebular hypothesis, a term that Whewell, ever ready with a new word or a new connotation for an older one, appears to coin[87] in his *Bridgewater Treatise* of 1833, *Astronomy and General Physics: Considered in Reference to Natural Theology*. By this time and particularly by the study of "Encke's comet," the motions of which were explained by positing a resistive medium, the ether, a consensus had emerged in favor of the belief that the solar system is ultimately unstable.[88] In his reply endorsing stability, Whewell enlists Laplace's own probabilistic arguments against the idea that the organization of the system could have arisen by chance. He asks whether there is any contradiction between cosmological stability and the argument that, by the existence of a resistive medium, the system must decay. That argument turns on his privileging of human scales of time. The system could have been unstable during the course of human history and futurity, but it was constructed to be stable enough so far as the human horizon is concerned: "It may perhaps appear to some, that this acknowledgment of the tendency of the system to derangement through the action of a resisting medium is inconsistent with the argument which we have drawn in a previous chapter, from the provisions for its stability. In reality, however, the two views are in perfect agreement, so far as our purpose is concerned. The main point which we had to urge, in the consideration of the stability of the system, was, not that it is constructed to last for ever, but that while it lasts, the deviations from its mean condition are very small. It is this property which fits the world for its uses."[89] The proper measure of stability, then, became not the existence of some time in which the stable solar system might decay—this Whewell took to be assured, and in a manner that defied Herschel's and Somerville's ahistorical astronomy, he robustly envisioned the prospect of an end—but rather whether that time makes any historical sense to humanity. And, claiming "not only that the past duration of the present course of things is finite, but that it is short, compared with such periods as we have had to speak of," he more emphatically declares that the resistive medium offers evidence for what the nebular hypothesis presumes, that the solar system had a common beginning, and from that beginning "to a condition still earlier than the assumed beginning;—to an origin of the original state of the universe."[90]

Whewell gave himself some room to expand on his views in his two multivolume undertakings, *The History of the Inductive Sciences: From the Earliest to the Present Times*, first published in 1837, and *The Philosophy of the Inductive Sciences: Founded upon their History*, first published in 1840. Herschel had already alluded to Whewell's work as underway in the *Preliminary Discourse*, declaring hopes that an analysis would soon emerge of the history of the different branches of science. In his dedication, Whewell returned the compliment, testifying to the importance of conversation with Herschel in the refinement of his own thoughts.[91] These are extensive and much (if still far from exhaustively) analyzed texts; we will briefly consider two matters of natural historical conceit and organization.

The first refers to Whewell's apology for treating only the physical sciences in his history. What recommended them for particular historical attention, he claimed, was their much-discussed connection. In the *History*, Whewell helped bring the English language ever closer to identifying science and natural knowledge, asserting that the physical sciences he treated were more definitively inductive than "Morals, or Politics, or the Fine Arts." There, he limited himself to those subjects that "appear to me to form a connected and systematic body of knowledge."[92] However, in the *Philosophy*, through his vision of scientific knowledge as the bridging of ideas and facts—as analogous to form and matter, the "Fundamental Antithesis," as he would come to call it—he repeatedly returned to the a priori and a posteriori limits of science. Concerning the ultimate matters standing at both the logical and chronological origins of things, he argued that an immanent knowledge of origins is impossible. And there, Whewell implicitly argued *against* the possibility of synthesizing the sciences according to any history, either: there was no potential narrative vista that could occupy them all in such a way as to disclose a principle of connection between them, beyond the type of history of scientific discovery he himself provided.

Individually, each of the sciences was supposed to follow a historical arc, with Whewell putting forward a conception of the maturation of the sciences. By this light, the physical sciences were more mature, and astronomy was the most complete of them. He structured that historical arc on the basis of a three-part sequence in which sciences moved from preludes, where the discipline under formation developed its central ideas and facts, to inductive epochs that established a framework for the discipline, and finally to sequels, a period of further articulation of and debate over the science. But as Geoffrey Cantor has observed, Whewell did not or could not apply this to the majority of the sciences he considered in the *History*.[93]

Then, in the first edition of his *Philosophy*, which begins with a set of, as it were, Baconian aphorisms, his final three aphorisms (111 to 113) conclude in relation to the "palaetiological sciences" that the original first cause is necessary to assume, but inaccessible, since historical chains of cause and effect point to, but are unable to reach, the highest origins.[94] These sciences rely primarily on historical rather than mechanical causes and the sole example of which up until that point was according to Whewell geology. And in the second volume, attention to the entanglement of origins and classifications persists, the boundaries of knowledge not yet formed or impossible to form looming at the edges of current knowledge programs, suggesting future work on morality, among other subjects.[95] There, Whewell organizes the sciences more explicitly by their fundamental ideas—the fundamental idea of time, for example, grounding the science of arithmetic, that of force grounding mechanics—but not in such a way that fundamental ideas could not also combine in the production of a discipline.[96] Once again, these inductive sciences naturally invoke history as a category, entangling historical and logical first causes and contrasting first and final causes. Here, unlike the biological sciences aligned with final causes, the historical sciences look to the first in a chain of causes to determine their underlying structure. Different fundamental ideas thus hold their own conceptual centers of gravity.

The nebular hypothesis potentially promised a narrative arrangement for all natural history. However, the idea of history as a final narrative was not in a position to supply an ultimate consilience of inductions, and on this different, a priori ground, geology and astronomy did not converge.[97] For Whewell as much as for Herschel and Somerville, historical narrative could not provide a synthesis. Natural history in the middle third of the nineteenth century was not primarily historical, even as it launched a process that many did not accept, the reconfiguration or narrativization of natural history as a historical synthesis. The scientist's universal history could not yet count as a natural history in Herschel's second sense of the term.

NARRATING HISTORY AND THE UNITY OF SCIENCE

The concept of a natural history thus almost immediately invoked the question of *narratable* history and the ultimate origins and limits or ends of things. Was the world meaningfully and measurably different in the past? Did every branch of science reveal a past that it could empirically access? If there were a robustly traceable linear history, and effects were produced through historical causes, could all of natural history be arranged historically? Would it follow therefore that the world would be different in days to come? Could first causes, logical and/or historical, be determined? Did the first logical cause point to the first historical cause and vice versa? Could such a modern history provide a sense of meaningfulness? Along such lines, the varying nebular hypotheses of Kant, Laplace, and William Herschel had been invoked and debated, often alongside attempts to conceive organizations of knowledge.[98]

But in a period of dissensus over which branches of science revealed a past, narrative history itself could not easily provide a framework for all of knowledge. Although history was beginning to emerge as a candidate final structure, natural history was not itself understood primarily as a history of the world. A concern with the organizational power of history can be gleaned at times as much from silent or received assumptions, or from apparent consensus, as from overt debate. The attested disjunction between geological and astronomical accounts of the origins of things demonstrates the potential appeal of achieving a historical synthesis as eloquently as any acceptance of the historical evolution of the world. As we will see in the next chapter, the appeal of the notion that the sciences were both tending toward synthesizing unification and positioned to tell the final story of the world was evident both in the enthusiastic reception of the work of systematizers such as Auguste Comte, Robert Chambers, and Herbert Spencer and in their critiques. The critiques worked to demonstrate the shortcomings of these totalizing accounts, without, however, fundamentally challenging the process of reconfiguring or narrativizing natural history as a historical synthesis and its weaving into broad-based pedagogical reforms seeking new audiences for new sciences, invoking notions of unity.

CHAPTER TWO

Dogmas of Unity and Questions of Expertise

As for other classifications, they have failed, through one fault or another, to command assent: so that there are almost as many schemes as there are individuals to propose them. The failure has been so conspicuous, that the best minds feel a prejudice against this kind of enterprise, in any shape.

AUGUSTE COMTE, in *The Positive Philosophy of Auguste Comte*, trans. Harriet Martineau, 1830

This book, as far as I am aware, is the first attempt to connect the natural sciences into a history of creation.

ROBERT CHAMBERS, *Vestiges of the Natural History of Creation*, 1844

The classification which this system implies may be said to be transverse to all ordinary classifications.

ROBERT CHAMBERS, *Explanations*, 1845

Among writers and scientific workers such as Herschel, Somerville, and Whewell, a discourse attempting to connect conceptions of the totality of things with ultimate limits (origins and ends) and with the relationship between the branches of knowledge explored the different and problematically related senses of Herschelian natural history, both treating nature as an array of particulars, a starting point of an open ended process, and analyzing it as a goal or end to be realized. This discourse entailed the idea of universal history and the representation of disciplinary knowledge, flirting with, encountering the obstacles to, or outright declaring a final objective for all these efforts together: a better understanding of humanity's relationship to its creator or the refinement of the human sciences on the pattern of the natural sciences, in progress toward an ideal world and in redemption of a bloody past.

But it was not this select company that, within this discourse of synthesis and unification, tended to narrate a synthetic history of the world. Instead, systematizers such as Auguste Comte and Robert Chambers, figures whom Herschel and Whewell might regard as impostors or upstarts, whose ideas

were judged by established men of science as conducive to both social upheaval and scientific incompetence, were those who most conspicuously set
themselves to establish conceptions of unity for this more narrative-based
agenda, synthetic visions predicated on different ideological commitments to
unity, social and scientific. Despite the criticisms that their systems ultimately
received and the faults they exposed, especially regarding the question of the
origins of the world and the actual unity of the sciences, some version of those
commitments to unity and synthesis nevertheless continued to inform and to
influence the thought and writing even of their critics. Herbert Spencer was
perhaps the most prominent synthetic thinker in these exchanges, developing both his evolutionary vision and his understanding of the relationship
between disciplines in critical dialogue with Comte. The principal question
for these commentators was not the desirability of such systematic accounts
of origins and unity, but who, if anyone, might be competent to achieve them
and the methods, if any, by which they could be achieved.

COMTE'S "NATURAL HISTORY"
AND DOGMAS OF UNITY

Like several of his English counterparts, Auguste Comte was no stranger to
conjectures concerning natural historical origins. He was for several years
one of the foremost advocates of Laplace's solar nebular hypothesis. Already
by 1826, he was giving the lectures in his home that would in the coming
decade result in the publication of the different volumes of his *Cours de philosophie positive*, lectures that attracted a number of luminaries, including Alexander von Humboldt.[1] From 1830 into the next decade, Comte gave annual
lectures dedicated to the workers of Paris on "positive astronomy," eventually
published in the *Cours*. By 1835, he was ready to present his own contribution
to natural philosophy/positive astronomy in an attempted proof of Laplace's
solar nebular hypothesis.[2] It was a proof that assumed that the planets had
emerged at a point of time when their centrifugal forces matched their gravitational attraction. From this fact, Comte derived Kepler's Third Law and
confirmed the observed periods of the planets.

But Kepler's law, as a mathematical consequence of Newton's gravitational law, did not represent new knowledge by virtue of this derivation and
the degree of its agreement with observation. Comte's proof of planetary
origins was circular: the consequence that the edge of the condensing solar
nebula or cloud orbits like the planet forming at the nebula's edge—and therefore conforms to Kepler's law—is embedded in the assumption that the edge

of the nebula orbits like the planet conjectured to form there. As a commentator at the time put it: "The vicious circle that Mr. Comte falls into may be summarized as follows: '*I assume in my formula that the sun rotates like the planet; and I find, after my calculations, that it rotates like the planet.*'"[3] Silvan S. Schweber, in excavating this history, has shown that this circularity was immediately perceived in France and Comte's memoire summarily dismissed. The critique of these same arguments a decade later in the United Kingdom would affect the reception of his ideas.

It was not this ultimately embarrassing work for which Comte was widely remembered, however. Rather, he was celebrated and more enduringly defended as a result of his theoretical and historical systematization of the sciences. Comte did of course have predecessors and articulated his extensive frame of scientific or "positive" knowledge following his former mentor, Henri de Saint-Simon.[4] Comte's indebtedness was a fact underscored by many, including Whewell and later still by Friedrich Engels, both of whom found Saint-Simon the far more cogent of the two. But Comte's theory also at times inspired and was promulgated by its more respectful detractors, Herbert Spencer among the most prominent of them.

In no small measure, Comte's effort was to establish a new social science. His system, like the works of the British synthesizers, suggested new areas for science to confront, and like them, Comte was deeply exercised by the question of the unity and disunity of knowledge. In his work of intellectual revitalization, he tasked himself with the elaboration of the positive or scientific philosophy through in part a delineation of what he regarded as the fundamental sciences and their relations to each other.[5] In this final positive/scientific age of knowledge, capturing the whole of the sciences in a system would itself, to Comte's mind, be a demonstration of the unity of the entirety of true knowledge, initiating a return to the more holistic ways of knowing of the ancients, but on the basis of a surer knowledge.

In such systematizing, Comte distinguished between a "Dogmatic" method and a "Historical" one. The latter displays the results of the sciences on the basis of how they had emerged. The former finds conceptual schemes ordering science from first principles, and as with Herschel's natural historical ideas, as a matter of an individual understanding: "Every science may be exhibited under two methods, the Historical and the Dogmatic. These are wholly distinct from each other, and any other method can be nothing but some combination of these two. By the first method knowledge is presented in the same order in which it was actually obtained by the human mind, together with the way in which it was obtained. By the second, the system of ideas is presented as it might be conceived of at this day, by a mind which, duly prepared and placed at the right point of view, should begin to reconstitute the science as a whole."[6] The Historical and Dogmatic might also therefore be seen as sharing a strong affinity with Herschel's two senses of natural history, as beginnings and ends.[7]

In Comte, there is a smooth elision in which these methods first apply to an individual science—the history or conceptual form of physics or physiology alone—but then address the sciences taken together: "In the main, however, our classification agrees with the history of science; the more general and simple sciences actually occurring first and advancing best in human history, and being followed by the more complex and restricted, though all were, since the earliest times, enlarging simultaneously."[8] As such, Comte's Dogmatic and Historical each suggested approaches to revealing a potential unity of knowledge.

Taking a cue from his vocabulary,[9] applying his terms less to the project of the individual sciences themselves and more exclusively to their overall relationships to each other, we can call assertions of belief in a scientific unity, according to whatever principle, "dogmas of unity." That belief need not assume or require some attempt to enact or even thoroughly explore its assertion of unity; such a dogma may indicate nothing more than the existence of such a belief, but may also include provisos and disclaimers relating to that belief.[10] And in these terms, historically minded dogmas of unity are what systematizers such as Comte (and Chambers and Spencer) held in common, constructing their own developing syntheses in relation to them.

Comte himself complained of the plurality of classifying dogmas, of as many dogmas of unity as there were people proposing them, and warned of the artificial, if not arbitrary, element of any such assertion, including his own.[11] However, the conspicuous failure of past attempts did not prevent Comte from attempting such a system himself. In his own work, he canvassed several. His preferred dogma identified the basic and independent sciences as those treating basic and general phenomena, the most complex and dependent as those treating the most complex phenomena. This hierarchy was predicated on the degree of complexity of subject phenomena and their decreasing age.[12] History and thought generally moved from the simple to the complex and equally from those things farthest from human scales (the astronomical), to those things nearest at hand (the social), to the fundamental divisions in between, given by physics, chemistry, and physiology. Comte saw the simplest matters as the furthest from the human and the most complicated as the most proximate. It is, in itself, a remark with an almost antique ring, since it affirms that the simplest things exist beyond the sphere of the human, or beyond what is beneath the moon. This simplicity also resonates with the view that history applies more to geology than to astronomy, a claim again no less redolent of older cosmologies.

By midcentury, if not earlier, Comtean positivism had been making its way into English society. George Lewes, John Stuart Mill, and David Brewster each reviewed Comte's work and translated it to the English context.[13] The translation of his lectures was taken up still more aggressively later by Harriet Martineau, whose popular condensation was intended to reach as broad an audience as possible. Those efforts helped give Comte's reflection on the different possible relations between the sciences a lasting fame, partly as a name for the tendency to introduce hierarchy in the sciences. And while the history of the disputation of Comtean dogma, not surprisingly, was nearly coincident with its articulation, the disputes extended chiefly to the shortcomings of what Comte attempted, or its untimeliness, and less to the idea of such an attempt at systematic synthesis itself. For his critics and for those who were unimpressed by his work, Comte and those who had embraced his system at some point were figures to define science against, not exemplars of the problematics of any vision of a unified and synthetic knowledge of the world.

Charles Darwin illustrates both such adherence to and repudiation of Comtean dogmas of unity. According to his own late report, Darwin claimed that he had not read Comte, feeling freed from that obligation as a result of the passing attack T. H. Huxley made in a much republished 1868 Edinburgh address, "On the Physical Basis of Life."[14] However, Darwin's program was constituted not only by templates derived from the work of those he acknowledged, such as those of Herschel and Whewell, but by templates derived from the work of those he dismissed. As Schweber puts it, "Darwin expressed the highest compliment that one thinker can extend to another by suppressing his indebtedness to Comte."[15] Many years earlier, Darwin had read more approvingly David Brewster's extended and largely sympathetic 1838 review of Comtean ideas on the nebular hypothesis and the relationship between the sciences.[16] This reading of the review, according to Schweber, was crucial to Darwin's appreciation of Malthus in a light essential for his discoveries. Despite then his later disavowal of the Comtean system, the ideological commitment and aesthetic attraction to such a prescriptive methodological unity of natural history and natural philosophy was part of the historical framework that Darwin inhabited and that Huxley also was deliberating over late into the century.

Huxley's own engagement with Comtean ideas was also more extended and ambivalent than his jabs suggested. Huxley never appeared to find much of worth in Comte's system, but while he regarded him as a flawed prophet of sociology, it was not the attempt at prophecy that drew his criticisms, but Comte's status as a scientific dilettante and that of Comte's disciples. Years before, in 1854, he had reviewed his acquaintance Lewes's account of positivism. By this time, Huxley had worked through medical apprenticeships and a scholarship to Charing Cross Hospital and had served in the navy between 1846 and 1850, as an assistant surgeon on the *HMS Rattlesnake*. Outside the Oxbridge elite and in an educational system that did not privilege his knowledge base, he struggled in a variety of scientific occupations before the mid-1850s saw his situation materially improve.[17] At the time of the review, Huxley was still struggling for security and as part of that effort contributing a column to the *Westminster Review,* which allowed him to pronounce on what science is and what it is not.

Largely praising, but also largely bypassing, Martineau's condensation of Comte's "six wearisome volumes of indifferent French," Huxley attacked the errors he found in Lewes's attempt to modernize Comte's science, rather than Comtean goals of synthesis as such, blaming the errors on the fact that Lewes had not done any scientific work himself.[18] Huxley noted: "We are taking advantage of no accidental mistakes, although those already cited would suffice to show, if demonstration were needed, how impossible it is for even so acute a thinker as Mr. Lewes to succeed in scientific speculations, without the discipline and knowledge which result from being a worker also."[19] The majority of the essay was spent in canvassing the work of scientific practitioners whose writings were more reliable gauges of their sciences. And in reference to work more directly related to Comtean philosophy—Comte's "Hierarchy of the Sciences"—Huxley gave priority to John Playfair, in his "admirable 'Dissertation on the Progress of Science,'—a sound and healthy essay on the philosophy of science."[20] In a letter to Charles Kingsley in 1869, his private assessment of Comte remained consistent, if brusquer and harsher than his public, while his views of Comtean classification were perhaps caustic enough to conflict with his earlier praise of the similar ideas he found in Playfair:

You are perfectly right in saying that Comte knew nothing about physical science—it is one of the points I am going to put in evidence.

The law of the three states is mainly evolved from his own consciousness, and is only a bad way of expressing that tendency to personification which is inherent in man.

The Classification of Sciences is bosh—as Spencer has already shown.[21]

ROBERT CHAMBERS AND THE QUESTION
OF EXPERTISE

Four decades ago, Frank Egerton observed that "the amateur synthesizer invariably chooses an area of knowledge having a strong social relevance in which the professionals in that field usually have not been willing or able to produce their own synthesis."[22] Such an argument is consistent with Huxley's views of Comte. But in their synthetic work, Somerville, Herschel, and Whewell themselves leapt over disciplinary boundaries in such a way as to problematize the conceptual integrity of an obverse category of "expert synthesizer."[23] Their own works, views, and status may have made that leap in some ways harder to make, but their publications suggested the extent to which neither popularization nor synthesizing was in any case contrary to their wishes or beyond their scope.

Each of these synthesizers, whether received as more amateur or more expert, was willing to synthesize on different grounds and with a different vision. One of the most sweeping syntheses was advanced by Robert Chambers. Four years after Whewell completed the first editions of his signature works on the history and philosophy of the sciences, Chambers's sensation-making *Vestiges of the Natural History of Creation* was published anonymously. As explicitly noted in its day, by its own author among others, it was a wide-ranging narrative synthesis connecting the different sciences through an expansive universal history. In putting forward what he believed to be a new induction of the unity of all the natural sciences, Chambers attempted in effect to become the Herschelian "one" to the scientific and nonscientific many.[24] The many responded with both praise and disdain.

As we have noted, a historical synthesis of some disciplines was already apparent as a project, even for some who greeted Chambers with the greatest derision, Whewell and Herschel among them.[25] In his *History of the Inductive Sciences*, Whewell had treated geology last of all the sciences. He critiqued the "antagonistic doctrines" of catastrophism and uniformitarianism, competing visions of the formation of the earth emphasizing either a series of violent events or the slow working of ongoing processes (and having coined "catastrophist" in this very text). In doing so, he noted that if geology thought to model itself after physical astronomy (as Whewell read Lyell reading Hutton), it could not do so by imagining geology as essentially ahistorical, as yielding the "perpetual uniformity" of Hutton and Lyell. This mistook the nature of astronomy, which in its aspect "as the science of cyclical motions has nothing in common with geology."[26] Instead, when treated "as a palaetiological science," astronomy suggested through the nebular hypothesis that there was progress in the heavens over time, and so, by analogy, progress in the history of the earth.

Such a position appeared almost tortured, particularly given his other claims concerning the nature of astronomy. Be that as it may, Whewell's classifications treated geology as essentially historical, so that a strict Huttonian position, that "we find no vestige of a beginning,—no prospect of an end," was ruled out. As a result, for a historical synthesis, geology must look to other disciplines for its historical lessons, even when those other disciplines (such as astronomy) seem essentially *a*temporal.[27] Chambers ran roughshod over any such fastidious reasoning and adopted the possibility that Whewell opened up, but would certainly not have recommended: to use the nebular hypotheses to render astronomy "palaetiological," linked to an equally historical geology.[28]

There were well-known intellectual resources for such a move close at hand. The astronomer J. P. Nichol had himself hoped to go beyond his own popular *Views of the Architecture of the Heavens*, to develop a fuller universal history. Judging by claims in *Views* concerning the phenomenon of the "ultimate dissolution of the solar system," his intended history extended a nebular progression into a sequential universal process. In broad terms, this is a view recalling Hutton's vision for earth's history: the solar system moved from arrangement to arrangement, from scene to scene, perpetually. Though Nichol's solar system had an origin, stipulated by a nebular hypothesis, how his comprehensive system was sustained once it began, and then what fueled its re-arrangements, was less clear. Nevertheless, for him, any indication that universal history was coming to a conclusion was itself proof that there *had to be* "unevolved wonders" securing a deeper permanence. Within the current cosmic scene, this "mighty cycle," Nichol theorized a progression shaping the physical and political world.[29] He was not the first to accept such an idea. A young Kant, for example, writing in the middle of the eighteenth century, argued for what he termed a "phoenix of nature": successive creation rippled out from a universal center into an endless spatial expanse, where the dissolution of any preceding worlds was followed elsewhere by a newer creation emerging from natures' inexhaustible treasures.[30] Kant's more mature expressions of universal history and of the future of civil societies and between states in turn invoked the promise natural philosophy held for understanding and forecasting the political world.

The extensive historical work on Chambers has revealed much of the reason for and the complexity of the sensationalism of *Vestiges*. Chambers was a scientific autodidact, and like Lord Henry Brougham and John Murray, involved in the construction of an expanding audience. A bookseller and author, in the early 1830s, he joined his older brother William in publishing the *Chambers Edinburgh Journal,* which (profitably for a future synthesizer) canvassed a variety of materials, including excerpting Herschel's *Preliminary Discourse.*[31] The brothers had a more extensive publishing firm, and as Sondra Miley Cooney has observed in her study of the firm, from the journal's launch in 1832, "the name of W. & R. Chambers of Edinburgh was synonymous with cheap, popular literature."[32] With a long intellectual and publication-based commitment to history and an interest in phrenology, in the 1830s, the younger Chambers became attracted to other branches of the sciences, such as geology and, particularly through his acquaintance with Nichol, astronomy[33]—and through such acquaintance, Nichol felt cheated by *Vestiges* of his own project for universal history.[34]

Vestiges itself underwent significant changes and varying receptions.[35] Across this changing production and context, Chambers's narrative began with the structure and evolution of the solar system and their relation to the broader cosmos. In the initial editions, the nebular hypothesis was unapologetically the initiating and central dynamic of that narrative, the gradualist principle inspiring the book's overall developmental model: the refinement of forms on more or less decreasing scales, where organic, progressive development followed gravitationally based inorganic formations.[36]

From the solar system, Chambers's history moved on to the emergence of the earth and its moon and to successive geological eras and their eventual concomitant structures of life. That origin story embraced the sparking of life from a spontaneous original creation of the organic from the inorganic, on to the development of the varied living forms, progressing to successive stages of greater complexity as the result of a postulated biological principle of development. Along with speculations on and assertions of life elsewhere in the solar system and throughout space, this organic trajectory reached up to and beyond human life and human difference (in which difference is mainly explained by different stages in development toward the highest Caucasian type), with a concomitant study of language, culture, and morality (and along with morality, criminality).[37] Against the background of this thoroughly historical natural history, asserting explicitly and repeatedly the unity of nature, Chambers concluded with the meaning of life itself and the future of humanity. As with the universal historians and the synthesizers who preceded him, he found that meaning in a narrative of successive human perfection. This overall narrative, in its broad structural and thematic strokes, bears a strong resemblance to the scientific evolutionary narratives that followed it, helping establish a generic template for evolutionary synthesis.

To many of its readers, the narrative of *Vestiges* instead appeared to be all too clear: a materialist progression that extended or transgressed the limits of science, while laying out the different roles of the different disciplines. The position advocated by the work was also generally regarded as aligned with radical visions of reform associated with Nichol or J. S. Mill.[38] At the same time, past and present-day commentators saw the question of the origins of life as the scientific heart of *Vestiges*, that itself inclined to specific radical social positions.[39] Upending the natural and the social order coincided.

But *Vestiges* also reinforced the structure of that evolutionary synthetic template in more fine-grained ways. For example, in reference to the first edition, in arguing that the earth is a generic body subject to universal laws, Chambers focused on known elements, their properties, and their distributions. In this, he connected that distribution to the initial state of the universe as "one mass," if presumably a diffused one, and he emphasized the importance of determining the relation of the moon to the earth and the importance of the earth's original state and life thereafter to the question of the sources of its heat.[40] Among his treatments of past life and the origin of life itself, he indicated the importance of past atmospheres and inorganic phenomena resembling life (such as crystallization), tying this to different contemporary experiments, underscoring both the shared, small-scale elements of life and electrical-chemical origins.[41] He mooted the possibility that life did not continue to originate, but did so in the remote past, while also looking for intimations or vestiges in the present, and he argued for the resemblance of life elsewhere to life here: "where there is light there will be eyes and these, in other spheres, will be the same in all respects as the eyes of tellurian animals," a claim that would seem strongly overstated by contemporary views, but is more important for the fact of the speculation in this history.[42] Together with the idea that cultural and political progression formed part of natural history, Chambers treated the question of human difference, itself no progressive element in the history of race and natural scientific speculation.[43] Throughout, he invoked domestic and democratic analogies, which he gave all the more substance by his final, socially targeted chapters.

Chambers claimed little originality, except in the overall narrative structure itself. But the originality and promise of that structure was a point upon which contemporaries were willing to defend him. In 1845, the Cambridge scholar Frederick Myers wrote to his friend Whewell in response to

Indications of the Creator. Whewell's *Indications* was a set of extracts from his *History* and his *Philosophy of the Inductive Sciences*, with the exception of an early chapter on the nebular hypothesis extracted from his *Bridgewater Treatise*. It was a work which Myers expected to be regarded as a response to *Vestiges* and in his letter to Whewell, he commented directly on the synthetic power of Chambers's book. Granting that its contents were "rather behind than before the highest science of our time," Myers confessed nevertheless that "I cannot but recognise a certain superiority in the book which will keep it alive, and am disposed to think that after all the scientific and other errors are deducted there will remain such a residuum of originality and truth as will demand and obtain for it a protracted consideration."[44] Endorsing Chambers's own beliefs, Myers clarified one important form of originality that *Vestiges* to his mind exhibited: he saw it as "the first sustained attempt within my knowledge to combine the highest results of astronomy, geology, physiology, and chemistry, into such a series as may serve to indicate, if not to demonstrate, the existence of a law of creation more general than any yet announced, a great law of life which shall bear the same relation to organised nature that gravitation does to mere matter . . ."[45]

As might be expected, Whewell in his response, though granting that it was published "with some reference to the Vestiges," resisted the idea that *Indications* was primarily a refutation of the still anonymous Chambers.[46] Whewell still more dramatically resisted *Vestiges'* synthetic power, comparing this to belief in Astrology and concluding: "If you were to point out to me palms, and lions, and crocodiles, and giants, in the masses of Western cloud-scenery, and tell me that you could not help believing them because the scene was so grand, I should never try to talk you out of your belief."[47] Not every synthetic image was successful.[48]

Chambers communicated the grandness of that frame through prominent rhetorical figures. The chapter headings of *Vestiges* were only the most conspicuous of these, often corresponding explicitly to different "eras" or "epochs."[49] The putative communal history that the book narrates is represented as written by the different natural scientific disciplines themselves, with each having a different share in the chapters in the tale.[50] The language of event and epoch structures the work, establishing processes and moments crucial to understanding a history that sought to capture such "sublime" scales of time and space. As the narration moves toward its own retrospective self-assessment and a review of all past history, it economically describes the shaping principle of the narrative as a whole: "The masses of space are formed by law; law makes them in due time theatres of existence for plants and animals;

sensation, disposition, intellect, are all in like manner developed and sustained in action by law. It is most interesting to observe into how small a field the whole of the mysteries of nature thus ultimately resolve themselves. The inorganic has one final comprehensive law, G R A V I T A T I O N. The organic, the other great department of mundane things, rests in like manner on one law, and that is,—DEVELOPMENT. Nor may even these be after all twain, but only branches of one still more comprehensive law, the expression of that unity which man's wit can scarcely separate from Deity itself."[51]

Among the significant changes the book underwent in subsequent editions were first, as Mary Ogilvie has emphasized, Chambers's attempts to argue explicitly against the necessity of the nebular hypothesis as that hypothesis came under increasing observational scrutiny and rear-guard attack.[52] Chambers responded in different ways to increasing statements of doubt about the nebular hypothesis, maintaining throughout its viability and at the same time claiming that the wider stakes of his argument were larger than the truth of any one hypothesis.[53] That the course of universal history followed immanent laws required a developmental principle, but not necessarily the specifics afforded by the conjectures of either the senior Herschel or Laplace.

Second, in a shift that Secord refers to as "revolutionary," Chambers began to move from the weakening circular classification structures of species advanced by the Scottish theorist William Sharp Macleay, which divided each successive classificatory group of the natural kingdom into groups of five closed circles based on species "affinities," to genealogical classification.[54] The significance of this shift, well before Darwin declared the imminence of natural historical genealogy, was not only to render the circular classification scheme outmoded, but to treat *any* nonhistorical classification as essentially provisional or pragmatic.

Across these succeeding editions, *Vestiges* was attributed to an evolving set of authors.[55] Brougham and Nichol were among the many names put about, as well as Charles Babbage (whose reasoning in reference to his computing engine and its implications for induction played a prominent role in *Vestiges*), Augustus de Morgan, Ada Lovelace, and Martineau.[56] This was an array of historical actors with varying expertise, social location, and status. That this variation itself included figures widely held to be expert is a reminder of the extent to which the possibility of such a synthesis was regarded as continuous with previous scientific work. But it is also a reminder of the problematic nature of expertise, particularly when the path to achieving it and the ways of demonstrating it were not so firm as they were constructed to be thereafter.[57] Putative amateurs could demonstrate expertise as a function of their writings, despite differences of formal education and social background: *Vestiges could* have made Chambers into an expert. In turn, someone regarded to be as expert as Somerville could suspect Babbage as the author, showing the extent to which it was difficult to decide the matter as an intrinsic function of the text, particularly so adaptive a text as *Vestiges*. It follows that Chambers's own status as an amateur was partly retrospective, a correlative in no small measure of *Vestiges* failing to resonate with elites and their ultimately closing ranks against it.[58]

We thus begin to see the difficulty of applying "expertise" in any way that assumes a certain uniformity of background or professional trajectory. Chambers did not and perhaps could not respect the limits that encumbered more discipline-specific practitioners and less synthetic popularizers. But this was not as any a priori matter a function of lack of training or experience and was not in itself an indication of amateurism. *Vestiges* was one among a field of ambitious synthetic options that belied claims to unwillingness and inability on the part of natural philosophical practitioners. What Whewell had claimed in reference to Somerville was perhaps still more aptly said of Chambers: that the synthesizer did not have to be judged an expert in all the disciplines she or he synthesized, and likely could not be.

CHAMBERS'S *VESTIGES*
AND HUMBOLDT'S *COSMOS*

There were other synthetic visions not entirely dissonant with that of Chambers, extended by those considered more expert. Nichol was himself likely a frustrated example, but a more resonant case was the iconic naturalist Alexander von Humboldt. The contrast between Herschel's response to Humboldt's synthesis and that of Chambers is instructive. For Herschel, the end of the sciences, and of Humboldt's efforts, relates to the promise of a natural history in his second sense, in a non-dogmatic, causal, ordered account of the unity of the whole of nature. Only the means to that end and the qualifications of those who sought it were at issue. And unlike Chambers, Humboldt could, as Herschel presented matters, more fairly lay claim to being one surveying the work of the many, with both the background and the vision to produce a synthesis.

The first volume of the English translation of his *Cosmos*, containing Humboldt's "cosmography"—the physical description not merely of the earth, but of the universe—was published in 1845, the same year in which Chambers had come under direct attack by Herschel in the latter's address to the British

Association for the Advancement of Science (BAAS) as its new president.[59] The massive endeavor appeared to a much less variegated reception than *Vestiges*, if not entirely to universal acclaim.[60] By that time, Humboldt was already in his late seventies, distant from the South American travels of his youth. His synthesis included in its multivolume ambit the history of science, literary and visual representations of the cosmos (from Hellenic to contemporaneous poetry, from antique landscapes as "accessories" to modern panoramas), as well as the natural history of the world. Like Chambers's synthesis, the narrative history it projected began with the combined nebular hypothesis of William Herschel and Laplace, resolving down in scale to the formation of the earth and its succession of geology, life, and its variations, including, ultimately, humanity.[61]

Herschel drew the contrast between Humboldt and Chambers in an 1848 review of Humboldt's work. As in his review of Chambers, what Herschel severely criticized was not the idea of a totalized account of natural history as an end, but imposing synthesis through a dogmatic vision of unity rather than by empirically grounded synthetic methods. Herschel's *Cosmos* review praised Humboldt's efforts at unification and his qualifications for making the attempt and excoriated other synthesizing doctrines, attacking Comtean positivism as "a dogma, which has of late been placed in prominence . . . to the detriment of sound philosophy." He dismissed it as a misconceived empiricism, overemphasizing "Law" and deemphasizing explanation. In his exposition, it stood beside two other "destitute" dogmas: the first, the mechanical or "occult wheelwork" view, which posited the moral and physical order as the action of a mechanism wound up and left alone, the second, the intrinsic belief that all motion results from "mechanical push and pull," that all phenomena are reducible to lawlike mechanical motions.

To Herschel's mind, Humboldt was throughout all of Europe the person "best fitted" for the task of discovering the synthetic principles underlying the unity of nature. It was not that Herschel rejected imaginative synthesis, as Chambers charged in his response to Herschel's 1845 BAAS address, claiming in *Explanations* that "it is to the chilling repression of all saliency in investigation, which characterizes the scientific men of our country and age, that I object, not to a due caution in selecting proper paths in which to venture."[62] It was only that hardly anyone could achieve it: "It is, therefore, by no simply clever writer, by no mere man of vivid imagination and fluent command of language and imagery—least of all, by any ideal speculatist who may have devised a system of philosophy spun from the abstractions of his own brain, and resolving all things into some single principle, some formula embodying all possible knowledge, that such a work can be entered upon without the certainty of utter and disgraceful failure."[63] For Herschel, Chambers and Comte were examples of such utter, disgraceful failure. Humboldt was at least qualified to hazard the attempt to be the natural historical "one."[64]

In fact, as Herschel judged the matter, Humboldt's own failures were not primarily a result of overreaching, but in Humboldt not being as ambitious and inclusive as his subject demanded.[65] In comments having a distinctive retrospective ring, coming just prior to the 1848 revolutions, Herschel starkly observed: "A great and wondrous attempt is making in civilized Europe at the present time: neither more nor less than an attempt to stave off, *ad infinitum,* the tremendous visitation of war." To this, the material conditions to allow for perpetual peace had to be in place, chiefly that nature supply increasing human wants in increasing abundance. His remarks brought more clearly to mind then the "civilizing" efforts of imperial science, the impact of Europe elsewhere in the world, and the debates over the virtues and vices of trade. "To no other quarter than to the progress of science can we look for the least glimpse of a fulfillment of the first of these conditions.... Science must wave unceasingly her magic wand, and point unceasingly her divining rod."[66] Evaluating Humboldt's synthesis, Herschel posited a strong vision of the ultimate target for the sciences, which remained close to his second sense of natural history, the "assemblage of phenomena to be explained; of effects to be deduced from causes; and of materials prepared to our hands, for the application of our principles to useful purposes." Ultimately, as with Chambers, Comte, and others, Herschel testified to the ambitions of scientific synthesis to capture and help advance human life. In this advance, he stressed the need for care and "long, calm, enduring patience."[67]

SPENCER'S PROGRESSIVE HETEROGENEITY

That the approach to natural history and philosophy was changing with the onset of the 1830s was something to which past and present scholars have testified.[68] For some, such as Herschel, Somerville, Chambers, and Comte, those shifts might have suggested that knowledge was on the path to being whole again. To others, such as Whewell, it might have suggested that knowledge was losing the wholeness it once possessed. For Herbert Spencer, it was possible to embrace portions of both positions.

In careful ongoing contradistinction to Comte, Spencer crafted his own historical view both of the natural world and of knowledge, articulating his own classification of the sciences working both to prove the fact of essential and deep divisions present between the sciences and to integrate this view into his own system.[69] However, this synthetic treatment did not, to his mind, suggest any dogmatic unity.[70] Synthesis and unity occupied distinct ideas in his work. When Spencer did emphasize unity, he ultimately related it in terms of identical patterns of progress in the sciences and identical ends.

Despite the attacks on the narrative synthesis that had been suggested by it, the nebular hypothesis remained a viable cultural concept. In the face of this extended record of critique, Spencer defended the hypothesis in 1858 in the *Westminster Review*. His "Recent Astronomy, and The Nebular Hypothesis" was headed by the citations of Laplace's complete works, Herschel's *Outlines of Astronomy* and *Results of Astronomical Observations at the Cape of Good Hope*, and Humboldt's *Cosmos*.[71] Despite the best of Herschel's efforts to debunk them, the article still mentioned Comte's mathematical proofs: Spencer was aware that "doubts have been cast on M. Comte's reasonings," but still concluded that "as a professor of mathematics, his authority is of weight."[72]

But other authorities were less imposing, to Spencer's mind, than the direct empirical evidence weighing against their own reasoning. Four years earlier, in "The Genesis of Science," also in the *Westminster Review*, Spencer reveals no sympathy whatsoever for the definitions and systems of science put forward by Lorenz Okun in his *Elements of Physiophilosophy* or by Hegel, both of which he dispenses with fairly quickly. The review is in fact a much more extended and respectful dissent from Comte, appealing to Whewell for the examples in the history of science needed for that disputation,[73] but also denying that the sciences could proceed according to their own internal logic without looking at how each affects each.

His differences with Comte allowed Spencer to craft his own historical view both of the natural world and of knowledge. In his work, the parallels between state of nature arguments, the beginnings of the world, and the evolution of ideas, are at their most explicit. He invoked the claim that the notion of identity/equality so necessary in his view to the rise of science was itself essential to the formation of groups and classes and to a conception of justice, each in turn required for the advance of knowledge.[74] This emphasis on the priority or coevalness of self-conscious sociability and science was articulated strategically, as part of Spencer's disagreement with Comte. It enabled the conclusion that sociology was as old as or older than the oldest sciences.[75] This was not simply a perverse reversal of Comte, but part of Spencer's holistically conceived rejection of Comtean or any other serialism.[76]

How each science affects each was not, to his mind, a process tending toward unity. Quite the opposite. In a third *Westminster Review* article, in 1857, "Progress: Its Laws and Cause," Spencer challenged one of the primary articles of synthetic thought, that humanity is the proper end of knowledge and the proper measure of progress. He begins by listing Humboldt's *Cosmos*, Lyell's *Principles of Geography*, and William Carpenter's *Principles of Comparative Physiology*, a stacking of texts the successive subjects of which suggest that Spencer was responding not only to Comte here, but to Chambers.[77] Consonantly with Chambers, he proposed that "the Law of organic progress is the law of all progress," with the nebular hypothesis as the ultimate illustration.[78] However, Spencer's anti-serialism did not prevent him from presenting a broadly serial, discipline-based, differentiating and branching history of the universe. He presents a concise history of the world, from astronomical to geological to biological to sociological formations; from the differentiation of the planets, to the differentiation of the human races, to the formations of church, state, the divisions of labor, to cultural formations such as language, art, music, and dance. In all, as he sums it up, "from the remotest past which Science can fathom, down to the novelties of yesterday, that in which Progress essentially consists, is the transformation of the homogeneous into the heterogeneous."[79]

Spencer thus claimed to uncover a deeper law behind the tendency toward heterogeneity from homogeneity in the plurality of changes and effects produced by a single force or cause.[80] But as with Whewell, he also claimed that what stood behind such knowledge was ultimately unrepresentable in principle: "under all things there lies an impenetrable mystery."[81] Spencer therefore saw the progress of the "sincere man of science" in the increasing appreciation of the fact "that the Universe is an insoluble problem." Spencer noted that the scientific man "sees himself in the midst of perpetual changes, of which he can discover neither the beginning nor the end," neither in the physical world around him nor in the developments of his own mental states. But resonant with *Vestiges*, Spencer allowed that scientist nevertheless to invoke the conjectural beginning of the universe, "that all matter once existed in a diffused form." Regardless, "inward and outward things he thus discovers to be alike inscrutable in their ultimate genesis and nature." The beginnings, the ends, the ultimate causes were all at the margins, regardless of how the relationships between the sciences and between the sciences and history were conceived. Whatever the suggested synthesis or the proposed unity of the sciences, the margins declared the ultimate limits and hopes simultaneously, distinguishing the contingently unknown from the necessarily so, underpinning despite that marginality the structures of natural history and knowledge together.

Natural history, as conceived of by Herschel and those who followed him, was understood, in its second sense, as a systematic plan for ordering manifold experience. In that light, as a plan, it did not have to be presented as a historical narrative. More than that, for Herschel and Somerville in the 1830s, it *could not* be presented as a history from first to last. As a conceptual matter, the schemes in place for surveying that whole were far from clear, and for Spencer, unifying first causes were inherently unknowable. But these thinkers also presumed that collective scientific labors needed projection onto the single (abstract or concrete) mind of the individual. The individual synthetic thinker was presented as key to the development of the structure on which that collective labor was knowingly or unknowingly directed. Natural history as a concept persistently involved the presentation of a collective labor of knowledge in such a way as to be translatable to an individual mind.

Chambers, and in a different manner, Comte, attempted to articulate such a synthesis, the one emphasizing the historicization of the world, the other the historicization of the sciences. Their vision of a totality did invoke ideas of first and last, of the order of knowledge and of historical time together, of the natural and social world, of cosmogony and human state of nature, of the origins of life and the material world. Their discourse placed different disciplines, seemingly ancient or apparently modern, in connection and in tension, particularly in the historical sense supplied by geology and resisted by astronomy. These dimensions were joined across alternative appeals to different conceptions of unity and therefore different conceptions of diversity, all in relation in this earlier part of the nineteenth century to varying levels of acceptance of nebular and genetic hypotheses.

In a review published in the final year of his life, an assessment of Comte's positivism, Whewell had dismissed synthetic efforts to classify and hierarchize. Like Herschel in reference to Chambers, it seems he was partly drawn out on the subject of Comte by Mill: "The praise which Mr. Mill bestows upon some parts of his [Comte's] writings is to me quite marvelous. But my wonder is somewhat lessened when I come to perceive, in reading these praises, that they refer to performances in which I conceive the object to be of small philosophical value, such as the classification of the sciences, and the arrangement of sciences one above another in a certain order. These attempts, even if successful, seem to me to be of small value. No science is yet complete; and yet when we classify and derive them, we suppose it to be so."[82] Far from arguing for the merits of his own efforts of conceiving of the entirety of knowledge, Whewell at this late date dismissed Comte by shrugging off the exercise of classifications and arrangements.

But from at least the time of Somerville's *Connexions*, the hazy interrelations traced among the natural sciences held out the stark hope that Herschel's "magic wand" and "divining rod" could be forged. These were tools that, when associated with Comte, Chambers, or Spencer, pointed to new, sometimes unsettling social orders—orders that might overthrow faith, class, discipline, and propriety—and orders in which the category of a "gentleman" of science, as once exemplified by Herschel, would seem less relevant than that of an industrious scientific professional.

For all of the synthetic architects considered here, the question and nature of cause, effect, the laws behind them, or the force driving them were in one way or another central. Whereas Spencer had proclaimed that "the Law of organic progress is the law of all progress," Herschel, in his 1848 review of Humboldt's *Cosmos*, had questioned the assumption that all phenomena are reducible to lawlike mechanical motions. Phenomena such as heat and electricity to his mind challenged that view: the descriptions or forms of their materiality were unclear enough so as to suggest that new modes of material existence and explanation might soon arise. Herschel was so annoyed by the emphasis on straightforward mechanism that, he claimed, the use of the term "force" in contexts like these "grates with something approaching to a painful harshness on our ears."[83] He looked instead to the substitution of a more general conception of "physical powers"—it was a more general conception that the "German Kräfte might bear, we think, without violence," he said, nodding approvingly at Humboldt.[84] In this interrogation overall, he even questioned the character of cause and effect itself and the nature of the ultimate facts that would form the ground of scientific explanation. With the meaning of causality, force, and materiality all being questioned, Herschel was giving voice to a larger set of interrogations addressing the nature of the scientific worldview.

While Spencer was defending the nebular hypothesis and universal progress as the basis of "the Law of organic progress," the emerging scientific concept of energy was however deployed not only to challenge and refine concepts of heat and force, but to advance critiques of organic evolutionary accounts such as those that Darwin and Wallace were beginning to make public. The idea of an ultimate force from which all natural phenomena might emanate on the one hand and the idea of a transformation of (the products of) one kind of force into another as ultimately articulated in the conservation

of energy on the other were not always clearly distinguished. As the vocabulary to make this distinction emerged, the principles of energy came to
be seen as undermining organic evolutionary accounts. In addition, as we
will see in the next two chapters, they perpetuated the tensions already existing within natural history and natural philosophy between the historical
and the ahistorical, unity and diversity, unity and synthesis, and scientist and
nonscientist.

CHAPTER THREE

The Many Faces of Force and the Mutability of Energy

> Absolute permanence is visible nowhere around us, and the fact of change merely intimates, that in the exhaustless womb of the future, unevolved wonders are in store. The phenomenon referred to would simply point to the close of one mighty cycle in the history of the solar orb—the passing away of arrangements which have fulfilled their objects that they might be transformed into new.
>
> J. P. NICHOL, *Views of the Architecture of the Heavens,* 1837

The constraints of the universal history that putative or even chance universal historians collectively posited, enjoining the different disciplines to coalesce in a narrative representation of the totality of things, were themselves in flux as a result of changes taking place in the conception of natural philosophy, changes that might also point to a new kind of person that science could produce. Arguments about the unity or independence of natural forces, attitudes toward the new vocabularies of energy coined in the development of thermodynamics, and visions of what constituted culture and the place of science in it resulted in interprofessional debates that involved not only the onetime hiking companions T. H. Huxley and John Tyndall or opponents such as William Thomson and Peter Tait, but also those authors of historical science to whom they were connected, such as Charles Darwin and Hermann von Helmholtz (1821-1894).[1] These debates relied on, contested, and invoked the work of such varied practitioners as Herschel and Whewell, the jurist and inventor William Robert Grove (1811-1896), the German physician J. R. Mayer (1814-1878), and Helmholtz himself, through work he carried out in his years as an army staff surgeon and beyond. The views on the unity or independence of natural forces, attitudes toward new vocabularies of energy, and visions of what constitutes culture varied strongly among of all those engaged in these debates. Nevertheless, their intersecting and interconnected discourses bound their concerns together, often overtly. While the foundations for a universal history were being crafted collectively, the project thus was simultaneously slipping out of the hands of any one individual, whether as a result of muddy and politically inflected arguments, or comparative unawareness of related or relatable efforts, or because of the varied expertise and authority needed to produce and evaluate that history.

This chapter and the next examine the later nineteenth-century discourse on and debates over force and energy as they influenced the potential syntheses of knowledge that these debates suggested, debates that were embedded in and helped craft new conceptions of Herschelian natural history and its relationship to intellectual endeavor as such.[2] In the effort to work toward a fundamental physical theory, a final theory in the sense of requiring no revision, the practitioners of the nineteenth-century physical sciences advanced a variety of different conceptions of unity based on the operation of what they postulated as the fundamental principles of matter, energy, and force. Any synthetic account of the world would, to their minds, have to construct its synthesis on the basis of these fundamental principles, demanding that any natural history or any narrative account of the world align with the unity conceived through these principles. In that sense, however, such a final physical theory established basic principles operating unconditionally throughout the universe, effectively everywhere and everywhen, an atemporal permanence in apparent contradistinction to historical and narrative theories of the world, conceived or structured in turn as dependent on such theory.

The counterpart of a final theory, a final story about the world, could however establish priority despite its temporal, historical nature—by making a case for the historicity of fundamental physical principles, by the historical facts any final theory could be challenged to explain, or by the constraints that narrative in itself imposed on any theoretical account of the world. A final theory could indeed be challenged on the basis of the narrative implicit or explicit within it just as a final story could be challenged on the historical theory of the world it itself constituted. Later in the nineteenth century and on into the twentieth, such accounts of the world proved to be dependent on what was increasingly referred to not as some version of force, but as fundamental principles of energy—the constancy and mutability in the world as explained by the laws of thermodynamics and the implications they held for human life.[3] The transition from conceptions of unity founded on the idea of force to newer theoretical conceptions of unity founded on the idea of energy and its convertability—changing in turn the understanding of force—are the subject of the present chapter. The next chapter then examines various different final stories, the various different and differing candidate efforts at cosmogonies and chronologies advanced throughout that same period, reflecting these fundamental conceptual changes and both perpetuating and constructing an ongoing tradition of scientific universal history connecting all knowledge in the authorship and legibility of one story.[4]

DUALITIES OF FORCE

Historians and philosophers have addressed the effect of the pedagogical and conceptual concerns of elder generations on those engaged in midcentury age-of-the-world debates. Herschel, Somerville, Whewell, Comte, Chambers, and Nichol all figure among these, their influence evidenced in the small and the large. So, for example, universal-historical and synthesizing scholarship provoked Darwin from at least the 1830s, as it did Huxley thereafter. Darwin was still more explicitly inspired, if in different ways, by predecessors such as Lyell and Humboldt, Herschel, and Whewell.[5] At the same time, the efforts of Darwin and Huxley to distinguish their work and writings from those of Chambers and other popular theorists attests to a cultural space partly inherited through *Vestiges* and older and other natural historical narratives.[6] Somerville's influence on the project of a universal history was in a different register. James Secord has emphasized how Somerville set an agenda of unification across the sciences for younger generations to pursue. She helped make plausible the idea that universal expertise is achievable, even necessary, and that there was coherent and ever more convergent agreement across the new sciences.[7]

The emergent discourses of unity in the middle of the nineteenth century depended on changing conceptions of force, as Iwan Morus has emphasized in discussing the importance of the exploration of unity in laboratory practice across Europe in the early nineteenth century: "New vocabularies were adopted to express . . . views concerning the unity of nature. William Robert Grove argued for the correlation of physical forces. Michael Faraday argued for the mutual conversion of forces. James Prescott Joule suggested that force was conserved from one transformation to another. Force was the primary focus in these discussions."[8] As the primary materials and the scholarship on them show, local environments were struck through with broader, even international concerns and saw the negotiation of these vocabularies. Inherited concepts of force and causality were matters of contestation. Newer universal historical narratives thus were in dialogue with alternative accounts written other times and places.

Despite their differences, one clear indication of the shared concerns and vocabularies of these accounts was the proliferation of conceptions of unity that revolved around dualistic notions of force, notions that grappled with the way in which a constant underlying unity could be articulated in the temporalities of a natural history. Regardless of whether the proposed, investigated, and popularized unity ultimately tended toward a view of a plurality of independent, but interconvertible forces or of one fundamental force showing different faces in different moments and experimental frames, practitioners shared an understanding of the fundamental character of their inquiry as oriented toward producing a synthetic account of phenomena in general, an account, however, that was at best problematically related to any possibilities of the character of narrative history.

Here, in order to examine these modes of authorship and invocation, I want to trace briefly aspects of the discourses circulating among a smaller set of these figures, those particularly involved in attempting to clarify and popularize that discourse on force. The nature of forces—whether they were regarded as subject to conversion or to conservation, or as the faces of some deeper principle or concept—presents a telling example of divisions in deliberations over scientific vocabularies and syntheses. These divisions were woven into the differing universal historical narratives soon to be hazarded by younger practitioners.

Despite the array of different positions on force and universal history taken by figures as diverse as Grove, Mayer, Helmholtz, Tait, Tyndall, and the Thomson brothers, as well as predecessors Herschel, Somerville, Whewell, Comte, Chambers, and Nichol, who had fixed elements of this scientific discourse that were deliberated over in succeeding generations in overt and subtler ways, these writers largely shared a number of presumptions. Although the universal histories then being crafted were the work of specific hands at particular places and often, no doubt, for profit and not only enlightenment, if we address the historical record in relation to the structure of these histories, we find that the rhetorical strategy or argumentative structure they employed was in many ways broader than any individual idiosyncrasy, intention, or even local preference. All proceeded by drawing a dualistic distinction within the concept of force that enabled them to reconcile the atemporal with the temporal and fundamental unity with change.

Across William Robert Grove's efforts (as with those hazarded earlier in the century), for example, he aligned universal history with different visions of potential synthetic unity. In the late 1830s, while working as a barrister, he carried out research into electrochemical cells, inventing a nitric acid battery that "was to make Grove's initial reputation as a natural philosopher."[9]

In 1840 he was elected a fellow of the Royal Society and the following year became professor of experimental philosophy at the London Institution. In lectures as a professor in the years immediately following, Grove publicly presented work on the "correlation of physical forces."[10] There, he launched into an extended critique of the concept of cause in reference to natural laws, a critique in dialogue with Comte, Herschel, and Whewell, among others.[11] For Grove, the example of the unification of electricity and magnetism as celebrated by Somerville, Herschel, and others brought the idea of causality into question.[12] After unification, it could be said with equal justice that electrical effects cause magnetic ones or that magnetic ones cause electric ones.[13] Grove opted instead for a view that "the various affections of matter which constitute the main objects of experimental physics, viz., heat, light, electricity, magnetism, chemical affinity, and motion, are all correlative, or have a reciprocal dependence."[14]

This view, that any "force" could be the "essential or the proximate cause of the others" and "that a force cannot originate otherwise than by generation from some antecedent force or forces" allowed him to endorse a view of the enduring nature of force itself: "Now the view which I venture to submit is, that force cannot be annihilated, but is merely subdivided or altered in direction or character. First, as to direction. Wave your hand: the motion, which has apparently ceased, is taken up by the air, from the air by the walls of the room, &c, and so by direct and reacting waves, continually comminuted, but never destroyed."[15] In this Comtean-inflected critique of causality, a view of nature different than phoenix or womb was being emphasized by this convertability and by a different view of natural history understood as a goal. To observe that electricity causes magnetism causes electricity and nevertheless to continue to understand these forces as distinct was to undercut the possibility of a program of knowledge based on the hierarchy of apparent forces.[16] If one force can be transformed into another and then into another and so on, which "causes" the other and when?

To deal with the problem such a vision of convertibility posed to the identity of forces and the causal sequences in which they were implicated, Grove distinguished between a "subjective" idea of force and an "objective" one. The subjective or qualitative usages referred to force as an entity or subject producing effects. Newtonian gravity could be treated subjectively as an agent acting in the world. The objective use related instead to effects themselves, as in the case of the coming together of different masses under gravity. The subjective identity of forces persisted; their objective products were never destroyed. But forces were also never newly created.

At roughly the same time, J. R. Mayer, in a very different intellectual and cultural context, was proposing another, different distinction in relation to force and causal concepts, between "quantitative" and "qualitative" determinations. His was a distinction between magnitude and form, and in contrast to Grove, he suggested an *Urkraft* standing behind all forces.[17] As P. M. Haimann has analyzed, one force causing another does not, however, imply that one can be reduced to another: the sequence of cause and effect need not contradict the underlying, atemporal unity of force, and different forces might be different faces of that *Urkraft*. Regardless of such possible distinctions, John Tyndall was to become firmly convinced that Mayer had captured the essential conservation principles and their applications to matter and life.

In his own 1862 lecture explaining the new laws of "force" and the explanations it gave (to which we will return), Tyndall asked, "To whom, then, are we indebted for the striking generalizations of this evening's discourse? All that I have laid before you is the work of a man of whom you have scarcely ever heard. All that I have brought before you has been taken from the labours of a German physician, named Mayer."[18] Presumably unlike other better-known contributors who had both a community of invested practitioners and clearer provocation, it was Mayer who, "without external stimulus, and pursuing his profession as town physician in Heilbronn ... was the first to raise the conception of the interaction of natural forces to clearness in his own mind."[19] Tyndall represented as crucial an evocative clue: "It was the accident of bleeding a feverish patient at Java in 1840 that led Mayer to speculate on these subjects. He noticed that the venous blood in the tropics was of a much brighter red than in colder latitudes; and his reasoning on this fact led him into the laboratory of natural forces, where he has worked with such signal ability and success."[20]

Through this distinctive history, Tyndall claimed that Mayer had preceded Joule by a year on the question of the mechanical equivalent of heat and had preceded John Waterson and Thomson by six years on the theory of the generation of the sun's heat by meteors—conceptions stitched deeply into the emergent universal histories.[21] "Here was a man of genius working in silence, animated solely by a love of his subject, and arriving at the most important results some time in advance of those whose lives were entirely devoted to Natural Philosophy."[22] Others then and now, however, saw Tyndall's heroizing account more as a move to claim the newer theorizations for others further from the British scientific elites, whether located in Oxbridge or North England.[23]

Be that as it may, each such contestation of force or cause asserted one kind of unity and argued against other kinds. Potential distinctions between views of those such as Grove or Mayer were to play no small part in the interpretive virtuosity displayed in priority battles over energy as the basic laws of thermodynamics were achieving status as fundamental doctrine. However, from the perspective available at the time, there existed a partly productive confusion about the meaning or the usage of "force." In an address of 1862, "The Relation of the Natural Sciences to Science in General," for example, Hermann von Helmholtz distinguished between "intensity" of force [*Kraft*]" and its "quantity."[24] It is the latter, the quantitative aspect of force, to which, for Helmholtz, the "conservation of force" applies.

Helmholtz had turned to such questions from a different standpoint still. Like Mayer (and T. H. Huxley), Helmholtz began his career with medical studies, first in Berlin at the Königliches medizinisch-chirurgisches Friedrich-Wilhems-Institut. Given his family's modest resources, Helmholtz accepted a stipend the terms of which exchanged his educational costs there for eight years of service in the Prussian military.[25] While still a student at the institute, Helmholtz also attended courses at the University of Berlin. As Arleen Tuchman has suggested, differences in training and ethos at these institutions—one having tended to the practical and regimental, the other to the "neohumanist"—may have affected Helmholtz's views on the nature of the university itself and therefore on institutionalized relationships between disciplines.[26]

As a student, he carried out an anatomical dissertation on the nervous system in invertebrates under Johannes Müller, professor of physiology at the university. His studies were followed by a year-long internship at the Charité in Berlin and thereafter by the term in the army. He continued his scientific work during his service, examining "animal heat," among other lines of research, and vowing with Emil du Bois-Reymond, Ernst Brücke and Carl Ludwig to create a fully "organic physics." Intimately related to such efforts was his articulation of *Erhaltung der Kraft*, or "conservation of force" in 1847, a work demanding broader intellectual training, competence, and synthesis. He focused on *Spannkräfte*, or "tensional forces"—"forces which tend to move the [material] point, before the motion has actually taken place" (what his counterparts in the United Kingdom would call "potential force") and concluded that "the sum of the tensional forces and vires vivae [the sum of what we would now express as the kinetic energies] present is always constant."[27] Helmholtz, like Grove and Mayer, saw at least two faces in the nature of force, and the fact of that duality brought with it the temptation to identify the terms of their vocabularies as much as to distinguish them.

The different aspects and notions of force raised the question of how historical programs of knowledge based on one or the other side of the different distinctions (subjective or intensive versus objective or quantitative) could relate to one another. Was nature fundamentally to be seen as matter and motion in which distinct forces remain distinct, as Grove might have it, or could there be an *Urkraft*, as Mayer emphasized? What should be the theoretical object of science as a whole in relation to these potential new fundamentals? Did the refinement of concepts of force ground all the more the possibility of a synthesis by fundamental laws (as Comte had already decried), or did the principles linked to conservation suggest the possibility of new synthesizing histories? And was there a synthetic thinker who could discern answers to these questions?

The efforts of these very diverse practitioners involved struggles for clearer articulations of force, cause, or convertability engaging these overlapping questions. These nascent conceptions provoked the possibility of synthesizing the disciplines as predicated on different notions of force or law, even while they problematized notions of causality. On the one hand, the conservation of energy and the potential determination of a final law when seen through the very different lens of expositions such as that of Grove or later a Tait or Tyndall, were intimately related: each suggested a deep, tangible, or measurable connection in the diverse phenomena of nature, a connection related somehow to force and perpetuation. But on the other hand, belief in the ability to transform one force into another or in the ability to capitalize on the effects of forces did not demonstrate exclusively the fact of an ultimate force lying behind all the others, as Mayer suggested. Such a belief also, as in Grove's treatment, could undercut a hierarchical view of the forces and inveigh against any scientific synthesis predicated on that view.

These common rhetorical and conceptual patterns extended beyond the dualities and fundamental character of the phenomena of force. The emergence of conceptions of energy and correlate newer understandings of force invoked totalizing representations of the natural world other than only the different possible evolutionary narratives. As with Herschel's natural histories, these totalizing representations related to other possible visions of scientific synthesis—to different possibilities of a Herschelian natural history of the second kind and therefore to different schemes of the nature, representation, and dissemination of knowledge as it was then conceived. In a period of always uneven professionalization, a common pattern in these reflections was the explicit theme that a sound "criticism of life" would need to turn on a wider appreciation of the new natural scientific disciplines and the realms of experience they disclosed.

"SO-CALLED EDUCATED MEN" AND
THE LIMITS OF NATURAL HISTORY

In these accounts, differing visions, disciplinary schemes, and representations of the limits of knowledge were all contested together in terms of the current state and status of science in the wider culture. In 1866, Grove addressed the British Association for the Advancement of Science as its president, arguing for the need for greater emphasis on natural science education. He lamented "the number of so-called educated men who, travelling by railway, voyaging by steamboat, consulting the almanac for the time of sunrise or full-moon, have not the most elementary knowledge of a steam-engine, a barometer, or a quadrant; and who will listen with a half-confessed faith to the most idle predictions as to weather or cometic influences, while they are in a state of crass ignorance as to the cause of the trade-winds or the form of a comet's path."[28] He hoped that improved "scientific studies" would soon occupy a still larger part of the curricula—and presumably produce truly educated men who would stand a better chance of resisting faith in unsound influences.

Taking stock of the state of science, Grove at the same time articulated a doctrine of continuity—a conception of nature and of the disciplines to know it, interconnected and in slow, gradual, naturalistic development.[29] With this doctrine in mind, he canvassed the "recent progress of some of the more prominent branches of science," connecting thereby astronomy, geology, chemistry, optics, magnetism, heat, and physiology to the question of life, the Pasteur and Pouchet debate, and the storm over Darwin's claims. Tactfully, but openly indicating his preferences for continuous development throughout and for a holistic view of knowledge,[30] Grove extended his doctrine beyond the physical sciences: "The same modes of thought which lead us to see continuity in the field of the microscope as in the universe, in infinity downwards as in infinity upwards, will lead us to see it in the history of our own race.... Our language, our social institutions, our laws, the constitution of which we are proud, are the growth of time, the product of slow adaptations, resulting from continuous struggles."[31] Through the physical sciences, the doctrine of continuity had treated subjects overlooked by antiquity. But not so "in ethics, in politics, in poetry, in sculpture, in painting," where knowledge remained nearly in its antique state.[32] These studies had yet to be remade into the mold of the new sciences, which were increasingly connected in historical, evolutionary formation. A new education and new people were needed to appreciate the industrial changes affecting the world, along with the physical laws that made those changes possible and simultaneously constrained them. At stake was the mastery of first nature by the conceptions, industry, and technological products of second nature. At stake was correcting a too narrow development of culture by virtue of the advances that natural science had made.

Both for such mastery and for the needed development of contemporary culture, Herschel, Whewell, Thomson, Grove, Helmholtz, and many others believed the question of scientific education was of central importance. But how general that education (scientific or otherwise) should be and what specifically should be taught were debated, even as trends and legislation moved toward universal reforms.[33]

Grove's continuity thesis made evident the linkage of scientific knowledge with its potential virtues in resisting popular falsehoods and with concomitant requirements for redesigned pedagogy and new articulations of the relations between the disciplines of knowledge. But this linkage was not itself new. British scientists had been deliberating for decades by that point over their own intellectual syntheses in relation to pedagogical reforms. In his *Philosophy* in 1840, Whewell had dedicated a chapter to the question and to the potential structure of natural scientific training in "intellectual education." It was a chapter that Herschel, in turn, underscored in an 1841 *Quarterly Review* piece. Expanding on Whewell, Herschel had argued for greater "individual and national effort" devoted to studies he found conducive "not only to physical . . . but the moral well-being of man."[34] What was conceptually at stake for these practitioners were not only the different possibilities for synthesis revealed in the classifications of knowledge and educational visions; at the same time, such potential shifts would ensure new scientific and social futures.

In the same years, natural philosophers of another generation were calling attention to similar questions, but recast in terms of the new vocabularies of energy that they were in the process of producing. Inaugurating his 1846 election to the chair that his Glasgow teacher William Meikleham had once occupied, Thomson began his address with the warning that "in entering on a new branch of study it is natural to look for a distinct statement of its subject. But nothing in science is more difficult than definitions. Attempts to give sharp and complete definitions, especially to define branches of science, have generally proved failures."[35] Thomson divides the world into "mind" and "matter," assigning to the former "mental science," and to the latter "natural science." It is to "mental science" that he first addresses the question of history, in reference to what he terms "mental history," in which he includes politics and (civil) history. Mental history, in turn, along with "personal experience and a knowledge of men and manners," grounds "mental philosophy."[36]

From these distinctions, he draws an analogy with natural history and natural philosophy. Here, in reference to a largely ahistorical natural philosophy, Thomson divides up the branches of knowledge in a way familiar to the disciplinary landscape of the period. Thomson's natural history is reminiscent of Herschel's in its first sense, relating to a description of the natural world based on the collection of facts. At the same time, his natural philosophy is reminiscent of Herschel's natural history in its second sense, where the subject is the determination of laws induced from those facts.[37] The former acts as the experiential and descriptive ground of the latter. Natural history, then, should precede natural philosophy. From this perspective, natural history is part of any science and can only by a light abuse of terminology be identified with its connotations in describing and classifying the natural world of vegetables, plants, and animals.[38] However, Herschel's divisions relied, in his view, on the primacy of force (in the Newtonian conception) for their underlying unity. As he and Tait worked to reconstruct Newton in the language of energy, this meant that this Herschelian scheme was now constructed instead on the basis of the posited fundamental character of thermodynamics.

As with Whewell, for Thomson, the limits of science implicitly pointed to the sacred, but the emphasis here, following explicitly on Herschel's *A Preliminary Discourse*, was more on how the disclosure of the mechanisms of the world endowed humanity with greater reverence for the creator. Science ultimately pointed to humility in the face of the totality of things, a sentiment resonating with the stream of religious thought asserting humanity's insignificance: "When we comprehend the vastness of the dimensions of that part of creation of which we know a little, and yet consider what an infinitesimal portion this is of the whole universe, how insignificant a being we must feel that man is, and how grateful ought we to be that God should still be mindful of him and visit him, and for the gifts and the constant care bestowed on him by the Creator of all."[39] As in one strain of European religious tradition, humanity was "anthropoperipheral," but in the spiritual, rather than the material order.[40] For Thomson, as for Herschel, humanity was in a sense doubly marginal, its material/spatial centrality was displaced, and its spiritual importance only weakly asserted.[41]

One year after his 1853 appointment to the Royal Institute, Tyndall likewise contributed a lecture on education to a series on the subject. The series also included lectures by Whewell and Tyndall's new senior colleague, Faraday, whom Tyndall deeply admired.[42] Whewell, in the preceding lecture of the series—arguing for "dependence of the progress of education on the progress of science"—found in Greek and Roman traditions the importance of mathematics and, as a science of moral conduct, jurisprudence.[43] But these, he observed, were deductive sciences, and it was by induction that the moderns had advanced beyond the ancients. So he argued for the introduction of inductive logic into the curriculum. By an education that includes "one or more departments of inductive knowledge, the mind may escape from the thralldom and illusion which reigns in the world of mere words."[44] Students of natural science stood better proofed against illusion, whether of the falsely supernatural or of mere opinion in the world.

There was a heroism in the vision of history of science that both Whewell and Faraday put forth in these lectures, relying on the greatness of founding figures such as Bacon or Newton. Tyndall celebrated in turn the heroes in his immediate world, in Faraday and the romantically minded Thomas Carlyle. As Frank M. Turner and Crosbie Smith have shown, however, the construction and choice of such heroes was hardly a harmless or apolitical act. It could be itself part of a strategy in the opposition of one scientific center to another (London to Edinburgh) or of one natural philosophy to another (North England energy physics to scientific naturalism).

There may have been an element of the Romantic hero in the construction of such figures, and Tyndall's own path to scientific prominence might have seemed to Tyndall himself a demonstration of the power of singular self-will. The child of a poor Irish family, he had moved from surveying work in Ireland and England to teaching mathematics in the Hampshire Queenswood College. With his friend and Queenswood colleague Edward Frankland, with inadequate preparation in either science or German, Tyndall went to Marburg, to Robert Bunsen's laboratory. Opting ultimately for the study of physics, he worked with Hermann Knoblauch and, then, briefly, in Berlin, in the laboratory of the Gustav Magnus. Among his circle in Berlin were Helmholtz's friend and fellow Müller student Emil du Bois-Reymond and Rudolf Clausius, another important figure in the sciences of energy.[45] Tyndall's movements likewise allowed him to position one scientific center and educational tradition against another, in this case the German against the English systems and approaches. His own more proximate reasons for emphasizing the German tradition in his own popularization were likely in part a function of this education and background. His early role as a translator of Helmholtz's foundational work on the conservation of energy also helped further cement a permanent attachment to German scholarship, and he repeatedly turned to figures such as Helmholtz as authorities for his history and for his science.

In the course of his lecture "On the Importance of the Study of Physics as a Branch of Education," Tyndall, like Whewell, adopted views that accepted periodizations, but he was more emphatic in attributing to each period a stamp that gave primacy of place to modern knowledge. He placed in soft opposition Greek learning—still a part of an English Oxbridge education—and that of the new sciences. For the latter, he offered a story of the alienation a student taught more exclusively about the classical world might feel toward modern culture and nature: "A few days ago, a Master of Arts, who is still a young man, and therefore the recipient of a modern education, stated to me that until he had reached the age of twenty years he had never been taught anything regarding Light, Heat, Magnetism, or Electricity: twelve years of his life previously had been spent among the ancients, all connexion being thus severed between him and natural phenomena."[46] Such a master of arts might have been willing to accept Grove's later dismissive characterizations of his own education.

But Tyndall did not advocate eliminating classical study: "We cannot yield the companionship of our loftier brothers of antiquity,—of our Socrates and Cato,—whose lives provoke us to sympathetic greatness across the interval of two thousand years. As long as the ancient languages are the means of access to the ancient mind, they must ever be of priceless value to humanity." But he argued against their thoroughgoing universality or relevance: "it is as the avenues of ancient thought, and not as the instruments of modern culture, that they are chiefly valuable to Man."[47] This gave him the opportunity to stress the particular importance of his own chosen field in its ability to appreciate and to construct the *modern* world: "We have conquered and possessed ourselves of continents of land, concerning which antiquity knew nothing; and if new continents of thought reveal themselves to the exploring human spirit, shall we not possess them also? In these latter days, the study of Physics has given us glimpses of the methods of Nature which were quite hidden from the ancients, and it would be treason to the trust committed to us, if we were to sacrifice the hopes and aspirations of the Present out of deference to the Past."[48] A synthesis of knowledge was still valuable, but it required a reconfiguration of knowledge that asserted the particular place of recent discoveries of physical law in any system of knowledge. It was therefore in the present and future of science, and above all of physics, that Tyndall invested his hopes in education.[49]

Tyndall was not alone in contrasting his own German scientific background with English scientific genealogies and education. From at least the early 1860s, as discipline-based debates over the history of the world were being launched in Britain, Helmholtz, as well, directed himself to the broad divisions of knowledge as reflected in the faculties of the university system. Historians have used Helmholtz as a central example of academic and intellectual developments across different fields of science in Germany in the nineteenth century. Helmholtz was contributing to a widening variety of natural philosophical subjects, whether the science of the eye, different aspects of the physical sciences, non-Euclidean mathematics, or musicology.[50] The interpenetration of different branches of knowledge had already enabled his contributions to different fields, including his invention of the ophthalmoscope, which did so much to establish his reputation, and his articulation of the conservation of force.[51] In an 1862 lecture, "The Relation of the Natural Sciences to Science in General," Helmholtz considered the consequences of what he regarded as the growing independence of the natural sciences: "One of my strongest motives for discussing today the relations of the various sciences to one another is that I am myself a student of natural philosophy and that recently natural philosophy has been criticized for striking out on a path of its own, for separating itself more and more sharply from the other sciences which are joined by common philological and historical ties."[52]

Earlier in the century, German "natural scientists" had objected to the minor place afforded to contemporary studies of nature. Within the philosophical faculty, there was much less emphasis on natural sciences than on classical philology, at a time when the philosophical faculty itself was generally considered the least important of the remaining three faculties of the university, law, medicine, and theology.[53] Likewise, and of central importance to Helmholtz the student and medical practitioner, shifts were taking place in medical training toward a greater emphasis on practice. Helmholtz made these observations at a time of increasing disciplinary gains and confidence on the part of natural scientific practitioners. But the blame for any ruptures with past systems of knowledge did not to his mind lie with natural philosophy.

In the 1862 essay, he drew a sharp line between Kant and Hegel, blaming the rise of Hegelian philosophy for a schism between the sciences that in his

view could hardly have taken place among Kantians. (Ironically enough perhaps, Hegel, in his own discussion of forces in the *Phenomenology of Spirit*,
also proposed a dualistic aspect of forces, of soliciting versus solicited forces,
however different the underlying conceptions of force at play.)[54] Helmholtz
advocated the preservation of the sciences within one university against the
introduction of specialized schools.[55] He canvassed a number of arguments
for maintaining that unity. On the individual level, knowledge of the different sciences safeguarded the intellectual health of the individual against any
"one-sided development."[56] Moreover, the sciences had supplied each other
"mutual services," advances in one depending on knowledge of the others.[57]
He connected these claims to the "enormous mass of information accumu-
lated in the sciences" so that the better its "order and systematization are,
the more extensive the accumulation of knowledge can be without the system suffering." The first part of that organization, he argued, "consists . . . in
rather mechanical arrangements of materials, such as are to be found in our
catalogues, lexicons, registers, indexes, digests, scientific and literary annu-
als, law codes, systems of natural history, and the like."[58] And though he observed that the work of such organization requires "a thorough understanding both of the purpose of the compilation and of the methods and content
of the science," he nevertheless argued that "our knowledge . . . cannot lie
dormant in the form of catalogues." The knowledge there must be folded
into "their laws and their causes."[59]

Pace Grove, and drawing different lessons from his own work on energy,
natural history in its second sense would for Helmholtz have to be causal.
His analysis in the essay is given in part to the different kinds of laws in
the different kinds of sciences, but ultimately tends to his final point: "The
sciences all have one common aim: to establish the rule of intellect over the
world."[60] Helmholtz conceived the struggle to establish this rule as an active
and hands-on one in which the welfare of humanity is at stake—regarding
the scientist of whatever type as "a fellow laborer on one great common
work, upon which the highest interests of humanity rest."[61] As Anson Rabinbach has argued, labor and energy were for Helmholtz and for the period
more widely concerns that went hand in hand, identifying human labor and
theoretical conceptions of energy and what the scientist worked on with the
work done to accomplish it.[62]

Practitioners such as Grove, Tyndall, and Helmholtz, then, were not free
of the urge for natural philosophical unity and synthesis. But in different ways,
they distinguished the basis upon which a new synthesis would be drawn
from different past moments and differently minded efforts. They pursued

these values through diverse avenues: by appeals to ultimate causes, ultimate ends, or visions of continuous development; through a focus on physical phenomena or what might lie beneath them; in the principles grounding systems of classification; and in the class-inflected, inductive, and industrial ethos of the new (middle-class) scientist. But they did so from a point of view predicated on the developing understanding of causes, ends, and change in terms of shifting understandings of the identity of force in relation to emergent conceptions of what was coming to be known as energy.

In Grove's early treatment, for example, correlative or reciprocal causality tended toward a unity of nature predicated on endless interchangeability. For him, nature's shape-shifting belied any unity adumbrated by the search for fundamental laws while intimating a fundamental unity in the fact of endless movement itself. And if Helmholtz or Tyndall asserted humanity's marginality, they did so on different grounds than Herschel, Whewell, or Thomson. Thomson's discussion of the concept of entropy asserted a historical progression that threatened any belief in endless convertability such as Grove's. Thomson's own assertions revealed still other rifts. The proclaimed universality of energy concepts, in debates over Thomson's analyses, in particular, was productive of disunity across disciplines as energy and entropy found application to evolutionary thought. Different commitments to unity and conceptions of natural history therefore could reveal philosophical disunity.

FROM FORCE TO FORCE AND ENERGY

While the practitioners discussed here all shared a sense of the necessity of disseminating an understanding of science and its implications throughout contemporary culture, they held no one shared vision of what the implications of such an understanding must be. In the labors of different scientists and commentators, the conservation of energy and the unity of forces would come to signal very different ways of potentially finalizing the projects of new visions of education, natural science, and natural history. The laws of the convertibility of forces and the conversation of energy suggested ideas as diverse as the universal development of all things and their ultimate dissolution. As we have seen in previous chapters, these contestations over synthetic visions of knowledge were expressed perhaps most visibly in relation to narrative histories of the world and the universe. These narratives emerged more keenly among natural scientific practitioners at midcentury, including representatives of several different disciplines, William Thomson chief among those speaking authoritatively if not unobjectionably across disciplinary lines.

Changes of inflection taking place in the conception of natural philosophy prior to and along with Thomson's own interventions and those of his collaborators led toward a possible reconciliation between the postulates of a general theory of things and the possibilities of a narrative account of their origins and ends. The transformation of the dualities of force into ideas of "energy" in work done by the Thomson brothers, along with Macquorn Rankine and J. P. Joule, building off the work of Emile Clapeyron and Sadi Carnot, among others, collectively established thoroughgoing conservation principles: whatever drives motion can be quantified more exactly and its quantity fixed. These practitioners followed out more vigorously the trajectories of comminution that Grove traced and concluded that the products of such conversions would not all be as useful as Grove's wave of the hand, "continually comminuted, but never destroyed." They deduced in turn that the universe must have had an absolute origin and must have a final, formless arrangement, even if such an origin must itself be ungraspable by scientific inquiry. In concert with P. G. Tait and stimulated by the work and interpretations of Helmholtz, Thomson, in particular, used the conservation of force/energy to maintain and refine lessons drawn by his predecessors and applied them to universal history.

Their predecessors had fixed elements of this scientific discourse that were deliberated over in succeeding generations in overt and subtler ways,[63] and Thomson was heavily invested in the work of his elders, an investment he demonstrated in his acceptance as much as in his engaged repudiation of their work.[64] Specifically, in relation to the issue of the age of the world, Crosbie Smith and M. Norton Wise have noted that Thomson's arguments about the age of the sun built on the analyses of both Herschel and Whewell. In his *Treatise on Astronomy*, Herschel had emphasized that the primary source of motion on the surface of the earth is the sun's heat. This was an observation Thomson would wield coercively, drawing out from it a historical template fixing cosmological concerns to more mundane ones.[65] Based on a new thermodynamic theory that Herschel, for one, did not accept, Thomson more firmly rejected the possibility of perpetual generation of solar heat.[66]

Thomson and his northern British colleagues saw the principles of that hard-nosed thermodynamics at play in the engines and industrialization surrounding them, binding together industry and science. Allied with Helmholtz, they extrapolated the implications of those scientific laws to every place and time physically conceivable. The idea of a universal history animating Whewell's *Bridgewater Treatise* had tied together the sciences of the heavens and of the earth, geology and astronomy.[67] But Thomson implicitly rejected the classifications of Herschel, Somerville, and Whewell that had pronounced geology a largely historical discipline, astronomy a largely ahistorical one.[68]

Thomson's assiduous efforts in dating the world may constitute the first extensive, secularized attempt in the era of "deep time" to date with precision the age of the earth.[69] Like the nebular hypotheses upon which it partly built, Thomson's vision involved efforts that in principle allowed him to give a historical account of both the earth and the heavens. Despite any overt attempt on Thomson's part, this dating practice played a role in the construction of the secular account of the universe as it emerged after Thomson's day. But it was a history of the universe that, in stipulating the contours of a natural history of the world as a representation of the totality of things, rejected still vaster ranges, embracing both longer times and biological and geological research.

Thomson was far from the first to make an extensive effort at dating.[70] If there was a difference in the character of Thomson's dating techniques, it was more as a matter of insisting on definite quantitative limits based on presumably unassailable natural and universal laws, on final theories/principles allowing him to establish both lower and upper limits. Furthermore, despite their theological inflections, apologetics, and roots, the technical details of his arguments and the intervals of uncertainty he established could easily be, and were, read as standing apart from natural theological or scriptural justifications. Unlike the case of some earlier chronologies, such as Buffon's undisclosed estimates of chronology in the previous century, the chronologies of Thomson, as with those of Darwin or Helmholtz, were not the assertions of an individual writing and speculating in relative isolation or even secrecy. His and their work called upon the efforts of a community of researchers, debating in public forums and popular media.[71]

Contrary to the apparent secularizing implications of Thomson's theories, however, as Smith and Wise note, his long-standing interest in the age of the earth was linked to his religious convictions.[72] These were by no means orthodox, Thomson inheriting the nondogmatic and tolerant latitudinarianism of his father, but his faith was firm. While the tone of his remarks on this question certainly seemed to betray an irritation with Darwin, the long-standing irritant in his considerations on the age of the world appeared to be uniformitarianism. Uniformitarian positions and explanatory methods were based in varying measure on convictions that past geological phenomena were the result of processes still active in the present and, as with Lyell, that they were active with similar intensity—positions that defer the question of beginnings and ends.[73]

Thomson's early biographer Silvanus P. Thompson noted that already by the early 1840s, Thomson had come to believe in the absolute origins of things. And by the 1850s, following out a direction Helmholtz had charted, Thomson extrapolated from his more terrestrial physical considerations, deducing the end of everything.[74] He developed thermodynamics in a way consistent with a religious standpoint rejecting any scope of time not taking sufficient account of the necessary energies willed available to sustain a given process. Uniformitarianism held open the possibility of a world, as Hutton saw it, indefinitely old, possibly eternal, and thus it tended to foreclose the possibility of any narrative account of its history. Certainly well before Darwin published his more controversial theories, Thomson had circulated work explicitly targeting uniformitarian principles. As the terms and focus of theoretical concerns shifted from force to considerations of both force and energy, an opening thus appeared for linking a fundamental theory of the natural world with historical narratives. A chronological pattern linking the creation, age, and future death of sun and earth thus was beginning to be more overtly delineated across Thomson's work. While Chambers had met with the derision of the natural philosophical elite for an analogous effort in a single volume on the basis of less technical reasonings, across a set of papers, Thomson, along with Helmholtz, began placing successive origins into a historical framework.[75]

As noted by the biographer Thompson, in 1842 Thomson's convictions about absolute origins were revealed as the result of his analysis of the linear motion of heat in an infinite solid. Seeking to make sense of negative values of time, weighing therefore the possibility of previous states of the system, given certain initial conditions of an infinite solid, Thomson concluded, "There *must* have been a beginning."[76] Expanding on his reflections, he went on to treat this subject in his inaugural dissertation as a professor at the University of Glasgow in 1846.[77] It was a set of ideas to which he would return as Thomson refined the vocabulary for these views, along with his North British colleagues, over the course of the next decade.

By 1852, in his short but important paper, "On a Universal Tendency to the Dissipation of Mechanical Energy," Thomson introduced a distinction between static and dynamical energy, focusing on the impact of Carnot's conclusions regarding the operation of engines. On the basis of this distinction and the consequences that flowed from it, he articulated his entropic view of the world, concluding that "within a finite period of time past, the earth must have been, and within a finite period of time to come the earth must again be, unfit for the habitation of man as at present constituted, unless operations have been, or are to be performed, which are impossible under the laws to which the known operations going on at present in the material world are subject."[78] Here, in contrast to the earlier Newtonian-inflected work of Nichol (or Kant), the emphasis is less on the endless fertility of nature and its unanticipated energies and instead on the possibility of unanticipated natural processes or supernatural interventions. The concession is that the historical accounting of the natural laws may be incomplete. As Smith and Wise remark, by the middle of the century, Thomson was beginning to see that "the changing arrangements of matter and energy could be treated mechanically (or thermodynamically) between their origins and endings, although those endpoints themselves, for example the creation of matter or energy *ex nihilo*, remained quite beyond mechanical or any conceivable kind of scientific reasoning."[79] After the absolute beginning of the world, its history and what should constitute it was fair game.[80] Physical science was free to extrapolate back to that beginning, with the evolution of the universe understood through different configurations of matter and energy, but not extending to the origins of matter and energy themselves.[81]

Thomson's attack on uniformitarianism demanded a reconciliation of astronomical and geological accounts of the world, and in attempting to treat both together through energy considerations, it undercut the Lylellian bedrock on which Darwin specifically and unrelentingly presented his argument.[82] Indeed, Thomson's mature and popular statements appeared to be prompted in part by the publication and extensive reception of Darwin's *Origin of Species* in 1859. While conceding what physical theory could not yet determine, Thomson prosecuted his arguments in the early 1860s, on the verge of his fourth decade, with a tone of confidence and even certainty, a tone at times further underscored by those who championed his views.[83] In this regard, Joe D. Burchfield has gone so far as to claim that "almost single-handed [Thomson] overthrew the mid-century uniformitarians' demands for unlimited time, and in the process stimulated a growing awareness of the necessity for quantitative measurement in geology."[84] And the emphasis on the quantitative was important to resistance to the uniformitarianism of naturalists who, as Cuvier put it decades earlier, "rely greatly on the thousands of centuries which they pile up with a stroke of the pen."[85]

In 1862, Thomson chose to publish a popular article entitled "On the Age of the Sun's Heat" in *Macmillan's Magazine*,"[86] his first extended publication on the meteor-contraction theory of the sun's heat. According to the theory, the hot sun formed through collisions and coagulations of meteors, as "was first discussed on true thermodynamic principles by Helmholtz,"[87] and the contraction of the sun was what primarily accounts for the present-day supply of heat outward, with the sun "an incandescent liquid mass cooling." Regarding the sun as in this way, as Helmholtz showed, would translate to a certain amount of available heat for the earth, the amount of which depended on the details of the changing density of that liquid across its body. By appealing to bounded estimates of the sun's specific heat (its heat capacity, or the amount of heat required to raise the temperature of the sun by a fixed amount) and its "expansibility" (the change in the volume of sun with heat changes), interpreted through the newly formulated laws of thermodynamics, the conservation of energy, and the increase of entropy, Thomson was able to compute ranges for both the sun's age and its life expectancy.

The choice of *Macmillan's* for the publication of these views is a marker of the publishing and disciplinary landscape in the period,[88] suggesting the extent to which physical or philosophical reasonings were not always unwelcome in the popular press.[89] Thomson at certain points did omit mathematical deductions, but he nevertheless gave a fairly thorough physical argument, enough so that he could cite the piece in more technical works. Moreover, in prosecuting this argument, he appealed to the public lectures of Helmholtz, themselves suggestive of shared public and international tastes for physical reasoning.

Tyndall, by then a prominent natural philosophical voice, offered a further example of such popular voices and tastes. Months after the appearance of Thomson's *Macmillan's* piece, in June 1862, Tyndall gave a lecture on force (as opposed to energy), the lecture in which he championed Mayer. Dropping a lead ball allowed him to demonstrate and explain basic mathematical relationships governing "mechanical effect" or "mechanical force." By noting further that a projectile striking a target produces a flash, Tyndall also drew out the connection between work and heat. Turning to the sun, and relying on the computations of John Herschel and Pouillet as to how much heat the sun uses up annually, he asked his audience: "How is this enormous loss made good? Whence is the sun's heat derived, and by what means is it maintained? No combustion, no chemical affinity with which we are acquainted would be competent to produce the temperature of the sun's surface."[90] Returning to his lead ball, he ultimately asked his auditors to consider how much more heat would be generated by its being dropped from a distance where gravity was negligible, and how much more still if more massive objects were dropped on a still more massive sun. "That lenticular envelope which surrounds the sun, and which is known to astronomers as the zodiacal light, is probably a crowd of meteors; and moving as they do in a resisting medium they must continually approach the sun. Falling into it, they would be competent to produce the heat observed, and this would constitute a source from which the annual loss of heat would be made good."[91] Given his own extended commitments to these questions, Thomson no doubt would have found both Tyndall's treatment as well as his attributions of credit vexing.[92]

Part of Thomson's motivation (and that of *Macmillan's*) for making such a public statement was however the ability to name Darwin in particular, to answer in as public and apparently as rigorous a manner as possible theories that gained so much attention through works understood as at least in some sense generalist themselves.[93] Referring to Darwin's uniformitarian estimate that it would have required some 306,662,400 years for the wearing away or "denudation" found in the Weald Valley in southern England, Thomson asked: "What then are we to think of such geological estimates as 300,000,000 years for the 'denudation of the Weald?' Whether it is more probable that the physical conditions of the sun's matter differ 1,000 times more than dynamics compel us to suppose they differ from those of matter in our laboratories; or that a stormy sea, with possibly channel tides of extreme violence, should encroach on a chalk cliff 1,000 times more rapidly than Mr. Darwin's estimate of one inch per century?"[94]

"It seems, therefore," Thomson concluded, "on the whole most probable that the sun has not illuminated the earth for 100,000,000 years, and almost certain that he has not done so for 500,000,000 years." And, "As for the future, we may say, with equal certainty, that inhabitants of the earth cannot continue to enjoy the light and heat essential to their life, for many millions years longer, unless sources now unknown to us are prepared in the great storehouse of creation."[95]

In the introduction to the *Macmillan's* article, Thomson thus declared that the configurations of matter and energy all point to the same fate: "The second great law of Thermodynamics involves a certain principle of irreversible action in nature. It is thus shown that, although mechanical energy is indestructible, there is a universal tendency to its dissipation, which produces gradual augmentation and diffusion of heat, cessation of motion, and exhaustion of potential energy through the material universe. The result would inevitably be a state of universal rest and death, if the universe were finite and left to obey existing laws."[96]

The new thermodynamics was not necessary for the narrative form of this argument—it was a view shared with Nichol. The new thermodynamic considerations nevertheless shaped the narrative depiction of the way in which that rest is approached, in particular, in the emphasis on the dispersion of heat: "But it is impossible to conceive a limit to the extent of matter in the universe; and therefore science points rather to an endless progress through an endless space, of action involving the transformation of potential energy into palpable motion and then into heat, than to a single finite mechanism, running down like a clock, and stopping for ever."[97] The contrast between a delimited and a limitless world is at this point a technical one: contradicting the finiteness of the universe still yields a bleak conclusion. The final sentence however deepens the contrast between the actual future and immanent thermodynamic predictions, refusing any future condition of being "left" on the basis of the need for a transcendent power in the past and present: "It is also impossible to conceive either the beginning or the continuance of life, without an overruling creative power; and, therefore, no conclusions of dynamical science regarding the future condition of the earth, can be held to give dispiriting views as to the destiny of the race of intelligent beings by which it is at present inhabited."[98]

The first and final words of the article imply calamity in the absence of an "overruling creative power"; the concluding rhetoric of the foreboding "unless" appealing to the storehouse of creation appears ambiguous about the likelihood that sources immanent to nature, but unknown to us, may be found that could prevent calamity. The overall suggestion in the article is that left to its own devices, physical reasoning alone would be insufficient to account for the creation of the universe and could not avoid the bleak future of an ever more dissipated universe. Natural philosophy is constrained by the postulation of an initial, but not necessarily original, state and a completely enervated end in time that is not the end *of time*. Contemporaries such as Helmholtz and future practitioners such as Ernest Rutherford (1871-1937) read the answer in Thomson's analysis as an unflinching refutation of the possibility of a brighter future.

However, Thomson's religious convictions and the scriptural resonances of his choice of words suggest that he believed that scientific conclusions are themselves necessarily limited, and that even if it appeared unlikely that science would transcend its current horizons to discover immanent sources of energy sufficient to keep the sun illuminated indefinitely, the fact that empirical science points to an inconceivable fate must be tempered by faith in the "overruling creative power" of the article's introduction.[99] But ambiguities and tensions persist, and with them a specific construction of the secular. Was, after all, the "destiny of the race" in any case in matter, in the material world? The tensions in the oscillation between a calamitous end and a redemptive one are a reflection of Thomson's attempts to reconcile the extrapolations of physical science and the limits set by the belief in a transcendent power. Extrapolating forward in time according to natural philosophical reasoning conjures up the image of a bleak future state, dismissed only by appealing to something beyond physics.

Thomson and Tait gave some further clue of these concerns in another popular venue, the Presbyterian *Good Words* magazine. Their article "Energy" appeared later the same year as his *Macmillan's* piece, in December 1862.[100] In each of the popular 1862 papers, the end of time is characterized by how long something interesting could last, and in each instance, the narrative often boils down to those features of the world in space and time that would allow humanity to emerge and that would fail to let humanity persist. Whatever optimism could be found in such an account depended on being able to celebrate the possibility of, in effect, divine intervention to produce or disclose new energies.

In "Energy," Thomson and Tait thus emphasize the implications of the first, rather than the second law of thermodynamics. The primary overt purpose is to ready a corrective to a number of presentations to "the interested but unscientific public" concerning the "ONE GREAT LAW of Physical Science known as the Conservation of Energy."[101] And they accomplish this by correcting the vocabulary of "force." On the understanding that "force is defined as *that which produces, or tends to produce, motion,*" the authors point out that force that doesn't produce motion is not conserved, hence dismissing the phrase "conservation of force" associated with practitioners such as Grove and Tyndall. They allude to a "host of errors" resulting from "confounding Force with Energy."[102]

And indeed, the article was in no small measure an attempted corrective to Tyndall himself.[103] The authors took notice of the fact that Tyndall's "On Force," in popularizing the new thermodynamic principles, drew from Helmholtz and Clausius and explicitly gave priority to Mayer over and above Joule and Thomson.[104] They faulted other presentations for their science, but Tyndall's account was explicitly faulted more for its history.[105]

In the article, the two authors endorse the cometary/nebular origin of the sun—that diffuse matter "endowed with the attractive force of gravitation" came together through collisions, eventually producing hot bodies cooling, including among them future planets.[106] They claimed that the energy lost by the sun radiating its heat is not in turn compensated by the infall of new comets—and this points to the decline of solar energy impinging upon the earth. Importantly, the sun for them provided a key to extrapolating to what is perhaps the very beginnings of things: "This view of the possible origin of energy at creation is excessively instructive. Created simply as difference of position of attracting masses, the potential energy of gravitation was the original form of all the energy in the universe." Eventually, then, all this energy will become kinetic, and once this happens, "the result will be an arrangement of matter possessing no realizable potential energy, but uniformly hot—an undistinguishable mixture of all that is now definite and separate—chaos and darkness as *'in the beginning.'*"[107]

The coauthors assert together the vision that Thomson had emphasized alone, his cosmogony and his correlate belief that to rely on science without accepting a transcendent power is to produce a grim picture of humanity's destiny. They find hope in another reference to scripture, in "that light which reveals 'new heavens and a new earth.'"[108] But whereas in some popular works of Herschel or Nichol an immanent science continually discovered and approached the workings of God and reflected therefore a more robust intellectual theodicy, in Thomson, as well as in Thomson and Tait, that more abstract divinity is largely relegated to the chronological margins and to the prologue and epilogue of a popular universal history. Their vision is suffused by something transcendent that is kept implicit, except at the margins, but future readers often saw instead only the bleakness of their conclusions about immanent natural phenomena. And for contemporaries such as Helmholtz, heat death was an ultimate truth against which no transcendent storehouse of energy could preserve humanity.

LINKED ENDS, DIFFERENTIATED MEANS

Contentious and broad as was the problem of writing universal history—broader still, when including all the popularizers of the time, regardless of their putative authority—the problem of many authors was also an epistemological virtue. Even despite contestations, different putative universal historians could observe that the question of history and the disagreements it entailed extended beyond any one apparently delimited theory, even as that history came to be shaped according to a few narrative principles drawing from the new theories of heat to emphasize the sustaining or destruction of worlds.

However, the questions of natural and universal history presented the problem of having no one author to which to attribute priority or work. In a sense, such a discourse was itself already a system that no one fully understood, maintained and distributed by many hands working toward shared *and* idiosyncratic ends. The new discourses of energy and force produced different pieces of a universal history that several hazarded, a history that deployed notions of energy to demarcate the times and places in which the universe could be sustained without transcendent intervention. At the same time, however, as the next chapter will explore, the new discourses of energy and force set different practitioners at odds. And within and beyond the horizons of these conflicts, the unity of forces and the transformability of energy could begin to stand for distinct modes for attempting to grasp the joint promises of a unity of nature and synthesis of knowledge.

CHAPTER FOUR

Schisms

Men of warm feelings, and minds open to the elevating impressions produced by nature as a whole, whose satisfaction, therefore, is rather ethical than logical, lean to the synthetic side; while the analytic harmonises best with the more precise and more mechanical bias which seeks the satisfaction of the understanding.

JOHN TYNDALL, Address Delivered before the British Association Assembled at Belfast, 1874

Natural Knowledge tends more and more to the conclusion that "all the choir of heaven and furniture of the earth" are the transitory forms of parcels of cosmic substance wending along the road of evolution, from nebulous potentiality, through endless growths of sun and planet and satellite; through all varieties of matter; through infinite diversities of life and thought; possibly, through modes of being of which we neither have a conception, nor are competent to form any, back to the indefinite latency from which they arose. Thus the most obvious attribute of the cosmos is its impermanence. It assumes the aspect not so much of a permanent entity as a changeful process, in which naught endures save the flow of energy and the rational order which pervades it.

Let us understand, once and for all, that the ethical progress of society depends, not on imitating the cosmic process, still less in running away from it, but in combating it.

T. H. HUXLEY, "Ethics and Evolution," 1893

The unity of forces and the laws of thermodynamics, though deeply intertwined, came to signal different modes of potentially finalizing projected visions of education, natural science, and natural history. Whether through a discourse of continuity, as with William Grove, or of energy, as with William Thomson and Peter Tait, or on the sublime and humbling scales emphasized by John Tyndall, but no less by John Herschel and William Thomson, theories built on the laws of the convertibility and dissipation of energy suggested new ways to narrate a final story about the world, a story telling of both universal development and potential dissolution.

These different natural historical narratives drew from many different sources, arising by conflict and confluence, by explicit, disregarded, and disavowed debt. The synthetic pictures of the world as a whole and its origins that they offered were disturbing or liberating, sublime or reductive, consoling or woeful, stirring the varied audiences reacting to them to a spectrum of responses. Deliberations over material and organic origins also remained linked to theoretical and pedagogical discussions over the interrelation of the domains and disciplines of knowledge—over the real or potential unity of the sciences as disciplinary and institutional practices and their claims to cultural priority and authority. As with the synthetic efforts and pedagogical claims discussed in the previous chapters, emergent structures of disciplinary authority and education were the site of arguments in relation to the relative importance of different kinds of knowledge for society as a whole and in relation to the question of which was the most authoritative of the sciences.

As a political matter, there was no consensus over which of these universal historical narratives could unify the variety of actualizing institutional disciplines or even over whether any narrative could do so. At the same time, many commentators, educators, and authors debated the potential of these narratives to construct interdisciplinary bridges, both criticizing previous theorizations and in doing so setting elements in place to write fuller histories, sometimes established via those critiques.

But Herschel's concern remained: who could responsibly attempt a synthesis of all the new scientific materials at hand? In the latter part of the century, scientific synthesizers such as T. H. Huxley and John Tyndall publicly criticized existing broad-strokes narratives, but along with Hermann von Helmholtz, central among others, hazarded such narratives of their own.[1] Like Kant and Hegel, they, too, sought to redeem the "slaughter bench" of history by the discernment or heroic assertion of a resplendent organizing and meaning-making idea, an immanent wonder in itself awaiting fuller disclosure. However, as we will see here, this putative disclosure proceeded as much through the process of dissensus as through any consensus. The narratives offered by Thomson, as we examined in the last chapter, and of Helmholtz, Huxley, and Tyndall, as we will examine here, were moral in character, but their narratives each pointed to a future secured on very different moral grounds—on that of a transcendent being, on "an overruling creative power," for Thomson, or on a transcendent order of immanent life and world, for Helmholtz, or on a new sublime wholeness and truth for Tyndall, or on immanence and ethical struggle, for Huxley. Their narratives all looked to the enlargement of the domains of natural science into human culture and thought, but the unifying principles on which they founded their hopes for the all-encompassing promises that science might offer as a ground for an enduring human culture could not be made by them to agree.

HELMHOLTZ'S "HIGHER MORAL PROBLEMS" AND HUMAN "DESTINY"

With Helmholtz less under the shadow of Chambers and less disturbed by Darwinian evolution, he claimed to stand apart from the anxieties about new scientific truths and their broader cultural acceptability or seemliness that so many commentators have seen as dogging the work of his British counterparts.[2] Regarding what he saw as the distinctive contributions of German scientists, he was willing to state hyperbolically that "among us there is a total lack of fear over the implications of the knowledge of the complete truth. In both England and France there are excellent investigators who are capable of working along scientific lines with complete dedication. Hitherto, however, they have almost always had to bow to social or ecclesiastical prejudice, and they have been able to express their convictions openly at the expense of their social influence and usefulness."[3] He saw himself, by contrast, as able to benefit from these investigators and at the very least to a draw a link between the emergent evolutionary history and moral science while advocating an integrated educational system.

Impassioned exchanges over the sustainability of nature provoked by the formulations of thermodynamics did not prevent Helmholtz from crafting epic histories in short essays. Helmholtz persistently took his case to the public, connecting the conservation of energy with universal historical concerns, writing separate articles on both, and making his debt to the nebular hypothesis still clearer than some of his British contemporaries.[4] Helmholtz established conciliatory and sweeping universal historical principles before any controversies or intellectual disputes emerged between Thomson, Tait, Darwin, and Tyndall, or Thomson and Huxley—before natural selection was established as a public matter of universal history, before Darwin was willing to declare publicly the synthesis taking shape in his private notes. It was a template that Helmholtz then extended and modified with Darwinian natural selection and in the light of new thermodynamic considerations, but it was not one he departed from significantly, despite all the debates waged between such friends and fellow travelers as Tyndall and Thomson.[5]

Like Darwin, Helmholtz sketched a natural historical synthesis well before its varied elements had met with any fuller articulation or consensus.

In his 1854 Königsberg address, "On the Interaction of Natural Forces," the popular lecture to which Thomson referred in his *Macmillan's* paper, Helmholtz was unhesitant and concise in putting forward a sweeping vision of universal history. Many of the ultimate principles of his history he shared with Thomson, most centrally the insistence on the universality and shaping principles of the first and second laws of thermodynamics. The first law fixes and makes fundamental the amount of energy or "mechanical force" in the universe. The second law charts the gradual transfer of that store into unusable heat. Keeping open the question of how foundational a category this conception of *Kraft* was, Helmholtz nevertheless gave voice to what Anson Rabinbach termed "transcendental materialism": "As Helmholtz tirelessly pointed out: energy was a transcendental principle, 'the groundwork of all our thoughts and acts.'"[6] Here, too, Helmholtz explored the question of whether the organic could be reduced to the mechanistic: whether the supposition of vital forces violated the new thermodynamics, whether the solar system and universe more widely could be a perpetual motion machine—in short, whether Kant's phoenix ever could reemerge from the ashes.

Perpetual motion was a possibility that he ruled out. He began his argument with automata. Helmholtz saw in automata links to more modern attempts to achieve perpetual motion, to produce work and money from nothing.[7] Contrasting the care taken over automata in the previous century with the industrialism of the day, he observed (in the language of the translation provided by Tyndall): "We no longer seek to build machines which shall fulfill the thousand services required of one man, but desire, on the contrary, that a machine shall perform one service, and shall occupy in doing it the place a thousand men."[8]

As Helmholtz related his stylized history, the recognition of the impossibility of this "philosopher's stone" led to the question: "If a perpetual motion be impossible, what are the relations which must subsist between natural forces?"[9] Here, through an extension of the second law of thermodynamics, Helmholtz presented a vision of the end of history, attributing the conception of "heat death" to Thomson, in an extrapolation that Helmholtz himself may have been the first to make in print.[10] Noting the general validity of this law, Helmholtz concluded that "if the universe be delivered over to the undisturbed action of its physical processes, all force will finally pass into the form of heat, and all heat come into a state of equilibrium. Then all possibility of a further change would be at end." This process would include the end of organic life. Helmholtz had given a mechanistic account of all living beings, laying stress on sunlight as producing the energies necessary to sustain life on earth. Though "we are all, in point of nobility, not behind the race of the great monarch of China, who heretofore called himself Son of the Sun," that position is shared by "the frog and the leech," the "whole vegetable world," and "even the fuel which comes to us from the ages past."[11] Consequently, "The life of men, animals, and plants could not of course continue if the sun had lost its high temperature, and with it his light. . . . In short, the universe from that time forward would be condemned to a state of eternal rest."[12] Though he did not foresee a counterindication, Helmholtz did concede that the law in question had yet to be proven "universally correct."[13]

But if anything, these are the more conservative elements of the piece. In the same address, he made short work of the nebular hypothesis, crediting it to Kant (a hypothesis that "in our own home, and within the walls of this town, first found utterance"[14]) and thereafter to Laplace. The elements of Helmholtz's "nebulous chaos" were equipped with forces/attributes of attraction, falling toward one another and in collision producing heat—and he used this as one plank to compare the new cosmogony to the Mosaic, the latter the narrative for which to supply a scientific reading.[15] In comparison to Thomson, at least in the context of a more popular address, Helmholtz did not emphasize the narrowness of time but its breadth, embracing the idea of older forms of life gone by and newer forms of life waiting to appear.[16]

Helmholtz nevertheless betrayed anxiety about the heat death of the universe. Indeed, his claim that the conclusion was "unavoidable" was more an admission, a deduction rendered through "physico-mechanical laws" that "are, as it were, the telescopes of our spiritual eye, which can penetrate into the deepest night of time, past and to come."[17] That spiritual eye, according to Helmholtz, brings us in view of these final truths: "As each of us singly must endure the thought of his death, the race must endure the same. But above the forms of life gone by, the human race has higher moral problems before it, the bearer of which it is, and in the completion of which it fulfills its destiny."[18] What these new problems were Helmholtz did not say in this essay, but they would appear linked to the historical process of the apparent re-creation of second nature, to the industrialization that permeates the essay, and thus to a world he believed to be reconstructing itself through natural science and producing by doing so new aesthetic and moral dilemmas.

This early narrative was a template for later natural histories to which he also actively contributed. Science intimated the origins and disclosed the destiny of the race, connecting nebulous astronomical beginnings to the rise, transformations, and decline of organic forms, projecting forward to the ultimate state of the universe itself. Though he declared the need to face the species and universal destiny, Helmholtz shared with Darwin, and, as we shall see shortly, Huxley and Tyndall, the sense that the new sciences disclosed wonders giving balance to the dispiriting conclusions Thomson and Tait excluded only by appeal to an overruling power. At the end of 1862, the year Thomson published "On the Age of the Sun's Heat," and "Energy" with Tait and the year Tyndall lectured "On Force," Helmholtz introduced his own series of lectures in Karlsruhe entitled "On the Conservation of Force." Here, he emphasized the almost "artistic satisfaction" of being in a position "to survey the enormous wealth of Nature as a regularly-ordered whole—a kosmos, an image of logical thought in our own mind."[19] As with Tait and Thomson, the surveying instrument of choice was the first law of the conservation of energy, an exercise that could bracket the heat death held in the view of the physico-mechanical, spiritual eye. As with Herschel, the ambition was for an individual mind to be in position to conceive the whole.

Helmholtz tied this conception of synthesis not only to a division of labor, but also to the figure of a single synthesizer and synthesis. And like Herschel, he focused on Humboldt as an exemplary figure—a bearer of the responsibilities for the higher moral problems confronting humanity, represented by a single mind. As knowledge of the details of "the enormous wealth of Nature" increased, Helmholtz observed that the task of synthesis and the burden of the synthesizer became more difficult. In a lecture of 1869, in the opening address to the *Naturforscherversammlung*, the Congress of German Scientists and Physicians, he recalled the synthetic work of his patron and observed that "it is obviously very doubtful whether this undertaking could be carried out in a similar way [today], even by a mind with gifts as peculiarly suited for the purpose as Humboldt's were and even if all his time and work were devoted to the task."[20]

Unlike for his earlier 1862 address, the intended audience now was natural scientists. In this comparatively restricted context, where "it is appropriate . . . to take account of the progress of science as a whole," and where "such a survey is desirable," Helmholtz nevertheless professed that "it is beyond the capability of any one man to carry out."[21] The admission however has an almost formulaic ring to it, all the more so in an address developing a vision of a comprehensive law embracing all "natural phenomena." Since "we experience the laws of nature as objective forces," this vision resolves into a "search for the forces which are the causes of phenomena."[22]

The structure of the address suggests then that between conservation of energy, evolutionary studies, and physiological advances in the study of life, a Humboldtian synthesis was indeed possible and implies that Helmholtz himself was perhaps the Humboldt of his age. Helmholtz connected the conservation of force to vitalism, taking them to be in contradiction. In turn, in relation to the question of the "fundamental principles of the theory of life," he offered an extended defense of Darwinian evolution.[23] This in turn raised the question of individual as opposed to species adaptation and brought Helmholtz squarely to physiological research. "Ophthalmic medicine," the focus of his own famed physiological studies, he argued, was the "one branch of medicine in which the influence of scientific methods has been perhaps most brilliantly displayed."[24]

But if a universal synthesis might be possible, possibly to be achieved by the powers of a single mind, there were obstacles to that achievement—obstacles, in Helmholtz's account, posed by differences in intellectual culture between Germany and the English context on which he himself drew. By the 1870s, Helmholtz was using the new cosmology he was proposing as proof of the relative intellectual freedom in his own country while noting that popularization had been a more effective enterprise in England.[25] In a preface to a German translation of popular work of Tyndall, *Fragments of Science*, Helmholtz singled out Tyndall's particular strengths as a popularizer and defended Tyndall against recent attacks by the astrophysicist Johann Karl Friedrich Zöllner, which he diagnosed as an attack on the Baconian inductive method, a method that "was followed earliest and has been followed most seriously by his [Bacon's] fellow Englishmen." Though Germany was the "last to break away" from a priori reasoning, it was nevertheless "the most resolute" in doing so.[26] Any new synthesis or system of education would have to be on the basis not of what Helmholtz termed the "metaphysical" (contrasted with proper philosophy, as astrology could be contrasted with astronomy), but on the basis of facts. To side with those such as Zöllner, who, Helmholtz suggested, espoused metaphysics, "would only set back the hope of a final solution of the problem in our present program of education."[27]

Whatever the truth of the symbiotic relationship between German and British men of science and the wider symbiosis it suggested for German and British natural philosophy, however, the congruity between points of view foundered on the questions of scientific narrative and priority. Helmholtz had come to draw not only from Thomson, but from Darwin, despite Thomson's critiques of Darwin. He also appeared to align himself with Tyndall in the latter's defense of the centrality of nineteenth-century German science. In Darwin's evolutionary work, Helmholtz further found the materials for a grand evolutionary synthesis like what Darwin himself had perceived through the accounts of Herschel, Whewell, Brewster's Comte, and others. However, his was a different account of cosmic evolution, proffered at roughly the period when Huxley would pronounce on the advent of the "evolutionary" school of geological and indeed universal history and a few years preceding Tyndall's most famous defense of scientific materialism and evolutionary history.

But the debates in the British context were not always the central axes of concern in the German context during the 1870s.[28] Instead, during this period, Helmholtz was lecturing on a vision of a more complete evolutionary history that unapologetically relied on the nebular hypothesis. As we have seen, he asserted that his own intellectual and cultural environment was better able to deal with the lessons of that history than the more fractious British context, and from the 1850s on, consonant historical, cultural, and pedagogical considerations ran throughout his public addresses. By 1871, well-polished refinements of Helmholtz's themes coalesced in a lecture delivered in Heidelberg and Cologne, "On the Origin of the Planetary System." There, Helmholtz erected the figure of Kant as representative of both philosophy and natural philosophy, enlisting him once again as a predecessor on the question of universal history. He appealed to Humboldt for more thorough physical geography. By this time, he could invoke spectral analysis to testify to the composition of the stars and the present-day nebula, asserting that their composition shares in terrestrial elements. Here, he alluded to William Herschel's nebular considerations[29] and to the newest estimates of average distances between nearby comets by William's grandson Alexander. The connection of these origins to the prominent features of the solar system, familiar from Kant and more thoroughly Laplace, he treated with refinements and corrections, maintaining the predominance of the sun for sustaining life.

And across this shaping universal history, Helmholtz once again, as with his 1854 address, returned to eschatology. He noted that as there exists evidence of the death of suns, so our sun, as it loses energy, dies: "This is a thought which we only reluctantly admit; it seems to us an insult to the beneficent Creative Power which we otherwise find at work in organisms. . . . But we must reconcile ourselves to the thought that, however we may consider ourselves to be the centre and final object of Creation, we are but dust on the earth; which again is but a speck of dust in the immensity of space"—just as, Helmholtz asserted, the life span of humanity and life itself is minute in comparison with that of universal history.[30] Humanity, he argued with Herschel, Thomson, and others, is marginal in space, time, and significance.

Nevertheless, from "Darwin's great thought" he concluded both hopefully and rhapsodically that "considering the wonderful adaptability to the conditions of life which all organisms possess, who knows to what degree of perfection our posterity will have been developed in 17,000,000 of years," the period he gave for the endurance of the sun's current intensity, given his contraction theory. As the past of the earth might seem inhospitable to the present, so might the present seem to the future.

But subtly thereafter, in a claim with Kantian resonance, Helmholtz moved away from what might easily be regarded as *our* posterity: "even if sun and earth should solidify and become motionless, who could say what new worlds would not be ready to develop life?"[31] Life else*when* also potentially invoked life elsewhere, at least in some form: "Meteoric stones sometimes contain hydrocarbons; the light of the heads of comets exhibits a spectrum which is most like that of the electrical light in gases containing hydrogen and carbon. But carbon is the element, which is characteristic of organic compounds, from which living bodies are built up. Who knows whether these bodies, which everywhere swarm through space, do not scatter germs of life wherever there is a new world, which has become capable of giving a dwelling-place to organic bodies?"[32]

Helmholtz went so far as to question the future state of an individual living creature, comparing it to flames and waves, which retain their identity despite being continually reconstructed. Noting that "the observer with a deaf ear only recognizes the vibration of sound as long as it is visible and can be felt, bound up with heavy matter," he asked: "Are our senses, in reference to life, like the deaf ear in this respect?"[33] In regard to the lecture, the question primarily relates to the future state of humanity. Helmholtz was not stressing the limitations of scientific knowledge, as his friend du Bois Reymond would in a sensationalizing address of the following year—an address that some such as Ernst Haeckel took as an attack on Darwinian evolution and to which Helmholtz himself apparently offered no reply.[34] But given the entire context, near and more extended, of Helmholtz's synthetic narrative of natural history, his query introduces the further questions of how exotic life might be, of how exotic are the places it occupies, and of how badly humanity might be constituted to know that other life. In this narrative, life in the universe might be much more exotic and prevalent than humans can sensibly appreciate.

THOMSON CONTRA HUXLEY

Judging the nature of the entanglement between human-centered and universe-spanning history exercised Huxley in scientific debate and in political thought. The often contentious activity in which Huxley was a principal actor clarified the terms of how cosmic evolution might function as a natural scientific expansion and refinement of Kantian universal history. Themes of civil history rooted in cosmic processes, of thermodynamics and human labor, and of the relation of the state of nature to human cultivation, were ones to which he returned continually.

At times, almost in mirror image to Helmholtz, Huxley contrasted the failings of the British educational system with the emergent strength of the German universities. He repeatedly defended universal education and, the year after his election to the London School Board, undercut arguments against the benefits of this universality in his 1871 address, "Administrative Nihilism." Arguing that a gifted but disadvantaged worker was more likely to take up radical socialism, Huxley saw general education as a way of providing such a worker (and therefore the state as well) with better horizons.[35] This pedagogical demand in turn undercut the "astynomocracy"—government limited to police action—he regarded as too prevalent an ideal among his contemporaries, Spencer first among them.[36] His critique brought him in explicit dialogue with Kant's "Idea for a Universal History with a Cosmopolitan Purpose."[37] Through Kant's exposition of the origins of the state, Huxley found tools to attack unrestrained laissez-faire liberalism. The "unsocial sociability" of Kant's (and Montaigne's) humanity pointed to the demand for a compromise between individual freedom and state regulation: individuals strove to be free of each other's constraints while seeking out the company that formed those constraints.

Universal historical narrative thus was growing outward from several different interdependent centers, geographical and discursive, as it brought more events and disciplines in its scope. Despite the early position on their reconciliation taken by as respected and connected a practitioner as Helmholtz, the most authoritative positions in Britain set physics, geology, and biology at odds over the question of geological uniformitarianism. Thomson's intervention in this debate and in particular his exchange with Huxley has been repeatedly analyzed, during his life and posthumously.[38] The debate was a prominent site for a self-conscious articulation of an evolutionary vision of both the material and organic world, and as such, for the possibility of writing a sweeping immanent universal history.

In the debate over uniformitarianism, the terms of Thomson's position with which Huxley took issue were well in place before a first response to Thomson in 1868. In his 1862 paper "On the Secular Cooling of the Earth," Thomson opposed uniformitarianism in broad terms as contravening the principle of conservation of energy, freely citing the findings of his *Macmillan's* article "On the Age of the Sun's Heat" and earlier papers. Thomson took a lyric attitude: "eighteen years it has pressed on my mind, that essential principles of thermo-dynamics have been overlooked by those geologists who uncompromisingly oppose all paroxysmal hypotheses, and maintain not only that we have examples now before us, on the earth, of all the different actions by which its crust has been modified in geological history, but that these actions have never, or have not on the whole, been more violent in past time than they are at present."[39] As evidence of this enduring concern, he attested his Glasgow inaugural dissertation, as well as his 1844 paper "Note on Some Points in the Theory of Heat"—a paper two years after his "On the Linear Motion of Heat." The ultimate form of the argument that he pressed relied on the current gradient with which the earth was losing heat from its center to its surface. On the supposition that the earth is in large part a cooling body from an originally hot state, Thomson extrapolated backward to determine how much time must have passed for that gradient to have come to be as contemporaneous measurements suggested. This geological work clarified in detail why, to Thomson's mind, the uniformitarians among the geologists—"those geologists who uncompromisingly oppose all paroxysmal hypotheses"—needed to reconsider their theories.

In a concise paper read before the Edinburgh Royal Society in December 1865,[40] Thomson further observed that "the heat which we know, by observation, to be now conducted out of the earth yearly is so great, that if *this* action had been going on with any approach to uniformity for 20,000 million years, the amount of heat lost out of the earth would have been about as much as would heat, by 100° Cent., a quantity of ordinary surface rock of 100 times the earth's bulk"—enough heat, that is, to melt a rock at the earth's surface larger than the earth itself. From this observation, it follows that the upper crust of the earth could not have remained as it currently is in present-day, refuting the uniformitarianism "held by many of the most eminent of British Geologists."[41]

The views in these publications alone give some context to the claims that Thomson made in a February address, "On Geological Time," delivered before the Geological Society of Glasgow in 1868, regarding his overall views countering uniformitarianism and the need for a geothermal survey, to which he had "long endeavoured, in vain, to call the attention of geologists."[42] Thomson commandingly began his address, "A great reform in geological speculation seems now to have become necessary."[43] This is the register that captured Huxley's attention and against which he launched his objections.

Thomson began by presenting an extended quote from Playfair's *Illustrations of the Huttonian Theory*. Connecting planetary and terrestrial cyclical stability, Playfair claimed that "the Author of nature has not given laws to the universe, which, like the institutions of men, carry in themselves the elements of their own destruction," that "He may put an end, as He, no doubt, gave a beginning to the present system, at some determinate time; but we may safely conclude that this great *catastrophe* will not be brought about by any of the laws now existing, and that it is not indicated by anything which we perceive." To the whole of which Thomson objected, "Nothing could possibly be further from the truth than that statement."[44]

Thomson argued that the fact that the tides "diminish the velocity of the earth's rotation round its axis, and lengthen the duration of the day"[45] points toward the conclusion that the current state of the solar system reveals its ultimate instability. On this basis, Thomson charged that "here is direct opposition between physical astronomy, and modern geology as represented by a very large, very influential, and, I may also add, in many respects, philosophical and sound body of geological investigators, constituting perhaps a majority of British geologists. It is quite certain that a great mistake has been made—that British popular geology at the present time is in direct opposition to the principles of natural philosophy." Regardless of how errant a celestial clock the earth is—the calculations vary—"the principle is the same. There cannot be uniformity. The earth is filled with evidences that it has not been going on for ever in the present state, and that there is a progress of events towards a state infinitely different from the present."[46] And once again invoking his *Macmillan's* article, Thomson also denied the possibility that the sun is a "perpetual miracle." That view was to his mind an assumption at work in much geology. Apart from the tides and the consequences of a delimited availability of solar power—each part of the celestial/astronomical argument—the last plank of his critique was the heat-conduction argument within the earth itself, returning to his earlier refutation of the doctrine of uniformity and his secular cooling papers.

But throughout his critique, Thomson was able to deduce the origin and destruction of the present order of the universe because he took so little issue with belief in the uniformity of the *ultimate* laws of the universe.[47] On the basis of this, he asserted (in line with discussions in the previous chapter) the need for a divine intervention to prevent the gradual destruction that thermodynamics promised. In this sense, the debate over the age of the world could be understood as not concerned with age in itself, but as a debate over *which* principles the natural scientist should accept as uniform.

The next year, 1869, in his presidential address to the Geological Society, Huxley took the opportunity to reply to Thomson, granting Thomson the audience among geologists that he had apparently been craving, if not precisely in the way he had been craving it. Huxley began his address by quoting Thomson's declaration that British popular geology had made a "great mistake." Whatever would follow Huxley's choice to initiate his address this way would all but guarantee Thomson's geological views further publicity. Given Thomson's status and care, that attention was also likely to make geological uniformitarianism appear more susceptible of criticism than it might otherwise have been. Taking up the defense of geologists in the role of their "attorney general," Huxley defended those he took to be targeted by Thomson, Hutton most of all.[48] In his storm of a response, he even asserted the priority of geology over biology, in the sense that (unlike geology with respect to natural philosophy) biology took its timelines from geological history.[49]

His most straightforward argument was based on professionalism: Thomson was simply wrong about the extent to which geologists were uniformitarians, because Thomson, not a geologist himself, was not in a position to characterize their views. Uniformitarianism was merely one option, and indeed, it represented an end point in a spectrum of positions, all of which could be combined. There were no modern geologists who were pure uniformitarians.[50] Huxley traced three "systems of geological thought." The first two were those that Whewell had characterized decades earlier: catastrophism, of which the traditional Mosaic geology was an example and of which geological persuasion Huxley numbered "some of the most honoured members of this society"[51]; and uniformitarianism, for which he adduced the examples of Hutton and Lyell. Moreover, from Huxley's point of view, the geological move away from examining origins was less a denial of the fact of those origins than an exclusion of the subject as a matter of geological methodology. He quoted Hutton and particularly Lyell to this effect.[52]

But it is to neither of these systems or poles of thought to which Huxley gave his allegiance. He subscribed instead to a geological "evolutionism," a school/system/speculation "destined to swallow up the other two."[53] To explicate it, Huxley drew parallels between the categories for studying the living forms—anatomy, development, physiology, distribution—and those of "earth knowledge"—stratigraphy, geographical succession, physical geography/meteorology, and the astronomical position of the earth.[54] In each, biology and geology, categories must be referred to causes, to biological etiology on the one hand and to geological etiology on the other. The explicit decision to bracket etiology altogether on the part of those geologists associated with the Geological Society was part of a foundational consensus, but a consensus that now had become an unnecessary hindrance. It also demonstrated the extent to which a Herschelian synthesis, in Huxley's view, had to remain firmly causal.

To Huxley's mind, Kant was the founder of geological etiology, a subject that in turn stood at the heart of geological evolutionism. He allowed himself an extended rhapsody on Kant—the Kant of broader cosmology: his nebular hypothesis, but also his physical geography. In weighing its strengths, Huxley articulated what he would also come to term "cosmic evolution": "Nor is the value of the doctrine of Evolution to the philosophic thinker diminished by the fact that it applies the same method to the living and the not-living world; and embraces, in one stupendous analogy, the growth of a solar system from molecular chaos, the shaping of the earth from the nebulous cubhood of its youth, through innumerable changes and immeasurable ages, to its present form; and the development of a living being from the shapeless mass of protoplasm we term a germ." And it is this school of thought that, contrary to Thomson's beliefs, Huxley believed to be more representative of contemporary geology, to be "assuredly present in the minds of most geologists."[55]

The remainder of the address is given to more specific criticisms of Thomson. He declares an indifference to Thomson's physics that, from Thomson's point of view, conceded the very point of contention and frustration. To Thomson, geology did rely on thermodynamic considerations, perhaps in a fashion similar to how biology, according to Huxley, took its timelines from geology. Here, however, Huxley accepts the possibility of reforms, whether for the substance of Thomson's critique or for the sake of argument.[56] Whatever the case, Huxley seems more offended by Thomson's attempt to set terms in such a way that biology and geology were subordinate to Thomson's physics and less scandalized by any idea that reconciliation across disciplines was ultimately necessary. For now at least, geology could stand apart.

A final history demanded reconciliation with theories also regarded as final. The events that structured the chronology of that history—where the rules bracketed transcendent intervention except as expressed in establishing laws as such—evoked stronger agreement than the structure of the chronological time itself. Thomson had in effect argued that his theories had the most thoroughgoing finality and most comprehensive applicability to those events. Huxley balked at that idea, resisting the claim that Thomson's chronology should supply the ultimate principle for reconciliation across the disciplines, except to the degree that it already agreed with Huxley's. From Huxley's point of view, *strict* chronology was not the ultimate arbiter, since even if adjustments would have to be made and dates more accurately determined, "it is not obvious, on the face of the matter, that we shall have to alter, or reform, our ways in any appreciable degree."[57]

Despite their differences, Thomson observed that at core, there was much agreement between Huxley and him. Most importantly perhaps, he believed that they each accepted Huxley's evolutionary geology, the geological school that Thomson claimed was the one in which he had been taught his geology decades earlier.[58] For Huxley, this vision included evolution by Darwinian natural selection; for Thomson, it did not. But neither disputant was in fact hostile to some form of transmutation.[59] Both by the next decade were advocating a naturalist evolution of life.[60]

TYNDALL'S SCIENCE CONTRA THE PROVINCES OF RELIGION

In 1874, John Tyndall delivered his most famous address, to the Belfast meeting of the British Association for the Advancement of Science as its president—an address that Ruth Barton refers to as "the most controversial session of the British Association in the nineteenth century."[61] Like Helmholtz's address before the *Naturforscherversammlung*, or Grove's before the BAAS in the decade preceding, Tyndall chose to present a synthesis that made universal history prominent, but in a way that, then as now, provoked debate over the nature of Tyndall's scientific materialism. The sensibility revealed in the address had roots not only in Tyndall's own long developing thought, but in German and Anglo-American Romanticism.[62] Tyndall's scientific naturalism was inflected by his embrace of the philosophy of Thomas Carlyle, helping produce what Stephen S. Kim has characterized as Tyndall's own form of "transcendental materialism," however in a different, if related, sense from the term applied to Helmholtz by Rabinbach.[63] As Ursula DeYoung economically notes, Tyndall "considered himself to be blending the truth about matter with the transcendentalism of Carlyle's philosophy."[64]

The structuring theme of the Belfast address was a defense of materialism, both in terms of its pedigree and inductive success. In reference to its historical roots, Tyndall celebrates the atomic philosophy over and above the less weighty philosophies of Plato and Aristotle.[65] He appeals to the varying forms of materialism as found in Democritus, Epicurus, and Lucretius—and from Lucretius, Tyndall draws a line to Kant through the nebular hypothesis. Approaching the modern period, Tyndall also invokes Copernicus and Bruno, Bacon, Descartes, Boyle, and Newton, using these figures not only to trace the advance of natural philosophy, but also to characterize different casts of mind, analytic or holistic—those given to dissecting the universe into parts and those susceptible to an impression of the whole. Tyndall argues that each has a place in natural philosophical work. Among his own scientific contemporaries, Tyndall gives pride of place to Darwin, praising his character, along with his science. Among the few others he singles out for more extended mention are Spencer for his speculations on the evolution of sensation and psychic life, and Carlyle as a mind that emphasized poetry, but could have pursued science.

But although it was Darwin who received Tyndall's most exclusive attention, it was not natural selection, but "Conservation of Energy," as he now called it, that Tyndall nominated as the greatest modern scientific principle. The planks of scientific naturalism—the conservation of energy, evolution by natural selection, and atomic theory—were all heavily in evidence, but Tyndall represented energy principles as "of still wider grasp and more radical significance" than even evolution.[66] Unlike in his "On Force" address nearly a decade before, Tyndall now attributed the discovery of the law to J. R. Mayer, James Prescott Joule, and the Dutch engineer Ludwig Colding equally. But if he avoided controversy with regard to the priority of the discovery of the conservation of energy, any controversy he might have avoided among immediate scientific colleagues was overshadowed by the controversy that the wider reception the address ultimately provoked. That controversy was not merely over the nature of Tyndall's materialism, but over the reaches of the scientific domain. In contrast to his old acquaintance du Bois-Reymond's "Ignorabimus" address of 1872, which sparked an enduring controversy in its declarations of ineliminable ignorance in relation to the nature of fundament of the universe and the origins of consciousness, Tyndall's address treating not inconsonant concerns was widely received as a provocative proclamation of scientific ambition.[67] Tyndall made explicit his contention that physical science takes precedence over religion in questions concerning the physical state of the nation and the world.

However, as Ruth Barton summarizes Tyndall's argument, "Science, because it satisfies only the understanding, cannot be a complete philosophy of life. In spite of this dualism Tyndall sought unity. 'The human mind . . . will still turn to the Mystery from which it has emerged, seeking so to fashion it as to give unity to thought and faith.' . . . Tyndall found this unity not in materialism but in a cosmical life that is manifested in both matter and mind, thought and emotion."[68] The pantheism that Barton thus identifies in Tyndall's address welded together and interlaced different emphases at different points, the search for "unity" also thus provoking different potential conflicts.

The Belfast address suggested that the relationship between life and matter could be approached in three possible ways, three possible modes of synthesis: as a problem of "logical sequence" (What derivation can science provide of life from the fundamental properties of matter?), as a problem of scales (At what scale does life emerge?), and/or as a problem of historical sequence (When in the past did life come about?). Each approach would rely on physical laws, but the accounts they suggested could not all be written out as easily through those laws. Tyndall admitted that "if we look at matter as pictured by Democritus, and as defined for generations in our scientific text-books, the notion of any form of life whatever coming out of it is utterly unimaginable." But he qualified that "those who framed these definitions of matter were not biologists but mathematicians, whose labours referred only to such accidents and properties of matter as could be expressed in their formulae."[69] The false definitions, Tyndall suggested, might have been the cause of a false severing of life from matter: "Divorced from matter, where is life to be found? Whatever our *faith* may say, our *knowledge* shows them to be indissolubly joined." And that knowledge in turn required hewing down to the molecular level.[70]

In a refocusing of the question of universal history that, as will be seen in the next chapter, was to become a characteristic of early twentieth-century work on the effort to construct a story of everything, Tyndall turned to the scalar as a basis for an approach to the historical. Thus, invoking Bishop Butler in an imagined response to an unnamed Lucretian disciple, Tyndall dismissed the possibility of a logical, foundationalist derivation of life from nonliving molecular materials. The denial of that possibility was for Tyndall the denial of a false and overextended materialism, the opposite of a life infusing pantheism. Accepting the argument that Butler voiced, Tyndall rejected one potential version of a linear synthesis projectively bridging contemporaneous biological, chemical, and physical lines of research. Mathematical foundationalism was a distraction. The knowledge at stake here was, to his mind, not a matter of deductive schemes.

This rejection was not a denial either of the alternative scalar projection that life is always associated with matter or the related historical possibility that life emerged from matter at a certain point in history in specific historical sequence. Dramatizing a potential scalar synthesis, Tyndall reminded his auditors and readers that a magnet split in two yields two more magnets. "We continue the process of breaking, but, however small the parts, each carries with it, though enfeebled, the polarity of the whole. And when we can break no longer, we prolong the intellectual vision to the polar molecules."[71] Like the persistence of polarity across scales, matter at even the smallest scale might have some kernel of life: "By an intellectual necessity I cross the boundary of the experimental evidence, and discern in that Matter which we, in our ignorance of its latent powers, and notwithstanding our professed reverence for its Creator, have hitherto covered with opprobrium, the promise and potency of all terrestial [sic] Life."[72] Tyndall imaginatively asserted a form of protohylozoism in which even apparently inorganic matter nurtures a pale organicity that might in time or in confluence with other matter at other scales yield more robust life.

This is not to suggest that Tyndall clearly sustained any extended distinction between a synthesis based on scales or one based on history. Moreover, his or Huxley's rejection of spontaneous generation, revisited in the Belfast address, could be linked to the failure of a scalar synthesis of a certain kind, one that attempted to find the key of life in some easy transition across scales. Whatever the case, what the imaginative splitting exercise revealed at bottom was an impenetrable "Power": "As little in our day as in the days of Job can man by searching find this Power out."[73] Without this mystery at the root of the natural scientific view of the world itself, it followed that "man the *object* is separated by an impassible gulf from man the *subject*. There is no motor energy in intellect to carry it without logical rupture from one to the other."[74]

This brought out the more controversial and distinctive dimension of his pantheistic reasoning—the claims it made for the grammar of secular reasoning: "All religious theories, schemes and systems, which embrace notions of cosmogony, or which otherwise reach into the domain of science, must, *in so far as they do this*, submit to the control of science, and relinquish all thought of controlling it."[75] He staked out this territory despite the fact that in scripture, Job was reminded that God, as opposed to Job himself, had this knowledge. In a reissue of the address, with prefaces "to the First Thousand" and "to the Seventh Thousand" reprints, Tyndall both narrowed and expanded the reach of this assertion: "We claim, and we shall wrest, from theology the entire domain of cosmological theory. All schemes and systems, which thus infringe upon the domain of science, must, *in so far as they do this*, submit to its control, and relinquish all thought of controlling it."[76] The shift between cosmogony and cosmology might have reflected Tyndall's own deliberations over the reaches of science, whether it could penetrate to universal laws and to the origins from which those laws ostensibly may have stemmed. Equally, it may have been a hesitation over the assertion of any origin whatever. The molding of the secular here was not without its twists, to which scientists would return.

In these published prefaces, Tyndall responded to some of the earlier criticism against him. "To the Seventh Thousand," included the observation that "the weightier allegation remains that at Belfast I misused my position by quitting the domain of science, and making an unjustifiable raid into the domain of theology."[77] Tyndall denied the validity of the charge: "Is it not competent for a scientific man to speculate on the antecedents of the solar system? Did Kant, Laplace, and William Herschel quit their legitimate spheres when they prolonged the intellectual vision beyond the boundary of experience, and propounded the nebular theory?"[78] By this point in history, Tyndall suggested, the nebular hypothesis had become uncontroversially secular and seductive: "Accepting that theory as probable, is it not permitted to a scientific man to follow up in idea the series of changes associated with the condensation of the nebulae; to picture the successive detachment of planets and moons, and the relation of all of them to the sun? If I look upon our earth, with its orbital revolution and axial rotation, as one small issue of the process which made the solar system what it is, will any theologian deny my right to entertain and express this theoretic view?"[79] In this rapid-fire defense and historical representation, Tyndall argued that the further step he took in Belfast merely had extended natural history and the natural domain of the secular: "As regards inorganic nature, then, we may traverse, without let or hindrance, the whole distance which separates the nebulae from the worlds of to-day. But only a few years ago this now conceded ground of science was theological ground. I could by no means regard this as the final and sufficient concession of theology; and at Belfast I thought it not only my right but my duty to state that, as regards the organic world, we must enjoy the freedom which we have already won in regard to the inorganic."[80]

To explain the resistance to secular grammars of the past, Tyndall returned to the theme of education. Binding together the historical and scalar, he conveyed a sense of the inevitable extension of the domain of natural scientific inquiry: "Have these forms, showing, though in broken stages and with many irregularities, this unmistakable general advance, been subjected to no continuous law of growth or variation?" Evolution to increasingly complex forms, Tyndall argued, was suggested by the geographical record. What then frustrated the possibility of the realization of that law? "Had our education been purely scientific, or had it been sufficiently detached from influences which, however ennobling in another domain, have always proved hindrances and delusions when introduced as factors into the domain of physics, the scientific mind never could have swerved from the search for a law of growth, or allowed itself to accept the anthropomorphism which regarded each successive stratum as a kind of mechanic's bench for the manufacture of new species out of all relation to the old."[81] But as a result of the biases of their training, "the great majority of [past] naturalists invoked a special creative act to account for the appearance of each new group of organisms. Doubtless there were numbers who were clear-headed enough to see that this was no explanation at all, that in point of fact it was an attempt, by the introduction of a greater difficulty, to account for a less."[82] However hylozoistic Tyndall's materialism, he staked for that scientific hylozoism a domain that could not be judged by religious dogma except in error. Thus Tyndall, like Helmholtz and Huxley, speculated on life and matter, history, logic, and scales. Like the others, he did so with distinct nuances, emphases, and points of interest.[83] Tacit agreement on the project for synthesizing a universal history produced instead disagreements and differences about the grounds and implications that such a history could have.

HUXLEY'S COSMIC EVOLUTION

In an address in 1880 in honor of the opening of a Science College at Birmingham—of a more enduring fame than Tyndall's Belfast address—Huxley had rejected any educational system that would produce "lopsided men," just as Helmholtz had feared one-sided development.[84] However, in contrast to Helmholtz, he advocated for the specialized school, holding scientific education to be crucial to the state, to the individual, to the "criticism of life which constitutes culture."[85] He constructed his argument in dialogue and debate with the opinions of the famed poet, critic, and inspector of schools, Matthew Arnold. In response, Arnold emphasized what he believed natural scientific education could not satisfy, that knowledge had to be "put for us into relation with our sense for conduct, our sense for beauty."[86]

In 1893, in the second Romanes lectures at Oxford, Huxley attempted to articulate how "our sense for conduct" was set against the cosmic process. "Cosmic evolution," one of the names by which that process was coming to be referred, "may teach us how the good and evil tendencies in man have come about; but, in itself, it is incompetent to furnish any better reason why what we call good is preferable to what we call evil than we had before."[87] Even more than this, however, Huxley claimed that "the practice of that which is ethically best . . . is opposed to that which leads to success in the cosmic struggle for existence." That practice "repudiates the gladiatorial theory of existence."[88] In published notes for the address, he invoked his "Administrative Nihilism" essay specifically on the question of Kant's universal history and ultimately posed the human state as standing against the political and historical actuality of the "state of nature"—the "state of nature" functioning both as a political philosophical and a natural historical category.

By the time of the lecture, Huxley seemed to share more ground with Arnold. In tension, though not contradiction, with his previous emphases, Huxley attempted now to see a space for humanity that somehow set itself against and therefore stood apart from the cosmic process. In a sense then, he also began to emphasize what he had shared with Thomson's ethical and universal historical viewpoint. The vision here was consonant with the emphasis on incompleteness of science, an emphasis that had been part of the debate over the virtues of natural historical narrative itself. Without reference to something outside itself, science to those such as Thomson and now perhaps Huxley offered a bleak picture of humanity's future, and even its present.[89]

To clarify his position, Huxley authored a "Prologomena" to the lecture relying on an analogy with the cultivation of a garden out of the state of nature. The garden, which compared to human society, demanded the gardener wage an insistent battle with nature outside and ultimately within the garden. To fail to do so was to compromise the garden's integrity, to let it slip back into first nature. There was no contradiction in human society warring with the natural world (with its own nature and the wider cosmos), just as there was none with the gardener warring with the tendencies of plants and the external assaults on the garden. Humanity might be an outgrowth of the cosmic process, but it sought to curb the universality of natural selection in the field of human life and society, just as the gardener produced conditions where natural selection was kept in sustained abeyance. Humanity succeeded to the extent that civilization succeeded. Ultimately however the cosmic processes would win out—an entropic vision would be realized, a future that Huxley mixed with the poetry, and perhaps to his mind the possibility, of the cyclical Great Year.

Huxley's analogy had clear limitations. The strongest was that there could be no gardener for human society. No administration could take advantage of artificial selection in order to surpass Kant's unsocial-sociable animals in the success of their emergence from the state of nature and in their advance of civilization. Those who advocated the suppression of the weak and infirm were as delusional as those who imagined that no administration at all was necessary. The balance between individual freedom and social restraint would perpetually need to be restruck to insure an ethical progress coordinated with a technological one.[90] Humanity's duty was to stave off the victory of the cosmic process indefinitely, not only the victory for entropy that Helmholtz had been announcing for decades but also that of natural selection.

CONCLUDING A CENTURY: METAPHYSICS, ENTROPY, ESCHATOLOGY—AND RADIUM

Hermann von Helmholtz offered a concise natural historical vision that set in coordinated motion a number of principles that would persist into future universal histories. The vision he promoted in his own work took less issue with the interdisciplinary battles of the natural sciences and accorded greater respect to the work of the nonphysicist, but in regard to the relations between the natural and humanistic sciences, in 1874, in a preface to a German translation of Tyndall, he admitted that there seemed to be a new, broader "schism" in play: "It has often been said—and even brought as a charge—that through the natural sciences a schism, formerly unknown, has been introduced into modern education. And, indeed, there is truth in this claim. A schism is perceptible."[91] In this regard, he maintained the virtues of natural science, finding "an important element missing in an exclusively logical-literary education": "I refer to the methodological discipline of the activity by which we reduce the confused materials we find in the real world, at first sight apparently ruled by chance rather than reason, to ordered concepts and thus make them suitable for expression in language. Such an art of observation and experiment, systematically developed, has as yet been found only in the natural sciences. The hope that the psychology of individuals and of nations, as well as the practical sciences of education and of social and political government which are based upon them, will attain the same goal, can only be realized in the distant future."[92]

This larger, cultural schism not only suggested a role for popularization, to bridge that gap and to underscore the need for education in the sciences. Popularization also embraced the ongoing, progressive vision that all social and political knowledge would be recast in the mold of the natural sciences. Helmholtz's universal historical template placed this program of knowledge and civil life within an energy-inflected, entropic narrative that hoped to trade pedagogical and axiological schisms for the sublime, if sober, synthesis the new cosmology offered.

However, the new potential syntheses of Helmholtz, Huxley, and Tyndall might themselves appear, by other lights, not only to differ from each other in important ways, but to be overly steeped in the assumptions of a fading nineteenth-century metaphysics. The routes to a recalibration of the vocabulary of the scientific history and synthesis were broad. The spread of the natural sciences into education, the new academic disciplines, and the reconfiguration of socioeconomic identities (rising social classes benefiting from the shift in pedagogical emphasis), the links between new educational models and industry, all were constitutive of that change, but at the same time fail to capture it. If the emphasis on the natural sciences held any special role among all these factors, it did so in part because in the latter half of the nineteenth century, natural science was increasingly regarded as holding a privileged, even threatening place in critical understandings of perceived shifts in worldviews. Detractors and celebrants of the new sciences alike placed particular importance on its implications for living. Nietzsche diagnosed the problems of his era by indicting the attempt to render history scientific, Dostoyevsky's underground man railed against the cultural imposition of the truth that twice two equals four, Arnold held out the hope for the continuing importance of classical education against the tide of Huxley and his ilk, while Huxley attempted to maintain a place for evolution in independence from Thomson's physics.

In the last decades of the nineteenth century, Ernst Mach (1838–1916) had also engaged in the political debates over education reform, debates that were seen as contrasting humanism with the natural sciences.[93] His own pedagogical views, articles, and textbooks were inflected by his critiques of classical natural scientific concepts with an emphasis on sensations, a point of view stamped with the Comtean name "positivism," despite his resistance to the term.[94] Along with other researchers of his period, such as Helmholtz, du Bois-Reymond, Brücke and the experimental psychologist Gustav Fechner, from whom he learned or upon whose work he built, Mach confronted the question of what relation physiology might bear to physics. His research in these subjects was partly based on the necessity of his initially spare means and on political circumstance. The Franco-Prussian War had helped initiate the contours of late nineteenth-century political alliances; it also elicited investigations with explicit scientific, technological, and political dimensions. In large part as a result of debates over whether the French had contravened international conventions by using bullets producing cratered wounds, Mach was ultimately led to his famous experiments on shock waves and Mach numbers.[95] The counterargument, which Mach had first heard articulated in

1881 at the First Electrical Exhibition in Paris, held that the injuries resulted from compressed air traveling with and ahead of the bullet. Mach had already been working on "spark waves," as well as on photography and ballistics, but it was not until the 1880s that he attempted to photograph shock waves produced by bullets—and ultimately decided against theories blaming compressed air for the wounds.[96]

In this period, Mach's research program increasingly sought to root out what he regarded as the metaphysical assumptions steeped in the worldview, naturalistic and cultural, in which he had been educated. With an emphasis very different from a Tyndall or even a Helmholtz, he made the case that physiology didn't just learn from physics, but had something to teach it. His would be a subjectively concretized physics in which the ego, understood as a "mental-economical unity," was itself a "practical unity," an epiphenomenon of the primary sensational elements.[97]

From one perspective, Mach advanced a line of thought that bound together sensation and physics more widely, with a Comtean emphasis on description made more empirically resonant. He intensified questions that would press on a more "classical" natural science, militating toward a view of knowledge as forever subjective and potentially subject to shifting needs from one historical moment to the next. At the same time, he emphasized the necessity of an economy of thought in natural science that shared in Herschel's very different natural historical "the one and the many." The uncomfortable cost of his demands that natural scientific knowledge should be thoroughly sensation-based undercut views of humanity as a disentangled product of processes and histories that human knowledge reveals. His work in turn contributed to resistance to the picture of a simple train of logical deduction that placed biological/physiological disciplines as the more or less distant predicates of the candidate axiomatic truths of physics.

At the end of the century, further contestations of any clean-edged physical worldview were in progress, including among them claims advanced by those who argued in very different terms from prevailing strains of more physical-physiological thought. The stability of the physical world had gone through a number of assaults since the ether was established as fundamental to nineteenth-century science. Even as Huxley was declaiming against an evolutionary ethics tethered to the nebular hypothesis, the cosmic stability that Laplace had been celebrated as achieving—the culmination of a Newtonian program that Newton did not endorse—was eroding in still other ways.[98] By contemporaneous mathematical standards, Laplace's demonstrations were no longer seen as decisive.[99] In the 1890s, the French polymath Henri Poincaré

(1854-1912), looking over the history of the successive attempts to demonstrate the stability of the solar system, emphasized that these were really only "successive approximations."[100] His view of this history—that the succession of demonstrations entailed no contradiction, that they involved successive deployments of mathematically "fictitious bodies" (of mental, logical constructions) and, implicitly, that their claims were adequate in relation to the conventions of their times—evinced his own philosophical views of scientific truth.

With an eye to proving the stability of the solar system according to his own methodological innovations, Poincaré had found himself in the late 1880s accidentally proving the enormous potential complexity of its motions and resigning himself only to probabilistic claims regarding unstable orbits.[101] He was forced to modify prize-winning views by advancing an evocative mathematical analysis that won him just as much admiration. His analysis of that instability was itself not dependent on the question of the ether. Whereas a Laplacian solar system could be destabilized by the existence of such a medium in space, a Poincarean solar system displayed its sensitivity and its improbable potential instability regardless of the existence of the ether.[102]

In the late 1890s, Poincaré used this work to articulate the mathematical possibility that even if the world should end in the heat death articulated by Helmholtz and Thomson, his own recurrence theorem might suggest that the universe could "awake after millions and millions of centuries."[103] Nevertheless, in the *physical* world, Poincaré's model "tends to a state of final repose."[104] This tendency would exist independent of the "hypothetical medium" of the ether or resolutions of questions about the ether's existence provoked by the behavior of Encke's Comet—partly as a matter of the behavior of real, rather than ideal bodies, partly, too, as a matter of the consequences of forces besides gravity.

But with the turn of the century, the specifics of Thomson's or Helmholtz's positions on such final repose and the shape that their claims gave to universal history were becoming increasingly tenuous, with more radical apparent possibilities for the future of the physical world than even Poincaré allowed. There was promise of stronger disciplinary reconciliation on grounds that had shifted beneath a now aging universal history. Thomson's own chronologies for the age of the earth and the sun were handily dismissed. George Darwin, Frederick Soddy, and perhaps most prominently, Ernest Rutherford, all relying on each other's work, were already prepared to place starkly different views of the history of the world in view of generalist or popular audiences, views that expanded the scale of such accounts, linked to new discoveries regarding the nature of energy and matter.

By 1905, Rutherford presented his case concisely in *Harper's Magazine*, casting the chronological debate as one of connection and conflict between disciplines: "The mutual interdependence between the sciences is illustrated in a very interesting manner by the controversy which has raged intermittently for more than half a century, between representatives of physical science on the one hand and of geology and biology on the other, with regard to the important question of the age of the earth."[105] Soddy and George Darwin had similarly cast the question as a matter of disciplinary conflict. Rutherford represented the backdrop of the controversy in terms of the major geological schools of catastrophism and uniformitarianism, quoting the same passages of Playfair as Thomson and omitting Huxley's evolutionary school. More or less equating their positions, he revisited the entrance into the "geological arena" of Helmholtz, Tait, and Thomson, now Lord Kelvin, whose victory as a historical matter he was less persuaded to accept. "Although Kelvin in this movement has gained the support of the majority of physicists, his conclusions have been subjected, at different times, to a considerable amount of adverse criticism, not only by geologists and biologists, but also by some representatives of astronomy and physics."[106] As he had done the year before in addressing the Royal Institute,[107] as George Darwin had done in a 1903 note to *Nature* upsetting Thomson's claims, and as Soddy had done in his 1904 book *Radio-activity: An Elementary Treatise* in reference to a similar claim of Tait's concerning the sun's energies,[108] recalling Thomson's statement in his 1862 *Macmillan's* article "On the Age of the Sun's Heat" that "inhabitants of the earth cannot continue to enjoy the light and heat essential to their life, for many millions years longer, unless sources now unknown to us are prepared in the great storehouse of creation,"[109] Rutherford declared: "This last remark seems almost prophetic in the light of the discovery of a body like radium which emits an enormous quantity of energy."[110]

Against a field of scientific literature and culture that had so often emphasized the limitations of energy, translating the energies of the sun into equivalents of energy/force calculable via materials such as coal—material that to researchers such as Tyndall already held tremendous stores of energy[111]—claims and speculations concerning radium must almost have seemed to be graced and even made suspect by a hint of magic.[112] This was perhaps especially the case for Thomson, who reacted by marginalizing the importance of radiation in reference to universal historical concerns. As Crosbie Smith and M. Norton Wise observe, Thomson's views overall not only involved theological commitments, but indicated "a profound attachment to men and machines"[113]—a transcendental materialism that by this point he could not allow to be transcended by new material. And Rutherford underscored the fantastic quality of radioactive materials all the more, which almost seemed to flirt with upending conservation of energy and the second law of thermodynamics.[114]

But it was Soddy, Rutherford's young English collaborator in the Chemistry Department at McGill University, who went that far. As Soddy observed, "Since the relations between energy and matter constitute the ultimate groundwork of every philosophical science, the influence of these generalisations on allied branches of knowledge is a matter of extreme interest at the present time." They were of so much interest that he believed such relations would precipitate "sooner or later, little short of a revolution in astronomy and cosmology. They will certainly be eagerly received, for it is only fair to add that they have been long awaited, by the biologist and geologist."[115] For these scientists, radiation was less magic than expected redemption.

Asking whether the second law did indeed apply to "sub-atomic change," Soddy suggested the possibility of a "cyclic scheme of evolution" where "heavy elements, on one side of the cycle, may yield a continuous supply of available energy, while the lighter elements are continuously growing, by the gradual accretion of masses, possibly of electronic dimensions, and at the same time storing up the 'waste' energy produced in the opposite process."[116] Soddy's phoenix of nature was not on the cosmic scale of Kant, but at the scales of the fundamental elements of the universe, light and heavy. Finding this admittedly speculative vision "much more complete and satisfactory," Soddy concluded his book by relating a progressive account of the evolution of knowledge: "Each new advance increases the period of time over which the laws of Nature can be regarded as having been in continuous operation without external interference." "The limitations, with respect to a beginning and an end, which one era imposes, disappear in the next. So that it is not unreasonable to anticipate that ultimately these laws will be recognised to operate, not only universally with regard to space, which has long been admitted, but also consistently with reference to time."[117]

If Soddy's progressive history of science is any measure, nineteenth-century uniformitarianism had already been uprooted from the scientific consciousness. But his views on universal history occupied only one position on the question. Pace Soddy, Rutherford concluded his Harper's article by reiterating the claims of the second law: "Although such considerations may increase our estimate of the probable duration of the sun's heat, science offers no escape from the conclusion of Kelvin and Helmholtz that the sun must ultimately grow cold and this earth must become a dead planet moving through the intense cold of empty space."[118] For Rutherford, Thomson's prophecy was not a statement of religious conviction. He read Thomson and Helmholtz together as straightforwardly declaring the ultimate end of all creation. In an immanent, causal natural history, Rutherford concluded that only nature, and not God, could produce new energies, and these would never be enough to perpetuate the world indefinitely. How universal this eschatological vision was at this point was open to question, as Soddy's alternative vision demonstrates. What was inaugurated in their work was a period where there was no dominant scientific universal historical narrative at all, even as the "matter" of the chronology was less in dispute.

However much Rutherford, George Darwin, and Soddy contradicted Thomson's story and potentially even Helmholtz's more conciliatory and synthetic account, they did not assert that the scientific disciplines could be divorced. Even as the historical chronology of the universe was being opened to significant challenges, sometimes to the point of undermining the possibility of any linear chronology whatsoever, these scientific laborers continued to treat chronology as a privileged site for the illustration of the connection of the sciences. If universal history did not appear definitively linear, it did remain a potential natural history, in Herschel's second and still very causal sense.

Much of Thomson's and particularly Helmholtz's, Tyndall's, and Huxley's later narratives remained compelling. They had established or extended a synthesis in very broad strokes. Although their chronologies included wider creation, they centered on the solar system, on the sun and its planets. From there, they each deliberated over organic origins, and then some form of transmutation to the present, ultimately forecasting a grim end in the far distant future unless something could prevent it. By the different lights of these accounts, humanity's meager beginnings and despairing ends could be redeemed in just a few ways: by divine intercession providing new stores of energy, or by the appreciation of an aesthetic vision of the grand connections of life across all time and perhaps space,[119] or by an unknown future fate or progeny, or by an ethical struggle on the side of politically unified humanity against the cosmic process. For all, the slaughter bench of history—what as a matter of predation alone had led Erasmus Darwin to proclaim "one great Slaughter-house the warring world!"—constituted no small part of the universe, past, present, and future. The unyielding evolutionary and entropic vision of much nineteenth-century science exposed its pervasive reach.

Natural history had been and was being reconstituted in deliberations over force, work, and energy, causality and correlation, dissipation and rejuvenation, pedagogy, disciplinary synthesis, and cultural unity. As a result, even as the slaughter bench that universal history took as its task to confront became the still more indiscriminate slaughterhouse of a cosmic process, how to recount the past and forecast the future remained under dispute.

PART II

CHAPTER FIVE

Undoing and Reassembling Scales and Histories

By the first decade of the twentieth century, old, now-stale syntheses were fragmenting; new syntheses were emergent. The connections and convergences that Mary Somerville had predicted to be imminent three-quarters of a century earlier, in their fuller development across not only the physical, but all the natural sciences, were stretching and being stretched by the epistemological and cultural fabrics out of which they arose. John Herschel's characterization of science as a natural historical synthesis of knowledge produced by many and digestible by one now could not be produced through the Humboldtian sciences, or even the sciences of energy. At the same time, the Laplacian nebular hypothesis was under critical scrutiny by astronomers and interested scientists in the United States, who soon regarded it as having little truth.[1] By the 1920s, accounts of the solar system were being advocated by those such as English astrophysicist James Jeans in which the birth of the solar system occurred not with the gradual resolution of a solar gas or fluid, but in a violent near-collision of stars. And it was not just the nebular hypothesis that was being contested. Other historiographic and scientific principles orienting the world were open to challenge, as well, particularly through a focus on and deployment of the new phenomenon of radioactivity. As with the earlier accounts of the origins and history of the universe, these views, too, were susceptible to dissensus and fissioning that, in a later outcome of that intellectual and technical exploration, would recast them anew and establish conditions for a potential coalescence of humanistic and scientific concerns—in effect, to a mythologization of science.

To the extent that philosophers and scientists such as Ernst Mach and thereafter Einstein could argue that the behavior of measuring apparatus is all that can safely be said of those quantities that are measured, an embodied or concretized physics could be posited, doing away with the abstractions, the "metaphysics," of the century before, partly on the basis of "physiological" emphases. In their treatments, time was increasingly identified with the behavior of clocks, space with measuring rods. This was work carried out by "new men," those educated in or educating within pedagogical systems increasingly recognizing the natural sciences.[2] The grounds for Thomson's, Tait's, and Helmholtz's universal history were eroding. And even though thermodynamic formulations remained in many respects crucial to the shape of that new history, the ultimate role and truth of such formulations were themselves cast in doubt.

The grounds on which a universal history might be constructed were shifting as a result of an expansion of time scales and the revision of central physical principles that underlay that scalar expansion. The notion of the whole of knowledge remained explicitly at stake. In this regard, at least, the concerns of Humboldt or Herschel, of Whewell or Comte, were still active. The terms of the relationship of the branches of knowledge to each other and the imagined conceptual structure embracing the sciences were explicitly theorized and contested. But scientific activity in the period that was being inaugurated in the early years of the twentieth century was undoing and recasting many of the preceding century's attempts to achieve a final theory and a final story. Thomson's overall universal history was effectively suspended, Darwinian evolution met with renewed scientific opposition, and syntheses were now sought out on the basis of principles jarring with the frameworks of preceding generations.[3] The natural sciences thus offered only an uneasy guide to what coherent, synthetic history could be told of the world and even to what overarching synthesis of knowledge was desirable. A linear or progressive universal history, in particular, looked like it might be an absurdity, and possibly any universal history whatsoever.

The emergence of the vocabulary for a scalar account of the history of the universe within the diversity of views in 1910s and 1920s, wrapped up in the dismissal of dominant nineteenth-century scientific-historical claims, is the subject of this chapter. It was a period when there was no universal historical consensus so prominent that to break with it would cast scientific reputations into doubt, when there was no accepted narrative structure governing the claims that science made about the universe's origins and future, when the universe that had been so long understood as resolving nebulously inward started instead exploding outward as the physical and temporal scales that any history of the universe would have to account for were expanded dramatically. At the same time, contemporaneous historical discourses and potential epistemological syntheses turned on strong agreements as to what such a history would entail.

To demonstrate this, I will focus on a small cluster of those figures who helped construct or maintain the scalar configuration of a new universal history in the first decades of the century and who at the same time began to craft a new popular scientific history. Apart from the foundational work of Einstein and thereafter the mathematical physicists Arthur Stanley Eddington and Alexander A. Friedmann, whose contributions are briefly reconsidered,[4] the American observational astronomer Harlow Shapley is of central importance to this cluster. Shapley attempted to demonstrate the greater extent of the universe and the liminality of the earth's position in it and to focus "borderland" research on the question of a naturalistic emergence of life from nonliving materials. These claims, he argued, had revolutionary metaphysical correlates.[5] Shapley persistently advocated a universal evolutionary history at a time when Einstein maintained his arguments for a stable universe and prior to the articulation of A. I. Oparin's or J. B. S. Haldane's views on the materialistic underpinnings of organic life.[6] Shapley posited as well an ahistorical scientific synthesis, mapping the world and scientific disciplines together by appealing to the scales of all objects of natural inquiry. These historical and scalar concerns were, in his treatment, mutually reinforcing.

In the 1920s, Shapley drew to Massachusetts the mathematically adept Georges Lemaître, a second central focus of the chapter. Lemaître was then a newly minted Belgian priest and soon became the theorist most commonly associated with spearheading arguments for the physical expansion and then the beginning of the universe itself. By 1931, Lemaître's and Shapley's efforts were reestablishing the shape of a universal history along the lines of newly written cosmic evolution: an evolutionary account independent of the nebular hypothesis with limited dependence on natural selection, an account in which the lessons of nineteenth-century energetics in a broad sense played an important, but contestable part. Each understood their account as metaphysical, axiological/aesthetic, and objective all at once. It was not a universal history that embraced all the different lines of research it could weave together, nor was it universally embraced. The assembly of new syntheses, historical and otherwise, was as much a matter of advocacy as it was a matter of awareness of the active research supporting it. While this universal historical assemblage was a kind of synthetic mosaic in progress, its composition was undertaken by crafters at times unaware of their peers at work on consilient areas.

EINSTEIN, THE STABILITY OF
THE UNIVERSE, AND HISTORY

In the nineteenth century, the apparently irreversible and continuous dissipation of useful energy—the increase of entropy—had seemed to require a transcendent intervention to prevent the heat death of the universe. And despite renewed considerations that challenged the nature and the ultimate ineluctability of that law—challenges aired by a Poincaré or a Soddy, as we've seen—dominant voices such as Rutherford and Poincaré's own nevertheless saw thermodynamic principles still forecasting a final doom. But even aside from any contemporaneous reasoning provoked by the study of radiation, some few challenges began to arise to the idea of an ultimate catastrophe. At the turn of the twentieth century, the idea of a perpetual universe was itself subject to renewed investigation, and the ultimate nature of entropy and the persuasiveness of Helmholtz's and Thomson's extrapolations of the second law of thermodynamics were open to question.

By the first decade of the new century, understanding of the second law was dominated by a statistical reinterpretation of it. So, for example, Max Planck (1858–1947), in the last decade of the century, moved away from a belief in the "absolute" nature of the law, embracing increasingly Ludwig Boltzmann's statistical interpretation. Planck's own famed research into black-body radiation furthered a line of physical interrogation dissonant with classical models of energy, helping set the terms for discontinuous or quantum conceptions.[7] Similarly, by 1904, as a result of considerations of Brownian motion, Poincaré was willing to accept the statistical, rather than absolute nature of entropy.[8]

Thereafter, however, it was the self-proclaimed Mach disciple Albert Einstein who delivered what became the consummation of a line of thought extending back to theorists such as Boltzmann and James Clerk Maxwell, through his own analysis of Brownian motion and the experimental confirmations thereof.[9] His suggestions of empirical arguments for the existence of atoms had implications for the history of the universe: if atoms actually existed, then Maxwell and Boltzmann's statistical mechanics became more compelling, probability appeared more intrinsic to the nature of natural laws, and Maxwell's famous demon could potentially be set free to produce violations of the second law.[10]

With Einstein, the nature of time and space themselves came under increasing interrogation, opening the possibility of new cosmologies bringing into doubt an idea of an "age of the universe" as such[11] and putting in question conventional principles of unity, synthesis, and history related through natural scientific knowledge. Indeed, if space and time were subject to questioning, along with the ultimate interpretations of the laws of thermodynamics, the nature of science itself and the nature of its validity were also no less open to interrogation in the period, before and after Einstein's contributions.[12]

Einstein's initial work with the quantized character of energy, on Brownian motion, and on the nature of the relativity of space and time appeared in 1905. After a series of academic posts in the following years in Bern, Zurich, and Prague, in 1914, months before the outbreak of the First World War, he moved to Berlin, where his work set in play the conditions for a reexamination of scalar and historical syntheses: the gravitational structure of the universe writ large and, soon after that, universal cosmogony—or rather, from his own preferred point of view, perpetuity and stability.[13] In that work, Einstein took a largely stable universe as an empirically warranted matter of fact and wanted to make sure that his physical model allowed for this.[14] In this sense, Einstein's general cosmological reflections are not cosmogonic, resisting a history of the world as such, despite nineteenth-century thermodynamic arguments. Regardless, they stand at a turning point in the language of cosmology, if not cosmogony, a brief transition point at which it seemed the present order of the world might be endless. Einstein thus articulated a conception of the universe as ahistorical, a conception with clear implications for the possibilities of authoring its final story.

In Berlin, Einstein advanced his work extending relativity from inertial to accelerating frames of reference. His emphasis on symmetries, so much a part of his original 1905 papers, was also in evidence in his analysis of the identity of gravitational and inertial mass and, from this, a physics identifying accelerations with gravitational forces.[15] Since according to general relativity, mass determines the geometry/curvature of space, and the geometry of space likewise determines the paths of free particles, the overall distribution of mass determines the potentially changing structure of the universe. This principle could come into conflict with favored models of that structure, as for example one concentrating all mass in a small area in infinite but otherwise empty space, suggesting that the universe would be Euclidean in all but a finite space surrounding that mass.[16] Having articulated his fuller general theory of relativity by 1915, Einstein considered these consequences in lectures given in 1917, thereafter reproduced as "Cosmological Considerations on the General Theory of Relativity." Here, he held that such stability "is logically consistent, and from the standpoint of the general theory of relativity lies nearest at hand."[17]

Although in the same work he declared that "whether, from the standpoint of present astronomical knowledge, it [his theory] is tenable, will not here be discussed," Einstein concluded with the concern that the theory accommodate a possible "quasi static distribution of matter," a condition he saw as required "by the fact of the small velocities of the stars."[18] He explained the introduction of the cosmological constant, the mathematical modification of his field equations allowing his cosmology to admit stable solutions, as a conclusion to be reached with some difficulty: "In the present paragraph I shall conduct the reader over the road that I have myself travelled, rather a rough and winding road, because otherwise I cannot hope that he will take much interest in the result at the end of the journey."[19] Introducing simplifying assumptions in discussing the overall structure of the universe, Einstein argued that "if we are concerned with the structure only on a large scale, we may represent matter to ourselves as being uniformly distributed over enormous spaces, so that its density of distribution is a variable function which varies extremely slowly. Thus our procedure will somewhat resemble that of the geodesists who, by means of an ellipsoid, approximate to the shape of the earth's surface, which on a small scale is extremely complicated."[20] As with Poincaré or Thomson, Laplace, and Nichol, stability was a prize to be won out of a set of equations or physical worldviews that by most appearances, on the contrary, evocatively suggested decay and death, or to be received as a gift to be given by God or from a universe preparing future energetic disclosures.

Einstein made his notion of stability more precise by beginning with an extended analogy with Newtonian mechanics, demonstrating that a "Newtonian stellar system" could not exist at all. Statistical mechanics suggested to him that like the propensity of radiation to dissipate, if the universe is itself endless, all matter would eventually be "lost in the infinite."[21] Unlike Poincaré, however, Einstein did not allow the equations to dissuade him, nor did he attempt to determine a kind of recurrent stability from them, but instead he appealed to force majeure. By what he termed a "slight modification" in Newton's equations, he demonstrated that he could posit a solution where the universe sustains a mean density of matter—a distribution of matter in the universe, that is, that by and large stays put. It is in this sense that Einstein understood stability and in this sense that the universe he posited and believed he lived in is, on a large temporal and spatial scale, static.[22]

However, there is an implicit tension between the autobiographical (that is, historical) moment detailing his own journey to these realizations and the geodetic mathematical language charting and asserting the effectively ahistorical distribution of matter in space and time. Einstein composed these thoughts at a moment of great apparent instability in the world around him, during a period in which he, like his British colleague Eddington, held pacifistic views that kept them not only out of the conflict, but outspokenly against it. Why Einstein was committed to such cosmological stability is not completely clear.[23] Nevertheless, it could certainly be regarded as consonant with his wider aesthetic and physical sensibility and consonant as well with a full-blown relativity principle demanding a physics independent of one's position in space and time.[24]

Viewed from this perspective, there is personal and local history, but not universal history, though consonant language pertains to each. What links them is an affinity between a vision of mathematics that distinguishes between local and global (where the local means the familiar and intuitive, or the Euclidean) and a vision of history in which the familiar concepts are all on the local side of the ledger—there are local histories, but there is no ultimate history of the universe as a whole. To ensure against that global history mathematically is why Einstein introduced the cosmological constant in his fundamental equations, a modification that in effect seeks to show that whereas Einstein is proposing a cosmology, a system of the world, he is not proposing a cosmogony, a beginning to the world.[25] A universe where the density of matter remains largely the same and where the stars remain slow moving from the perspective of the observer on this world is a universe without meaningful, distinctive, far-reaching events.[26] In this sense, and though his physics had no place of absolute rest, it was not Einstein who unfixed the fixed stars. The cosmological constant is a move against history.

Given the few scientists actively engaged in cosmological inquiry, Einstein could not easily be said to have established a consensus around a new eternalist natural history. The question of heat death as itself a marker of a kind of universal future also remained. Further, as Helge Kragh has observed, "at first the astronomers did not pay much attention to the theories of Einstein and de Sitter"—Willem de Sitter, a Dutch astronomer, found solutions to Einstein's equations for a massless world, a world stable only in the absence of matter.[27] Nevertheless, as we will see, the language for cosmological narrative was increasingly consonant with Einstein's own, both mathematically and metaphorically. In each case, it focused attention on universal history and the scales of the universe to question what unified picture of the world might be possible and how broadly and temporally the picture of the world could apply to the universe as a whole.

Einstein concluded his remarks with disclaimers as to the state of observational astronomy, and with the reservation that his modification to the gravitational field equations with the cosmological constant "is not justified by our actual knowledge of gravitation."[28] And almost immediately, the theoretical footing for its inclusion was displaced. While he was formulating his cosmological considerations, Einstein was in the midst of correspondence and debate with de Sitter. It was in fact in the same year that Einstein published his cosmological formulations that de Sitter showed they admitted the possibility of a universe unstable only if empty. And within a few further years, observational astronomy would produce challenges to Einstein's static conception. It would seem that like Thomson, Einstein indicated, anticipated, and perhaps helped to focus attention on the very points to which his theory was ultimately regarded as susceptible to criticism.[29]

NEW SCALES, NEW ENERGIES, NEW HISTORIES

Harlow Shapley pursued observational astronomy aggressively, and his concern with scales in the universe and its cosmic evolutionary history is unmistakable. Perhaps unsurprisingly for an astronomer, the question of the magnitude of things appeared to animate him from the time of his dissertation work in the early nineteen-teens with Henry Norris Russell at Princeton, estimating the size of eclipsing binary stars.[30] Subsequent to the completion of his dissertation in 1913, he took up a post at California's Mount Wilson Observatory that would last him through and just beyond the decade. From there, Shapley produced a sweep of single-author and coauthored articles. A large part were published in two series, "Studies Based on the Colors and Magnitudes in Stellar Clusters," in *Contributions from the Mount Wilson Solar Observatory*, and "Studies of Magnitudes in Star Clusters," in *Proceedings of the National Academy of Sciences*. As one of his former students, the astronomer Helen Sawyer Hogg, noted, over half of the total output, over eighty papers during his tenure at Mount Wilson, examined globular clusters.[31]

Through this work, he and his collaborators began drastically revising the size of the known universe, decentering at the same time the place of the sun, which, by dominant astronomical models, had been placed near the center of the Milky Way.[32] This, as well as Shapley's criticism of the so-called island universe theory of the spiral nebulae—the theory that they are giant and distant galaxies in their own right—set the terms for a famed debate with Heber D. Curtis of Lick Observatory at Mount Hamilton in California. The protagonists differed primarily on the scale of the galaxy (Shapley holding for the much larger size) and the distance to those nebulae (Shapley arguing against Curtis's much more distant placement). A much-revised version of the debate appeared in print in the *Bulletin of the National Research Council* as "The Scale of the Universe." Here Shapley puts forward arguments to which he returned repeatedly, that "the significance of man and the earth has dwindled with advancing knowledge of the physical world."[33]

The year of his debate with Curtis was also a year when the time had come to stage the scientific funeral of Kelvin-Tait-Helmholtz universal history. It only remained for a prominent advocate of Einstein and a Shapley correspondent to deliver the coup de grâce. Not only conceptions of the physical scale of the universe, but accounts of the time scales of its existence, were being expanded, even apart from the ahistorical world Einstein proposed. Arthur Stanley Eddington was the chief popularizer of relativity in the early part of the century and an architect of the 1919 eclipse expedition to confirm Einstein's general theory. Resisting violent upheavals in his astronomy and in politics, a pacifist claiming conscientious objector status during the First World War, Eddington treated the Kelvin-Helmholtz theory of the source of the sun's energy with disdain.[34] In 1920, he characterized the theory with evident frustration as "an unburied corpse." In this same address, "The Internal Constitution of the Stars," given to the British Association for the Advancement of Science in Cardiff, Eddington bridled against those astronomers and physicists who continued to accept Thomson's belief that the energy source is gravitational, not paying attention to the time scales that this demanded. And like Rutherford fifteen years earlier, he doubted that Lord Kelvin's account had ever been as dominant outside of physics: "What is the source of the heat which the sun and stars are continually squandering? The answer given is almost unanimous—that it is obtained from the gravitational energy converted as the star steadily contracts. But almost as unanimously this answer is ignored in its practical consequences. Lord Kelvin showed that this hypothesis, due to Helmholtz, necessarily dates the birth of the sun about 20,000,000 years ago; and he made strenuous efforts to induce geologists and biologists to accommodate their demands to this time-scale. I do not think they proved altogether tractable. But it is among his own colleagues, physicists and astronomers, that the most outrageous violations of this limit have prevailed."[35]

Eddington heaped scorn on the theory, partly as a belated intervention in the debate over who has the authority to pronounce on such matters, partly as a timely intercession in the reemergent inquiry into the relationship between the scientific disciplines. From the point of view of science as a whole, of a potential historical synthesis, he claimed that "if the contraction theory were proposed to-day as a novel hypothesis I do not think it would stand the smallest chance of acceptance. From all sides—biology, geology, physics, astronomy—it would be objected that the suggested source of energy was hopelessly inadequate to provide the heat spent during the necessary time of evolution."[36] With a new physics on the horizon and new sources of energy continuing to be explored, organic evolution could now testify against that older physics. The new reservoirs of energy Eddington appealed to were "sub-atomic," sources that, he believed, given the constitution of the sun, would provide enough energy to fire it for fifteen billion years. This perhaps puts in some perspective his derisive claim that "Lord Kelvin's date of the creation of the sun is treated with no more respect than Archbishop Ussher's."[37] In the 1920s, there certainly was room for a new history of the universe, one on time scales far exceeding those dominating the discourse only twenty years before.

In fact, there was room not just for a new universal history, but for several new universal histories, because not only were conceptions of the physical and temporal scale of the universe being expanded, but the possibility was being maintained that a plurality of universes could satisfy Einstein's equations. In 1922, the Russian mathematical physicist Alexander Alexandrovich Friedmann demonstrated decisively that Einstein's field equations admitted nonstable solutions: that is, there are universes that either expand uniformly over time or oscillate between periods of contraction and expansion, and all conform to Einstein's equations.[38] From the perspective of those actively interested in crafting or assessing a new universal history in Einsteinian language, such outcomes could be disturbing. Einstein himself found them so, his initial resistance and dismissiveness of such results suggesting the extent to which Friedmann's oscillatory solutions could be assimilated neither to a static nor to a more straightforwardly historical universe.[39]

The first survey article on Einstein's general relativity wasn't published in Russian until the year before Friedmann published his work on Einstein's field equations, in 1921. As Friedmann's biographers Eduard Tropp, Viktor Ya. Frenkel, and Artur D. Chernen observe, "Soviet scientists got acquainted with the new theory"—Einstein's general or, as it was known to them, "strong" theory—"only after several years delay."[40] The lag in years was less a matter of intellectual gestation and more of political chill. But the reception, once established, was an energetic one.

In recollections of his teacher, the Russian physicist George Gamow, to whom we will return in depth in coming chapters, presented Friedmann as "basically a pure mathematician." This is an image of Friedmann corroborated by the recollections of Vladimir Alexandrovich Fock, who attended a regular seminar given by Friedmann, as well as by Vsevolod Konstantinovich Frederiks, the author of that first 1921 survey article—it was Frederiks who emphasized the physics, Friedmann the mathematics.[41] Nevertheless, Friedmann "also had a vast interest in the application of mathematics in various branches of the physical sciences."[42] In that regard, Friedmann organized what Gamow described as a "fleet of manned and unmanned balloon flights" to study the atmosphere. It was as a result of one of these flights that Friedmann contracted the pneumonia that led to his death in 1925.

Friedmann's intervention, from Gamow's perspective and those of future commentators, was to approach Einstein's work "from a purely mathematical point of view." Doing so demonstrated a mathematical error on Einstein's part, effectively the error of division by zero, which in turn led Friedmann to realize "that this opened an entire new world of time-dependent universes: expanding, collapsing, and pulsating ones."[43] And with his characteristic anecdotal humor, Gamow described Einstein's famed reaction to Friedmann's work, when Einstein was first alerted to it through the intervention of theoretical physicist Yuri Krutkov: "As a result of his conversation, Einstein wrote Friedmann a short and somewhat grumpy letter, agreeing with his argument."[44] The agreement, however, was on the question of the mathematics, not on physical plausibility. After trying and failing to refute Friedmann's computations, Einstein conceded that there were unstable solutions to his model, but not the physical possibility of these solutions.

It is intriguing to note that before his death, even the "pure mathematician" Friedmann wrote a popular book entitled *The World as Space and Time*. The description of the work as popular is not Friedmann's, who it appears had little regard for popular exposition. Relativity, in particular, Friedmann argues, is "impossible to popularize," a position that Einstein, a relativity popularizer himself, might have contested.[45] According to Friedmann's biographers, "He would like to make his book generally comprehensible, although he preferred not to call it popular, so as not to create a too easy attitude in the reader: 'so as to avoid a wrong conception of the present article in the reader's mind.'"[46] The text does suggest a serious expectation on Friedmann's part that the layperson will be given some grasp of the materials laid out in mathematical terms. In this regard, Friedmann gives mathematical treatments of the meaning of space, of measure and curvature, of time and dimension, more elementary than those presumed in advanced treatments of Einstein's theories.

What has to be kept in view here and also as a continuing question is how the vocabularies of popularization have changed, along with changes in the role played by popular works in forming a culturally resonant universal history.[47] That Friedmann's book was at least quasi-popular is itself demonstrated by the very need to explain its mathematics in order to explain at some length notions of time and space already articulated in more technical journals. But such basic treatments can be found at the birth of relativistic physics, for example in the work of Einstein's 1905 papers, and continued on through the popular work of those such as Eddington. A popular piece, as for example Helmholtz's 1854 address "On the Interaction of Natural Forces" or Thomson's short 1862 *Macmillan's* article, framed the debate over the possibility of universal history in terms of the first and second laws of thermodynamics and the credibility of uniformitarianism. As will be seen directly, the cultural force of popular works was not only a nineteenth-century matter, Friedmann heightening the mathematical demands of that terminological framework.

But mathematical treatment (and the early Soviet context) still permitted religious allusions in Friedmann's work. Even in his more technical papers, Friedmann did not shy away from terms with religious overtones and, in the margins at least, the religious functions as an anticipation of the truths to come. In what was to become a major topos in the genres of synthesis under development in the twentieth century, suggested already in the previous century through the work of Helmholtz or the arguments of Tyndall, science should be the consummation of earlier, philosophical or religious questing for truth.[48] The epigraph to the first chapter of *The World as Space and Time*, for example, is verse 21 of chapter 11 of the Book of Solomon, "Thou hast arranged all things by measure and number." The epigraph captures the spirit of Friedmann's claims, but also suggests a sublime register not dissimilar to the allusions of a Thomson. In a more extended discussion of Augustine's philosophy of time, asserting the possibility of different modes of gauging it, Friedmann approves of the critique of measured time offered by the saint. Here the sacred is at stake, as well, in the aura of the epigraph and in the idea of reaching across historical time to find those past figures arguing for the correct philosophical views of the world.

With his work, Friedmann opened up different possible histories all at once, without on the current state of knowledge clearly endorsing any, as if to suggest by contrast to Einstein that his technical readers may take their pick.[49] Einstein of course preferred the version in which the universe is stable—the global nonhistory. The beginnings of a universal *history*, one that asserts a beginning to the world and a process that moves from beginning to present, did not emerge in force until later that same decade, in the later 1920s, in a context that was even more entangled with religion by the historical observers of the day.

TOWARD SCALE AS GENRE

For Harlow Shapley a universal history had to contend with and synthesize the new conceptions of the scales of the universe and new techniques of measuring it. That to his mind involved questions beyond the ken of any particular scientific discipline or even beyond the domain of natural science itself. In November 1923, Shapley, then director of the Harvard College Observatory, took part in a three-person symposium on the question of the origins and distribution of life in the universe. The symposium seemed less a discussion than a prepared debate between Shapley and his colleagues, the professor of plant morphology Edward C. Jeffrey and the professor of ecclesiastical history Kirsopp Lake. Their positions were put forward as emblematic, a drama staged between astronomy, biology, and theology.[50] In introducing his contribution and the symposium, Shapley observed invitingly: "We may show to each other, and to you who listen in, that a subject such as this gains by synthetic treatment."[51]

To the audience and his interlocutors, Shapley stressed that the problem of the origins of life was too large to involve any one person or discipline, requiring instead "collective thinking" and a "broader outlook"—including presumably, and perhaps surprisingly to some, the central input of an astronomer.[52] For scientists like Shapley looking for bridges to other disciplines, this interdisciplinarity was a virtue of the question.[53] The theologian Lake criticized Shapley's view of life as being too geocentric or anthropocentric. In Lake's view, Shapley's argument was predicated on the strong assumption that life everywhere must resemble material life around us. For those like the biologist Jeffrey, who saw in Shapley's interest the endorsement of a belief in life elsewhere in the universe, the shape of the question of life could be overstretched. No decisive empirical evidence as yet existed for life elsewhere. Despite popular scientific arguments for the possibility of life on Mars, astronomical reports of the possibility of built structures there, and popular stories of assaults by its inhabitants, no Martians had been observed or yet invaded.[54]

Insofar as he treated life's emergence as a question for natural scientific investigation and historical inquiry, Shapley's position was in certain ways consonant with the earlier positions of Huxley and of Tyndall, and by the

1920s, it was difficult to find a Shapley work that didn't allude to natural history and evolution. There are hints of this even in the unpublished and published versions of his address in the "Great Debate" with the fellow astronomer Curtis. More thorough early evidence is found in a May 1923 *Harper's* article entitled "The Universe and Life." The paper is notable in sharing aspects of the programmatic outlooks of the later published works of J. B. S. Haldane and A. I. Oparin, in imagining the sciences as a whole to address the question of life, in setting the terms for the emergence of life in the earth's distant past, and (in ways roughly evocative of Tyndall) in challenging the division between the living and the "lifeless."[55] However, it did not suggest specific mechanisms for life's emergence (as Oparin or Haldane did), or attempt to find its essence (as the physicist Erwin Schrödinger might be read to do in his later famed reflections on life). Rather, it looks to the conditions for its possibility, an approach that later in the century became characteristic of the search for life elsewhere.[56] And here, Shapley made an overt distinction between astronomical and human scales, observing that "the difference between the evolution of the Galaxy and the evolution of the political state is to him," the "impersonal cosmogonist" conversant in astrophysics, "the difference between the cosmic and the comic. The one approaches the eternal and infinite; the other approaches the infinitesimal and evanescent."[57]

Political commentators began to take stock of these claims to changes in understanding of the history and scale of the natural and civic world. To track the way in which these treatments of universal scales and the popular accounts of them were part of contemporary discourse is to see how they circulated as generic forms, imaginatively assembling, reconciling and reorienting the work of different disciplines for varied ends, projected unifications precipitating further contradictory or opposing stances. A playful case came from the hands of the relatively young journalist and author Elmer Davis. Davis was a frequent *Harper's* contributor, a future head of the World War II Office of War Information, appointed by Franklin Delano Roosevelt, and an eventual critic of Joseph McCarthy.[58] Responding back in 1925 in published correspondence to claims by Shapley that the universe is much larger than had previously been thought, Davis found it "a great consolation . . . to learn that [quoting Shapley] 'the future of the Galaxy may involve much growth.' This solaces my civic pride," and Davis explained what "civic" meant to him: "If I apprehend your Professor rightly, our universe is at present only some twelve hundred quadrillion miles across, and our visible and cognizable portion of it, if not exactly in the suburbs, is at least in the 181st Street neighborhood, or perhaps even farther out. Depressing thought, that one lives in the Bronx, the Inwood, the Bath Beach of the universe." But given the growth that Shapley held out as a possibility, "there is every hope that our universe may yet annex not only suburban star clusters but some which are still remote. It makes one expand the chest, does it not, to realize that one's universe displays this much public spirit, that it is following the example of Los Angeles and annexing not only everything in sight but much that is out of sight." Marking the winds of astronomical change and prominence, Davis warned that "we can't get excited too soon. I hear that my friend Ed Hubble of Mount Wilson has just discovered a new nebula that is bigger than the whole galaxy of which we have hitherto been so proud. To meet this threat to our preëminence I can think of only two measures: (a) take a new census as they do in Los Angeles or (b) pass a few resolutions of the Galactic Chamber of Commerce as they do in St. Louis."[59]

The specific article Davis lightly parodied was a piece commissioned by the editors of *Harper's* entitled "How Big Is the Universe?: Notes on Recent Measurements of the Milky Way." In a preamble to the piece, the editors sung praises of the importance of Shapley's work: "One of the most significant scientific events of the present day has been the discovery that the universe—the visible, measurable universe—is immensely larger than it has ever previously been thought to be. No astronomer has played a more important part in this discovery than Professor Shapley." In the article itself, Shapley related how different kinds of stars—variable stars, eclipsing binaries, and Cepheids—each offered methods to determine distances where triangulation was impossible. The editors therefore believed theirs to be "the first authoritative and comprehensive article written in terms intelligible to the layman, on the new methods of measuring astronomical distances and their staggering results."[60]

By 1926, having already contributed popular articles in periodicals such as *Harper's* and the *Nation*—also focused on the origins of life, the magnitude of the universe, and the place of humanity—Shapley's first popular book, *Starlight*, was published in an imprint pertinently entitled "The Humanizing of Knowledge Series." Here he conceived of the universe in terms of a grand evolutionary (if not natural selective) history.[61] He is at his most explicit in this regard in the conclusion: "We are, or should be, impressed by the general scope and dignity given to the evolutionary conception by the recent studies of astronomy and physical chemistry. Evolution is not limited chiefly to the relation of man to his anthropoid forebears. That phase is one of the minor steps in the development that pervades the whole universe. In truth, we cannot restrain the feeling that the whole of organic development, from the earliest one-celled protozoa to human consciousness and the higher hereditary instincts, is trivial and transient from the standpoint of the development of the material cosmos."[62]

As documented by historians Joann Palmeri and James Gilbert, ideas such as these were not superficial ones for Shapley. Across the 1920s, he was invested in the work of institution building in the hopes of establishing an ultimately unsuccessful interdisciplinary "Institute of Cosmogony."[63] The word "cosmogony" was for him intimately related to the natural history proceeding out of that first universal creation and to the integration of the sciences.[64] Palmeri quotes him from a 1923 letter: "Sooner or later there may be a central institute in cosmogony in general, combining the cognate phases of physics, chemistry, astronomy, biology, and other sciences."[65]

Some sense of the ways in which Shapley's more popular presentations resonated or failed to do so was given by published readers' responses, as with the case of Davis. A more conspicuous response was the 1927 article "Black Science" in Harper's. It was written by the then chief editorial writer for the Herald Tribune and future Pulitzer Prize winner Geoffrey Parsons. Parsons had clear stakes in the questions Shapley was raising, given that he was becoming a universal historian himself, with his 1928 The Stream of History soon to be published—a work that prompted a glowing review in the New York Times claiming that Parsons "has done something which nobody else has ever done. He has succeeded in writing a history of the world."[66] Parsons' Harper's article specifically targeted the pedagogical and scientific visions of H. G. Wells—who himself had authored an extensive universal history already by then repeatedly revised and reissued[67]—and of Shapley and seemed in part to be provoked by the series title for Starlight: "The humanizing of knowledge is resulting in gross misconceptions. It is possible to wonder whether the present era of popularization may not yield more darkness than light. A recent volume Starlight by Harlow Shapley offers a flagrant example."[68]

The journalists Parsons and Davis were reacting to these apparently outsized claims of significance. Part of their own responses was a challenge to the sensibility of taking measure of the universe at all—either in terms of what it could demonstrate axiologically, aesthetically, or metaphysically (as with Parsons) or even physically (as with Davis). To speak of humanity's universal marginality was, in a sense, to treat the universe as something so understandable as to be relatable in mundane, even municipal terms. Such civic pride in the universe, or injury from it, was not only a response, but also part of the more thorough vision at work in the world-expanding astronomy of the early part of the twentieth century.

Shapley devotes the second part of his "flagrant example" *Starlight* to explicit treatment of "two of the most fascinating subjects of modern astronomy: the evolution of stars, including the origin of the earth and other planets; and the measurement of the dimensions of the stellar universe."[69] Parsons was rankled by Shapley's treatment of both subjects. The keynote in regard to dimensional measurement was that through recent innovations—where Shapley's own technical work played no small part—the universe had grown much larger and humanity much smaller and more marginal.

In the face of such claims, Parsons called for a different kind of book: "If anyone could find the time amid the rush of books telling what is known, he might draft a new kind of textbook, a single slender volume recording what is not known, a Grammar of Ignorance." Parsons wrote this in response to what he saw as a lack of humility on the part of certain scientists and popularizers in the 1920s. Instead of a "Credo" for the volume, he suggested a different kind of testimony, "an Ignoramus," that would declare: "From science and by reason we as yet know neither whence the universe came nor where it is going; what I am that read this, nor what it is that I read, nor whether there is an I; nor what is energy or space or matter; nor the explanation of any force or thing, whether heat or light or electricity, or thought or imagination or love."[70] More than a half-century before, Emil du Bois-Reymond had praised the "manly resignation" with which the natural scientist pronounced his enduring "Ignoramus."[71] Ironic though this later invocation might be in light of the earlier explicit confession of ignorance, Parsons might have appealed to du Bois-Reymond to claim a now lost "Ignoramus" and recall a still more exigent "Ignorabimus."

Mooting the possibility that future knowledge might point to the uniqueness of human life and thought, Parsons asked his readers, "suppose the sonnets of Shakespeare were in one side of a balance, would the Milky Way tip the other side, or not?" For Shapley, who in his debate with Lake and Jeffrey two years before had referred to the "place of humanity in space" as "a small but boisterous bit of cosmic scum," such a question must have seemed absurd.[72] But this lay critical questioning, fair or not, was evidence of critical resistance to Shapley's arguments for the significance of his history and scales.

The year 1930 saw the publication of two revealing books from the standpoint of Shapley's own contributions, *Flights from Chaos: A Survey of Material Systems from Atoms to Galaxies*, and the technical monograph *Star Clusters*.

In each, the question of scales was in the ascendancy.[73] The latter work was intended to communicate the state of research on star clusters at its time of composition and became a standard reference for decades thereafter. In the former, based on a series of lectures he had delivered the previous year in New York, Shapley sought to extend the classificatory language he helped coin for stars to the classification of all "material systems" in the universe. It is a system of classification that appears repeatedly in his papers and across his works.

As he observed, these material systems were not merely all the materials in the world, but the different orders of organization were a function of their different sizes.[74] He took this to be an innovation of the book, one that also implicitly laid out the proper scales of the different sciences, even while clarifying that the order of the whole was important for each.[75] Within this, he also found an opportunity to reinforce in passing the question of the material basis of life.[76] And here, as well, he observed that the scales set up a system by which the unknown itself could be shaped, so that both what counts as progress and what counts as what remains to be discovered could be constructively elaborated.[77] By 1930, then, Shapley had advanced programs for future research based on potential scalar and historical syntheses, each shedding light on the other.

THREE ACCOUNTS OF TIME: SHAPLEY, LEMAÎTRE, EDDINGTON

Implicitly and often explicitly at play in accounts of the scales of a universal history are the limits of those scales and of the account itself. The limits provoke the question of the history of time and space, courting paradox in the possibility of history before time and outside of space. Wrestling with Kantian proscription, the authors of such accounts have often bracketed these conceptually murky questions, pointing to what is beyond the universality of their history and to the concomitant limits of theory, while nevertheless attempting to address that beyond. Einstein's efforts to maintain a stable universe worked to undermine the pertinence of such questions straightforwardly posed, where final beginnings and ends of time are not meaningful or relevant to an account rejecting global history. But at the same time, Einstein's attempt to posit a cosmology unqualified in its range did not simply concede that a universe without a beginning is, in Kantian terms, "too large for our concept." But then questions of how a law-like universe exists and more generally of why the theory is valid themselves indicate the limits of understanding, qualifying the scope of the cosmology. By contrast, as with varied accounts of unstable universes, if the universe has such a beginning launching the everything that is the universe, including its dimension of time, then its beginning works to foreclose further inquiry into the origins of these beginnings. Such origins would also appear to concede Spencer's "impenetrable mystery," or to demand the play of time that only exists with or through the beginning, or to suggest that the structure of the universe has itself a prehistory in a temporal dimension prior to and apart from the universe it structures. Scientists who argued for universal beginnings also generally made such concessions, granting that a historical universe becomes a concept itself now "too small" to include its own origin without recourse to some "higher temporal condition." Likewise, they confronted the limit of time understood as its future rather than its origin, as the ends of such histories, entailing conceptual quandaries paralleling those posed by origins, and also demanding the revision of entropic forecasts in appreciation of a widening universe and changing physical understandings of space, time, energy, and matter. However the history of time was to be confronted, whether directly or in the attempt to bracket or negate it, cosmogonic considerations repeatedly dwelled on that history, attempting to reconcile it with new conceptions of the energy sources, scale, and structure of the universe. Such was the problem confronting Shapley and the sometimes conflicting candidate accounts that he, Lemaître, and Eddington worked to produce.

In the year following the symposium on the origin of life, Shapley found himself advising a researcher who would offer him the tools to extend his scalar views and his evolutionary history to a very beginning of time and, perhaps paradoxically, before time. In the year Shapley was making his first very public forays into the question of life on earth and elsewhere, 1923, Georges Lemaître had been ordained a Roman Catholic priest. Directly following Germany's invasion of Belgium in August 1914, Lemaître had enlisted with his brother in a volunteer corps, serving with distinction, though apparently not without some difficulty with his superiors.[78] Whether his action there confirmed him in his choices of profession is debatable. It did not of course dissuade him from a life in religion and science. Completing a course of study in mathematics and physics in his native Belgium, the young priest traveled to Cambridge to study astronomy with Eddington.[79] He attended Rutherford's lectures on radioactivity, as well, and so was close to two men who had knowingly set themselves counter to the chronology dominant at the turn of the century.

For the yearlong period in which Lemaître took up residence in Cambridge, he and Eddington both made lasting impressions upon each other. During the preceding year, 1923–1924, Lemaître had prepared a submission on relativity for a competition sponsored by the Belgian Ministry of Sciences and Arts, which he won and which, with a fellowship from the committee for Relief in Belgium of the Belgian American Educational Foundation, provided him the travel allowance necessary for research at Cambridge.[80] In the following year, Lemaître took his work and study to the United States, researching spiral nebulae and variable stars with Shapley at Harvard while learning the tools of spectroscopy and simultaneously enrolling in doctoral study at MIT. He also made excursions to Flagstaff, Arizona, to visit the astronomer Vesto Slipher and to California to visit Edwin Hubble at Mount Wilson, attempting to glean first-hand new observational astronomical data. By the end of the year, as Kragh has observed, Lemaître "partially transformed himself from a mathematical physicist to a theoretical astronomer."[81]

Given their personal histories and varying expertise, the association of the American observational astronomer Shapley and the Belgian mathematical physicist Lemaître is somewhat surprising, speaking more in the first instance to Shapley's prominence at the time than to anything that tied these two together. His scientific achievements did not turn on the mathematical sophistication associated with Lemaître or Eddington—he collaborated with or often relied on his wife, Martha Betz Shapley, in order to carry out more complex mathematical computations, his own strengths lying elsewhere. For reasons of varying strengths and cultural location, Shapley thus might perhaps have seemed to offer comparatively little to Lemaître—more data than ideas. However, if Lemaître had sought to learn more about observational methods and connect with the networks of observational astronomers, Shapley was a compelling choice for an advisor.

There is little available comment as to how their professional associates understood their interaction and very little about it in their limited correspondence. Writing to the relativist Theophile de Donder, Eddington expressed the hope that Lemaître would do well with Shapley—expressing no particular surprise at the association—though it is difficult to read too much into a few sentences.[82] There remains scant historical work uncovering their connection, partly the fault of the paucity of relevant papers. What there is could easily reinforce the picture that Shapley looked to trusted others, including Lemaître, for help in making sense of and rigorously shaping his astronomical observations. Nevertheless, a number of firmer intellectual connections suggest themselves.

Despite the more "classical" and less "mathematical" flavor of Shapley's own work, and the more irreligious tone of his early claims, the continuities between Shapley and Lemaître's outlooks become, in broad strokes, apparent, including perhaps the outlines of a social and perhaps political internationalism: for each of them, from the point of view of the 1920s, the prospect that the universe looked to be growing larger was a more sensible one than Davis, for example, presented with his lighthearted parody of a municipal sensibility. And in good Spencerian fashion, for Shapley, in particular, there was a way to look backward from ever-increasing complexity to a minimum of simplicity, a simplicity that might be dated in years.

Lemaître spent 1924–1925 in Cambridge, Massachusetts, briefly returning in 1927 in order to complete course requirements for his thesis. In the years in between, he formulated his thoughts on the idea of a dynamic universe, and in 1927, he wrote his first paper detailing these thoughts,[83] reprinted in March of 1931 in the *Monthly Notices of the Royal Astronomical Society* as "A homogeneous universe of constant mass and increasing radius accounting for the radial velocity of extra-galactic nebulae" and first published in the relatively obscure *Annales de la Société scientifique de Bruxelles*. Positing an expanding universe, he derived what later became known as the Hubble Constant, connecting spectral data (red shifts) to cosmic expansion. The history of the world depicted in this paper is one that begins with the static universe of Einstein, a state from which the universe has since evolved, and approaches in the future a de Sitter–like empty universe. In a strong sense, then, Lemaître's work was a reconciliation of the de Sitter-Einstein debate and their mathematical/physical positions within that debate. It was a reconciliation he explicitly sought to achieve, working on a possibility Eddington had foreseen years before. Even though it did present the grounds for a global history, suggesting how the contours of the universe as a whole varied over time, this was not a theory that posited any ultimate origin to the physical world as a whole. But in historical retrospect, as we will see, Lemaître appeared to be edging toward the concept of a growing universe that could be spoken of as once having been not only smaller, but younger. At minimum, he helped to lay the groundwork for such a theory, even if unknowingly.

Over the course of the same years, the astronomer and polymath Edwin Hubble had been carrying out a set of observations at Mount Wilson suggesting to him that certain nebulae were outside the galaxy, an endorsement of Curtis's position in the Great Debate. Then, in collaboration with Milton Humason (an extraordinary autodidact), he articulated a law concerning the red shifts of galaxies. A potential implication of this work was the idea that the galaxies are receding from each other, and moreover that the farther the galaxies are from each other, the faster they are flying away one from the other. In the context of the cosmology of Einstein, in which Hubble and Humason's results were viewed as problematic, Eddington took it upon himself to devise a solution, an announcement he made at the meeting of the Royal Astronomical Society at the start of 1930. Upon reading the publication of Eddington's intent, Lemaître forwarded his former instructor a copy of his 1927 paper. Eddington, realizing the paper's significance, forwarded it in turn to de Sitter, who shared the view of its importance, then published a treatment of it in the *Monthly Notices of the Royal Astronomical Society* in May 1930 and arranged for its republication in the same journal the following year.

At the end of the same year, Shapley in a session of the Council of the American Association for the Advancement of Science put forward evidence associated with Lemaître's nonstatic vision of an expanding universe (but not yet a universe with a birth) above that of Einstein's static, in effect ageless universe or that of other contenders such as de Sitter or Richard Tolman. A report from the *Herald Tribune* on New Year's Day 1931, present among Lemaître's papers in Louvain, noted Shapley's comparison of the distribution of the US population to the distribution of the galaxies, as found in Shapley's survey of some thirty thousand of them. Though the full implications of Shapley's survey were not entirely clear (whether his vision supported any contending theories, including Lemaître's), in a subsection of the article titled "Einstein Doctrine Under Fire," the newspaper reported that Shapley's results "are said to push to the forefront a theory of cosmogony put forth by Abbe Le Maitre [*sic*] . . . who has published a paper which constructs a universe in which matter is scattered and is radiating outward." The article reported in turn a supposed early exchange between Shapley and Lemaître: "It seems to me that you are trying to reconcile your religion and my science." In response: "No, I am only trying to work out an equation."[84] The reportage itself might have been discerning what Lemaître had yet to posit: a beginning to the universe.

The overarching ways in which Shapley's and Lemaître's positions were consonant is revealed by contrast with others with whom their sensibilities, along these same dimensions, differed. In 1930, connecting Lemaître's recent cosmological solutions to "behavior of the spiral nebulae," Eddington decided "to review the situation from an astronomical standpoint"—this although he took his prior attempt to study the stability of "the Einstein world" to be preempted by "Lemaître's brilliant solution."[85] In his published remarks, he assumed that the mass of the universe is constant and then compared it with the mass of an Einstein universe, M_e, which itself is determined by the cosmological constant λ.[86] If the mass of the universe M is greater than M_e, then "it seems to require a sudden and peculiar beginning of things"; if M is less than M_e, then given estimates for universal expansion, "the date of the beginning of the universe is uncomfortably recent." However, the final, limiting case of equality, $M = M_e$, yields the "philosophical satisfaction" of a world "as beginning to evolve infinitely slowly from a primitive uniform distribution in unstable equilibrium." Opting for this case, Eddington noted that "we allow evolution an infinite time to get started," but "once seriously started," the rate of that evolution is similar to the second case.[87] Eddington further emphasized that for $M = M_e$, Einstein's equations could be integrated, as already demonstrated by Lemaître in his 1927 paper. Nevertheless, despite this apparent consonance in their treatments, Eddington's reasons for recommending the initially static and gradualistic "Eddington-Lemaître model" over and above the other cases already hints at the oppositions between the sensibilities of the two scientists that soon became explicit.

In 1931, within days of the press reports of Shapley's survey, perhaps stepping back from many recent theoretical and astronomical discoveries—the world was after all much bigger and much less stable than it had been only a few years before—Eddington reflected on the ideas of a past and a future for the universe, conceding as well the force of cosmogonic speculations. In his presidential address to the Mathematical Association on January 5, 1931, republished in the semipopular forum of *Nature*, he explicitly contrasted the working of evolution, which produces great complexity and organization, with that of entropy, which promotes disorganization. And he observed that it is ultimately entropy that must win out: "Evolution teaches us that more and more highly organized systems develop as time goes on; but this does not contradict the conclusion that on the whole there is a loss of organisation . . . the high organisation of these [organic] systems is obtained by draining organisation from other systems with which they come in contact. A human being as he grows from past to future becomes more and more highly organised—at least, he fondly imagines so. But if we make an isolated system of him, that is to say, if we cut off his supply of food and drink and air, he speedily attains a state which everyone would recognise as 'a state of disorganisation.'"[88] From the point of view of natural history, both the entropic and the evolutionary play a role, even if in organic history, it is evolution that tends to be emphasized. Evolution here does not apply to the universe as a whole, for which entropy is the rule, though it does involve the material formation of everything leading up to the emergence of life and humanity.

In the far future, as well, entropy according to Eddington must defeat evolution: "Ahead there is ever-increasing disorganisation. Although the sum total of organisation is diminishing, certain parts of the universe are exhibiting a more and more highly specialized organisation; that is the phenomenon of evolution. But ultimately this must be swallowed up in the advancing tide of chance and chaos, and the whole universe will reach a state of complete disorganisation—a uniform featureless mass in thermodynamic equilibrium. This is the end of the world."[89] But the end of the world is not the end of time. The latter goes on even after the former concludes. Time outstrips history: "Time will *extend* on and on, presumably to infinity. But there will be no definable sense in which it can be said to go on. Consciousness will obviously have disappeared from the physical world before thermodynamical equilibrium is reached, and dS/dt [the rate of change of entropy] having vanished, there will remain nothing to point out a direction in time."[90]

Though the universe continues to exist in this state of thermodynamic equilibrium—"heat death"—it is not by human or conscious lights a universe that can be said to have an interesting history. It was the end of the world posited by Helmholtz and that had presented a dilemma to Thomson, a future he himself forecast and from which he withdrew. The apparent fate of the world remained remarkably similar, to Helmholtz, to Thomson, and to Eddington, even as Eddington tried his best to bury his predecessors. But Eddington also suggested a "loophole," not unlike that suggested by Poincaré, that chance fluctuations could also produce new order out of the heat death—a possibility in which he could express only as little confidence as Poincaré: "Nevertheless, I feel sure that the argument is fallacious."[91]

The question of entropy also appeared to lead Eddington, if despite himself, to the conclusion that there was also a *beginning* to time, though almost certainly a beginning that presupposed the static world preceding cosmic evolution as he articulated only months before: "Following time backwards, we find more and more organisation in the world. If we are not stopped earlier, we must come to a time when the matter and energy of the world had the maximum possible organisation. To go back further is impossible. We have come to an abrupt end of space-time—only we generally call it the 'beginning.'"[92] But this is clearly not a happy conclusion for Eddington: "I have no 'philosophical axe to grind' in this discussion. Philosophically, the notion of a beginning of the present order of Nature is repugnant to me. I am simply stating the dilemma to which our present fundamental conception of physical law leads us. I see no way round it; but whether future developments of science will find an escape I cannot predict."[93] Ultimately, then, the question of the maximum state of organization and its dissolution dictates the parameters of what is at least a human time. As expressed by Eddington in this address, space-time may have indeed existed prior to the "beginning" and may exist after the "end of the world," but this is neither a human nor a scientific concern. This beginning and end both yield dilemmas, each inevitable, each aesthetically unsatisfying.

Eddington appealed to the cinematic as a metaphorical resource in articulating his argument.[94] Here, he is resisting a subjective vision of science: "The view is sometimes held that the 'going on of time' does not exist in the physical world at all and is a purely subjective impression. According to that view, the difference between past and future in the material universe has no more significance than the difference between right and left."[95] As Eddington sums up that claim, "The fact that experience presents space-time as a cinematograph film which is always unrolled in a particular direction is not a property or peculiarity of the film (that is, the physical world) but of the way it is inserted into the cinematograph (that is, consciousness)." But he declares that "if this view is right, 'the going on of time' should be dropped out of our picture of the physical universe," and if so, "we must, of course, drop the theory of evolution, or at least set alongside it a theory of anti-evolution [evolution in reverse] as equally significant."[96]

On the contrary, for Eddington, time here—empty time—is a property of the film, not the viewer. The metaphor functions as an intellectual resource for manipulating time—or at least for representing that manipulation. And the film can be completely rewound, to a beginning, the start of the film. To ask what happens before the film begins is to ask a question that strains the metaphor, a question that may seek to understand the propensity of the film to be rolled exclusively in one direction apart from its insertion in the cinematograph and its projection.

It is an irony of Eddington's declaration—that a beginning to "the present order of Nature" is both ineluctable and repugnant—that spurred on Lemaître to propose a universe with an origin. Whatever his actual motivations and inspiration, it was Eddington's nontechnical article in *Nature* that Lemaître credited as inspiring his theory of the primeval atom.[97] And what might fairly now be called a tradition of discussing avant-garde cosmological theories across a set of public texts from Helmholtz and Thomson forward was instantiated with this exchange. In the very first articulation of his originary account, in what some commentators see as the first expression of a Big Bang Theory, Lemaître paraphrases Eddington's language in putting forward what he takes to be the more natural alternative:

> Sir Arthur Eddington states that, philosophically, the notion of a beginning of the present order of Nature is repugnant to him. I would rather be inclined to think that the present state of quantum theory suggests a beginning of the world very different from the present order of Nature. Thermodynamic principles from the point of view of quantum theory may be stated as follows: (1) Energy of constant total amount is distributed in distinct quanta. (2) The number of distinct quanta is ever increasing. If we go back in the course of time we must find fewer and fewer quanta, until we find all the energy of the universe packed in a few or even in a unique quantum.[98]

Now the reel is played backward, the second law of thermodynamics in reverse combined with the quantization of matter producing a single homogenous state/particle. Lemaître concludes: "If the world began with a single quantum, the notions of space and time would altogether fail to have any meaning at the beginning; they would only begin to have a sensible meaning when the original quantum had been divided into a sufficient number of quanta. If this suggestion is correct, the beginning of the world happened a little before the beginning of space and time. I think that such a beginning of the world is far enough from the present order of Nature to be not at all repugnant."[99]

For Eddington, time outlasts the end of the world. For Lemaître, the beginning of the world precedes the beginning of time. Lemaître here not only inverts Eddington's language and sensibility in order to make his own theory more attractive, he also puts forward an alternative vocabulary for the world. For Eddington, the beginning of the world is the (perhaps) atemporal state of highest organization; for Lemaître, the beginning of the world precedes the beginning of space-time itself. Lemaître's beginning, then, is different in kind from Eddington's. Positing such a point of externality outside space and time can only with great difficulty avoid a transcendental viewpoint to that world order. This is a position Eddington may have been avoiding.[100]

To Eddington's cinematographic film, Lemaître counters with the phonograph: "Clearly the initial quantum could not conceal in itself the whole course of evolution; but, according to the principle of indeterminacy, that is not necessary. Our world is now understood to be a world where something really happens; the whole story of the world need not have been written down in the first quantum like a song on the disc of a phonograph. The whole matter of the world must have been present at the beginning, but the story it has to tell may be written step by step."[101]

The declaration that something really happens suggests what Lemaître here took to constitute the denial of a global history: a universe where no event takes place at all. But the events that proceed from it are not necessarily like the unrolling of the film reel or the turning of a phonograph disk. The true story of the world, the final story, emerges *in* time, in itself and in humanity's composition of it.

It is perhaps at this point that it is hardest to keep the question of Lemaître's religious commitments outside of this account. In this context, the Lemaître biographer Dominique Lambert sheds some light. He observes that in an early manuscript, Lemaître had a final paragraph with the words: "I think that every one who believes in a supreme being supporting every being and every acting, believes also that God is essentially hidden and may be glad to see how present physics provide a veil hiding the creation." Lambert observes that it had been Lemaître's intent to show that the idea of an origin to the world was not "théologiquement répugnante," either.[102] God here appears to stand behind the veil of space, time, and the creation and dissolution of that primeval atom, in a realm inaccessible to empirical inquiry.

Lemaître's theory soon emerged in a fuller form. In a 1931 article of the *Revue des Questions Scientifiques*, Lemaître again posited a huge, superdense atom, what Gamow later says is better termed a "primeval nucleus,"[103] whose disintegration produces the elements in the universe. He gives the following much-quoted précis of his own theory:

> The world has proceeded from the condense to the diffuse. The increase of entropy which characterizes the direction of evolution is the progressive fragmentation of the energy which existed at the origin in a single unit. The atom-world was broken into fragments, each fragment into still smaller pieces. To simplify the matter, supposing that this fragmentation occurred in equal pieces, two hundred sixty generations would have been needed to reach the present pulverization of matter into our poor little atoms, almost too small to be broken again.
>
> The evolution of the world can be compared to a display of fireworks that has just ended: some few red wisps, ashes and smoke. Standing on a well-chilled cinder, we see the slow fading of the suns, and we try to recall the vanished brilliance of the origin of the worlds.[104]

For Lemaître, the question of the origin of the world became wrapped up in the question of the origin of cosmic rays, a much-debated problem in his day. Lemaître argued that cosmic rays are indeed primeval fireworks falling over the world, released by the splitting of that primeval atom, a point of view that Einstein, among others, ultimately found both aesthetic and intellectually satisfying.[105]

It is the point of his words about the increase of entropy and the related increase in the number of light quanta—the particles of light—that Lemaître repeatedly cited as the fact compelling him to theorize a universe begun in some simplest state.[106] In giving so much pride of place to the second law, in a universalizing spirit, Lemaître applied statistical considerations without limitation of space or time, as did Einstein before him, but now asserting what Einstein denied and Friedmann proved mathematically possible: that galaxies might fly off forever, losing themselves in the infinite in a way apparently unanticipated by Einstein.

Lemaître's work by this point was strongly distinguished from the previous theoretical work of Friedmann, linking the primeval quantum to the origin of cosmic rays, the fact of radioactive decay—itself suggesting some earlier period of higher abundance of radioactive elements—and predictions for the red shifts of galaxies. It was the growing consensus that these are best explained by the hypothesis of an expanding universe that soon brought Lemaître's theory some favorable reception.[107] Lemaître speculated that the dissolution of the primeval atom can be understood as a matter of quantum mechanics, but did not pursue this extensively.[108] Whatever the reasons, he contented himself with raising the question. Thus, in the same 1931 article, "L'expansion de l'espace," published in the *Revue des Questions Scientifiques*, he asks: "Does the table of elements really end with uranium? Have we not come too late to know heavier elements which were almost completely disintegrated before our birth? Are not radioactive transformations a faint residue of the original evolution of the world and did they not take place, on the stellar scale, several billion years ago?"[109] In a flourish of counterfactual history, Lemaître adds: "If man had appeared on Earth a hundred billion years later, he would not have known radioactive bodies, long since exhausted; he would not have known the extra-galactic nebulae, withdrawn beyond the range of his telescope; he would really have come too late into a world too old."[110] In effect, we are lucky enough to live in the radioactive era of the universe, suggesting that the universe will age beyond this. The age of the universe then—its youth—is gauged by its scrutability to people.

Lemaître directly states as much while at the same time introducing a history of the universe: "Our universe bears the mark of youth and we can hope to reconstruct its story. The documents at our disposal are not buried in the piles of bricks carved by the Babylonians; our library does not risk being destroyed by fire; it is in space, admirably empty, where light waves are preserved better than sound is conserved on the wax of phonograph discs. The telescope is an instrument which looks far into the past. The light of nebulae tells us the history of a hundred million years ago, and all the events in the evolution of the world are at our disposal, written on fast waves in internebular ether."[111]

It was not an account that appealed to everyone. Eddington's response to the primeval atom of 1931 is different from his response to Lemaître's dynamic universe of 1927, and his reservations are aesthetic in a way also reminiscent of his politics. In 1933, Eddington published *The Expanding Universe*, a popular text based on a public lecture given the year before to the meeting of the International Astronomical Union in Cambridge, Massachusetts. Here, Eddington observes: "Since I cannot avoid introducing this question of a beginning, it has seemed to me that the most satisfying theory would be one which made the beginning not too *unaesthetically abrupt*. This condition can only be satisfied by an Einstein universe with all the major forces balanced"[112]—a universe, that is, that is neither contracting nor expanding, but nevertheless unstable.

The Lemaître-Eddington model may indeed provide an escape from the dilemma of a universe conceived of as static, but unstable. On this basis, he contrasts his theory with Lemaître's. As Eddington describes it, Lemaître's world begins with a "violent projection," inclining Eddington to the view that "I cannot but think that my 'placid theory' is more likely to satisfy the general sentiment of the reader."[113] The resonance of this phrasing echoes not only Eddington's broader pacifism,[114] but other popular responses to Lemaître's theory, such as the characterizations of "violent birth." For example, a *New York Times* review of Eddington's book in the column "Topics of the Times" and evocatively entitled "Nebulae and Nations" writes of Lemaître's theory: "Plainly, this is the kind of universe to contain the kind of terrestrial world we are now living in. It is a world in which the nations are desperately bent on getting away from each other as fast and as far as they can. The unity of civilization and the brotherhood of man were good enough notions for Newton's ordered universe living under a single law. But we know better now. The cosmos is relativist and centrifugal. Charity begins in your native galaxy. There is milk enough in the Milky Way to sustain us comfortably in complete independence of foreign nebulae."[115]

The political echoes of Eddington's language are taken to parodic lengths here—the cosmos itself now, *pace* Shapley, is rendered comic—but nevertheless, the question of the unity of humanity was taken to be at stake in cosmology and cosmogony. This becomes increasingly linked to the exploratory language evident in Einstein's cosmological paper, but is also appealed to throughout Eddington's work, here and elsewhere.[116] Matthew Stanley has argued that Eddington's 1919 eclipse expedition was represented as an adventure in scientific testing and international reconciliation. Correspondingly, the casting of cosmological truths as universal also in the sense of requiring international effort was increasingly implicated in the explicit political message of popular cosmology.[117] And despite his greater readiness to accept or at least treat Lemaître's "violent" beginning, Shapley shared in the universalism underlying Eddington's views.

"AT THE BEGINNING" . . .
"SOMETHING REALLY HAPPENS"

In the early 1930s, Lemaître captured the attention of a wider audience than the small group of cosmologists in a position to evaluate the breadth of his theoretical work. His travels across the United States were covered by central newspapers (the *Los Angeles Times*, the *New York Times*, and the *Washington Post* among them) attracted by a cosmogony written by a scientist and priest and by Einstein's interest in the affair. Many of the articles were straightforward reports announcing Lemaître's new theories and relating the possibility that they may explain the origin and nature of cosmic rays. Throughout, however, even in the more perfunctory coverage, there is an appeal to biblical language like that invoked by Friedmann and that at least implicitly connects Lemaître's vocation, his theory, and broader popular theology.

Thus, for example, a *New York Times* article in May of 1931 entitled "The Exploding Universe" focuses in particular on Lemaître, discussing the cosmological constant in relation to other constants of nature: "What the relation of these fundamental constants may yet be to one another no scientist has yet divined. In them the heavens now declare the glory of GOD. Verily, as JEANS maintains, the cosmos is the work of a mathematician."[118] In December of the following year, a similar spirit is evident in the concluding paragraph of a *Los Angeles Times* article, but with a still tighter weave of allusions: "It is an awesome conception—that mother atom, as lone as God, giving violent birth to everything."[119]

Several commentators share the view that the conception is, if not awesome, certainly radical or revolutionary.[120] But the question of its radicality seems to depend in large part on the perspective of both the historical actors and those of us who return to the period as analysts. Was the idea radical in the cultural context from which the scientists emerged, a cultural context many tried to address? And in the apparently narrower sphere in which practicing scientists voiced their opinion (whose work, as we've seen, always spoke to and from positions broader than an institutional base), the nineteenth century established strong precedents among scientists for the idea of an origin and for a history of the world. Though physics famously forgets, the major protagonists of this history did not forget Thomson's work; certainly neither of Lemaître's instructors, Rutherford or Eddington, did so, even as they dismissed it.[121] The Kelvin-Hemholtz theory still needed burying in 1920 after all—a corpse, perhaps, but one in plain view.

Throughout the early part of the century and until well after, the question of origins was a contested and contestable matter. Apart from being an innovative work of mathematical physics, to posit the idea of an originary universe may have seemed a radical professional act. And it was certainly provocative to those such as Einstein and Eddington who found the idea of universal origins a repugnant notion. But to the extent the idea of a beginning might have been radical among a scientific elite, this stands in some contrast to a larger public conception or even just a wider scientific assumption, given, perhaps, the evolutionary bent of Shapley.[122] The literate classes educated in science may not have been ready for a cosmogony, certainly not without great reservations. The broader public however appeared ready to accept the radicality of the idea of an originary universe, a public excited by a figure such as Lemaître and ready to connect his conceptions to theological ones.

Even granting the straightforward aesthetic appeals of Eddington, the reservations of the professional scientists arguably found some justification in distinguishing sharply between the mathematical-physical formulation of the origins of a cosmos and the broader cultural conceptions of the origins of a world. But the valence existing in the larger culture receptive to a cosmogony and a universal history may well have exerted a force working to overcome those reservations. Lemaître's first word on the subject was in a nontechnical article, one inspired by another generalist article written by his instructor Eddington on the same subject of universal import. And if there is any truth to Lemaître's own sense of the matter, those very same views of Eddington's inspired Lemaître to reflect on and produce a relativistic and quantum conception of the universe, a universe with an origin and a history. Though no doubt Lemaître's opinions helped, by some lights, to substantiate a creationist view of the world—God having created and split apart the primeval atom behind Lemaître's "veil hiding the creation"—his nontechnical articles did not explicitly endorse a religious creationist story.

Instead, they play with the religious terms, presuming that their public will make all the associations, associations that drove popular interest in any case. Neither newspaper readers, nor the authors of the articles they read, simply misread Lemaître as promoting a traditional view. An excitement at a *new* cosmogony is evident, even if at times understood as a reinterpretation of the old.

It is also not entirely clear whether a beginning to the world was a *conceptually* difficult idea, even if a radical one. Certainly the cosmogonists considered so far, from Thomson to Shapley to Lemaître, believed they could address a broad public. Though some, such as Friedmann (if in a different cultural context) might have scoffed at the popular expositions, there's little to suggest that they believed the subjects impenetrable to the public. They published such treatments themselves, connecting them to other, earlier, nontechnical thinkers (such as Augustine), and they certainly appreciated the broad appeal of their scientific work. If Lemaître's theories were radical, then, their radicality was not primarily predicated on the difficulty of conceiving of such an idea, which itself was repeatedly alluded to as simple.[123] Nor was it radical in provoking significant offense among the scientific lay public. Instead, its radicality would have to be gauged against scientific opinion in its own day. As far back as 1917, Einstein's explicit rejection of unstable universes turned on what he perceived as a contradiction with empirical facts, even though the conceptual alternatives suggested by his own theories—unstable universes, if not necessarily origins—were evident to him in those very reflections. His rejection might be regarded as a shying away from radical alternatives, or the fact of his recognition of such alternatives might instead be regarded as how far from radical they were as conceptual matters.

Whatever the resolution of the question, the move from Thomson's history, through Einstein's eternity, to Shapley's scalar discoveries and universal evolution, and "back" to Lemaître's history *did* involve reorganizing the scale, and the overall conception of the cosmos. A universe that in some senses had been resolving inward toward the solar system's and humanity's formation now raced wildly outward, fragmenting from an unimaginably massive point, a point that was nowhere and everywhere and before time, spilling its fireworks over an uncertain and unstable and accelerating present that itself had potentially been produced through evolutionary forces.[124] With this vision came some promise for the resolution of chronologies long in conflict, even as the scalar and historical "facts of the matter"—the size of the universe, the "speeds" of evolution and entropy, the character of matter and energy themselves—shifted.

From Thomson on, organic (life) and material (physical) history had been decisively separated, promoting different timelines, with only one having anything that merited the name history: the local and the living. Whatever the differing views regarding the question of evolution, few were arguing that *life* might have existed in perpetuity, except if it were divine. By contrast, in the 1910s and 1920s, few physicists were arguing that the universe had not existed in perpetuity, even if partly because it was not a central question of discussion for any large group.[125] In that period, the organic—life—had an eventful history, but the material universe as a whole did not, certainly not one that produced any significant consensus. But now, in Lemaître's almost relieved declaration: "Something really happens. . . . The whole matter of the world must have been present at the beginning, but the story it has to tell may be written step by step." Thereafter, as Shapley increasingly seized upon the possibilities of synthesis, addressing a larger public by his systems of classification and his evolutionary history, the history of life dovetailed increasingly with the physical history of the world, and a fuller, ordered chronology became possible.

Curiously then, Shapley's work, in drafting frameworks by which to judge the merits of scientific synthesis, shared something of the spirit of the journalist Parsons's vision. In the sketch of his "Ignoramus," Parsons suggested how chapters of the book "might analyze the sciences in turn, ending in a crude estimate of charted as against uncharted areas, according to the progress achieved in each." He posited an organization based on levels of completeness or finality moving from the certainties of mathematics down to the comparatively ignorant human sciences, history last of all—a quiet deference to nineteenth-century synthesizing impulses. Parsons extended the question of ignorance further, to the measure of *all* knowledge. As such, the Grammar of Ignorance would be charged with the "harder task" of estimating "the total of all knowledge as against the unknown." This would not mean dismissing the knowledge that the sciences possessed: "The figures as to the several sciences would accept the present basis of the sciences, that they describe given fields of fact, observing, measuring, analyzing, formulating hypotheses of operation, without attempting as yet to deal with first causes or last ends. A figure representing all knowledge must necessarily pay heed to these vast frozen poles of thought."[126] Shapley and Lemaître, Einstein, Eddington, de Sitter, and others who had embarked on constructing new natural scientific worldviews, speculating on the first causes of history and the limits of the physical world, appeared to hover too close to the poles. And yet, in fine grain, these same researchers shared in Parsons's hopes, casting about for figures that might give them confidence in their speculations, distinguishing between transcendent and imminent beginnings, between human and sublime scales. The relationship between the Parsonian Grammar of Ignorance and the emergent synthetic schemes were far from opposites, instead forming between them a kind of Möbius band: the ignorance of the sciences was beginning to be measured by what scales of space remained uncharted, what passages of history remained unwritten. Their successes were increasingly judged by the same frameworks, establishing what counted as charted versus uncharted, written versus unwritten.

For the histories before the turn of the century, the ratio of known to unknown was now too low. Having decisively turned their backs on the time scales set by Thomson and the stability encouraged by Einstein, historically minded scientists such as Lemaître and Eddington, and in a different register, Shapley, began to set terms for what could constitute a final story of the universe. As we will see in the next chapter, the genres of synthesis to which their work on the scalar registers of the universe and temporal character of its history gave impetus was imbricated with and helped foster other genres, as well.

The conditions for the possibility of that final story had emerged by the early 1930s—not just the conditions, but the contours of that story, its genre—but the story itself had not emerged, nor had any accompanying theory of finality to ground it. Though it speaks to the possibility of penetrating that first primeval nucleus, Lemaître's account contented itself with speculations, not universal historical research programs. The biological and geological were gestured at, but not written into his account. The spread of the originary ideas of an Oparin or Haldane or the future reconciliations of organic evolutionary syntheses were not realized. Comparing different research enterprises of the day, historical actors could still find that the stars might be older than the universe.[127] The universe might not conform itself to the municipal analogies or organismic notions of age. For human conceptions of it, the world might still be both too large and too small, placing contradiction in the way of putative final stories.

In Lemaître's account, little was expressed about the composition of the world, little said about why so much of this element exists as opposed to another, though he invited speculation as to the distribution of fundamental matter. This was an issue soon regarded as necessary in the telling of a more complete history, at least once the relative abundance of the basic elements was more generally understood as a question about the emergence of materials *in time*. From the point of view of the cosmologists, judged by the way they related their popular stories, the sciences were compartmentalized in terms of time (such as the period under study, the duration of processes) and space (the size and location of the objects under study), apart even from the questions of techniques, goals, and languages. To the extent that uniting increasingly dispersed fields was an active question at all, history as full cosmogony and chronology—the idea of a final story—was too speculative an idea to unite the sciences and yet too seductive an idea to ignore.

CHAPTER SIX

Other Emerging Genres of Synthesis

From the Fabulaic to the Foundational

... So from the root
Springs lighter the green stalk, from thence the leaves
More airy, last the bright consummate flower
Spirits odorous breathes: flowers and their fruit,
Man's nourishment, by gradual scale sublimed
To vital spirits aspire, to animal,
To intellectual, give both life and sense,
Fancy and understanding, whence the Soule
Reason receives, and reason is her being

from Book Five of *Paradise Lost*[1]

In the early 1930s, George Gamow, already in his late twenties a well-known physicist, was trying to escape the Soviet Union. He somewhat stiffly described the circumstances years later. His recollection was in response to a questionnaire, an "interrogatory," sent to him by the Department of the Navy in order to inquire into his former associations and to "clear up questionable points without the filing of formal charges."[2] According to the description Gamow provided the navy, the first, it seems, he had ever penned of the episode, he had made several failed attempts to escape.[3] The most serious effort was an ill-fated kayak trip in the summer of 1932 with his first wife, Lyubov Vokhminzeva, known as Rho. The couple set off south across the Black Sea from the Crimea to Turkey, surrounded at one point by a school of porpoises. But on the second day, they met with bad weather, and the effort to stay afloat left them both in a delirious, even hallucinatory state. When the wind died for a time and the sea was calm, its surface looked to Gamow like a cathedral floor, and he "almost thought" to hazard walking on it. Rho and he ultimately made shore that night, seventy miles west of where they'd set off, and when questioned about it thereafter used the bad weather as a cover to explain why they had landed so far afield of their point of embarkment.

Gamow's efforts to leave the Soviet Union the next year were initially no more successful.[4] Prospects at that point were bleak, because he recently had been denied permission to travel on three different occasions: to attend the first Nuclear Conference in Rome in 1931, another invited conference in Copenhagen arranged by friend and mentor Niels Bohr, and a summer-school teaching appointment in the United States at the University of Michigan at Ann Arbor. The premise for his departure in 1933 was the Solvay conference, convening to discuss the structure and properties of the atomic nucleus. Gamow's earlier work on the nucleus had earned him his scientific reputation and an invitation to the celebrated meeting. Just prior to the turn of the century, Rutherford had noted the phenomenon of "alpha decay," the emission of a particle consisting of two protons and two neutrons (an "alpha particle") from an atomic nucleus. And indeed, not long after proposing the "liquid drop" model of the nucleus (a model understanding the nucleus as an incompressible fluid), Gamow had worked briefly in the United Kingdom with Rutherford in Cambridge. With regard to alpha decay, in particular, this was an emission from the nucleus that should have been impossible, at least according to conventional classical physical understandings. But the young Gamow in the late 1920s had realized that according to the new quantum theories, such a particle did have a chance to escape—a phenomenon that came to be known as "barrier tunneling."

The Solvay conference offered another opportunity for his own escape, but he was at first refused permission to attend. He learned only thereafter that it was through direct application to the Kremlin by the French physicist Paul Langevin that had secured his visa. Langevin was the conference chair, an enthusiastic Communist, and the then chairman of a French-Russian scientific alliance committee. He had himself been urged to the effort by Bohr, among others, who felt that obtaining permission for Gamow to attend would otherwise be impossible. But a visa was necessary for Rho, as well, who was at first denied permission to accompany her husband. To try to secure that permission, Gamow arranged a direct meeting with Vyacheslav Molotov, who, as Gamow puts it "was at that time vice-president (or something) of the Soviet Government," through Nikolai Bukharin, the Marxist theorist and Stalin rival.[5] The ensuing conversation between Molotov and Gamow, in Gamow's recollection of it, ended promisingly: "I walked out beaming."[6] Nevertheless, Rho was refused permission to travel. Only after Gamow categorically refused to go himself until the very eve of the conference did the Soviet officials relent: the Gamows were given visas to Brussels, allowing for their own unlikely escape from the political field that seemed at one point to have them "stuck in Russia for good."[7]

Gamow's escape was one flight among many. Einstein had left Germany in 1932 in uncertain circumstances. His friend, the physicist and Vienna Circle philosopher Philipp Frank, reported Einstein's prediction that he and his wife, Elsa, would never return.[8] Over that same period, in Pasadena for the winter of 1932–1933, Einstein met with Lemaître. In the following spring, while he was in Lemaître's native Belgium, his property in Germany was searched and confiscated. The year before, Einstein had already discussed with Abraham Flexner, the founding director of the Institute for Advanced Study, the possibility of going there. Regardless, prior to his permanent departure for the United States, an extended period of uncertainty had ensued.[9] In England and the United States, from the early 1930s, uneven efforts thus began to try to find new homes for scholars under threat.[10] As they did so, they helped create emerging networks of scholars devoted to aspects of research that would contribute further to efforts to develop and to popularize scientific accounts of universal history.

This chapter and the next focus on efforts of synthesis and popularization by such figures in the middle third of the twentieth century, just prior to the fuller articulation of a compact vision of the history of the world. Across this period, in particular, I examine the converging efforts of Gamow and Shapley, two energetic welders of larger networks of scholars and cultural actors exercised by questions of history, synthesis, and the place of science in the wider culture.[11] As prolific and widely read popularizers who articulated synthetic frameworks and parallel synthetic scientific programs across a range of media and in different institutional contexts, Shapley and Gamow merit particular attention. But their connections will also allow and require the examination of more diffused efforts of the larger groups of scholars to whom they became connected or upon whose work they based their own synthetic efforts, an analysis that will continue into the chapters to come. Such scholars include Haldane and others such as Julian Huxley associated with the so-called "evolutionary synthesis," an overt attempt in the years preceding and during World War II to find an evolutionary framework for biological research, including the advances of genetics.[12] But they further include members of the much-studied Vienna Circle, such as Otto Neurath and Frank. My treatment here takes advantage of the fine-grained analysis carried out in the historical scholarship on these movements. The focus is on further conceptualizations of scientific synthesis in relation to the emergence in the twentieth century of other meaning-making practices that I call "genres of synthesis" and to the theorization of and activism toward a practical ethics and governance in the broader cultural context in which these genres developed.

This chapter begins to trace the development of contributions of these genres in relation to a historical synthesis, focusing on fabulaic, foundational, and historical genres. Building on and sharing in the work of other travelers into the scientific "borderlands," including in different measure Einstein, Eddington, Lemaître, Frank, Huxley, Haldane, and Oparin, Shapley and Gamow played a particularly strong role in the refinement and dissemination of these technical and more popular synthetic templates for the natural sciences. Such synthetic efforts constructed social and conceptual frameworks to bring together scientists and scientific work, both in the scientific disciplines themselves and in popular conceptions of science and scientists, efforts that became only more urgent in the face of contemporary efforts to divide the sciences by nationality and religion.

As had been the case since the time of Thomson, any new universal history in the twentieth century would have to nest within it contemporary geological and biological understandings in their turn and balance the evolutionary and the entropic. In the wake of the new emphasis on the immensity of universal scales, the topics covered by the emerging popular version of universal history now were coming into focus: an explosive beginning, or effective beginning, a radioactive burst, the formation of galaxies and stars, the succession of the geological and then the biological, on earth and potentially on far distant elsewheres. They included, too, the question of the possible end of the world, evidence given in terms of accelerating galaxies and radioactive decays.

To these, we can also add the appeal to the cinematograph as a metaphorical resource, the casting of cosmological researchers as the heroes of a perennial quest for truth linking them to earlier figures from many cultures, the political affinities and lessons promoted through that research and its results, and the religious resonances of the entire tale. Finally there are the uncertainties as to what might constitute the first and final moments, the work of bracketing of those questions as either poorly defined, as nonscientific, or as a veil behind which to peer means little more or less than to speculate. The registers in which beginnings and ends (and particularly the ends) of the universe fall—peaceful or violent, destructive or redemptive, eventful or flatly and eternally eventless—also continued to invoke the question raised with Helmholtz and Thomson and persisting with Shapley of what this scientific view intimates about life's purpose.

As William B. Provine has emphasized in the context of his analysis of the "evolutionary synthesis," however, what the particular synthesis covered and what counted as its seminal texts depended on subdiscipline, institution, and nation. And what took place in each account and its significance depended on which historiography of science an analyst adopted. Important texts of the period for the evolutionary synthesis, depending on discipline (whether, for example, genetics or paleontology), included Theodosius Dobzhansky's 1937 *Genetics and the Origin of Species* and George Gaylord Simpson's 1944 *Tempo and Mode in Evolution*.[13] Ernst Mayr's *Systematics and the Origins of Species* appeared in 1942, based on lectures given the previous year. The humanistic, ethical, and political dimensions thematized in such texts varied with context and author—even as the idea that a new universal history and scientific understanding of life deeply affected the question of humanistic self-understanding and future was a common feature of them all.

Apart from popular accounts, the more technical experiments in synthesis I examine in detail here delve more into the practices of the physical than the biological sciences. But such a distinction is not always easy to sustain, since both the scientific roots of synthesis and the popular and generic expressions of them are far more tangled and widespread than the always uncertain and exposed disciplinary walls that often appear less porous than they are. As discussed in the previous chapter, for example, in addition to his focus on the scale of the universe, Shapley had already by 1930 begun to emphasize the question of the nature and origin of life in relation to "borderland" science. He had also begun to posit a scalar framework for synthesizing disciplines and scientific results together. As we have seen repeatedly, the topics plainly involved in any attempt at synthesis inevitably ramify to other topics, as well. And as will be seen here, the same may be said of the genres of synthesis that were emerging in the first half of the twentieth century: the scalar, fabulaic, foundational, and historical genres of universal history of necessity tended toward mutual imbrication, if only most evidently in the case of the genre of history itself. But, also of necessity, scalar accounts involved assumptions about fundamental laws whose workings were narrated as a story about the nature and shape of history. As what follows shows, the same was the case for the fabulaic, foundational, and historical genres as they emerged together with the scalar. Universal histories compassed a range of ruling figures upon which they individually relied in varying degrees, figures themselves tied to varying, interdependent visions of scientific orientation, organization, fulfillment, and finality.

MATERIAL ORIGINS AND FABLES OF SCIENCE

Well before resettling in the United States, Gamow was able to draw an audience and stir up colleagues. Some were moved so far as to make him a focus of institution-building projects. Karl Hall has excavated this early history, relating how Gamow founded the "Jazz Band," a group that included the physicists Lev Landau and Matvei Bronshtein. Though informal, the bonds linking this group to Gamow were strong enough that Landau and Bronshtein "had further embarked in the fall of 1931 on an impolitic campaign to get the twenty-seven year-old Gamow elected to the Academy of Sciences."[14] Gamow had hoped through these efforts to construct a theoretical institute as a kind of counterpart to Bohr's in Copenhagen. He was indeed elected to the Academy of Sciences, but the Jazz Band did not succeed in forming the institute.

From equally early in his career, Gamow developed a taste for giving popular lectures. One of these on stellar fusion was given with Bukharin in attendance. According to Edward Teller, this was in 1928, some six years before Gamow's arrival in the United States in 1934, at the Moscow Academy of Sciences. The subject of the lecture was the recent theories by the British astronomer Robert Atkinson and the multinational (Dutch-Austrian-German) physicist Fritz Houtermans, theories that claimed the energy of the sun is produced by reactions between atomic nuclei.[15] After delivering the lecture, Gamow was asked by Bukharin whether there might be a practical, terrestrial use of nuclear reactions. According to Teller, Bukharin "offered to turn over to Gamow the Electric Works of Leningrad for a few hours at nighttime if that would help in the job."[16] It was not a practicable possibility; Gamow declined.

Gamow's own recollection, published posthumously, differs in a few points[17]: the lecture is given three to four years later, in 1932, and in Leningrad. The offer is instead for the apparently more modest few minutes one night a week to the electric power of Moscow's Industrial District.[18] Regardless, the offer to harness fusion for the state, however impracticable, was one that appears to have given Gamow pause, making clear to the young scientist the potential power of lecturing to a mixed audience—the stakes of public science.[19]

Once out of Russia, Gamow continued to forge networks and to popularize. Following Solvay, Gamow spent several months traveling across Europe on different grants, with monies first provided by Marie Curie and then by Rutherford and Bohr.[20] Afterward, he immigrated with Rho to the United States, where he was able to take up the Ann Arbor summer-school appointment denied to him two summers before. There he reported briefly meeting a young graduate student named Tompkins, who triggered his imagination and sense of humor, a man he would meet only a handful of times over the coming years, but who played an important role in relation to his popular writings.

While at Ann Arbor, Gamow was contacted by the applied physicist Merle Tuve on behalf of the president of George Washington University and was soon installed in Washington, DC, as a professor. As a condition of his appointment, Gamow was allowed to select another faculty member, so shortly after his own installment, he arranged for the appointment of Teller. At the same time, Gamow published the first of what became a long, seemingly inexhaustible series of papers on the subject of and usually including the phrase "origins of the [chemical] elements," the first appearing in 1935. The original paper (and the ones thereafter) attempted to begin to account for the relative abundance of the different fundamental elements.[21]

To consider the issue of the origins of the fundamental elements of the universe as a research question, several planks upon which that agenda was based needed to be firmly in place. First, there had to be some agreement on the existence and identity of such elements. In the 1930s, when Gamow first started to pursue this line of inquiry, this consensus was only fairly recently established. Atoms were potentially the stuff of fiction before the first decade of the century, even if a useful fiction. Einstein's theory of Brownian motion, together with the contributions of Marian von Smoluchowski and Langevin and the experiments of Jean Perrin, helped establish the reality of atoms—though the ever independently minded Mach had remained unconvinced of their more than hypothetical character.[22] Likewise, the distribution of matter in the universe was also itself only coming to an accepted clarity in the 1920s and 1930s. Gamow's 1935 paper did not rely on distribution data, concerning itself primarily with the potential of transmutation for the explanation of element origins. But as distribution became a central question, the data in this regard relied on the work of the Swiss-born Norwegian geochemist V. M. Goldschmidt, later supplemented by the meteorological data of the American geochemist Harrison Brown.[23]

Second, a sense of some process suggesting elements can emerge was needed. The recent history of radiation and of atomic "artificial transformations" was instrumental. Contra the belief that the elements are "certain prime substances," "the discovery of radioactive phenomena and their interpretation, due to Rutherford . . . has shown, however, that chemical elements can be, under certain circumstances, transformed into one another."[24]

Third, and relatedly, it was useful to think in terms of a common origin, and for that, radiation again played a role: could all the elements emerge from some common material and then, through radiative and recapture processes, form in different combinations to produce some relatively stable shapes? But as seen in the last chapter, this latter question, a question of stability, could also be cast as a matter of the cosmological arena itself. And so there were already evolutionary conceptions of space-time as a whole suggesting that *all* things have an origin, and a common one, at that.[25] However, in his very first paper of this "origin" series, this was not a connection Gamow drew.

Gamow entitled his 1935 paper "Nuclear Transformations and the Origins of the Chemical Elements," publishing it in the *Ohio Journal of Science*. Rutherford was the point of departure: "Rutherford succeeded also, as early as 1919, in producing artificial transformations of light, usually stable, atoms, bombarding them with the intensive beams of fast α-particles. . . . Further investigations of Rutherford and his schools have shown that analogous processes can happen for the [*sic*] most of the light elements and that the probability of disintegration decreases very rapidly with increasing atomic number [Z, the number of protons in the atom] so that already for Z>20 no emission of protons can be observed."[26] Gamow turned to the capture of neutrons within stars as what allowed the possibility of the formation of heavier elements. This was not a Lemaître fissionlike model, but Gamow was considering element formation in the context of neutron-based reactions, reactions of prime importance to fission experimentation.[27]

At a time before the discovery of the fission of uranium, Rutherford's transmutation of elements was work closer to the heart of Gamow's expertise in nuclear physics. Gamow began by advocating the possibility of further transmutations, appealing to Rutherford's experiments in bombarding and "disintegrating" atoms with beams of fast-moving particles. Here, it was entirely on the question of stars that he looked for the production of the heavier elements, hypothesizing that they emerge as the disintegrated residue of fragments erupted from a large and dense stellar nucleus. He did not tie the creation of the elements to the broader timelines of the creation of the world itself. There was no set of central events to which this process was linked, tentatively or otherwise, no attempt to give a sense of when element formation itself began to take place. Narrative ingredients *were* introduced as potential contributions to an emerging genre of synthesis—the formation of different types of galaxies, or more immediately, the contrast between heavy and light element formation—but nothing more than an oblique chronology was suggested. Though Gamow did see the abundance of elements as connected to their origins as early as 1931, here, without the emphasis on distribution, history did not meet transmutation.

The study of the historical formation of the elements was not the only extended effort Gamow initiated in force in this period. His interest in popularization persisted, but was not limited to public lectures. After immigrating to the United States, he began his more popular publication career, first with his "Tompkins" series, a succession of humorous articles, books, and one completed film featuring a bank clerk, C. G. H. Tompkins, who journeys in dreams to and through a variety of different worlds as then understood by science. Gamow's first forays into popularization in English were through Tompkins, beginning in 1938, three years after his first "origin of elements" paper, but they extended well into the 1960s, in a succession of fables exploring the nature of the universe on a variety of scales, including the origins and nature of human life. In Gamow's hands, the synthetic possibilities of a scalar approach to the universe and of the fabulaic came to be strongly tied to each other, through depictions of journeys taken if not always toward then in view of the ultimate truth of things.

The literary genre of the Tompkins stories is that of the picaresque. Through Tompkins's dream voyaging across different media and scientific disciplines, Gamow portrayed science as offering adventures in other worlds, worlds of exaggerated special and general relativistic effects, of humans the size of subatomic particles, of bizarre domains within the human body. Although the picaresque is fictional work that prototypically follows the haphazard adventures of a rogue or *picaro*, a character very different from Tompkins, and although the traditional *picaro* provokes action, rather than having action visited upon him, Tompkins, like the *picaro*, follows a winding route through the domains he visits, and as in the traditional picaresque, there is little that *demands* he visit one place over another. His own experience through these worlds, seen from his own viewpoint, is a chancy one, where any knowledge gained is learned by way of the adventure itself.[28] This knowledge doesn't transform or reform the *picaro*/Tompkins, even if the reader may learn something new about what shapes that world. What can finalize the journey, or circumscribe this synthesis, is that there might be only so many worlds to visit, that some last world awaits. In the context of scientific synthesis, this possibility translates to the idea that with each new world, more knowledge of this world, the home world, is known, and that after traveling to a limited set of such worlds, a complete knowledge of the home world will be possible.

At the same time, in step with this learning by adventure, the relationship implicitly posited between the sciences in this form is more casual than strict. In capturing the cavalier nature in which at times Gamow depicted his own scientific path, Tompkins's adventures trace to a degree Gamow's own trajectory through different subdisciplines of the sciences, following him from the nucleus of the atom to the workings of the human body.[29] Retrospectively, then, as will be seen in future chapters, there is an affinity between Tompkins's scientific adventures and Gamow's narrating his own much-better-informed adventures through science, stitching together varied scientific research through his own work and his own life path. But not until midcentury (the 1950s), when Gamow began to prepare the Tompkins adventures through the human body and thereafter to attempt to put *that* Tompkins on film, does the Tompkins series offer a kind of totalizing synthetic vision, connecting different branches of science through Tompkins's explorations.[30]

The fables of the Tompkins series also became a kind of bridge for Gamow from technical work into more clearly generic scientific laypeople's texts. In the Tompkins stories, in his physics papers, and in his popular textbooks, it is unclear how self-conscious this work of synthesis was as Gamow constructed that bridge, but it was work being done, formulating the grounds of potential synthetic genres for science while deciding on what counted as the narrative *of* science.

The hero of these accounts, the bank teller named Mr. Tompkins, was inspired by that mathematics graduate student with whom Gamow crossed paths in Ann Arbor. As Gamow explained: "The initials of Mr. Tompkins originated from three fundamental physical constants: the velocity of light c; the gravitational constant G; and the quantum constant h, which have to be changed by immensely large factors in order to make their effect easily noticeable by the man on the street."[31] Mr. Tompkins, for Gamow, is such a man on the street, winding and dreaming his way through vivid displays of exotic physics and science.

In 1937, Gamow first attempted to place his Tompkins stories in the United States in a series of prominent magazines, including *Harper's*, the *Atlantic Monthly*, and *Coronet*, all of which rejected him. It was not until May of the next year, at a conference in Poland entitled "New Theories in Physics," that Sir Charles Galton Darwin, a physicist and the grandson of his namesake, suggested to him that he send the story to C. P. Snow, who then edited the popular science magazine *Discovery* for Cambridge University Press. The first editorial notes of the very first volume attempt to establish the mission of the magazine as reconciling science and the humanities, themes with which Snow's name was to become so firmly associated: "Our object, then, is to give readers an interest both in the Sciences and the Humanities by making the work of both as plain as possible. Whether we fail or not remains to be seen. We mean to try."[32] With the introduction of a new series in April 1938 under his editorship, Snow reaffirmed the mission. He wrote of the scientific results of his era as a "golden age" in which "every means ought to be used to explain these results. They ought to be within the reach of the interested laymen, even if he does not possess the time or training to follow all the technical details. The advance of knowledge, indeed, ought to be made as easy to understand as is humanly possible."[33] For such a mission, the Tompkins stories must have seemed a godsend. And Gamow took up Darwin's suggestion: "Thus, when I returned to Washington, I fished the old manuscript from my desk and mailed it to England. The next week I received a cable from Cambridge: 'Your article will be published next issue. Send more. Snow.' Well, this was the beginning, and one Mr. Tompkins article followed another in the pages of *Discovery*."[34]

The stories were first published in six issues over a six-month period, in each case as a series of dreams, from "Dream I: The Toy Universe" to "Dream VI: Tompkins' Final Adventure."[35] Prior to "Dream I," an explanatory letter of Gamow's was published above the story inset. There, Gamow explained that "if we imagine other worlds with the same physical laws, but different numerical constants, the relativistic and quantum phenomena would be easily accessible to observation. We may say that even the primitive savage in such worlds would be acquainted with relativity and quantum principles, using them for his hunting purposes and everyday needs."[36] In this way, Gamow tied the notions of physics to the quotidian of an exotic elsewhere. To underscore that Tompkins's knowledge therefore came more through experience than instruction, Gamow noted that "in the following stories a certain Mr. Tompkins, who does not know any mathematics or physics, being a clerk in a city bank, is brought in his dreams into different worlds of such types where he can directly observe the phenomena about which modern science concludes from very elaborate investigation." This adventuring into other worlds, however, was an introduction to the structure of this world, of what underlies apparently universal experience.[37] In this way, Tompkins becomes an accidental adventurer and anthropologist, visiting other worlds of physics on other scales to assimilate their knowledge and underscore the conventions/laws obtaining in his own home.[38]

The stories were from the first constructed as a kind of *Alice in Wonderland* of science, a connection Gamow drew explicitly in dedicating the first collected volume to Lewis Carroll and Niels Bohr. In one such dream, Tompkins attends a technical lecture on "the problems of Space, Time and

Cosmology," taking nothing from it but the phrase *"the space in which we live is curved, closed in itself and in addition expanding."*[39] That night he dreams he is on a tiny planetoid in a universe itself so small (and with much smaller c, and much greater g) that he can readily perceive the effects of its expansion in its cooling ambient temperature and the pink of distant objects. He perceives thereafter a shift in the color of these objects to violet and the increasing warmth of the universe, both effects of the universe now in contraction. He wakes feeling hot, but saved from the crunch that would have destroyed his dreaming self.

In another dream, Tompkins attends a physics lecture on "EINSTEIN'S THEORY of Relativity." Falling asleep during the lecture, he wakes "in a beautiful old city with medieval buildings lining the street." The speed of light in the city is only about 10 miles per hour. When he bikes there, the streets appear to contract. He sees others apparently flattened by their motion, and meets a grandfather younger than his granddaughter.[40] It is a city where relativistic effects, the dilations of time intervals, the contractions of lengths, are all immediate and unmistakable. "Evidently nature's speed limit is lower here"[41]—so much lower that when Tompkins leaps on a bicycle, the streets grow dramatically shorter.

At the same time that Gamow was working both on the origins of the elements and on different story worlds, he was also working on the question of stellar and interstellar formations—work he found ways to connect lightheartedly to Tompkins. In a *Physical Review* article drafted in 1937 and published in 1938, Gamow examined the relationship between the evolution/life of a star and the thermonuclear processes that give it its energy, a topos that itself would become an element of all the emerging genres of synthesis. In the course of this article, "Nuclear Energy Sources and Stellar Evolution," in line with his 1937 paper, Gamow observed, "In order to obtain the formation of heavier elements inside a star it is necessary that some of these thermonuclear reactions between light elements should produce an appreciable number of neutrons, which would then be radiatively captured by heavier nuclei, causing a continuous building up of heavy elements."[42] In a 1939 *Physical Review* paper, coauthored with Edward Teller, Gamow looked to the question of gravitational condensations of gases as what had formed the gaseous nebulae in the skies. The authors connected this to cosmological considerations, an evolutionary model of the universe offering an explanation of how matter could have condensed to form the nebulae.[43] Gamow and Teller ultimately concluded that this condensation must have occurred in a much earlier state of the universe. The final words of the paper are humorous testimony to Gamow's other ventures: "We consider it our pleasant duty to express thanks to Mr. C. G. H. Tompkins for having suggested the topic of this paper and to H. A. Bethe for valuable discussions."[44]

Gamow took only one year to popularize this joint work with Teller (and Tompkins). In 1939, he received offers to publish a collected edition of the Tompkins series, *Mr. Tompkins in Wonderland* (1939), from Cambridge University Press at approximately the same time that Viking Press contacted him with a book offer that became *The Birth and Death of the Sun: Stellar Evolution and Sub-Atomic Energy* (1940). In this latter text, Gamow's interest in the history of the world is announced from the outset: "In this book, the author, who has been closely connected with the progress of research on these problems, attempts to give, in the simplest terms of which he is capable, an outline of the fundamental discoveries and theories that now permit us a general view of the evolution of our world."[45] He placed his own work in a chronology that began with Helmholtz's contraction theory, through Eddington's work on stellar interiors, through to his work with Teller, as well as with physicists Carl von Weizsäcker and Hans Bethe. In a chapter of the book entitled "The Birth of the Universe," he more clearly than anywhere else in his work at the time announced his belief that the universe as a whole is historical: "If we now look backward in time, reversing the process of progressive expansion [the spreading apart of the galaxies], we are obliged to conclude that, a long, long time ago, before the galaxies or even the separate stars were formed, both the density and the temperature of the primordial gas that filled the universe must have been extremely high."[46]

And in line with the sensibilities of the day, he soon imagined a voice of protest: "'Enough! the reader has by now certainly exclaimed. After all, this book is supposed to be based on certain physical realities. But all this talk of the universe's being formed from a superdense and superhot gas sounds very much like metaphysical speculation!"[47] Nevertheless, without appealing to either Lemaître or Friedmann, Gamow asserted that there was enough research concerning these "seemingly metaphysical speculations" to provide evidence for the "early stages of the development of our universe."

ORGANIC ORIGINS AND THE QUESTION OF FOUNDATIONALISM

A decade after Lemaître's intervention, a metaphysical air indeed still prevailed in work on the origins of the universe. The atom bomb helped to change this, together with the work Gamow was busy doing in order to construct a

seductive, narrative synthesis of natural scientific research. And any synthetic narrative explicitly would have to embrace not only the origins of the
material world, but the origins of organic life, a topic with inevitable metaphysical and theological valences.

Though pursued by few, the research and speculation into the origins of
life itself persisted, represented as having potential for bridging scientific
endeavors. As with the origin of the material elements, the origin of life as a
research question now required the interplay of distinct scientific disciplines,
an interplay tied to the political, aesthetic, and epistemological seductions of
unity.

In emphasizing humanity's cosmic marginality, and promoting at the
same time "borderland" or cross-disciplinary science, Shapley saw the promise of new syntheses. The problematic that he promoted as central to that borderland work was the origin and nature of life. For his own part, he examined
the question from an astronomical perspective: What planetary conditions
were necessary for life to emerge and survive? Such an approach was broadly
consonant with the efforts of other scientists whose work he would come to
know and promote in later decades, with Haldane and the Soviet biochemist
A. I. Oparin the most prominent examples. Oparin's and Haldane's sketches
toward new syntheses of science were based on partly foundationalist reductionist claims about the operation of ongoing fundamental laws, laws seen as
driving history, views that were also imbricated in their own political and axiological commitments. The intricacies, modulations, and variations of these
views underscore the need to distinguish between the different possible conceptions of foundations and the different scopes for logical disciplinary reduction they can suggest.

In his *Origin of Life*, the young Oparin attempted to tackle the question of
the ultimate beginnings of life. Between 1924, in a first slim text, and 1936,
a second, expanded and revised one, he produced a treatment significantly
more extended than that suggested in debate over the same years by Shapley. Like Shapley, his problematic of life was less a focus on its character or
definition than a speculation as to its first emergence, whatever life might
be.[48] Unlike Shapley, Oparin, however, suggested specific chemical processes that might have resulted in life, and he investigated the relationship
between organic and nonorganic forms. In his research, he made no apology for throwing as many different kinds of science at the problem as he
could muster.[49] For Oparin, life as a research problematic required historical
work to situate and propose it properly, and an array of natural scientific
research to address it.

Both versions of Oparin's argument, earlier and later, took stock of a large set of subjects that were at least in part well beyond his more narrow fields of expertise, including the history of science, astronomy, cosmography, geology, and chemistry, tracing threads of inquiry relevant to his question. He moved from the spontaneous generation debate of the previous century to the more recent synthetic biology concepts of his own period. Indicting all past attempts, Oparin attempted to locate the problem in its proper natural historical context: life as such is not always being newly created, as spontaneous generation suggested, and is not coextensive with matter, as Tyndall softly speculated, but rather emerged in certain rarefied conditions that could obtain only far in the earth's past—a natural selection of prebiotic materials.[50] The past for him was more another world than another country. To illustrate its place in the universe, to characterize its own geological and biological evolution and its peculiar chemical conditions, Oparin's exposition compared the young earth with other planets.

So much was true of his problematic generally. But the differences between the first and second editions—the period over which concerns with the origins of material elements were being more fully articulated—provoke the question of whether Oparin's historical and political sensibility significantly informed his theories. The second edition is framed in terms of an overt dialectical materialism, a change that historians have treated as something of a puzzle. In Oparin's treatment, foundationalism in science came to be strongly implicated in politics and political philosophy: Could the sciences in principle be reduced to some set of fundamental laws in the universe, beyond which there would be no further need or further possibility of going, some ultimate structuring truths? In disciplinary terms, this was persistently linked to the questions of whether any science ultimately could be reduced to physics: could for example biology be reduced to chemistry and chemistry to physics, so that deductions from physical law would yield chemical and biological laws? Loren Graham has pointed to the fact that the earlier Oparin accepted what from the perspective of his later works would appear to be a naïve materialism, one that sounded much like a foundationalism in certain senses akin to Comte's complexity-based structure.[51] In this regard, Oparin appealed to Engels, who in his natural history rejected what by that time must have been associated with the Comtean point of view.[52]

Regardless of whether Oparin found such an argument in Engels, the dogmatic unity that such naïve materialism proffered—which could be straightforwardly associated with Comte only through a somewhat one-sided reading—led to a foundationalism that the later Oparin of 1936 could not support.[53]

Nevertheless, in both texts, through a problematic that deployed the sciences roughly according to which natural historical period they could enlighten, Oparin projected a historical synthesis in his account of the origins of life, even if it was less consistently a foundationalist one.

But in this work on origins, history, and materialism, Oparin was not entirely idiosyncratic. In the periods between Oparin's first and second treatments, Haldane advanced consonant, if far from identical, views.[54] Haldane and his own eventual embrace of Marxism has raised paralleling questions for historians: Did Marxism or the dialectical materialism of Engels's natural historical ideas act as a spur to a materialistic understanding of the origins of life? Given that in the case of both Oparin and Haldane, avowed dialectical materialism came after initial speculations on the origins of life, the more straightforwardly posed historical question might instead be whether their thoughts concerning the origins of life shaped their political views.[55]

The role of dialectical materialism as a far from straightforward foundationalist credo, suggesting the possible range of foundationalisms, or political philosophy more generally in its relation to research on the origins of life, is a reminder of the ways in which scientific syntheses and problematics arise in a broader cultural context. For those such as Haldane and the prominent origins researcher, essayist, and socialist/Marxist thinker J. D. Bernal, the meaning-making role of science was invested in a left-leaning political activism, informing their scientific and popularizing efforts. Haldane's speculations on the origins of life were increasingly wrapped up in synthetic scientific and cultural conceptions tied to questions of politics and ethics across a whole generation of scientific researchers. Theirs was a point of view that Haldane's friend Julian Huxley insisted was underwritten with the universalism of a full evolutionary process replacing older mythologies. These synthetic concerns embraced the relationship that politics, ethics, and axiology bear to science—questions that had interested as varied figures as Herschel and Comte.

Like Oparin, Haldane explicitly considered the relationship of scientific synthesis and political philosophy. Weighing the question of socialism and a technologically enabled state, he focused on the ideas of scales and of history. His 1926 essay "On Being the Right Size" concluded on the question of the scale of democracy and socialism as biological and technological matters.[56] In the piece, different political institutions are themselves so many different animals and have to be at the right scale to function. To get that scale wrong, as a political choice, would be to produce at best a very awkward animal, at worst, one that could not survive. "To the biologist,"

Haldane claimed, "the problem of socialism appears largely as a problem of size." And at this time, socialism did not appear to him as empire sized: "But while nationalization of certain industries is an obvious possibility in the largest of states, I find it no easier to picture a completely socialized British Empire or United States than an elephant turning somersaults or a hippopotamus jumping a hedge."[57]

Be that as it may, by the time of his contribution to the question, in the late 1920s, Haldane was still in Cambridge, writing essays and science popularizations in the course of more technical research. Haldane's image of a "hot dilute soup," articulated in his 1929 essay "The Origins of Life," provided a resonant and enduring figuration for the scene of that life's emergence.[58] Haldane's is a much slimmer and scantier sketch than either of Oparin's, a "short but brilliant essay of eight pages," as Bernal came to refer to it.[59] In adjudicating the question of credit, Bernal argued that the essay contained "the two leading ideas which have run through the whole development of the subject." These, as Bernal saw it, were the fact of "intermediate forms" between living and nonliving matter and that of the "reduced" or oxygen-free or oxygen-poor state of the early atmosphere.[60] This retrospective endorsement points out what were later to be the central elements that universal historical narratives drew from Haldane, elements of an emerging genre of synthesis that were prominent in such a brief essay: materialist accounts of the emergence of life as we know it via intermediate forms from nonliving matter in an environment starkly different from that of the present.

Haldane worried that this view would be dismissed as materialistic, precluding the possibility of spirit.[61] To this, he added: "This answer [to the origin of life] is compatible, for example, with the view that pre-existent mind or spirit can associate itself with certain kinds of matter." Nevertheless, he emphasized that in his opinion descriptions of the relationship of mind (and presumably soul) and matter were all "rather clumsy metaphors." He concluded therefore that "the biochemist knows no more, and no less, about this question than anyone else. His ignorance disqualifies him no more than the historian or the geologist from attempting to solve a historical problem."[62]

HISTORY IN SCIENCE, POLITICS, AND CULTURE: "EVOLUTIONARY HUMANISM"

The new generation of politically and culturally active scientific researchers married cultural and pedagogical traditions, and the results also formed part of the emerging genres of synthesis, connecting concerns more specific to scientific practice to the larger culture in which scientific research circulated and was embedded. For Julian Huxley, for example, like those such as the Polish-English mathematician and litterateur Jacob Bronowski, a new humanism was to be combined with a wider natural scientific apologetics and formed an important component of the genre of universal history that they advocated. The traditions they invoked were perpetually provisional, formulated in a cultural context where the tendency was more to contrast than to unify the natural sciences with the humanities. Huxley's own extended family had argued for a balance across humanistic and scientific studies, but a balance weighted toward different sides of disciplinary scales. His grandfather Thomas Henry had emphasized the new sciences and modern languages, his grand-uncle Matthew Arnold, classical studies. The newer attempts to make the fields of knowledge whole appropriated the terms of those older debates, but the terms themselves were weighted with very different twentieth-century meanings and political dilemmas.

Julian Huxley and Jacob Bronowski made plain that there was no necessary logical contradiction between humanism(s) on the one hand and science(s) on the other. One of the ways for forming their reconciliation, they argued, was through a recuperation of a humanistic poetics in the context of a scientific framework.[63] As such, for Huxley, evolution was overtly constructed as a progressive development in terms of a universal history celebrating the human as current and crucial pinnacle of a cosmological process coming to consciousness of itself. His defense of the language of evolutionary and human progress was in overt contrast with many of his scientific colleagues, and yet (as he himself observed) tapped into currents of their own writings. Bronowski would also represent the progress of humanity in not altogether different terms—and even Shapley would eventually depict the human as a local peak in a broader evolutionary landscape rising toward consciousness.

Huxley named this synthetic turn "scientific humanism" and later "evolutionary humanism." These were varieties of humanism not only in the sense of depicting human beings as dominant, culminating, and therefore central to any account of natural history. In this new "psychosocial" phase of evolution, the rise of humanity signaled a decline in a prior phase of physical evolution, a decline resulting from the advance of consciousness and the conscious control of evolution itself. Huxley had begun to articulate these ideas well before, in the nineteen-teens, but he refined them in the context of 1930s economic depression and political instability.[64] Awareness of humanity's emergent ability to take control of evolution, Huxley argued, meant taking concrete steps on a social level, demanding extensive coordination and planning among scientific experts and other political actors. To resolve crises of planetary social and psychological habitats—of domestic and of international politics—required integrating disciplines (natural scientific and humanistic) and scientific training. Only in this way could humanity assume control of all these environments to positive rather than cataclysmic effect. At the same time, Huxley argued for the need to create space for generalized individual freedom in order to insure the fruition of individual promise, a promise that he generally held to be plentiful and diverse. Across the 1930s and 1940s, he advocated for planned state interventions that would maximize individual freedom, what he regarded as the promise of scientific evolutionary history. But that promise to his mind required some form of the increasingly charged proposals of eugenics, a strong element of the wider ethics discourse among his colleagues.[65]

As a result, Huxley deliberated repeatedly over the proper form of governmental action, what would be nationally and globally necessary to confront the dilemmas of a modern age. He carried this into active involvement in various political groups[66] and throughout the 1930s wrote a slate of essays that spoke directly to or affected these and related scientific and political themes. Though regarded as politically moderate in comparison with an increasingly Marxist-leaning Haldane, Huxley went so far as to publish a treatise on scientific social planning entitled *If I Were Dictator*. It was a book that took seriously its fanciful premise, one that doubtless played into the fears of the mounting critics of social planning and contemporaneous political experiments in scientific and technological control. Huxley held "as part of my dictatorial philosophy" that on the basis of the advice of a central, coordinating "Science Council," social improvements would be rationally enacted.[67] This planning ethos was part of a wider advocacy in the 1930s and 1940s for expertise-based governance. But however much Huxley the dictator hoped "to achieve the socialized state without socialism," socialist or fascist dictatorships were precisely what a number of critics and colleagues feared.[68]

At the same time as he pursued this scientific humanism, Huxley posited a strongly related (if not, to him, identical) "evolutionary synthesis," explicitly articulated during World War II, a reconciliation of Darwinian natural selection with post-Mendelian genetics: "biologists may with a good heart continue to be Darwinians and to employ the term Natural Selection, even if Darwin knew nothing of mendelizing mutations, and if selection is by itself incapable of changing the constitution of a species or line." So he would advance a "reborn Darwinism," a "mutated phoenix" from the old Darwinism before it.[69] The foundations and features of this synthesis, as well as its actuality, have long since been disputed.[70] As presented by Huxley, however, he was merely naming trends that owed much to the work of R. A. Fischer, Haldane, and Sewall Wright, among several researchers, and sharing in viewpoints of others such as Theodosius Dobzhansky and C. H. Waddington.[71]

And these were far from the only contributions to the emergence of the diverse and multinational configuration of scientists and topics thereafter celebrated in the evolutionary synthesis. If the historiography of the evolutionary synthesis has leaned too much to the British accounts in this period, however, it is possibly because Huxley and his British colleagues stitched together the scientific, cultural, and political so thoroughly and so overtly.[72] Within his synthetic scientific work, Huxley argued for his views of scientific ethics and progress. Thus, in his seminal *Evolution: The Modern Synthesis*, he claimed that research into intraspecific selection—or selection between members of the same species—confounds the belief that selection necessarily promotes "constant evolutionary progress."[73] He states the political implications of this clearly: "It disposes of the notion, so assiduously rationalized by the militarists in one way and by the laissez-faire economists in another, that all man need to do to achieve further progressive evolution is to adopt the most thoroughgoing competition." Indeed, "rational applied biology"—presumably, eugenic policy, in particular, but scientific planning in general—would do better than natural selection in promoting evolutionary progress.[74]

In correspondence to these commitments, Huxley emphasized the distinction between evolutionary and human progress, but these he saw as ultimately entangled with each other. Evolutionary progress was objective and based on degrees of control of the environment and independence from it, while human progress was based on human values and fulfillment.[75] But in a world in which evolution was increasingly psychosocial and in which humanity increasingly affected the environment, human values were affecting the further course of evolution, he argued, introducing subjective choice into objective development. His grandfather had announced similar ethical evolutionary dilemmas, and the junior Huxley's own formulation functioned as a blueprint for researchers to come.

The ethical and utopian visions of Huxley and others associated with the evolutionary synthesis, such as Haldane and Waddington, were rooted in the thick context of the early part of the twentieth century, but were also refined in a dialogue with the concerns of the late nineteenth century. For Huxley and his more immediate networks, this dialogue was often with the figure Thomas Henry.[76] The English and Anglo-American conversation more generally took up concerns articulated in earlier generations, restructured through the ideologies and imminent political concerns of their own periods.[77] Julian debated his grandfather's positions on questions of ethics during World War II in a way that made clear how much those earlier deliberations still penetrated the younger Huxley's own understanding of evolutionary history and humanism, even as they functioned as resources for the postulation of contemporary views.

In 1943, Huxley was invited to give a lecture on the fiftieth anniversary of his grandfather's famed Romanes lecture, "Evolution and Ethics." The lecture that he entitled "Evolutionary Ethics" gave him the opportunity to present his views, consistent with his articulations of evolutionary and human progress and in the midst of ideologically charged, widespread conflict. The earlier Huxley's arguments also played a formative role in the contemporaneous and like-minded views of Waddington, whose own position on science and ethics in previous years elicited a broad response from scholars, including Bernal, Huxley, and Haldane.[78] All these scientists shared with the elder Huxley the emphasis on ethics writ large—at the level of the community and the state—and gave less attention to the ways in which science might help detail individual codes of conduct. From this point of view, the fact of internalized individual conflict and choice mattered, and the spectrum of possible individual choices, but any structured guide to individual behavior was not the issue. They embraced the synthetic view at play in T. H. Huxley's claims—the latter's "cosmic evolution"—but Huxley found his grandfather's presentation too restrictive, his representation of the destructive force of cosmic evolution too one-sided. Waddington shared Julian Huxley's assessment and took from Freudian theorization the inscription of the ethical onto the individual mind by way of the superego.

Haldane favored the elder Huxley's views over those of Waddington (and presumably his friend Julian's), recasting T. H. Huxley's claim dialectically as "the cosmic process, which was responsible for human evolution, negates itself by generating the ethical process." From this point of view, Haldane advanced an ethical dilemma similar to Huxley's: "There is a real contradiction, which will be resolved when men not only realize, as eugenists do, that they ought to control their own evolution, but also possess, as they do not at present, the knowledge necessary for this control."[79] At the same time, Bernal looked to ground ethics in a practical-minded, nondualistic Marxist vision. On that basis, he found in Waddington's appeals to a Freudian superego and the direction of evolution a mythical shift: "he has thrown away old myths and sanctions, but has felt the necessity to introduce new ones for justification."[80]

These considerations, of natural history, science, ethics, and governance, were widely debated by scientific practitioners, often in the course of their involvement in political or military work. Haldane, Bernal, and Bronowski were all examples of this. Bronowski was himself no stranger to reflections on science, society, and politics through an interest in Marxism and an opposition to fascism, actively taking up government positions from the time of his operations research during the war years through to his appointment to the British National Coal Board and thereafter.[81] Haldane, a combatant and scientific researcher in First World War, carried out scientific research and served as a counselor in the second.[82] From at least the time of his optimistic *Daedalus* of the 1920s, he energetically reflected on the way science would and should shape society. But in debate with Waddington and others, he also insistently advised caution, protesting vigorously against any poorly reasoned attempt to refine social ethics or social planning on the basis of tentative scientific results.[83] Over the 1930s, his views on governance in relation to science were increasingly inflected by a strengthening commitment to Marxism, postdating speculations on the naturalistic emergence of life. Bernal exhibited a still more enduring leftist political sensibility, with a lasting commitment to Marxism, to theorizations of science in history and society, and to the history and basis of life. His inquiry into the latter, promoting and extending the work of Oparin and Haldane, was itself instrumental in the articulation of the origins of life as a research program.[84]

Both the practical political efforts and scientific ethics of these scientists were linked to movements and commentary in the United States. The attention they received is revealed, for example, in the famed sociological considerations of Robert K. Merton in the late 1930s and early 1940s. Merton attempted to define a scientific ethos partly on the basis of the proclamations of such leftist British scientists.[85] Conversely, the United States saw its own debate over the question of the integration of science, government, and society, from which extended process, for example, the National Science Foundation emerged: the conservative Vannevar Bush opposing the most liberal reforms championed by policy makers such as Senator Harley Kilgore or by the National Science Foundation committee formed by Shapley and the celebrated chemist Harold Urey.

But the positions taken by these scientists on the relations between science, politics, and culture are not always easy to capture, both in themselves and in their evolution over time—a point Jessica Wang demonstrates through the example of Urey.[86] Urey, who had been awarded the 1934 Nobel Prize for the discovery of "heavy hydrogen" or deuterium, led efforts during the war to find practicable methods for separating out U-235 from U-238. Prior to the war, Urey had been a member of progressive left organizations in the United States and, directly following it, argued publicly for internationalism in relation to atomic energy.[87] But in the course of the decade, he became increasingly hostile to Communism and the possibility of international cooperation. In the postwar period, he would also become increasingly involved in the question of planetary formation and the origins of life.[88] Many of those actively forming the new histories of the world and considering the new tools of war thus were also actively at work framing new, historically minded syntheses of knowledge in the broadest sense, stretching past conventional domains of the scientific disciplines. These moments in their extended projects cannot easily be held apart. Regardless of their points of view, these scientists demanded a historicization of ethics through a new universal history, establishing an inquiry into the origins of ethical values and its most effective social or political inducements. And with the advent of World War II, ethical questions and their relation to the knowledge and practices of scientists assumed a new urgency.

EXPLOSIONS: STARS AND BOMBS

Whether tending to scalar, fabulaic, foundationalist, or historical genres, contemporary syntheses depicted an explosive beginning to the universe. Work on the potential military use of thermonuclear reactions in World War II and thereafter structured and saturated this topos, in scientific accounts and in popular imagination.

As Shapley, Haldane, and Frank devoted their energies to their universalist-minded scientific and political activisms, at the onset of war, George Gamow was employed as a consultant to the Division of High Explosives in the Bureau of Ordnance of the US Navy Department. Though as he himself noted, his expertise on the atomic nucleus would have recommended him for Manhattan Project, Gamow was not cleared for work on it. Instead, as he recalled, the problems upon which he advised were "mostly concerned with the propagation of shock and detonation waves in various conventional high explosives, and the discontinuous transitions from shock into detonation."[89]

What specific role the imagery of explosive impact had on the emergence of his concept of what came to be known as the Big Bang is an open question, but there are intriguing links. During this consultancy, Gamow remained at George Washington University, working only on Tuesday and

Friday afternoons for the navy offices on Constitution Avenue.[90] The war effort connected Gamow to other High Explosives Division consultants such as the mathematician and game theorist John von Neumann. Gamow particularly valued the connection to Einstein, who was by then settled in Princeton. Einstein asked that given his age, special arrangements be made to present him with the problems on which he was asked to consult and Gamow was delegated with the task of taking the train to Princeton to confer with Einstein on behalf of the navy.

On his own report, one of Gamow's wartime efforts suggests how close his work on conventional explosives came to aspects of the atomic bomb project. "My other part-time activity for the High Explosive Division," other than his consultation with Einstein, "had an experimental character and pertained to the 'dent sensitivity' of various explosives. . . . These experiments were done in the Navy yards on the Potomac River, in collaboration with a high-explosives expert." In the course of this work, Gamow realized that "one could achieve a very high compression by producing a detonation wave that converged at a point, and it became clear that such a detonation wave could be formed by combining two explosives with different propagation velocities of the detonation process." Receiving the okay from Einstein, Gamow and his colleagues were set to test a simplified version of the idea at Indian Head, the Potomac River navy proving grounds. But to test a more realistic case, they needed to look elsewhere, and for this they chose a Pittsburgh high-explosives factory on contract with the Bureau of Ordnance. "But when I showed the drawing to a man from Pittsburgh, he made a long face and said that his company could not undertake that task and refused to answer my question, 'Why not?'" The project was buried at the bottom of the priority list and, Gamow claimed, "I suddenly realized what was being worked on at a mysterious place in New Mexico with the address: P. O. Box 1663, Santa Fe"—the mailing address, that is, for the secret site for the building of the A-bomb. "Years later, when I was finally cleared for work on the A-bomb and went to Los Alamos I learned that my guess had been correct."[91]

During this consultancy period, however, some of Gamow's work was less restricted. He returned to the question of the origins of the elements while at the same time publishing a second natural historical text, the 1941 *Biography of the Earth*. Here, Gamow connected that book to his earlier work: "In his previous book *The Birth and Death of the Sun*, the author endeavoured to present a picture of the wide world of stars and giant stellar families, among which even our Sun is but a small, inconspicuous body and the Earth an insignificant speck, merely a tiny 'observation platform' for fantastic excursions into the endless universe." For this text, Gamow proposed that he and his readers take "an imaginary lens and study somewhat more closely the dark little globe carrying our destiny."[92]

Most of the popular text does not focus on Gamow's expertise, treating as it does primarily geological natural historical questions. But the book does address the question of what fuels the sun, and as with the work of Helmholtz and Thomson/Kelvin, forecasts a grim end, though in very different terms from those of a century before—a death by fire, rather than frost. Though there will be enough "alchemic fuel" to power the sun for "another 10 billion years," "a more detailed study of the processes taking place in the solar interior also leads to the conclusion that the steady decrease of the amount of 'hydrogen fuel' will only cause more violent 'combustion' of what is still left, so that, contrary to ordinary expectations, *the Sun must become more and more brilliant from century to century.*"[93] The oceans will evaporate, the atmosphere will become so hot that it will escape into space, and humans themselves "will either perish of the heat or be forced to emigrate to some more distant planets, provided, of course, that they are highly intelligent beings and have solved the problem of interplanetary communication."[94] Here he invokes his own work to explain that "investigations carried out by the author at the time he was preparing this book indicate that *even when 'walking the last mile' our Sun will once more demonstrate its might and will burst into a brilliant display of fireworks.*"[95] So to Lemaître's fireworks evolution of the world, Gamow answered with a fireworks death.

But as with Helmholtz, Thomson, Eddington, and Lemaître, Gamow wove together ends and origins in both his popular generalist and technical scientific accounts. So in an address before the Washington Academy of Sciences in 1942, Gamow made the question of explaining the abundance of the different chemical elements central: "Considering the known abundances of various elements from the point of view of possible nuclear transmutations, we should ask ourselves first of all whether these abundances are due to some nuclear processes taking place *at present* in various parts of the universe, or whether the abundance-curve should be considered as a 'frozen-distribution' corresponding to some unusual conditions that existed in the early creative stage of the universe?"[96]

Gamow expanded on these two broad possibilities for explaining the element abundance curves, an equilibrium state and an explosive, fragmentation theory, a Lemaître-like fission of some initial, more homogeneous state. In answering the central question of the address, Gamow rather surprisingly asserted (by the lights of his arguments prior to the war, and from a more contemporary point of view of the question) that the abundance of light elements may indeed be explicable by the appeal to stellar sources, but not so much the heavier ones.[97]

Ruling out equilibrium theories by appealing to recent astrophysical research suggesting that the theory could not reproduce the abundances of the heavier elements, Gamow thereafter focused on the other possibility, "to see which kind of distribution could be obtained from the hypothesis of a rapid breaking up of the original superdense nuclear matter. We must remember that, according to our present knowledge of nuclear-fission-processes, all nuclei that are heavier than uranium would be immediately broken into two or more approximately equal parts."[98] The approximately equal distribution of the heavier elements would, he claimed, naturally result as a statistical matter from the splitting up of the original, superdense nuclear fluid—but would unfortunately mean an absence of the lighter elements. These latter however might be accounted for by the emission of neutrons, which would further decay into protons, as well as to the possibility of as-yet-undiscovered fissioning processes with regard to the heavier elements. He concluded: "It seems . . . that the main features of the abundance curve may be explained as the result of a complicated fission-process." To answer the question of the distribution of the elements from the theory of their origin "in a more definite way it will be necessary to study the stability of superheavy nuclei with respect to multiple-fission and to investigate the statistical distribution of the fission-fragments in a successive breaking down process."[99] At this point, the birth of the universe was on the model of a not-yet-produced fission bomb.[100]

Gamow's fission model as expressed in 1942 acted as a kind of enabling hypothesis for his considerations, as evidence of how much of this thinking went hand in hand with considerations also informing more applied work. As both the transmutative and destructive potential of fission were being understood and explored, Lemaître and then Gamow were exploring the original, creative, and explosive view of the fissioning of some first, simple primeval state.

During the same period when so much of the scientific war effort was centered on the nucleus, Gamow again sent Tompkins through his slumbers to the world of the atom. As with earlier work, the 1944 *Mr. Tompkins Explores the Atom* involved the manipulation of scales.[101] Here, he visits the atom to find himself swooping to and fro along a track, with other misty figures, the "gay tribe of electrons" circling the nucleus of a sodium atom. Tompkins is addressed by a "Father Paulini," who, in monk's frock, bears a strong resemblance to the physicist Wolfgang Pauli. Father Paulini, according to rules articulated largely by Pauli, explains that "it is my mission in life to keep watch over the morals and social life of electrons in atoms and elsewhere."[102] Given the contemporaneous work examining explosive atomic processes, there is a temptation to read the priestly subatomic Paulini as a figure who might also speak to the morals and social life of scientists.

Another dream of Tompkins occurs when, visiting a cyclotron with the physics professor whose lectures he had earlier attended, now his father-in-law, he comes too close to the apparatus and is knocked unconscious by the "generating electric tensions." In this dream, Mr. Tompkins meets a woodcarver and is "struck by his strong resemblance both to the old man Gepetto in Walt Disney's Pinocchio, and the portrait of the Late Rutherford of Nelson hanging on the wall of the professor's lab." The woodcarver is in the business of making nuclei: "Naturally it requires some skill, especially in the case of radioactive nuclei, which may fall apart before you even have time to paint them."[103] Though the woodcarver builds gold and shows Tompkins transmutative processes, Tompkins does not witness fission. Instead, Tompkins would have learned about it during the lecture he had attended "just before he was struck by a high tension spark," a lecture entitled "The World Inside the Nucleus." Here, the professor had described the discovery of nuclear fission by the German chemists Otto Hahn and Fritz Strassmann, omitting Lise Meitner, and how this makes access to new sources of energy possible.[104] Explaining how this could "produce very important changes in our economy," the professor cautions that "it is not an easy task to create the conditions necessary for such an explosive process, and it is very difficult to say whether the solution to this problem can be expected within a year, a decade, or a century."[105] Given the events of 1945, the answer was very well "within a year," of publication at least, as Gamow himself might have suspected.

And indeed, in that same year, 1945, Gamow prepared a manuscript entitled *Atomic Energy in Cosmic and Human Life: Fifty Years of Radioactivity,* the September preface of which states, "In 1945, the radiation of uranium . . . burned and obliterated a good part of two Japanese cities, and now the spectre of the 'atomic bomb' hangs heavily over humanity, threatening its complete destruction in case of another armed conflict between the major powers."[106] The book surveys the "Modern Alchemy" of transmutation, the fueling of stars, and the origins of that "atomic fuel" in the origins of the universe. Its last chapter focuses on the physics of the bomb, but concludes on a brief hopeful note for the potential peaceful uses of atomic energy. One of these is nuclear power plants, the other is the exploration of space.

Here Gamow envisions a "peacock-model" spaceship with a radioactive tail: "When it gets out into the empty interstellar space it opens its tail and sails proud as a peacock towards the stars." Gamow's final sentence aligns nuclear-powered exploration with peace: "Let us hope that the best important achievement of atomic energy will lie in planetary exploration, and not in human destruction."[107]

At the conclusion of the war, Gamow served on the scientific advisory board "to the Abel and Baker shots," the nuclear weapons tests conducted in July of 1946 at the Bikini atoll precipitating further tests and a lasting displacement of the Bikini Islanders.[108] "As a member of the board, I was asked to deliver to the 100 high-ranking navy officers (a few admirals, mostly captains, and a sprinkling of commanders) ten lectures on Nuclear Physics in the movie theatre of the Navy Building. The lectures had to be followed by examinations, the result of which had to be entered in their service dossiers. I have saved a few of these examination papers. (Everyone got [sic] A)."[109] The exam began with basic multiple-choice questions on the size of the atom, on the comparative size of the nucleus, and on the energy involved in nuclear transformations. Its first essay style question was on the differences between nuclear fission and fusion, a very basic question that by this time Gamow was also considering in relation to cosmogonies.

At the end of that summer, Gamow returned to the question of that distinction once again, in the pages of the *Physical Review*, in his September 1946 "Expanding Universe and the Origin of Elements." Here he confidently declared: "It is generally agreed at present that the relative abundances of various chemical elements were determined by physical conditions existing in the universe during the early stages of its expansion, when the temperature and density were sufficiently high to secure appreciable reaction-rates for the light as well as for the heavy nuclei."[110] He dismissed again the equilibrium theory as being unable to explain the abundance curve and went on to connect, perhaps for the first time in a technical paper, element production, expansion, and the formation of stars and galaxies, referring to his work with Teller, if only obliquely.[111] The principles of a historical account were suggested here more clearly than in the past—the unfolding through time of materials not on the basis of a massive fragmentation, but on that of successive aggregation and decays of some initial state.

There were no explicit words left for the fission model, no discussion of the primeval atom. Instead, Gamow speculatively concluded on an idea that perhaps owed something to the "primeval nucleus" of Lemaître, but in no way suggested either the primeval atom or its fragmentation: "The present high abundance of hydrogen must have resulted from the competition be-tween the beta-decay of original neutrons which was turning them into in-active protons, and the coagulation-process through which these neutrons were being incorporated into heavier nuclear units."[112] This "coagulation-process" is perhaps the first mention of a fusion-based model of the origins of the universe. And the *transmutation* of neutrons into protons through this process of beta-decay became his new center of focus—and soon enough, that of his student Alpher. Gamow concluded in reference to potential time-lines: "It is hoped that the further more detailed development of the ideas presented above will permit us to understand the abundance-curve of chem-ical elements giving at the same time valuable information concerning the early stages of the expanding universe."[113] As in the case of fission, the fusion model of the creation of the world preceded the construction of the bomb.

The association to the nuclear bomb is not fanciful, and Edward Teller would be far from the only person to knit together the new cosmologies and thermonuclear reactions.[114] It was a topos that Gamow at times exploited, both on the level of comparisons to nuclear processes and, much more plainly, to more conventional artillery. In another strand of his work, the popular, postwar effort of 1947, *One Two Three . . . Infinity: Facts and Specula-tions of Science*, Gamow directed the reader:

> Imagine now an artillery shell that has exploded in midair, sending its frag-ments in all directions. The fragments thrown out by the force of the explosion fly apart against the gravitational forces that tend to pull them back toward the common center. It goes without saying that in the case of shell fragments, these forces of mutual gravitational attraction are negligible, that is, they are so weak as not to influence at all the motion of fragments through space. If, however, these forces were stronger, they would be able to stop the fragments in their flight, and to make them fall back, to their common center of gravity. The ques-tion as to whether the fragments will come back or fly into infinity is decided by the relative values of their kinetic energy of motion, and the potential energy of gravity forces between them.
>
> Substitute for the shell fragments the separate galaxies, and you will have a picture of the expanding universe.[115]

This artillerylike representation neither turns on questions of fission or fusion nor illustrates them. It suggests instead the different ways in which cosmogo-nies and explosive fragments were linked in a period when synthesizing ac-counts were inflected by the postwar imagery inspired by the atomic bomb. We will shortly see further such representations and deeper connections.

But the plan of *One Two Three ... Infinity* itself is a reminder of the other topoi of scientific synthesis then at play, apart from the universal history that a new cosmogony promised. The book was also a combination of wonder stories in the manner of the Tompkins tales and the questions of scale that occupied Shapley and Haldane, as well: "The book originated as an attempt to collect the most interesting facts and theories of modern science in such a way as to give the reader a general picture of the universe in its microscopic and macroscopic manifestations, as it presents itself to the eye of the scientist of today. In carrying out this broad plan, I have made no attempt to tell the whole story, knowing that such attempt would inevitably result in an encyclopedia of many volumes." Nevertheless, Gamow qualified: "At the same time the subjects to be discussed have been selected so as to survey briefly the entire field of basic scientific knowledge, leaving no corner untouched."[116] An encyclopedic structure was not necessary to canvass the whole of the sciences.

The four corners of the books, its four parts, are on mathematics, on space and time (and Einstein), and then on the microcosmos and macrocosmos. The comparatively shorter macrocosmos section refers the reader to the more extended treatments in Gamow's *Birth and Death of the Sun* and his *Biography of the Earth*. Here "basic scientific knowledge"—a foundationalist approach—is linked to a broad distinction between microworlds and macroworlds—a scaled approach. But because it is not a book primarily of synthesis, these different possibilities are sketched together but are not extensively pursued. At that time, Gamow may have taken a synthetic approach further only by being still more controversial and speculative than was his habit. Whether the whole world has a history was an open question, just as whether the world could be comprehensively mapped through a multidisciplinary play of scales, reduced to a few foundational laws, or whether a figure such as Tompkins could ever visit enough other worlds to understand the one in which he lived. It was also an open question as to whether a world that included humanity would last much longer anyway.

CHAPTER SEVEN

Humanisms, Nuclear Histories, and Nuclear Ages

While it remained an open question whether a history of the world was coherent or possible, the advent of nuclear arms posed the question of its end, of whether the conclusion of a universal human history might not wait for the sun to cool or to blaze. The cultural response to the bomb itself following the war promoted a vision of a shared universal condition, the world living under nuclear threat. Likewise, that response, crediting scientists with the production of terrible wonders, simultaneously promoted their roles as political advisors, as teachers for anxious publics, as scientific sages, and deepened the possibility of establishing a new, scientific, culturally resonant myth: if scientists and technicians could produce such wonders then the deepest secrets and truths might soon be revealed. Emergent synthetic universal histories were shaped in light of thermonuclear processes and their human historical consequences, in the context of the emigrations of scientists and others, movements precipitated by conflict that the new weapons could seem on the verge of consummating. Networks of scientists and other cultural actors forged new communities as a function of others blowing apart. Scientific researchers, consultants, activists, social commentators from different disciplinary and institutional positions, at times distant from nuclear research itself, formed assemblages diverse in knowledge that some consciously worked to unify. In that work, associations between the ruling figures of different genres began to take shape, between the smaller, the more fundamental, the non-living, and the older/more enduring on the one hand, contrasted with the larger, the more dependent, the living, and the newer/more transient on the other. The varied potential syntheses at stake could nevertheless coalesce or collide, informing practical scientific and political efforts, passing different verdicts on them.

In November 1945, the Polish-English mathematician and litterateur Jacob Bronowski visited Nagasaki and Hiroshima as part of a team studying the impact of the atomic bomb strikes. Bronowski had been brought to the Research and Experiments Department of the British Ministry of Home Security by J. D. Bernal in 1943 for operations research on optimizing the destructive impact of bombing German and ultimately Japanese sites.[1] Near the time of the Hiroshima and Nagasaki strikes, this work was undertaken partly in liaison with an intelligence team in the United States.[2] Soon after the strikes, his department formed a survey group to take stock of the impact on the cities. As Bronowski recalled over a decade later, he entered Nagasaki with the ship's loud speakers sounding out a famous Louis Jordan tune, "Is You Is or Is You Ain't My Baby?" over the devastation.[3]

He portrayed the visit as both a personal turning point and a "universal moment," through which he confronted "the experience of mankind."[4] But the moment was *not* a turning point for humanity. To him, that experience included the ruins both after and before the strikes, "the industrial slum which Nagasaki was before it was bombed, and the ashy desolation which the bomb made of the slum." Each to his mind posed Jordan's question from the point of view of civilization as a whole: Did civilization give birth to both atrocities? He emphasized the continuity—that "we are not here fumbling with a new dilemma." The recent devastation had only increased the scale, urgency, and irony of the dilemmas inherent in the material conceptions of civilization.

Implicit in the structure of Bronowski's question was a muted apology both for the bomb and for the role scientists had to play in its construction. His observation suggested that the fault for the bomb, as for the slum, was a collective one. Moreover, his point tended to lessen the singularity of the bomb as such: civilization bore horrors before the nuclear age as well as after. The bomb only rendered more starkly a question that it had already been necessary to face. But the stark power of a single bomb renewed a demand for an adequate language and instruction in the face of the existential dilemmas the bomb intensified.

The consequent intensification of humanistic deliberations that nuclear conflict provoked in the United States and England gave added weight to the universal history those such as Shapley, Gamow, Bronowski, and Huxley were already articulating, focusing attention on the value of science in relation to the meaning of civilization and the progress of history. In order to be cultured and civilized in any sense, natural scientific knowledge was by their lights necessary; it was impossible otherwise to confront these "universal moments," existential, murderous, and mythopoetic.

Though distinctive in their apologetics, Bronowski's arguments were nevertheless part of larger midcentury transatlantic discourse privileging the place of science and scientists in society. The devastation that the bomb could and did cause was often, *pace* Bronowski, seen as new and singular. In turn, the building of the bomb was widely regarded as a pivotal point in human history, one that demanded new modes of national and global politics, with scientists given a special status as uniquely knowledgeable and farseeing.[5]

This larger, ethically inflected discourse, with its roots in prewar political and pedagogical efforts, propped up further the cultural platform from which scientists could begin to establish a newer, widely accepted history. After the war, a number of scientific social planners and advisors increasingly tied their practical political efforts and utopian visions to some form of "scientific humanism," articulating a humanistic vision of science pursued only most insistently by Julian Huxley. For those such as Huxley, Bronowski, and Shapley—and in different measure and spirit, Bernal, Haldane, and even Gamow—political action and poetic energy were tied to a scientific ethos and to a conception of the role of science and scientists in postwar culture, all of which tended toward the mythologization of science and its accomplishments, as they contended with the threat underpinning that status in the catastrophes which new energies posed for the fate of humanity, and in the devastation and displacement they already caused. The privileged place of the scientists, the contours of a new scientific universal history, and the many effects and views of the bomb were part of a tightening cultural configuration.

A new synthetic universal history was being constructed by the likes of Lemaître, Haldane, Oparin, Gamow, and Shapley, an account that was increasingly intertwined with the projects of big science. These synthetic efforts, particularly given the combination of popularization and network building, as with Gamow and Shapley, were presented to scientists and to interested publics neither as preliminary injunctions nor as conceptually taxing prescriptions. They were instead more insistently aggregative than reformative, pointing to changes of orientation and understanding in ways largely recuperative of past scientific labor. These efforts at syntheses were structured across research paper series, public and classroom lectures, private correspondence, and institution-building and network-building endeavors. They were not limited to any specific mode of intellectual production. This chapter traces the provisional structure and authorship of a new collective story, a new scientific epic, and the movements that furthered it—movements of scientific humanisms, of political action, of the atomic science of cosmology, and the cosmological science of the bomb.

The development of nuclear weapons was not only connected to a new universal history by way of the depiction of scientists manipulating mythic powers licensing them to tell mythic stories. The techniques, experiments, and theorizations of both nuclear and conventional bombs—and bombings— were the scientific fabric of that new history, patterning it in a number of ways, establishing it as an empirical project. Research programs determining what fuels the stars and the project of building the hydrogen bomb were intertwined. The former raised immediate questions concerning potential applications for new energy sources on earth, while the latter offered a new cosmogony resulting from the study of thermonuclear reactions. Both stars and bombs were bound up with the visions of the emergence of the world: the Big Bang, following a "Big Squeeze," was like a thermonuclear explosion. The heavy elements were ultimately accepted as being cooked up in the interior of suns. The data that informed the building of atomic bombs and reactors pointed to what had informed the construction of the universe. Gamow and his network shed particular light on the connections between the disciplines of the bomb and of cosmology as they advanced a compelling, empirically grounded universal history and at the same time fostered a scientific network to debate that history. As we have already begun to examine, the explosives consulting work of Gamow and his colleagues and students brought his efforts during the war close to the work at Los Alamos. Relatedly, he represented the first few moments of the world by appealing to conditions within the interior of an exploding thermonuclear bomb, a bomb that by the late 1940s he was employed in the effort to build. With Gamow, we see how the histories of cosmology and of the technologies of fission and fusion are different sides of a joint labor.

That larger story of the history of the world and nuclear history was at the same time an explicit attempt at the reconciliation of different branches of science and also (particularly for Huxley and Bronowski) of the sciences with the humanities. The reconciliation between different studies was not effected outside the writings of a few select scientists. Division instead was underscored in the midst of the Cold War by the highly contested categories of Gamow's erstwhile Tompkins publisher, C. P. Snow, who himself built explicitly on Bronowski's views.[6] For them, a properly balanced education would help orient the future of the nation and of civilization. In the context of intradisciplinary conflicts, where, in the immediate postwar period, extended debate was waged over whether the universe has a history, natural scientific timelines nevertheless moved toward convergence or were shaped so as to be conducive of convergence. Those timelines led back to a universal origin, corroborating a theory steeped in the empirical and theoretical legacies of conventional and atomic bombs and itself nursed in the interior of a hydrogen bomb.[7]

THE S IN UNESCO

In the same set of retrospective reflections on the nuclear strikes of Japan, Bronowski added his own views to the long-standing British deliberations over the nature of ethical development: "In the field of ethics, of conduct and of values, we think as Aquinas and Spinoza thought: that our concepts must remain unchangeable because they are either inspired or self-evident. In the field of science, four hundred years of adventure have taught us that the rational method is more subtle than this, and that concepts are its most subtle creations."[8] Like Bernal, Waddington, Huxley, and Haldane, he took as his theme the cultural evolution of ethics, revealed and propelled by science. "The subject of these essays is the evolution of contemporary values. My theme in them is, that the values which we accept today as permanent and often as self-evident have grown out of the Renaissance and the Scientific Revolution."[9]

Likewise, for Julian Huxley, the first director of UNESCO, the United Nations Educational, Scientific, and Cultural Organization, ethics was embedded in the evolutionary process in a way that Julian Huxley thought his grandfather had failed to appreciate. It made more sense to his mind to ground ethics in evolution—including cultural evolution—than to evaluate evolution by the imposition of any falsely transhistorical ethical view.[10] In contrast to T. H.'s position or Haldane's defense of it, ethics and evolution were therefore in no necessary opposition whatever, and, in fact, evolution generally clarified what ethics was and could be.

For Huxley, as for Bernal or Haldane, regulation in the form of scientific, rational planning outweighed political liberalism (even in an emphasis on individual promise and fulfillment), science and technology aligned with progress and optimism, and the dangers of the world came from ignoring science. And despite fears of air strikes, of emergent fears of nuclear conflict and irradiation, Bronowski, Huxley, and their interlocutors were all more optimistic about science. Though they did not all participate in equal measure or with fully consonant beliefs, they had helped give voice to and broadcast the "social relations of science movement" and the "era of meritocracy" of postwar Britain.[11]

But in an age of global conflict and nuclear arms, coordination was necessary beyond national borders. So, for his founding directorship of UNESCO (1946–1948), Huxley crafted the organization's civilizing/enculturing internationalism in terms of a scientific and evolutionary humanism that he saw as the only option for avoiding political, religious, or philosophical factionalism. In consequence, "the general philosophy of Unesco should, it seems, be a scientific world humanism, global in extent and evolutionary in background."[12]

The emphasis on planning and expertise of Huxley, Haldane, and others put them in the sights of critics such as John Baker and Michael Polanyi, as well as F. A. Hayek, who during World War II had already feared the future outlined by advocates of planning.[13] Hayek and Polanyi corresponded over fears of the "propoganda of the Haldanes, Hogbens, Needhams, etc."[14] The biochemist Joseph Needham, who was a friend of Bernal and also a "heterodox Marxist," became fascinated with China during the war and, like Huxley, had been involved with efforts to form UNESCO—and to have "science" in its name.[15] For him, as for Bronowski, a central question was what were the cultural preconditions for the emergence of modern science and why that science had evolved in the West and not elsewhere. To these individuals, national and increasingly global *scientific* planning was necessary for survival; for their opponents, that planning constituted the threat to survival, as so recently evidenced in German and Russian social and political experiments.

Like his counterparts in the biological sciences, whose popular work he followed, Shapley, too, combined his scientific research, his political activism, and his views on the origins and nature of humanity—a synthesis that also included ethical concerns and an engagement between science and religion. To some, his name would be suited to fall under the "etc." that Hayek had in mind. Like Huxley and Needham, Shapley played a role in the history of UNESCO. He described his involvement with the formation of the organization in his memoirs, giving himself no small measure of credit. As he recalled it, the former president of the New York City Board of Education, James Marshall, approached him in 1945, asking him to help lobby diplomats convening in San Francisco to include education in the United Nations Charter. According to Shapley, the State Department resisted even a straightforward resolution on education being brought before Congress: "They thought we were just internationalists acting prematurely."[16] But a resolution was passed, and thereafter Shapley took part in the UNESCO conference in London to play a role in crafting the organization's charter. Once again, he recalled opposition by the State Department, "because I was a 'dangerous character.'"[17]

Also like Huxley and Needham, Shapley was reputed to have put the *S* in UNESCO—or, if his memoirs are indeed accurate, to preserve the *S* from its excision at the hands of Archibald MacLeish.[18] But, as the previous chapter suggests, UNESCO was far from Shapley's only organizational effort on behalf of science in these years. He lobbied for the formation of what was to become the National Science Foundation, in opposition to the views championed by Vannevar Bush. Despite these differences, Shapley was a member of the Committee on the Discovery and Development of Scientific Talent, the findings of which were included in Bush's much studied report to the president, *Science, the Endless Frontier*, one of several committees that contributed to the report.

Another part of Shapley's activist efforts were linked to Philipp Frank, efforts in ways consonant with the continuing shape of their synthetic intellectual commitments to drawing together the sciences and concerns usually associated with the humanities. After his more secure installation at Harvard, Frank initiated an "Inter-Scientific Discussion Group" including both Frank and Shapley as committee members. Between 1939 and 1944, Shapley had been president of the American Association for the Advancement of Science, where, in 1947 (under the call for disciplinary unity made by Shapley's successor, Howard Mumford Jones), the Institute for the Unity of Science was formed.[19]

The second ongoing institutional connection between Shapley and Frank had begun in 1940, with Shapley arranging Frank's inclusion in an annual "Conference on Science, Philosophy and Religion."[20] Frank noted thereafter that "the precise aim of this Conference was to establish a common understanding of democratic principles that would help to overcome the high-pressure propaganda of totalitarian values." "Relativism" was an explicit target.[21] Frank argued that such relativism as modern science espoused "has not the slightest thing to do with agnosticism or skepticism, that it is in no way hostile to the belief in ethical or democratic values."[22] His defense that the shift in human self-conception instigated by science was an enrichment and aid to ethics strongly shared a spirit with Shapley's conception of what he called metaphysical "adjustments"—ideas that in broad strokes were consonant with those of their English counterparts.[23] These adjustments were to Shapley axiological shifts in response to increasing scalar displacements of a physical center: from localcentricity (for example, the village), to geocentricity, to heliocentricity, to galactic marginality, and then finally to the awareness of the certitude of life elsewhere in the universe—the earth as no longer the center of life.

But Shapley's spatial/scalar sensibility was married to a historical, evolutionary view concerning itself with ethics and social values as human life developed. For decades, Shapley had been attempting to erect a scaffold for a full-blown evolutionary history, including in its scope the formation of life from the material universe. From early on in such efforts, he had engaged scientists and other scholars with stronger religious sensibilities, goading to a degree that lessened over time. As seen in chapter 5, as far back as 1923, he had taken part in the symposium on the origins of life.[24] Shapley's interest in the interface between astronomy, religion, and other scientific disciplines appears not only in his overt and publicized synthetic efforts, but in his being awarded the Pope Pius XI prize for astronomy in 1941—if much to his own surprise.[25] Frank had helped to institute such "cross-cultural" interests for Shapley and others, more enduringly.

Frank's own contributions to the 1940s conference were ultimately shaped into the collection *Relativity: A Richer Truth*, a work making the case for the ethical virtue of relativistic thought in a broad sense. Frank deliberated over the particular role of philosophy, rejecting the idea of any superscience as he tended increasingly toward an emphasis on the sociology of science. Intriguingly, given his connections to Shapley and other spokespeople of science, the conference gave him an opportunity to critique the figure of the popularizer: "The specialist in a field of science works hard and is proud of his achievements, but sometimes he feels lonely because of the number of people interested in his work is small." As such, the specialist will use technical language (such as "energy" or "freedom") misleadingly, contributing to philosophy and general discourse "nothing but ambiguous expressions."[26]

However, syntheses such as those being constructed by Shapley, Gamow, and indeed Huxley and Bronowski were taking deeper root in both technical and unapologetically popular venues, and with them, a rough convergence of scientific and humanistic interests as then construed. In the field of more conventional astronomical work, Shapley had in previous decades maintained his continuing stewardship of the Harvard College Observatory, directing the building of the university's graduate program in astronomy and sustaining prolonged collaborative efforts, including seminal work on the Magellanic Clouds.[27] The political concomitants of this work in the postwar period saw him called before and blisteringly confront the House Un-American Activities Committee in 1946, investigated by the government for his participation in internationalist and putative Communist agencies and by 1950 one on the list of Joseph McCarthy's proclaimed communists.[28] These forms of intervention were in stark contrast to the governmental consulting work of the Gamow network and to the contemporaneous works of Gamow individually, even as Shapley and Gamow began to invoke each other and to correspond on the question of the origin and structure of the universe, while taking advantage of evolutionary humanisms tied to the evolutionary synthesis.

Regardless of the deep differences across varied efforts at disciplinary and cultural synthesis, and even including much of the criticism these syntheses invited, ethics, humanism, evolution, and science were all part of a putative universalizing narrative fabric, a narrative that, because it involved the universe as a whole and the quest for knowledge about it, tended to be cast as an epic. These elements of science and humanism were inseparable even as their division was being lamented and would come to be lamented still more. A cultural and political space was being established for a very practical mythology, one that allowed scientists to advise governments, to address questions of education, values, and cultural difference, to theorize on the origins of the world and life, and to work on applied problems as typified by atomic research, bombing strategies, and the effects of radiation.

BOMBS, BIOPHYSICS, AND ATOMIC COSMOGONIES: A BIG BANG

The overt connections between bomb physics and universal history became still more intertwined in the postwar period and gave greater standing to the figure of the sagelike scientist. Cosmological science and bombing technology emerged in parallel, dependent on each other for theoretical and empirical advances. As the formation of the original elements came together with the understanding of fission and fusion, so, too, did an understanding of what fuels the stars and, in turn, their origins and those of the elements and of living matter. Bomb theorists and strategists, as well as biologically trained scientists theorizing and observing the effect of war and of nuclear radiation, made the proclamations of a Shapley, a Bronowski, a Bernal, or a Gamow carry weight, despite their differences in background, training, and research.

A Prologue and Three Acts

Late in 1947, for the Guthrie Lecture at Oxford, J. D. Bernal attempted to outline a possible history of the world, a story of the emergence of the organic out of the inorganic. He began by apologizing for his subject, one that was distant from the work of the physicist Frederick Guthrie for whom the series was named: "physics has changed much since his time, and even then the foundations were being laid for its transition towards biology."[29] In the published version of the lecture, he noted how his title had been an "unconscious echo" of T. H. Huxley's 1868 lecture "On the Physical Basis of Life." But as Bernal saw it, the elder Huxley, "great evolutionist as he was," had nevertheless never attempted a historical account of the unity of life. It was only on this question of origins itself that Bernal felt a true departure from Huxley was possible.

Like Shapley, Bernal regarded the question of origins as essentially multi-disciplinary and collective: "It is probable that even a formulation of this problem is beyond the reach of any one scientist, for such a scientist would have to be at the same time a competent mathematician, physicist, and ex-perienced organic chemist, he should have a very extensive knowledge of geology, geophysics and geochemistry and, besides all this, be absolutely at home in all biological disciplines." It was a project the internal politics of which itself reflected the social-planning initiatives Bernal saw as possible and necessary: "Sooner or later this task will have to be given to groups rep-resenting all these faculties and working closely together theoretically as well as experimentally."[30] In a viewpoint similar to that of the physicist Edwin Schrödinger in his account of life, on whose ideas of "negative entropy" Ber-nal drew, and in words unknowingly paraphrasing Shapley's claims for a never-realized cosmogony institute, Bernal believed that understanding the nature of life must call upon the whole of knowledge.

But he believed there was still resistance to be overcome: "Until recently discussions on the origins of systems were in some way improper to sci-ence, but now even in physics itself the questions of origins are coming into discussion, as, for example, those of the nebulae, the solar system and the elements themselves." These were, of course, the elements of older histori-cal syntheses, but ones that over the preceding decades also had marked the work of Eddington, Lemaître, Shapley, Gamow, and others. For his own fo-cus on the material origins of life, Bernal was able to draw on a number of synthesizing authors, including the earlier work of Huxley, the physiolo-gist Lawrence Joseph Henderson, V. M. Goldschmidt, Haldane, and Oparin. To fill out his sketch more fully, he addressed the natural philosophical writings of Engels, asserting a historical progression inflected by Marxist historiographies.

A vivid historical synthesis emerged, one that he represented as a play of three acts with a long prologue. The prologue was all that preceded life on the earth, a part written by those physicists and geochemists such as Goldschmidt dealing with the origins and distributions of the elements. As for the body of the play, "The first act deals with the accumulation of chemical substances and the appearance of a stable process of conversion between them, which we call life; the second with the almost equally important stabilization of that process and its freeing from energy dependence on anything but sunlight. It is a stage of further synthesis and of the appearance of molecular oxygen and respiration. The third act is that of the development of specific organisms, cells, animals and plants from these beginnings."[31] Bernal proceeded briefly to sketch the play, expressly leaning on the researchers who preceded him, summarizing a drama of the world where the stage and its actors developed together.

It was a sketch he was sure would someday be regarded as absurd. But he nevertheless argued that his survey and the synthesis it suggested constituted progress, progress that was not overly mechanistic. As with other left-leaning scientific thinkers such as Needham and Haldane, Bernal treated social and cultural evolution as continuous with the physical evolution that preceded them. As with Engels, these different historical epochs were not simply reducible to one another: the emergent laws of a biological epoch could not be reduced to the physics that preceded life historically and logically. Given the unique emergent properties of individual epochs, an account of origins was itself split between the immanent possibilities of matter and their actual, contingent articulation, their "cosmic history."

Despite the dependence of the account on geophysics, geochemistry, spectroscopic analysis of stars, and physics more generally, Bernal therefore rejected a foundationalist, reductionist model. The biophysics that he saw as marking a turn in his discipline "does not mean any reduction of biology to simple physics. The complexity of the successive evolutionary levels mark definite orders of advance each with its own laws and each with its own internal unbalance leading forward to new levels."[32] The progress in understanding was marked "by elucidating structures, by following out mechanisms," methods by which "we may at least clear the air of mystical interpretations which serve simply to conceal ignorance."[33] In keeping with the nineteenth-century natural and social scientific tradition, Bernal saw the successive evolutionary levels as themselves embracing the culture of his own period, the final act of his play. What the maturation of the sciences still promised was a fuller orientation of humanity and history: "Only an understanding of the biological and social sciences can show how to make those changes of real benefit to man, and how to balance power with intelligence."[34]

Bernal had strongly tied origins of life to the origins of matter in order to justify and to endorse research into the latter, to concretize the turn to biophysics. He cited Darwin in correspondence with Hooker tracing a similar connection, but to the very opposite effect: "It is mere rubbish thinking at present of the origin of life; one might as well think of the origin of matter." As Bernal saw it, "We are now almost in a position to take him [Darwin] at his word. For the origin of matter, at least the origin of the forms in which matter presents itself on this earth, is at last becoming clear."[35] Though he offered no specific citations in this regard, the scientific results Bernal had in mind were clear: "The chemical elements, in their ordered multiplicity, and with their apparently arbitrary and enormous variations of abundance, are seen now as perfectly logical products of a process going on in some primordial hyperstar or concentrated universe some four billion years ago."[36] Attempts to understand the origins of matter were not rubbish thinking, and neither, Bernal suggested, were attempts to understand the origins of life.

Fusion versus Fission: The Prologue in Several Acts

In drawing the connection between the origins of matter and origins of life, Bernal suggested an awareness of Lemaître's earlier research. But he also noted that in brief contemporaneous debates over cosmogony centering on the differentiation of matter itself, "the physicists have reconstructed, somewhat provisionally, the temperatures and pressures of that time from the relative abundances of different elements, and in the process have discovered anomalies which would not have been without some such hypothesis as to origin."[37] These physicists were at work writing Bernal's prologue on the origins and distributions of the elements in greater detail. The question still remained of how to derive the current distribution from some original distribution or state, and to what degree that historical process was explained by primarily fission-based or fusion-based models.

For Gamow and his collaborator and student Ralph Alpher, one obstacle to refining and testing the theories they were forming was finding the appropriate data for carrying out potential cosmogonic computations. This would require information about the chances neutrons had of colliding with the nucleus of a given atom and being absorbed by it, the atom's so-called "capture cross-section." Since this information had been crucial to the Manhattan Project, it was not easily available at the time.[38] A first batch of data was supplied through the publication of work by nuclear physicist Donald Hughes, another Naval Ordnance Laboratory graduate, who was studying properties of materials that might be useful in nuclear reactors. What Hughes's measurements allowed was a computation of the processes Gamow had speculated at the end of his 1946 paper, the weighing of the neutron decay rates against the capture of neutrons by other atoms.[39] These considerations must have formed no small part of what Bernal had in mind when referring to the provisional reconstructions of physicists.

The computational and theoretical work done in collaboration with Alpher was published in a famous 1948 paper, "The Origin of Chemical Elements." As a letter in the *Physical Review*, Gamow and Alpher published what was thereafter and continues to be called the Alpher-Bethe-Gamow paper. Bethe had written no part of it, but was introduced (to Alpher's initial annoyance) by Gamow for "alphabetical" reasons. An expanded version of the paper was published in *Nature* later that same year. The original letter is little more than a page, the article in *Nature* a little more than three pages.[40]

In them, Alpher and Gamow opted for a fusion-based cosmogony in which they speculated on the formation of all elements, light and heavy. Referring to the earlier 1946 "Origin of Chemical Elements," Gamow and Alpher remarked in the 1948 article that "as pointed out by one of us, various nuclear species must have originated not as the result of an equilibrium corresponding to a certain temperature and density, but rather as a consequence of a continuous building-up process arrested by a rapid expansion and cooling of the primordial matter."[41]

In a *Physical Review* paper of that same year, Alpher coined the term "Ylem" pronounced "I-lem": "Very shortly after the beginning of the universal expansion, the ylem was a gas of neutrons only. These neutrons began to decay into protons and electrons." A footnote explains: "According to Webster's New International Dictionary, 2nd Ed, the word 'ylem' is an obsolete noun meaning 'The primordial substance from which the elements were formed.' It seems highly desirable that a word of so appropriate a meaning be resurrected."[42] And so it was resurrected, even if briefly.

In Alpher and Gamow's picture, the neutrons spontaneously decay into protons and electrons, and the protons collide with electrons and decay into neutrons. A rough balance is preserved. With the expansion of the universe, the temperature and density fall, and the collisions between protons and electrons aren't energetic enough for the fusion into a neutron. The Ylem will have few free neutrons, since nothing compensates for their spontaneous decay. At this lower temperature, atoms can and will form having high ratios of neutrons to protons. By further decay of the captured neutrons, atoms with additional numbers of protons are formed, along with freed electrons.[43] The heavier elements emerge.[44]

In 1949, however, the physicist (and later Nobel Prize winner) Maria Goeppert Mayer and Edward Teller together published a paper itself called "On the Origin of the Elements" in *Physical Review*. In it, they "explore the hypothesis that the heavy nuclei were formed by the disintegration of a 'cold' nuclear fluid containing a great excess of neutrons"—as opposed to the "overheated nuclear fluid" that Alpher and Gamow postulated as the initial stage of matter in their account. Mayer and Teller noted that "the breakup of a primordial nuclear fluid is in many respects similar to the fission process."[45]

The research was based on materials they had presented the year before to the 1948 Solvay Conference, research that debated the possibility of accounting for the production of the heavier and lighter elements together. They observed the relative difference in the ratios not only of lighter to heavier elements, but also of the isotopes of those elements. Whereas Mayer and Teller conceded that the lighter elements "may have been formed by thermonuclear reactions," they disallowed this in the case of heavier elements. Instead, they observed, "There is conclusive evidence that at the time of production of the heavy nuclei, the proportion of neutrons considerably exceeded that which is now present in nuclei."[46]

Ultimately, they focused on what they termed "the polyneutron model": "In order to obtain a model of a nucleus which can serve as the starting point for the break-up process we shall assume that an assembly of neutrons forms a nuclear fluid which will not spontaneously disintegrate into neutrons. The only limitation imposed on the size of this polyneutron is that its total mass should not exceed the mass of a star."[47] This is a limitation that could seem onerous only in the light of the primeval atom theory of Lemaître, upon which this theory indeed follows. Though they did not refer to Lemaître and could perhaps even posit the nuclear fluid as an initial condition of the world, Gamow clearly saw their work as advancing Lemaître's earlier hypothesis.[48]

In each instance, whether fission-based or fusion-based theory, the initial state of consideration was (as with Lemaître) not taken to be the start of time itself—or rather, the question wasn't even raised in the technical materials. Indeed, as Gamow made clear in popular work on his fusion-based theory, the time before the initial state might be infinite. If the abundance curve was, for Gamow, the "first document" concerning the history of the universe, then the time prior to the initial state out of which it was composed was a prehistory that might be endless. In his unpublished words, Lemaître had, for reasons of faith, welcomed "how present physics provide a veil hiding the creation." Lemaître's veil stood here, too, before the fission or fusion explosions that heralded the contemporary cosmological state of nature and that had disturbed the contemporary political state so strongly. In contrast to Bronowski or Bernal, the thematization of recent conflicts was more muted, even as the reliance on physical data from the war structured the research program.

Lemaître's veil, however, had become a curtain. At the end of January 1949, Alpher and the physicist Robert C. Herman presented a lecture on nucleosynthesis—their atom-building theory—to the American Physical Society at Columbia University. News of the lecture appeared the next day in an article in the *New York Times*, the final section of which is entitled "The Curtain Rises." Citing the lecture, the author reported that Alpher and Herman asserted that "no scientific data of any kind are available whatsoever on what went on 'behind the cosmic curtain' before it 'rose' on our present expanding cosmos three billion years ago." Thereafter, the article explicitly divides the history of the universe into a series of seven scenes, from the curtain rising to the present day, in a cosmic drama the first scenes of which are "strikingly similar in some respects to the story of Creation as told in Genesis."[49]

NUCLEAR AND RELATIVISTIC ORIGINS

It was during the period of this deliberation over the original state of the universe that both the hydrogen bomb effort and the fusion model were reaching maturity. The production of a universal history and the technical research spurred by the nuclear arms race remained entangled, partly as a matter of their joint studies in nuclear physics, partly as a matter of the researchers who authored these studies. Thus Gamow in the early 1950s explained that in the early period of expansion and "nuclear cooking"/nucleosynthesis "conditions throughout the universe closely approximated those existing in the center of an exploding atomic bomb."[50] And in addition to the bomb research of Gamow and Teller, Mayer had worked with Urey on U-235 separation in the Manhattan Project's Substitute Alloy Materials Lab and, from 1945, with Teller in relation to radiation from uranium explosives (a research context he apparently could not disclose to her), first at Columbia University and then at the University of Chicago's Metallurgical Laboratory.[51]

At about the time of his publication of the massive polyneutron paper with Mayer, Teller had relocated for a time to Los Alamos, having sustained his relationship with the laboratory there following the war. The September discovery of the Soviet Union's successful test of an atom bomb in 1949 made the debate on the question of a hydrogen bomb more pressing, and soon an effort toward its production had begun. Physicists such as John Wheeler and (less actively, according to Teller) John von Neumann and Enrico Fermi also joined the effort, as did Gamow.[52]

In 1949 and 1950, Alpher and Herman as a team, and Gamow separately, each published overviews of research in the field on the question of the origin of chemical elements. For the group around Gamow, a robust history of the universe across a set of events that includes initially the formation of elements and galaxies and the principles of their formation (neutron capture, beta decays, and the earlier principles of gravitational condensation and energy dissipation) was also a technical research program. One epic ratio of universal time to a short list of important events would have to embrace the work of many different disciplines.

In July of 1949, Gamow published a piece entitled "On Relativistic Cosmogony" in *Reviews of Modern Physics*, an overview that appeared explicitly as a defense of relativistic cosmogony, over and above different possible modifications of that theory that Gamow chose not to explicate. Establishing his historical account as "an attempt to understand the development of various characteristic features of our universe as the result of its expansion from the originally homogeneous state," he also presented this history as a research agenda, highlighting the role of general relativity in determining the further connections between element formation and the formation of galaxies and stars.[53]

Alpher and Herman also offered their own overview of the current state of analysis on the question of element formation in an April 1950 *Reviews of Modern Physics* article, "Theory of the Origin and Relative Abundance Distribution of the Elements," where they find previous overviews of the field insufficient, including Gamow's. They begin the article with an epigraph, a quote from Ben Jonson's *The Alchemist*: "Ay, for 'twere absurd, To think that nature in the earth bred gold, Perfect i' the instant: something before. There must be remote matter." They weigh the strengths and weaknesses of three dominant historical alternatives, each positing a different possible history for the world, only one of which they embrace as the dominant account: a "frozen-in" equilibrium of elements, Mayer and Teller's polyneutron fission theory,[54] and their own fusion-based neutron-capture theory. They mention as well "matter creation" theories, more commonly known thereafter as steady-state theories. But to these, they give only passing attention.[55]

The equilibrium theory of element production remained inadequate by the lights of their own research program, failing to reproduce the abundance curve with any accuracy. But their problem with Mayer and Teller's theory and with steady-state ideas was more an aesthetic one, more a matter of taste than a calculation of improbability: "The polyneutron fission theory has had considerable success in predicting relative isotopic abundances," but "with this theory and with theories of 'matter creation' [the steady-state theories] the supposed initial conditions seem improbable to us at the present time."[56] An initial state beginning with a "hot, nuclear gas" (the ylem) was more likely to their minds than one beginning with an enormous homogenous object fragmenting into small parts—and presumably more likely or compelling than an ongoing creation of matter.

It was against the background of these disagreements that in a catalyzing set of broadcast popular lectures given in 1949 and adapted into a text the following year, the English astrophysicist Fred Hoyle then coined the term "Big Bang," not as a descriptive appellation, however, but as a pejorative characterization of the work of the Gamow network. He contrasted the "Big Bang" with his own steady-state theory, which required the continuous creation of matter everywhere in the universe, a theory on which he collaborated with the mathematician and cosmologist Sir Hermann Bondi and the Austrian-born astrophysicist Thomas Gold. Gamow himself disavowed the name that Hoyle had given his theory and, given Gamow's preferences, the universal history that the Big Bang came to represent might have been understood differently. Its acceptance was itself the product of further efforts to achieve a synthetic account of the origins of the universe linked to the explosiveness of a nuclear age.

BIG BANG VERSUS BIG SQUEEZE

Fred Hoyle's biographers claim that the idea for his own theory came from a conversation among Hoyle, Bondi, and Gold inspired by and directly following the viewing of the 1946 English horror film *The Dead of Night*. The plot is a circular one in which the beginning wraps around to meet the end, but apart from this comparison of an endless structure, the relationship between the film and the theory seems strained.[57] Regardless, Hoyle's point of view asserted a uniformitarian immodesty: the idea that science must presume principles that are the same everywhere over time and space. He defended this view over and above those of the Gamow network (who go unnamed) by observing that in the evolutionary account, "the whole of the matter in the Universe was created in one big bang at a particular time in the remote past. On scientific grounds this big bang assumption is much the less palatable of the two. For it is an irrational process that cannot be described in scientific terms." Here, Hoyle could be read as targeting Lemaître's "beginning before time itself" that forever remained outside of any possible account. But this was not true of Hoyle's theory: "Continuous creation, on the other hand, can be represented by precise mathematical equations whose consequences can be worked out and compared with observation. On philosophical grounds too I cannot see any good reason for preferring the big bang idea. Indeed it seems to me in the philosophical sense to be a distinctly unsatisfactory notion, since it puts the basic assumption out of sight where it can never be challenged by a direct appeal to observation."[58] For a public audience that would include scientists, Hoyle attempted to naturalize continuous creation and to render the "Big Bang" a product of faith.

Gamow's response to the "Big Bang" was what he termed the "Big Squeeze," answering the criticisms directed at him by Hoyle by giving a fuller, generalist exposition of his theory and undercutting the advantages Hoyle believed the steady-state had over his Big Squeeze. As far as I can tell, the term was not one that Gamow coined for or used in the technical literature. Instead, it appears in his 1952 book *The Creation of the Universe*, where, as with his previous research on stars, Gamow took current cosmological ideas into the generalist arena in a popular text. In it, the technical and the popular narratives merge, and the historical account Gamow formed produces a synthesis with implications for his more technical work.

In the prefatory remarks, Gamow identifies *The Creation of the Universe* as the third and therefore consummating chapter of a cosmological trilogy, "being a sequel to *The Birth and Death of the Sun* and *Biography of the Earth*. While the first volume dealt essentially with the place of our sun among the stars, and the second with the place of our earth among the planets, this book is concerned mainly with the universe as a whole."[59] There is little either in the construction of his two previous books or in his papers to suggest that Gamow had in mind the trilogy prior to 1952. It seems more likely that this was a retrospective assessment, one endorsed by the increasing possibility and success of his historical agenda. The first printing was in April, and the book was already in second printing by September. It also helped promote cultural products that took as their object the "universe as a whole." It is clear that as he was writing the texts in succession, he saw each in rela-tion to the last as completing some part of the tale or emphasizing another part of it, where that tale was as complete a history of the universe as could be narrated by the cues of contemporary science. Moreover, his retrospective christening of them as a trilogy suggests the affirmation of the belief that he had completed a story.

In writing his history of the world, Gamow was not looking back to his "Origin of Elements" series alone, but to his work on stellar formations, as well as to that of his predecessors and collaborators. The continuous and concise historical picture that emerged in the text is that after the period when radiation density dominated the density of matter in the universe, the uniform gaseous material that filled it could begin to condense under the action of gravity. Individual clouds of gas could then form, what Gamow referred to as "protogalaxies." Gaseous eddies would develop at every scale, leading to the formation of galaxies and stars, and of stars within the galax-ies, to increasingly formed developments, and onward to the final words of the conclusion of the book: "... it took less than a hour to make atoms, a few hundred million years to make the stars and planets, but three billion years to make man!"[60] The prologue saw the major acts of the drama.

Like Bernal, Gamow was not merely summarizing past texts. He enacted a theoretical synthesis in his generalist work and in a more specialist con-text at roughly the same time. So a kind of synthesis through turbulence—through dimensional scales, as the scientific idea of turbulence suggested—seems to have been composed by him in the generalist text prior to or along with journal publication. Gamow used values of density and mass of the ylem (or, just more generally, "an originally uniform gas") and concluded that "since the values for density and temperature used in our calculations were derived on the basis of purely nuclear data, we may have here a bridge (or rather a viaduct) uniting the microcosmos of nuclear particles with the macrocosmos of stellar systems!" Nested in this evolutionary synthesis is a synthesis of scales predicated on the data provided by nuclear research. And in a February 1952 letter to the *Physical Review*, "The Role of Turbulence in the Evolution of the Universe," he posits galactic formation as linked to a possible primordial turbulence. The letter appeared contemporaneously with his appeal to the idea of turbulence as a bridge, an exposition permit-ted in a popular text.[61]

Gamow chose the *Physical Review* letter in which to cite from an address of Pope Pius XII to the Pontifical Academy of Sciences months before to make the mischievous point that the cosmological theories he defended "can be considered now as an unquestionable truth."[62] But in *The Creation of the Universe*, Gamow made a more sober appeal to St. Augustine in order to explicate a more self-consciously periodized part of that cosmological history, asserting that "it was St. Augustine of Hippo who first raised the question as to 'what God was doing before He made heaven and earth.'"[63] Augustine was perhaps a figure inherited from Friedmann, who earlier had referred to him in the context of modern scientific conceptions of time. Gamow calls the time before element formation the "St. Augustine era"—an era behind Lemaître's veil (behind which stands not God, but another material order) and behind Alpher and Herman's cosmic curtain. What takes place before this era is perhaps unknowable, but in this more generalist setting, Gamow allowed himself to speculate, asserting a belief that perhaps he himself would not regard as a physical theory. He termed the "creation" the "Big Squeeze" as a consequence of formal considerations: "We can now ask ourselves two important questions: why was our universe in such a highly compressed state, and why did it start expanding? The simplest, and mathematically most consistent, way of answering these questions would be to say that *the Big Squeeze which took place in the early history of our universe was the result of a collapse which took place at a still earlier era, and that the present expansion is simply an 'elastic' rebound which started as soon as the maximum permissible squeezing density was reached.*"[64]

In one sense, this is an unambiguous beginning: expansion from the highly compressed state outward. In another sense, it is clear that Gamow saw the history of the universe as consistent with a wider narrative that contained an unknowable prehistory.[65] The previous traces of earlier worlds may indeed have vanished entirely: "Most likely, the masses of the universe were squeezed together to such an extent that any structural features which may have been existing during the 'pre-collapse era' were completely obliterated, and even the atoms and their nuclei were broken up into the elementary particles (protons, neutrons, and electrons) from which they are built."[66] Still more assertively: "Thus nothing can be said about the pre-squeeze era of the universe," a point reasserted in Gamow's concluding summary.[67]

Gamow thus claims that there is the time that exists in the accessible universe and also another temporality intimated by logical priority, of events in some sense outside of accessible time that nevertheless may precede the effective beginning of the world. Gamow's account says almost nothing else of St. Augustine's Era except what has here been cited, just as he would lead us to expect from his observations. Anything physically interesting that could be said could be said only after the universe had begun to expand following the "latest" squeeze event. Gamow's Big Squeeze theorizes the beginning of a story that can be told, but unlike Hoyle's "Big Bang" rendering, it is not necessarily *the* beginning. The concrete story Gamow believed he could tell merely has an effective beginning, the placeholder that is St. Augustine's Era giving way to the time when the atom-building processes are set in motion and events can be told.[68]

In neither the steady state nor the Big Squeeze theory was there a simple beginning of things, as opposed to other popular conceptions of the Big Bang. The mathematical term that Hoyle used to modify Einstein's formulation in order to allow for the constant creation of matter—his "creation tensor"—did not so much explain the creation of new matter as posit it. And both theories produced physical predictions. But whereas Hoyle's theory precluded any beginning dating and giving sense to an age of the universe, Gamow's Big Squeeze was consistent with a story either with or without such a beginning. St. Augustine's Era "could have lasted from the minus infinity of time to about three billion years ago."[69]

It was in the contestation of the Big Squeeze by steady-state theorists that a more thorough picture of the Big Bang ultimately emerged. Whereas Gamow looked to the creation of all the elements of the universe in an early, exotic state, Hoyle and his confreres Gold and Bondi looked instead to stellar processes. And these explanations had opposite virtues. Gamow could account for the production of light elements by his one primordial oven, Hoyle for the heavier elements through his stellar one. But it was only through the combination of such explanations, with the framework of Gamow's theory having the clear advantage of being able to subsume the others, that a convincing theory emerged. The final story that did ultimately hold sway was not Gamow's, nor was it Hoyle's. In technical and popular journals and in many reviews in the period and thereafter, the Big Squeeze was forgotten, and the steady-state theory was rejected for the Big Bang and the synthesis it provided.

However, the extent to which Gamow doesn't affirm that the Big Squeeze is *the* beginning reveals how the Big Bang might have been understood differently than it is now. It might have been theorized and popularized just as Gamow elaborated in *The Creation of the Universe*, as a narrative that effectively but only doubtfully in truth begins with some first, explosive event. In fact, if anything, Gamow was pressing for this latter view: an empirically unreachable ("metaphysical") era, intelligible but not perceptible, that came in some sense—logical, temporal, or both—before the explosion, the first empirically accessible event.[70] But Hoyle's "bang" resonated with the explosive technologies then tied to scientific achievement and authority and played on the relationship between cosmological and (thermo)nuclear research.

The adoption of the Big Bang story instead of the Big Squeeze was a measure of how the story could very well have been understood as a different kind of history. So, for example, the term "prehistory," understood as the period before the Big Bang, might have been adopted more widely, and as suggested by his appeal to "archaeological records" here and still more extensively elsewhere, it was an idea put forward by Gamow as a potential conceptual resource.[71] But as Hoyle's ire makes clear, it is a knowledge-limiting metaphor, reasserting the unknowable, that does not appear much taken up apart from Gamow's use. Gamow's account demanded a certain epistemological modesty, an allowance for unanswerable questions, and its eclipse by the Big Bang theory shows how unappealing in principle such modesty might have been to contemporaries at the time. The notion that knowledge has its limits, that scientists pronounced "Ignoramus," did not accord with the contemporary image of science or of the scientist as privileged knower.

THE PLACE OF HUMAN LIFE
IN AN EXPANDING UNIVERSE

"What, then, in your opinion is the value of natural science?" Erwin Schrödinger imagined someone asking him in a series of lectures in 1950 entitled "Science as a Constituent of Humanism," "I answer: Its scope, aim and value is the same as that of any other branch of human knowledge. Nay, none of them alone, only the union of all of them, has any scope or value at all, and that is simply enough described, Γνῶθι σεαυτόν, get to know yourself."[72] Schrödinger, along with other architects of scientific humanisms and with theorists of evolutionary histories we have examined, argued that science writ large addresses ethics, if not on narrow questions of quantum mechanics and causality, then partly by assembling historical and other universalizing schemas allowing humanity to know itself—and destroy itself—more fully.

The text of Schrödinger's lectures was published the following year as *Science and Humanism*, and it was one of the books that Harlow Shapley recommended in a new university course he offered in 1952 called Cosmography—An Approach to an Orientation. It was based on the proposed new discipline of cosmography, a mapping of the universe according to modern coordinates,[73] and in it Shapley endorsed Schrödinger's synthetic vision of knowledge. Through the Harvard General Education course, Shapley integrated questions of cosmogony and anthropogeny into the curriculum, an experiment in educational reform he had been advocating for some years prior. The course's motivating questions were largely shared with those that had originally motivated his planned cosmogonic institute, but they extended further to philosophical, literary, and religious connections.[74] It was a course that attracted large audiences, attention from other institutes, and more popular notice—even though cosmography did not in his day emerge as the new synthesizing discipline he hoped it might become.[75]

Measured by its cultural products, the cosmography course was nevertheless successful. Among the succeeding generations of scholars and instructors, George B. Field and Eric J. Chaisson helped produce some of the most widely read contemporary accounts of "cosmic evolution."[76] The cosmography course also may have been behind Shapley and Gamow establishing a correspondence in 1952. In November of that year, Gamow gave a guest lecture in the course, intended to be an elementary presentation on "the origin of the elements." However, Shapley added: "beyond that slight indication, I make no predictions or restrictions. If the universe evolves along with the elements, and if the galaxies get created—that will be fine."[77] This was an invitation that Gamow must have taken up in a spirited way.

The notion of scales played a role in the historical narratives developed by both Shapley and Gamow. In the course, Shapley offered different maps or alphabets to the world, taking matter, energy, time, and space as basic entities. So, for example, the Periodic Table is "an alphabet for matter," but it is Shapley's own extended classificatory, scalar effort, the "Table of Material Organizations," that is an "alphabet for the universe." In that sense, the alphabets listed synthetic options, including the historical, and the course treated the scalar as the most expansive alphabet.

Shapley's best-known major publication effort, *Of Stars and Men: Human Response to an Expanding Universe* (1958)[78] took up this alphabetical structuring, emphasizing its implications for the orientation of humanity. By this point, Shapley added an important qualification underscoring the longstanding metaphysical implications of his work, the inconsequential position of humanity: "In one respect the animals and plants excel the stars. In the complexity of their molecules and molecular aggregates, living organisms transcend the atomic combinations of the inanimate world."[79] But it is clear that it was Shapley's intent to undercut this flattery by demonstrating the plurality of life in the universe. In this extended context, in a manner similar to Bernal, he advocated for Haldane's and Oparin's work on the origins of life, framing their ideas in terms of his historiographic concept of metaphysical adjustments. Thus he circulated more broadly his claim that every scalar physical displacement of the center provokes a metaphysical displacement of and adjustment in humanity's self-understanding and worldview.

As with Gamow's cosmological trilogy, Shapley's *Of Stars and Men* can be regarded as a case of a popular text revealing and constituting the consummation of a line of work.[80] In the book's attention to evolutionary history broadly conceived, including the emergence of civilizations, it helps establish major structural elements to future instances of a narrative of cosmic evolution, the scientific epic. Its own cultural impact can partly be gleaned from John and Faith Hubley's 1961 *Of Stars and Men*, narrated by Shapley, itself followed by a short film on technology and the environment, in thematic dialogue with the first underscored by the title *Of Men and Demons* (1968).[81] Faith Hubley observed that she and her husband (who himself had been blacklisted for refusing to cooperate with the House Committee on Un-American Activities) "had wanted to make a film reflecting on "the marriage of science and art. Then we read this book by Harlow Shapley, *Of Stars and Men*."[82] The Hubleys saw the book and their film as having an immediate political message: that "to stop destroying the planet and get rid of weapons, then we really had to work on a new development [the next stage of evolution] and he [Shapley] went through a kind of history of western science to make that statement."

In the book, in line with his cosmography course, Shapley expanded on his own modes of classification, the alphabets for time, space, energy, and matter. On alphabets in general, made prominent in the visual strategy of the Hubley adaptation, Shapley stresses the nature of scientific organization and knowledge together—grammars of the natural and social world. "Since their invention or emergence, alphabets have enabled men to coordinate better their knowledge and ideas, and to comprehend many phases of the surrounding complex world. They have served to reduce the seeming chaos and to lay the foundation for civilized cooperation among individuals and groups. The alphabets have also enabled men to advance their cultures and to build stable societies."[83] His alphabets were also a key to social harmony. It is the new knowledge of cosmography that demands the introduction of new alphabets to order both knowledge and knowledge makers.[84]

Unlike efforts to purify science of its metaphysical confusions and reveal its coherence, Shapley's scientific alphabets are presented as continuous with past endeavor, as already sufficient for their orienting tasks, linked in varying measure to historical, foundational, and particularly scalar syntheses. Two of these alphabets he understood as already established, a historical alphabet (geological ages) and a material one (the periodic table of elements).[85] In the remaining two alphabets, Shapley advanced his own innovations. The first was the "ether spectrum" or radiation sequence constructed *as* an alphabet and is the least developed grammar he offers.[86] The second is the scalar/spatial alphabet, or the "organizations of nature." This was his now long-standing classification of systems of material organization.

Gamow, too, continued to exploit the possibilities of scales—so much part of the turbulent syntheses envisioned in *The Creation of the Universe*—sending Tompkins inside himself in *Mr. Tompkins Learns the Facts of Life* in 1953 to explore the human body and its workings, extending the picaresque synthetic possibilities to biology. And this interest in the fundamentals of life returns conceptually and "genetically" to an interest in a language of nature, of its alphabets and grammars, a broad, mid-century concern apart from Shapley or colleagues such as Frank.[87]

The exploitation of scales and the appeal to alphabets is evident in Gamow's motivating work with the "RNA-Tie Club," a group of prominent scientists convened by him in the attempt to understand the relation of DNA or RNA to protein building. The governing discourse, according to Lily Kay, was shaped by Gamow, understanding the problem as one of translating or encoding DNA information expressed in an alphabet of four letters (the nucleotides) into a larger alphabet of twenty letters (the basic amino acids). How is the genetic information encoded in DNA translated through RNA to the building up of proteins, and then in turn to the functions of the cell? To attempt to answer this, Gamow organized the "RNA-Tie Club" to try to work out this scheme carefully, with twenty-four members in all: twenty regular members for the number of amino acids, and four additional honorary members, one for each nucleotide in a nucleic acid. Retrospectively, the cast of the twenty regular members conveys something of a natural scientific pantheon, including James Watson and Francis Crick, Richard Feynman, Max Delbrück, Erwin Chargaff, and Gamow's old friend Edward Teller.

Gamow had initiated contact with Watson and Crick in a letter of July 1953, invoking Tompkins in the very first line: "Dear Drs. Watson and Crick, I am a physicist, not a biologist, and my interest in biology can be justified, if anything, only by my recently published book 'Mr Tompkins Learns the Facts of Life' (Cambr. Univ. Press. 1953). But I am very much excited by your article in May 30th Nature, and think that this brings Biology over into the group of 'exact sciences.'"[88] Through an apparently random process, the scientists received their appellations: Yčas's was Cysteine (Cys) and Gamow's Alanine (Ala).[89] Given Gamow's role in forming the club, he referred to himself as "The Synthesizer" (or "Synthesiser"; spellings varied).[90] And, rounding out the account, a draft of Gamow's first paper in this regard, entitled

"Protein Synthesis by DNA Molecules," originally listed both G. Gamow and C. G. H. Tompkins as authors.[91] However, Gamow was prevailed upon to remove his coauthor prior to publication. He promoted the club's collaboration in both technical and popular work, and in attempting this alphabet of life, he explicitly embedded the club's research question in evolutionary historical context. In this way, he attached the origins of life to the origins of the material world and worked to glue together different disciplines.[92]

Gamow was installed as a professor at the University of Colorado Boulder in 1956 and he turned his hand to writing elementary and introductory textbooks on the sciences, works that again emphasized the generic synthesis of scales. His first of this series in 1958 was *Matter, Earth, and Sky*, a college-level introductory scientific textbook including such topics as "The Chemistry of Life," "The 'Book of Sediments,'" and "Cosmology." Concerning its structure, Gamow explained: "If one has a large library composed of books written by many different authors in many different languages on many different subjects and published over a period of many years, it is impossible to arrange them on the shelves in such a way that they will form a consistent sequence in all respects. This is the situation that faces any author who attempts to present as a continuous narrative any broad subject such as the entire field of the physical sciences. Thus, one has to resort to trickery and put things together in such a way that they will at least look like a continuous, uninterrupted presentation running from alpha to omega."[93] The trickery that Gamow had in mind was scalar: "The plan selected for this book is that of a trilogy dealing first with things of our size, secondly with things much smaller, and finally with things larger than ourselves."[94]

Gamow's trickery adds the complication of making time an additional element in scalar organization. That is, at different orders of spatial magnitude, different orders of existence are correlated to different characteristic durations. But this principle is not as central as spatial magnitude to the organization of subject matter. And as with the historical and its choice of central events, humanity is used as the basic measure for the account.

SCIENCE AND WISDOM

At the same time, such scalar syntheses, dependent on or linked to the materialist conceptions of Oparin, Haldane, Bernal, and others, suggested that the conditions for the emergence of life must exist elsewhere. Shapley argued against the idea that humanity is special by once again positing other, if rare, life-inducing chemical environments in the universe. For him, this making humanity modest allowed for an enlightened universal politics. His was an outlook in broad terms shared by a community of scholars who saw in science a common reference point for all humanity, with Shapley's cosmography course reflecting the work of some of those authors, such as Bronowski, Haldane, Schrödinger, and Frank. Nevertheless, critics of the period (the many who, for example, debated Shapley or Haldane) and the generations that followed, also accepting metaphysical beliefs as bound up in the physical arena of the world, saw in that history and scale the possibility of rendering the entire human condition farcical, its united enterprises as much as its conflicts. But despite Gamow's famously farcical humor and his provocative pranks, neither he nor Shapley, nor those who identified strongly with humanism such as Huxley or Bronowski, saw living in light of contemporary scientific wisdom as a travesty.

Reflecting on the implications of the atomic bomb, Bronowksi concluded that "science has nothing to be ashamed of even in the ruins of Nagasaki."[95] Scientists were so far from being to blame for actual and potential nuclear disaster that, to Bronowski's mind, they represented an exemplary community of truth seekers, uniquely able to diagnose, confront, and resolve dilemmas birthed by civilization. If there was blame to be had for the state of the world, it was on the part of any society that remained ignorant of science or didn't allow the conditions that furthered science to be cultivated.

In 1955, Bronowski, now the director of the Coal Research Establishment in Gloucester, England, took part in the century-old tradition of addressing the British Association for the Advancement of Science on the question of education.[96] Bronowski lamented the current state of general education, arguing that to be cultured and more properly oriented with respect to the modern world, a stronger understanding of science was necessary for both individual and social direction. To fail to cultivate scientific literacy among state subjects or citizens was to risk surrendering democracy to a dictatorship propped up by scientific expertise, as Hayek and Polanyi had feared. And to reject a scientific examination of the social choices was to entertain the irrational governance that Julian Huxley, for one, had indicted. Bronowski gave the "mischievous example" of Churchill, who claimed in a wartime directive to be "personally . . . quite content with the existing explosives"— evincing what Bronowski regarded an irrational attitude toward the bomb. "This bland phrase is a monument to a nonscientific education."[97] He believed that the virtues of a democracy, when constituted by an educated citizenry, are that the state would not have to be at the mercy of dictatorial and irrational preferences. Given a society "penetrated, as ours is, by technical skills and engines,"[98] facing decisions on atomic energy, genetics, and their interrelation, a general understanding of science was a way to escape the utopian or dystopian dictatorial visions Bronowski saw entertained in the works of H. G. Wells, Aldous Huxley, and George Orwell.[99] C. P. Snow was not far wrong when he observed that others such as Bronowski had been making points similar to his own, that the world was lamentably divided between the scientifically literate few and the others, and the latter were more blameworthy.

Sharing in such claims, partly based on addresses such as Bronowski's, Snow began to publish on his "two cultures" thesis the following year, culminating in his explosive Rede lecture near the end of the decade.[100] Snow's "two cultures" gave a name to the dominant and sometimes cloudy discourse for analyzing the changing divisions between the natural sciences and the humanities. Arnold had complained three-quarters of a century earlier that in the Anglophone context, the term "science" was increasingly being surrendered to the natural sciences. By the time of Snow's lecture, that war was lost.[101]

For Snow, the sine qua non of scientific literacy was the second law of thermodynamics, the law of the increase of entropy. This law remained essential to the physical history of the universe. But that new history had been set in renewed motion by the interior workings of a bomb, and its pronounced consequences, along with new theorizations of life increasingly in relation to nuclear thought and act. For later generations, however, cosmogony and natural history seemed to lose their associations with bomb physics and the strife of World War II. Instead, the history flowing out of the Big Bang became more exclusively associated with the political universalism evident in Huxley's vision of UNESCO or in Shapley's activism. This universalism was linked far more in the public eye to an ethos of peace than to one of hubris and to almost spiritual scientists, such as Shapley increasingly became and that Huxley and Bronowski already were. From the perspective of modern commentators, this shift was oddly indebted to the success of Gamowian cosmologies and the humbling reaches of Shapley astronomy: the historical and scalar syntheses they enacted appeared to take thought far from the earthly and the mundane, even as they taught new lessons for new humanisms and however much they were rooted in political calculations and conflicts. For scientific writers, researchers, and readers of a new scientific history in the immediate aftermath of the Second World War, the structure, reception, and confidence of their work were deeply intertwined with the successes of the bomb and the ethical dilemmas it created or clarified. While these ethical dilemmas remained for still later generations, however, the terms of those dilemmas were rendered in less explosive, yet still more cosmic and mythic registers. And the cosmogonies and universal histories that figures such as Gamow and Shapley continued to help institutionalize were soon rendered into mythic accounts and their narrators into larger-than-life questors after truth.

CHAPTER EIGHT

Scientific Myth and Mysticism

Here is the leading edge of panic in modern man which drags his society after him. It is his failure to make a unity of, as it were to make head or tail of, two sets of values. On one side, he is taught to take as his own a ragbag of second-hand traditions about human motives that are out of date. And on the other, he is taught to treat as gospel (a new gospel) the newsworthy speculations of scientists whose context he does not know.

JACOB BRONOWSKI, *The Identity of Man*, 1965

In 1967, Loren Eiseley edited a volume of Thomas Henry Huxley's *On a Piece of Chalk*. The publication and Eiseley's introduction to it were an affirmation that Eiseley was a cultural successor of Huxley, the "Huxley of his time."[1] Along with Huxley's markedly different counterpart Louis Agassiz, Eiseley credited Huxley with having made it possible for the "working man to enter and partake unhesitatingly in the delights and responsibilities of learning."[2] But Eiseley also used Huxley to make clear a reservation, expressed in reference to those biologists of Huxley's time who "tried to simplify the creation of life, and who unconsciously sought to combat religious superstition by reducing life's enormous mysteries to the gurgle of bath water running out of a tub." To Eiseley's mind, these reductive scientists "paid an unnecessarily high price for their knowledge of time and organic change."

Via Huxley, Eiseley was pronouncing on the hubris he thought characteristic of his own age, an age of "lightning-filled laboratories and cyclotrons." The reference is in part to the vivid image of simulated lightning filling a beaker, catalyzing reactions transforming inorganic molecules into the components of living material in Stanley Miller's successful experimental attempt to produce amino acids from nonorganic materials in conditions simulating a potential early earth atmosphere—and hence suggesting that scientists might soon be in the position to simulate the origins of life on earth.[3] However, science, he made clear, could not fully explain human experience. His own shadow—an avatar of himself and of the mysteries of life—remained "opaque to the clear rays Huxley had cast with such alert penetration into the evolutionary depths of the past." Eiseley speculated that this fact, that humanity, the world, and life cannot be entirely captured by science, explained Huxley's own declarations that "in ultimate analysis everything is incomprehensible," that the work of science is only the more modest effort "to reduce the fundamental incomprehensibilities to the smallest possible number."[4] To imagine otherwise was to obfuscate the limits of science and make false claims for the extent of its powers—to invite the construction of a myth of science.

But Eiseley used the term "myth" in two opposing senses—reprovingly, as the product of the mystification of science's limits, but also reverently, to denote the mysteries that lie beyond its reach. In an article entitled "The Secret of Life" published in *Harper's Magazine* in October 1953,[5] a year after Gamow's *Creation of the Universe*, Eiseley declared: "After having chided the theologist for his reliance on myth and miracle, science found itself in the unenviable position of having to create a mythology of its own; namely, the assumption that what, after long effort, could not be proved to take place today had, in truth, taken place in the primeval past." Eiseley was then taking stock of past failed efforts to determine the secret and origin of life. To his mind, these left science in the "half embarrassing position of having to postulate theories of living origins which it could not demonstrate."[6] In the article, later reproduced as the final essay of his popular 1957 collection *The Immense Journey*, he confessed that his characterization of the origins of life as "mythology" was "perhaps a little harsh," but the rhetorical contrition only underscored his use of the terminology of "myth" in this case to mean falsity.

However, Eiseley was chairman of the Department of Anthropology at the University of Pennsylvania and a curator of "early man" at the university museum from the late 1940s to the late 1950s, and from that disciplinary perspective, he also appealed to "myth" in a more reverential sense, drawing on Claude Lévi-Strauss, Mircea Eliade, and Joseph Campbell, among others, to oppose myths viewed positively to the myths that some science was offering contemporary culture. Unlike the myths of the human past, which had led to a respectful equilibrium of people with their environments, science, as he later expressed his claim, "as it leads men further and further from the first world they inhabited, the world we call natural, into a new and unguessed domain, is beguiling them."[7] Certain scientific disciplines may open a vista onto an unspoiled past closer to that first world, even while the logic and actuality of technology forever threatens to spoil further the lost worlds those sciences reveal and that science as a whole obscures. Eiseley was increasingly a poet of long geological times, and to him and others he drew upon, such as Rachel Carson, nature remained a storehouse of wonders. The technologically altered state of the present world opposed and threatened to desecrate the more resonant older vistas and ancestral roots imperfectly reconstructed by the more conservation-conscious disciplines of archaeology, anthropology, and geology. Scientific quests for universal truths threatened to constitute a myth leading humanity astray.[8] As a widely read essayist and commentator on scientific naturalism, Eiseley consistently critiqued what he regarded as pernicious elements of scientific ambition or ethos, elements that enjoyed increasingly popularity in contemporary culture.

And indeed, a wider receptiveness to science-based and "pseudoscientific" or mystical visions from at least the late 1950s onward, among scientists and the populace alike, helped make legible and attractive the prophetic registers adopted by some scientific popularizers and commentators.[9] As Eiseley wrote in the 1970s, "We constantly treat our scientists as soothsayers and project upon them questions involving the destiny of man over prospective millions of years."[10] With the scientifically enabled alienation of humanity from the "first world" of nature "the scientist is forced into a new and hitherto unsought role in society."[11]

There is no dearth of examples in the twentieth century of the call for science to supply new myths, overt or otherwise, understood by their proponents if not in a reverential sense of a term, then certainly in an appreciative, positive sense. To some, such as Haldane, Huxley, and figures such as Teilhard de Chardin, science was in a position to embrace prophecy. From the time of his *Daedalus*, Haldane had countered science to the scholarship of his Etonian and Oxonian education. Across the apparent gulf that Snow had yet to label, Haldane had declared that "I am absolutely convinced that science is vastly more stimulating to the imagination than are the classics but the products of this stimulus do not normally see the light because scientific men as a class are devoid of any perception of literary form."[12] But many prominent popularizers by that point, from Huxley to Huxley, Helmholtz to Shapley, had hazarded a literary voice, and the generations of scientific popularizers that followed Haldane would increasingly employ the literary forms of myth and epic.

Haldane had himself been making a point that applied equally to Bernal as much as it did to him, that "pictures of the future are myths, but myths have a very real influence in the present." Referring in the 1930s to H. G. Wells as "our greatest living mythologist," he saw it as the duty of such mythologists to disseminate the idea that humanity would evolve, a belief that Haldane thought would come to be as commonplace as the belief "in the idea of social and political progress."[13] The evolutionary myth was the literary form that Haldane hoped new mythologists would take up, whether they were literary-minded scientists or scientifically learned authors.

At the same time, myth in the more general negative sense of false belief was persistently invoked as what science worked to supplant. Bernal had expressed sentiments like Eiseley's above in another context, when he had critiqued what he saw as C. H. Waddington's scientific-ethical mythology.[14] But even from the time of his own 1929 *The World, The Flesh and The Devil: An Enquiry into the Future of the Three Enemies of the Rational Soul*, Bernal had argued that as religion is superseded by science, "the paradisiacal future of the soul fades before the Utopian future of the species"—licensing science-based speculations on that transfigured future.[15] It was for science alone to establish utopias on a mythic scale.

Others, such as P. B. Medawar and Theodosius Dobzhansky, agreed with Eiseley that prophetic science was overstated and flirted with mysticism, however much such scientists might display different attitudes toward that "extrascientific" dimension. But any of these perspectives recognized the phenomenon of science extending its claims into the realms of myth, whether understood positively or negatively, whether the scientific authors in question were accusing others of mythmaking or being accused themselves, whether they were despairing at circulation of scientific myth or instead trying to right all that was wrong in its absence.

Eiseley thus was far from the first or only scientist or science commentator to apply ambivalent notions of the mythic in relation to his or her contemporary science, notions with implications for the meaning of a scientific worldview.[16] Other spokespeople for science also distinguished between scientific mystification and myth in similar ways. Bronowski's insistence on the autonomous truth of literature and his defense of evolutionary humanism also avoided grounding conceptions of myth in science. Such critiques of scientific attempts to confront the mysteries of life and its origins targeted an increasing willingness on the part of scientists to fashion a new mythology and discerned in culture a broader receptivity to new scientific myths. But even through the essayistic individualism of Eiseley or the humanism and scientific historicism of Bronowski—itself essayistic even in the medium of television—prominent scientists regarded themselves as in position to provide new answers to what they understood as great and perennial questions.[17]

THE RHETORIC OF SCIENCE AND MYTH

Mythic vocabularies thus were lifted from varied traditions applied to mean either "grand lie" or "great truth." Often, as Bernal, for one, suggested, these vocabularies silently suffused scientific cultural discourse.[18] As science as a cultural presence assumed the character of mythology, whether in the senses of truth or untruth, science as a category also came to encompass many opposites. The different senses of myth all circulated around what science could mean—and, at times, as in the cases of Haldane and Eiseley, the different meanings of myth could all be applied to scientific work by the very same figures. The acceptance that scientific myth might find and the shape it might take were not forgone conclusions.

The possibility of a comprehensive evolutionary myth required not only individual, well-educated literary utopians, but terminologies provided by existing collectives and institutions in order to produce, reproduce, and consume that myth. Apart from Haldane or Bernal, as we have seen, Gamow, Shapley, Huxley, and their networks had invoked scriptural powers and antique figures.[19] Gamow's tone may have been as playful with religious terms as it was with scientific sobriety, but nevertheless, his scalar and historical syntheses found in the scriptural the aura of mythic scales and times. His discourse intertwined in structure and theme the terrible miracles of the atomic age, giving proof to his mythic invocations.

These discursive cultural resources formed a part of what Lily Kay has termed "scriptural technologies." Understood in the full play of their meaning, of script and scripture, these technologies were instrumental, as Kay saw it, for supplying "the potent imagery and discursive software with which that mythical object—the genetic code—became constituted."[20] As much can be said for the mythical object of a contemporary myth itself. Gamow recommended an evolutionary universal history, a myth, in Haldane's terms, that could be fit easily together with the cosmic evolution of Shapley or the historical evolutionary humanism of Huxley.

Eiseley himself was bringing those terminologies to the attention of a wider public. In the same years that he reproved theories about the origins of life as falsifying myths, he attempted to foster a reverence for universal history as a positive myth. As early as 1953, in an article entitled "Is Man Alone in Space?" he had taken up the Big Bang theory, running Gamow's and Lemaître's models together.[21] Eiseley had described a "monobloc," by which he might have been alluding to the Primeval Atom, "whose contracted density abolishes our everyday experience of space, time and matter" and in which "life is inconceivable."[22] He characterized both the alternatives to the Big Bang picture and the Big Bang itself in terms with heavy cultural associations, criticizing "recourse to the assumption that matter is somehow being created continuously in space, although this process has never been observed" by quoting approvingly "the apt words of Herbert Dingle of London

University, 'It exempts us from having to postulate a single initial miracle on condition that we admit a continuous series of miracles.'" Steady-state theories had only recently crossed the Atlantic. Helge Kragh notes that it was Fred Hoyle's popularization of them in 1950 that changed matters.[23] But even when steady-state notions had transcontinental reach, evolutionary theories enjoyed a consistent advantage. Eiseley sided with the majority, though he did so in part to critique the hope for other life in the exploration of space the myth might seem to support.

Into the later 1950s, before the Big Bang–steady-state issue was largely felt to be settled, Gamow and Hoyle remained the representatives of the opposing and unmatched sides. However, the scientific disputes that resulted remained disputes over disciplinary authority and the status of science and technology within the state and statecraft. But under the pressures of their own discourse, scientific popularizers also deliberated over the cultural preeminence of scientific truth and anxiety about the perversion of that truth by older mythologies. By the late 1950s and 1960s, reconciliation across scientific disciplines in fact seemed an increasing need *and* possibility, linked to the perceived dominance of the physical sciences and the promise of the biological sciences.

The synthetic genres turned on and helped to construct an epic narrative in the form of a final story: so much historical time made sense of through so few historical events. The Ylem, the formation of the elements, of the galaxies, stars, planets, humanity, and then outward to cosmic doom—these topics formed the major periods or moments. The event of humanity's emergence or "ascendance," whether as storyteller, cosmic visionary, or technology user (often all at once), was persistently constructed as a central event. By the early 1950s, many of the elements for this historical synthesis had been composed or announced, with the emergence of life and mind as increasingly prominent elements of that history.

As far as the material world was concerned, the contours of the account appeared to be hardening in a direction that differed only in details, as will be seen from other prominent universal histories of the time. So in 1961, Gamow did not do very much work to revise *Creation of the Universe* on the basis of research carried out over the decade since the first edition. There were few differences except in chronological scale of time between the old 1952 text and the new. Gamow's revision attested to the structure he had described in the first edition, the "somewhat hazy and fragmentary but in its general outlines quite definite" story. Five billion years were neatly substituted for the original three billion years since the St. Augustine Era.[24]

The fact that between the two editions, two billion years could be added without comment and so little in the nature of the story be changed is revealing of way in which the shape of the historical narrative, more so than its content, could determine how the story was to be told, that the story itself did not depend critically on chronological time.[25] By contrast, when a similar expansion of the scales of space had taken place in the decades before this, the conception of the universe had shifted from one of stability to one of explosive and even turbulent change. It thus is clear that Gamow considered the contours of his Big Squeeze story as largely established, regardless of the expansion of the time scale by so large a percentage of its original breadth. And in both editions, he explicitly allowed for uncertainty on the level of a few billion years while noting that all evidence pointed to a universe whose age was on that order of magnitude.[26]

But while the form of such stories increasingly seemed to be established, care was necessary in how to instantiate that form—to represent the story of the universe and humanity's place in it in a way that could be viewed as a myth in the positive sense of nurturing reverence to those truths it revealed and not criticized as a myth in the negative sense of overstating the ascertainment of those truths or their depth. To mistake how it was to be represented was to risk being seen as a lapsed scientist, one with a patently religious or mystic agenda, or by contrast, a sciolist, a deliberate or oblivious pretender to knowledge. This was part of the perceived danger in the production of a new mythology. Scientific historians shared Haldane's preferences for an evolutionary mythology, but their individual accounts and the cultural vision the accounts promoted might be starkly different. To produce putative scientific mythologies meant at the same time contesting alternative scientific mythologies. Such a creative effort meant courting the accusation not only of an irresponsible self-indulgence, but also of mistaking the task of science and the nature of the relationships between science, ethics, and society.

MYSTICAL UNIVERSAL
HISTORIES AND RESPONSES

When Faith Hubley was interviewed concerning the 1961 film adaptation of Shapley's *Of Stars and Men*, she recalled that "Dr. Shapley thought the next stage of evolution was psychic and that we had to start preparing for it." This was accurate: by that point, Shapley was arguing that humanity was entering a third stage of evolution, the stage of the "psychozoic kingdom." By this time, in fact, he was willing to consider quite a bit more: "the evidence is good that the forebrain . . . is something very remarkable in the animal world, and perhaps justifies the separate classification" of humanity.[27] On the basis of this, he confessed haltingly that "I have almost brought myself to the point of believing that man is important in the universe."[28] Any hesitation on this point was a sea change in the course of Shapley's arguments.

Shapley made a case for the singularity of the human mind. At the same time, by arguing that life elsewhere "should be similar to life here—similar in pattern and quality," he intimated that humanity could just possibly be of a more cosmic consequence.[29] Through such tentative self-understandings, "we begin to glimpse a cosmic goal for ourselves." It was a goal he saw as "angelic" or "spiritual," one that recommended three ethical principles. First, a "universal philanthropism." Second, a proper "cosmic orientation," an orientation demonstrating that the universe is "much more glorious than prophets of old reported." And third, that it was necessary in this era to be "mindful of the mind," to understand "our inner life" better.[30] With such a vision before humanity, "We see faintly a mystical light glimmering on a new horizon."[31]

His was, for all its apparent distinctiveness, a point of view shared with many, but in various hands, controversial. Shapley's was not the most "mystical" or the most "inner" of visions. For decades, Julian Huxley had been making a similar claim as a prominent element of his scientifically inflected humanism: through collective, conscious, mental effort, humanity was now in position to control evolution. Eiseley invoked a similar belief in a new age of cultural and mental evolution—itself turning on technology—dating to the ticklish figure of Wallace in the previous century.[32] Though the evolutionary synthesis was coming to an extensive consensus, not every version of it was welcome as a new orienting mythology.

The specific term Shapley chose, "psychozoic," was telling. It was linked to a broader international discourse, evidenced in the writings of Russian geochemist V. J. Vernadsky, the French paleogeologist and priest Pierre Teilhard de Chardin, and Julian Huxley. These views had had wider cultural attraction: Vernadsky and Teilhard had taken up and transformed in different, if broadly consonant, ways the nineteenth-century Viennese geologist Eduard Seuss's notion of "biosphere."[33] This work would also inspire the literary theorist Mikhail Bakhtin in crafting his notion of a dialogical sphere/logosphere and literary scholar and semiotician Juri Lotman's ideas of a "semiosphere."[34] In turn, Vernadksy had drawn not only on Seuss, but on, among others, James Dwight Dana and Joseph Le Conte, who had claimed in the nineteenth century that the evolution of living matter proceeds in a definite direction. This phenomenon had been called by Dana "cephalization"—"the brain, which has once achieved a certain level in the process of evolution, is not subject to retrogression, but only can progress further." That progression heralded what Le Conte had referred to as the "psychozoic era," the terms of which Shapley ultimately took up.[35]

Still other terms of art clustered about "the psychozoic," first among them "the noosphere." It signified the evolved and evolving sphere of mental patterning and activity, suggesting a historical progression of spheres precipitating spheres. The noosphere was said to have issued out of the biosphere, itself undergoing a process of cephalization, and out of the noosphere might issue the semiosphere, and so on to future biocultural evolutions.[36] This crisscrossing terminology of neologisms or near neologisms encouraged an identification of Vernadsky and Teilhard's ideas, even as later scholars have emphasized their independence.[37] The discourse as a whole signaled attempts to address the further conscious evolution of a meaning-making universe, the theorization of which was sometimes treated as mysticism. But as we will see, attempts to construct a holistic view of the universe and the role of mind in it were also often treated as legitimate intellectual inquiry. A stronger point of contestation was whether such an inquiry could be deemed a scientific one.[38]

Teilhard's universal history, in which the noosphere was a central target of explication, clarified these stakes. Trained as a paleontologist and geologist, he drafted his *Le phénomène humain* well before its publication, perhaps as early as in the late 1930s, though other sources date the writing to his isolation during the war years.[39] What appeared to nettle a number of his harsher critics of the late 1950s and early 1960s, when his views came to prominence, was the fact that in the very first words of his 1947 preface, he noted: "If this book is to be properly understood, it must be read not as a work on metaphysics, still less as a sort of theological essay, but purely and simply as a scientific treatise."[40] This claim was the sticking point for many responses to *The Phenomenon of Man*.

Teilhard's claim was rooted in his professional work. He had spent some twenty years in China, originally traveling there in the early 1920s as part of a geological expedition with the Muséum National d'Histoire Naturelle. Living in China from 1926 to 1946, he was connected to the discoverers of Peking Man and worked on fossils and artifacts stemming from the find. The Second World War prevented his return to Paris, where in his office as priest, his ideas met with enduring resistance by superiors. His manuscript was ultimately published the year of his death in 1955 and translated into English in 1959 with a preface by Julian Huxley.

Teilhard's work conveyed an equivocal historical synthesis, a progressive evolution toward a spiritually inflected consciousness. Science had failed to take account of the "inner" or "within of things," the facts of human self-understanding and meaning, attending instead to the "outer" or "without of things." The development of differing spheres of existence was founded on interrelated geneses: a Lemaître-inflected cosmogenesis, geogenesis, biogenesis, anthropogenesis, noogenesis, a christogenesis. These views he took to be supported by a scalar understanding of the sciences: "If there is one thing that has been clearly brought out by the latest advances of physics, it is that in our experience there are different orders of 'spheres' or 'levels' in the unity of nature, each of them distinguished by the dominance of certain factors which are imperceptible or negligible in a neighboring sphere or on an adjacent level."[41] Unlike Gamow, or for that matter Bernal before him,[42] he did not apologize for such levels as purely conventional or "fictional." Phenomenologically, these are presented instead as substantive, distinct realms.

The fundamental evolution of the universe, he claimed, is convergent, tending to some ultimate point of integration, the "Omega Point." This was made evident, to his mind, through the tendency of matter to produce consciousness, which was at the same time a tendency to union, fusion, synthesis, and symbiosis across all conceivable scales and differences. Binding the spheres, in the historical projection that developed them, Teilhard saw a "*mega-synthesis*, the 'super-arrangement' to which all thinking elements of the earth find themselves today individually and collectively subject."[43]

Among those who recommended Teilhard's work in the late 1950s and early 1960s, if with qualifications, were Huxley and Dobzhansky.[44] Others linked to the evolutionary synthesis, such as Ernst Mayr and George Gaylord Simpson, dismissed its scientific content, even while Simpson in an early review mentioned Teilhard's earlier paleontological monographs respectfully.[45] Peter Medawar, in 1961, went further, denouncing the work in style and substance, and unlike Simpson, offering no courtesies based on Teilhard's scientific research. Medawar dismissed as well the scientific discipline of the *Phenomenon*'s author.

Medawar had won the 1960 Nobel Prize in Physiology or Medicine, a prize shared with Frank Macfarlane Burnet, awarded for the discovery of "acquired immunological intolerance." The prize was tied to Medawar's experimental work on tissue grafts, an outcome of research he participated in through the War Wounds Committee of the Medical Research Council of

Great Britain. In 1959, the same year that Teilhard's *Phenomenon* was translated into English, Medawar had given the Reith Lectures, a series of BBC
radio broadcasts named in honor of the first director-general of the BBC.
Medawar entitled his lectures "The Future of Man," and they were soon
published as a book under that same title. In them, he argued views that
happened to be in strong opposition to Teilhard's and Huxley's.[46]

Medawar's Reith Lectures were broadcast from the middle of November
in 1959, beginning just prior to and extending beyond the centennial of the
publication of Darwin's *Origin of Species*, overlapping with the University of
Chicago commemoration bringing together Darwin's grandson Charles Galton, T. H. Huxley's grandson Julian, Shapley, as well as Dobzhansky, Ernst
Mayr, Hermann Muller, G. G. Simpson, C. H. Waddington, among other no-
tables.[47] "Previewing" the coming celebration, Darwin, Huxley, and Shapley took part in a panel discussion in the television show *At Random* on November 21. Their wide-ranging conversation emphasized the question of
the evolution of the mind and included a discussion of cosmic evolution
by Shapley, during which Shapley indicted Huxley as too anthropocentric.
Here, Shapley asserted the generic quality of intelligence, noting that "this
so-called mental disease, this neurotic complex we call 'intelligence,' is a
pretty common quality. It is not confined to man but goes down through
the whole animal world." Shapley endorsed the wondrous possibilities and
meaningfulness modern science and particularly astronomy offered in comparison to the "myopic predictions and concepts of the ancient Church Fa-
thers" and its demand once again for human adjustment. Huxley replied
with the observation that "the population experts are not very hopeful,"
which provoked Shapley to retort: "Again you are being anthropocentric."
But to Huxley, this amounted to no charge at all: "I should hope so! After
all, we are *anthropos*."[48]

Adlai Stevenson, the former governor of Illinois and soon to be ambas-
sador to the UN, who also took part in the exchange, had been comparing notes with Shapley on the problems they each had recently declared to
be the greatest facing humanity. Shapley had lectured on nuclear bombs
and the "population explosion" and Stevenson had been writing on nuclear bombs and the disparity between rich and poor nations.[49] Together,
these were the three "menaces" that C. P. Snow had called attention to earlier that year, in his famed two-cultures lecture. Huxley named them together near the conclusion of the program: "The three big problems are
preventing our civilization's being wiped out by atomic warfare; preventing
overpopulation from engulfing the world; and bringing the underprivileged

nations up to an improved standard of living."[50] Huxley returned to these themes in his charged Thanksgiving Day convocation address, emphasizing again humanity's significance despite its peripheral cosmic location, its evolution as psychosocial and its habitat as the noosphere, the idea of which he explicitly attributed to Teilhard. Huxley called overall for new "evolution-centered" thought recognizing "man's destiny . . . to be the sole agent for the future evolution of this planet," and the concomitant need to address not only Snow's menaces, but others including unrestrained material production and consumerism, the "democratic myth of equality," the spread of communism and the loss of cultural diversity, all contrary to the "evolutionary vision" he looked forward to as the basis of a new, future religion.[51]

Less than two weeks earlier, in his first Reith Lecture "The Fallibility of Prediction," Medawar confessed feeling so unconvinced by expert opinions on humanity's biological future or about the ability to make sound predictions on the basis of the state of science that he decided instead to focus on the reasoning that the learned applied to their forecasts. In the fourth lecture, "The Genetic System of Man," broadcast on December 13, he observed that in spite of the many uncertainties of science, listeners (and later, readers) "will have heard it said that the fall of nations can be traced to genetic causes—for example, to the persistent infertility of a ruling class; you have been alarmed by insistent declarations that we are declining rapidly in intelligence; you have been stirred by the pronouncement that man can now control his own evolution . . . and you have been taken aback by the confidence with which some experts will assert what other experts will just as confidently deny."[52] His point was that too little was known to substantiate these claims.

In the first of his lectures, "The Fallibility of Prediction," he emphasized this as a difference between biological and physical sciences: "I first thought of attempting a grand prophetic statement about Man's future as Homo Sapiens—of doing for biology something of the kind that physicists have done when they have written about the shape of the foreseeable future or of what might happen in the next million years."[53] But biology was not ready for this kind of confident prediction: "if I were to attempt anything of the kind on behalf of biology, I should be obliged either to weary you with endless qualifications and reservations and disclaimers, or else to try to disguise the thinness of the reasoning by taking refuge from time to time in apocalyptic prose." Thinness of reasoning or speculative "apocalyptic prose" Medawar took as characteristic of biological prophecy.

As with those who turned to universal history less apologetically, Medawar left the brain for last. Also entitled "The Future of Man," his overall thesis in this, as in other lectures, is that very little could be learned from the scientific understanding of nature. His points resonated far more with cautions expressed by T. H. Huxley in his later years than with Julian Huxley's perorations, and stood in contrast with the growing assertions that depressed him when reviewing Teilhard: "we can jettison all reasoning based upon the idea that changes in society happen in the style and under the pressures of ordinary genetic evolution; abandon any idea that the direction of social change is governed by laws other than laws which have at some time been the subject of human decisions or acts of mind. That competition between one man and another is a necessary part of the texture of society; that societies are organisms which grow and must inevitably die; that division of labour within a society is akin to what we can see in colonies of insects; that the laws of genetics have an overriding authority; that social evolution has a direction forcibly imposed upon it by agencies beyond man's control—all these are biological judgments; but, I do assure you, bad judgments based upon a bad biology."[54]

Medawar was disposed to dismissing the futurist, scientistic sketch Teilhard had composed and that was just then being widely circulated. And roughly a year after his Reith Lectures, Medawar did indeed dismiss it, as an assembly of incoherent judgments based on second-class and false biology. Far from the scientific treatise it claimed to be, it was instead a work of self-deceiving nonsense, produced by a "naturalist" who "practised a comparatively humble and unexacting kind of science," but one who "must have known better than to play such tricks as these."[55]

Medawar attacked Teilhard's thesis by reinforcing the differences between scientific disciplines and disciples: "Laymen firmly believe that scientists are one species of person. They are not to know that the different branches of science require very different aptitudes and degrees of skill for their prosecution." The designation of "naturalist" had bite, since Medawar further claimed that it was within "the intellectually unexacting kind of science in which he [Teilhard] achieved a moderate proficiency. He has no grasp of what makes a logical argument or of what makes for proof. He does not even preserve the common decencies of scientific writing, though his book is professedly a scientific treatise."[56] Confessing distress and despair through his study of the work and its reception, Medawar attributed its success to four factors: its antiscientific sentiments, an unintelligibility mistakable for profundity, its superficial characterization of a fraught human condition, and the validity conferred on it by Huxley himself. Supporting all these was a market created by a literate public "who have been educated far beyond their capacity to undertake analytical thought."[57]

These remained the terms of battle over a popular scientific synthesis: from the divisions of and between the sciences—among those associated with the evolutionary synthesis and beyond them—to the possibility of alternative conceptions of sciences, their genres, and their effect on the general public and its future welfare. Which synthesis, and which scientists from which disciplines were authorized to produce it? In criticism paralleling F. R. Leavis's response to the wide circulation of C. P. Snow's lecture, but preceding it by roughly a year, Medawar also characterized Teilhard as a portent of a kind: "I have read and studied *The Phenomenon of Man* with real distress, even with despair. Instead of wringing our hands over the Human Predicament, we should attend to those parts of it which are wholly remediable, above all to the gullibility which makes it possible for people to be taken in by such a bag of tricks as this. If it were an innocent, passive gullibility it would be excusable; but all too clearly, alas, it is an active willingness to be deceived."[58] For Medawar, Teilhard was a minor scientist peddling a false myth, but a myth that found an audience, learned and lay.

MINDFUL OF MIND

The intersection of the language and terms employed by those such as Teilhard, Vernadsky, Huxley, Haldane, or even Medawar, suggested a wider set of commitments. These views dipped into teleological water. They adduced varying, interrelated notions of progressiveness: direct expressions of "higher" and "lower" (as with Huxley); the progressive complexity of biological or molecular forms (as Medawar cautiously endorsed)[59]; an increasing ability to form more intricate solutions to problems of living (unconscious or not); a superior control over nature.

Progress was often an assumption at work in ideas of a transcendent reason, a faculty that also somehow evolved immanently. "A century ago Charles Darwin and his cooperators, especially Thomas H. Huxley, presented a strong case for biological humility." Nevertheless, argued Shapley by the 1960s, "All the new revelations should promote our respect for the universe and our pride in the human mind. Would that the mind were more powerful, more penetrating, more free of delusions, free of silly hopes and prejudices."[60] In this intellectual context, consciousness itself was rarely granted the possibility of great variation. Reasoning about nature and life was generally presumed to be tending in the same direction, toward some fixed pole of expanded consciousness, greater control of the environment and greater truth. Or, in the implied obverse of Shapley's terms, toward greater power, penetration, understanding, freedom, clarity. Consciousness and intelligence were regarded as commensurate not only on earth, but across the universe.

Shapley himself in the same years expressed a possible alternative to such a view. Rejecting any perceived pride in human domination, and in terms similar to his claim that intelligence was a "pretty common quality" made during the *At Random* discussion, he noted that "a close student of social insects, however, will not boast about the superiority of the human brain." In contrast to the belief in the uniqueness of human reasoning, he asked: "What evidence is there of thoughtlessness and unreason in the bird choosing its nesting site or a spider locating its web?"[61] But instead of the assertion that among such animals could be found "a different kind of thinking and reasoning," as might be suggested from a strictly immanent evolutionary commitment, Shapley asserted that between the minds of these creatures and those of humans, the difference was of "degree or intensity." What is shared across the entire set of "animal forms," he claimed, is a "generic mind."[62] In Teilhard or Huxley's more marked vocabulary, such a generic mind, measured in degrees of consciousness across the biosphere, was a characteristic of the nöosphere. What must have filled Medawar with such despair was that these views were so popular among professed scientists, as measured by the attention they garnered.

Views such as Shapley's, Teilhard's, or Huxley's were respectable enough to set the tone for international meetings of science. Similar to the putative subject of Medawar's Reith Lecture, the 1963 meeting of the CIBA Foundation took as its theme "Man and His Future." Huxley supplied the introductory address, and Haldane the concluding one.[63] Among the other contributors were the physical anthropologist Carleton Coon, the physiologist Albert Szent-Györgi, Hermann Muller—whose paper was read, though illness prevented his actual attendance—and Joshua Lederberg; among the discussants were Francis Crick, Bronowski, and Medawar. Covering thereby a variety of specializations, the talks were orchestrated to address the collective future of humanity and the planet: the physical resources and environment, the biological, political, and axiological welfare and prospects of life. The perils of ecology, as so recently clarified by Carson in *Silent Spring*, of apparent population trends, and of nuclear warfare were much on their minds. As Haldane observed in the final open discussion, "Ethical Considerations," "The trouble is that everything is happening too quickly with these atomic bombs."[64] Likewise, when Bronowski challenged Lederberg on the point of Muller's and Lederberg's eugenically minded papers ("what problem are we trying to solve?"), Lederberg replied, "In answer to Dr. Bronowski's question about our motivation, I think most of us here believe that the present population of the world is not intelligent enough to keep itself from being blown up."[65] In sum, a union of population and nuclear concerns, and a biological prophecy, one again taking up Snow's menaces.

Though he did not launch strong objections recorded in discussion, much in the tone of the conference would likely have frustrated someone as skeptical of "evolutionary control" as Medawar or someone as contemptuous of Teilhard.[66] Huxley's initiating statement turned again on the necessity of planning, coordination, integrated learning (with a special emphasis on eugenically minded genetics), and meaning making, in view of the necessity "to think of the future of man in the unfamiliar terms of psychosocial or cultural evolution."[67] In those remarks, he twice invoked Teilhard, in relation to "the convergence of the psychosocial process upon itself" and in relation of the new organic habitat of the noosphere—the "covering of the earth's sphericity with a thinking envelope."[68] Unsurprisingly, the belief in the human ability to control evolutionary processes suffused the address.

As much as the centrality of humanity had been and was being questioned by scalar and historical evolutionary syntheses, it was also persistently reasserted. "To me," Huxley noted, "it is an exciting fact that man, after he appeared to have been dethroned from his supremacy, demoted from his central position in the universe to the status of an insignificant inhabitant of a small outlying planet of one among millions of stars, has now become reinstated in a key position, one of the rare spearheads or torchbearers, or trustees . . . of advance in the cosmic process of evolution."[69] What Huxley repudiated by the statement, the demotion, might easily have been taken chapter and verse from Shapley, except that Shapley at that point also endorsed a position not far from Huxley's own, if at the same time finding Huxley's account anthropocentric in its valuation of the specifically human. Life and mind were invoked as the heroes of an evolutionary tale onto which mythic connotations continuously precipitated. Regardless of how the scientific knowledge was conceived as being assembled, a given synthesis was often construed as offering a flattering place for humanity in life and consciousness—whether in the biospheres and noospheres, or in the molecular complexity of human biology. The determination of that place was invariably treated as central to scientific inquiry, even when that place was represented in less exultant or more halting terms.

But as Eiseley suggested in his 1958 *Darwin's Century: Evolution and the Men Who Discovered It*, a book released in advance of the Darwin centennial, a concern with the place of the human mind in physical and cultural evolution was hardly new. Alfred Russel Wallace had turned to that problem in the 1860s, linking the comparative stability of recent human evolution to the emergence of the human brain. Eiseley focused on Wallace's contention that a second, nonphysical phase of human evolution had begun with that event, an evolution guided by a brain "whose primary purpose was, paradoxically, the *evasion* of specialization."[70] Later that decade, Wallace came to the conclusion that natural selection could not give sufficient account of the emergence of the brain—of the natural experiment, as Eiseley put it, "to totally encephalize the hand."[71] Wallace had concluded that a guiding intelligence may have shaped human development. As Eiseley saw the matter, the contemporary context was such that "to consider, or express approval of, certain of Wallace's ideas is occasionally to find oneself labeled, along with Wallace, as a 'mystic.'"[72] And indeed, in a 1962 address, his contribution to the John Dewey Lecture series, Eiseley complained of being labeled a mystic himself by scientific colleagues who dismissed such expressions of wonder at the universe and human conceptions of it. In the lecture, published as *The Mind as Nature*, he compared the entire evolution of the universe to the "smaller universe of the individual human brain."[73] The latter had an evolutionary past and a future that needed protection from repressive scientific ideologies.

FOUNDATIONS OF LIFE

In new idioms then becoming current, scientists and explainers of the larger cultural and spiritual implications of science were becoming public intellectuals and popular experts. At the start of the 1960s, ideas such as the psychozoic or the noosphere and the wonder expressed at the physical and mental universes could be employed for political and aesthetic purposes, as when John and Faith Hubley crafted Shapley's ideas and their own filmically, in *Of Stars and Men*, in aid of an artistic-scientific aesthetic that was at the same time politically minded. Eiseley, from his longtime Philadelphia perch, increasingly protested the liminalization of the natural world and the flatness of insensitive scientistic worldviews. Frank and Shapley, each by then in their seventies, had long been involved in efforts to bridge science and more overtly spiritual views, as Eiseley attempted to temper science with an awe of life and the demands of political universality.

By this time, in the late 1950s and early 1960s, Eiseley was in at least limited correspondence with figures such as Rachel Carson and Dobzhansky (and, earlier, with George Gaylord Simpson). His correspondence with Julian Huxley in the early 1960s, apart from academic papers forwarded by Huxley, concerned a minor misunderstanding linked to the television program *The House in Which We Live*. The program had been hosted by the landscape architect Ian McHarg, growing out of McHarg's "Man and Environment" course at the University of Pennsylvania. Announcing Eiseley's participation as the first guest of the second series, the *University of Pennsylvania Almanac* observed that prior guests included, among others, Carleton Coon, Erich Fromm, Julian Huxley, Margaret Mead, and Harlow Shapley[74]—figures of scientific authority. Later in the same decade, Eiseley hosted NBC's *Animal Secrets*, a nature show targeting school-aged children, and in the decade following was himself to be a subject of a documentary written and produced by Lester Novros entitled *Cosmos—The Universe of Loren Eiseley*.[75] He was sought out for a number of television and writing projects, and permissions were asked for the use of his work for varied educational groups, environmental organizations, and cultural endeavors, including dance and visual art.[76] One such book request from Charles Scribner's Sons toward the end of the decade would result in his response to space travel, *The Invisible Pyramid*. Over the course of the 1960s, he found himself among a constellation of those who addressed humanity's relation to the environment, becoming widely associated as a spokesperson for and poet of nature.

But if Eiseley came to be regarded as a bard for environmental concerns, he did not always feel himself to be appreciated by contemporary culture, nor was he appreciative of it. The elder masters of a scientific or scientifically inflected sublime rarely expressed comfort with prominent countercultural movements—even if these movements did not always assault the institutions and knowledge bases they represented. Eiseley called for an epistemological modesty and for an intellectual pluralism, but each steeped in tradition.

Eiseley ultimately excoriated what he perceived as an attack on the cultural role of science on the part of the young in the 1960s, and he even blamed contemporary efforts to explore outer space for causing this disillusionment with the scientific culture and institutions that he represented. Eiseley and other elder statesmen of letters reacted with consternation, at times with hostility, to their perception of campus unrest and a politically charged generation gap. In the midst of campus agitation, both Eiseley and Bronowski did not initially feel their messages were being heard. Witnessing unrest on his own Philadelphia campus while assuming an increasingly conservative posture, Eiseley argued that the value of education was being undermined.

The protests, he reported, had extended into his own classroom where he claimed to be challenged on the worth of learning evolution. In a volume to which both Bronowski and he contributed, Eiseley turned to the mythographer Joseph Campbell, that "very able student of folklore" for the terminology to describe the events: "the chronicle of our species from its earliest page, has been not simply an account of the progress of man, the toolmaker but—more tragically—a history of the pouring of blazing visions into the minds of seers and the efforts of earthly communities to incarnate unearthly covenants."[77]

Bronowski used similarly dramatic language, finding that "the generation gap is now a moral chasm, across which the young stare at their elders with distrust, convinced that the values that make for success are fake."[78] Bronowski wasn't entirely unsympathetic to what he saw as a criticism of an entire value system preceding them, nor did he want to side with the animosity and fear he found in published reactions to the youth protests. But conceding all this, he objected: "A whole generation of liberals and humanists, to which I belong, is bewildered at the discovery that the young include us in their charge of hypocrisy. We made liberalism respectable by our labors, and turned it into an intellectual faith; and now we are distressed to find that out heroic memories of the hungry thirties and the Spanish Civil War are dismissed as out-of-date mythology."

The historical synthesis Eiseley and Bronowski offered was, to their own minds, in strange relief against the perception that scientific knowledge was suddenly not to be trusted. Both were taken aback by their contemporary culture, and from their different points of view, thrust into difficult positions. Eiseley was offended by the disrespect shown to elders and the cultural artifacts those elders celebrated; Bronowski was upset that the young did not perceive the struggles of and their debt to their liberal elders. The elder men each stood in defense of the universities, in addresses and publications, and of the syntheses of knowledge that were accumulating, and stood for the scientifically inflected humanisms they each broadcast.

However, from the perspective of Eiseley, Bronowski, and others, if the evolutionary and life sciences were threatened, a more potent threat came not from the student protests themselves, but from dangerous scientistic dogmas. For Bronowski this would take the form of a kind of instrumentalism or "push-button philosophy" not to be confused with and contrary to the spirit of science. Eiseley for his part increasingly protested against the effects of scientific overreach, mechanistic ideology, and large-scale technological efforts. This led him to the topic of space travel as a central example of unfulfillable longings aroused by science, resulting in the disillusionment he saw driving the protest movements around him and predicated on mis-understandings of the nature of life.

Eiseley long had published skeptical views on the subject of space travel and the possibility of life elsewhere. In a 1953 essay, "Little Men and Flying Saucers," he judged the possibility of intelligence elsewhere against the re-cent fact that "the present theory of the expanding universe has made time, as we know it, no longer infinite." From this he concludes that "it is now just conceivable that there may be nowhere in space a mind superior to our own." The context he had in view was clear: "Since we now talk, write and dream endlessly of space rockets, it is no surprise that this thinking yields the obverse of the coin"—that rockets might arrive from space, that intel-ligences elsewhere might visit.[79] But he emphasizes the Darwinian lesson that though life "may exist out yonder in the dark," nevertheless "it will not wear the shape of man." With that, he couples what he regarded as more re-cent astronomical and cosmological lessons: "In a universe whose size is be-yond human imagining, where our world floats like a dust mote in the void of night, men have grown inconceivably lonely." He quietly ties the search for origins to the anticipation of life elsewhere, but was pessimistic of what might be drawn from either: "We watch the stars but the signs are uncer-tain. We uncover the bones of the past and seek our origins. There is a path there, but it appears to wander. The vagaries of the road may have a mean-ing however; it is thus we torture ourselves."[80] Modern science, to Eiseley's mind, could only reinforce humanity's isolation; the scientific quest inward and outward might carry meaning, but definitely brought pain.

The contemporary forms of the search for extraterrestrial intelligence (SETI) were in play by the end of the same decade, with the research of Gi-useppe Cocconi and Philip Morrison, the observational work of Frank Drake, and the Green Bank Conference of 1961—what Steven J. Dick has referred to as "three seminal events" for modern SETI.[81] A half-decade predating these attempts, in another 1953 article "Is Man Alone in Space?," Eiseley had claimed, thinking perhaps more of earlier attempts, rather than anticipating new attempts to come: "We want to radio Mars and get back the equivalent of 'How yuh, boys?'" How much he thought this possibility would be taken up is unclear, but his facetious characterizations also predated those that would come to haunt the efforts to maintain steady public funding for such attempts.[82]

A chronology prepared by the NASA history office of the "Astronautical and Aeronautical Events of 1962," reporting to the House Committee on Sci-ence and Astronautics, took note of a December 29th address Eiseley gave to the American Association for the Advancement of Science. According to the brief summary, he "predicted the massive emphasis on space and armaments might lead to downfall of modern civilization," drawing a comparison with "Ancient Egypt's tomb and pyramid building," and what he regarded as other self-destructive preoccupations of past civilizations.[83] And so in his 1971 *The Invisible Pyramid*, Eiseley lamented that "science endows the culture with great energy, so that far goals seem attainable and yet grow illusory. Space and time widens to weariness. In the midst of triumph, disenchantment sets in among the young."[84]

More broadly still, however, Eiseley and other advocates of either scientific plurality or—more controversially—the intractability or mystery of nature (or both) found it necessary to contend with foundational dogmas of unity, the overarching belief that all phenomena, including human life, should in principle be reducible to the operation of a single set of fundamental laws, if not with full-blown syntheses achieved on the basis of that belief. In the memory of future synthesizers such as Edward O. Wilson, the scientific idiom of Watson and Crick had helped advance an atmosphere inimical to ecological sciences and to field biology.[85] Knowledge less foundationalist in conception, from that perspective, became more hobby ("butterfly- or stamp-collecting") than science.

This was not merely an artifact of Wilson's memory. In the summer of 1964, in an address before the American Society of Zoologists, Dobzhansky, for example, argued against any monolithic approach to biology.[86] By that point, Dobzhansky had already adopted and promulgated Big Bang science, adopting the Gamow network's "Ylem" and using the term "cosmic evolution" to promote what he saw as the overall view of the sciences that nature "is the outcome of a historical process of development." He wrote this in his 1955 textbook *Evolution, Genetics, and Man*, where among his suggestions for further reading were Gamow's *The Creation of the Universe*, Bernal's *The Physical Basis of Life*, and Oparin's *Origin of Life*.[87]

His ideas were in dialogue with those of Mayr in a number of ways, whose own classifications of biological causality carried with them the implication that a multiplicity of approaches was necessary in the life sciences. Dobzhansky took this as the theme of his address. He noted how at the time, "the view has gained some currency that the only worthwhile biology is molecular biology. All else is 'bird watching' or 'butterfly collecting.'"[88] Dobzhansky argued that what sets biology apart from physics or chemistry is the hierarchical character of its subject matter, a sequence from "molecule, cellular organelle, cell, tissue, organ, individual, Mendelian population, species, community, ecosystem."[89] All the levels above the molecular he gave to the purview of organismic biology, giving some intrinsic weight to the other, nonmolecular subdisciplines of biology. But the larger point was that fruitful research takes place at all these levels at the same time. Even the example of Watson and Crick's successes of the decade before required the prior discovery of genes and of research on crosses that themselves took place far from the idiom of molecular biology. And the restriction to a reductionist program with claims to having determined the foundations of life and with the mistaken view that organismic biology had completed its project ignored that history of science demonstrating how persistent new discoveries were in well-trodden fields—putting aside the number of species that had yet to be understood and examined more fully.[90]

Moreover, molecular biology was to Dobzhansky's mind neither the most exciting nor the most important development in contemporary biological research: "Perhaps the most significant and gratifying trend during the last two decades or so has been the increasing unification of biology as a field of knowledge." It was not primarily through molecular biology that this unity was being established, however. Instead, it was the historical evolutionary synthesis that bound biology: "The unity is understandable as a consequence of common descent and of universal necessities imposed by common materials. The diversity is intelligible as the outcome of adaptation of life to different environments."[91] Dobzhansky noted that in the end, biology had to confront the question "What is Man?"

When looking to the future, he also abided by the projections of Haldane, Bernal, Huxley, and Teilhard that humanity would be able to assert greater control of nature and its own evolution: "Man must ... face the problem of adapting his culture to his genes, as well as adapting his genes to his culture. Man is being forced by his culture to take the management and direction of his evolution in his own hands. This is perhaps the greatest challenge which mankind may ever have to face."[92] By this point, the trope of humanity as master of evolution was becoming central to universal history.

In the lecture of the following April, in a symposium in honor of Mendel, Dobzhansky closed optimistically on the future dilemmas by invoking and endorsing the more holistic viewpoint of Teilhard. To do so, however, he dismissed Teilhard's claim that the *Phenomenon* was a scientific treatise and repudiated Teilhard use's of the term "orthogenesis," suggesting as it did a belief in a predirected evolution. What he endorsed instead was "Teilhard's basic insight that the cosmic, biological, and human evolutions are not only components but are developmental stages of a single process of universal evolution. This single process has a discernible direction. It has advanced from matter, to life, to thought."[93] But older notions of orthogenesis were inconsistent with Teilhard's basic view that this advance was not at all the "long strip-tease act" in which all evolution becomes meaningless. For Teilhard, history revealed instead a groping, creative evolution that experimented with new living forms and possibilities.[94] In contrast to the "twilight" of humanity forecast by scientists such as Muller, Charles Galton Darwin, and Julian Huxley, Dobzhansky saw Teilhard's views as embracing the future in a way inimical to the eugenics so many had been convinced was necessary to stave off that twilight. He concluded his address with the final words of Teilhard's "megasynthesis" section: "The outcome of the world, the gates of the future, the entry into the super-human—these are not thrown open to a few of the privileged, nor to one chosen people to the exclusion of all others. They will open to an advance of *all together*, in a direction in which *all together* can join and find fulfillment in a spiritual renovation of the earth."[95]

FOUNDATIONAL REDUCTIONS WITHOUT SYNTHETIC ASCENTS

The foundationalist dogma of unity that Dobzhansky attempted to overturn, a dogma he found especially prevalent among biologists, was, he suggested, also active more widely. From the late 1960s, the condensed matter physicist and future Nobel Prize winner P. W. Anderson voiced objections to such foundationalist dogmas among physicists and in the understanding of the relationship between the sciences as a whole. Anderson's own position by the late 1960s, as a renowned theoretician at Bell Labs who was also a permanent visiting professor at the Cavendish Laboratory in Cambridge, UK, and a recent inductee (1967) to the National Academy of Sciences, gave him at the very least a favorable vantage point over which to assess the social, formal, and substantive relations between the subdisciplines of physics.

Anderson was motivated to articulate his views not only by the "arrogance of the particle physicist" whose "intensive research may be behind us"[96] but also those more directly involved in the examination of life, for "we have yet to recover from that of some molecular biologists, who seem determined to try to reduce everything about the human organism to 'only' chemistry, from the common cold and all mental disease to the religious instinct"[97]—a point to which Dobzhansky might easily enough have assented.

Based on the Regents lecture given in the University of California in is 1967, Anderson's 1972 *Science* article "More Is Different: Broken Symmetry and the Nature of the Hierarchical Structure of Science" debates what Anderson takes to be a common understanding—that there is only one set of fundamental, foundational physical laws, the study of which was cornered by the subjects of "some astrophysicists, some elementary particle physicists, some logicians and other mathematicians, and few others."[98] He found such a perspective in the theoretical physicist Victor Weisskopf's distinction between "intensive" and "extensive" science, supporting the view that scientific disciplines such as solid state physics were simply applied elementary particle physics, molecular biology simply applied chemistry, social sciences applied psychology and so on. To this, Anderson exhibited physical systems arguing to the contrary, demonstrating that at new scales, new foundations emerged.[99] A central point was that in scaling up to the larger molecules, physical symmetries and the laws of nature were different than they were at lower scales, and unpredictably so from the point of view of those other scales. New emergent foundations had to be discovered and enunciated.

Anderson's lecture and article are, in a number of senses, evocative of other work that called for new conceptualizations of the relations between the sciences, Schrödinger's past reflections on life perhaps most trenchantly.[100] Anderson posited an interlocking, almost rigid framework of foundationalism, scales, and complexification: new scales were correlated to new foundations and to more complex phenomena. As a result, "we expect to encounter fascinating and, I believe, very fundamental questions at each stage in fitting together less complicated pieces into the more complicated system and understanding the basically new types of behavior which can result."[101]

But across any ladder of scales, Anderson all but ruled out the possibility of "synthesis," in the sense of moving from the building blocks of the previous stage/scale to an anticipation and/or construction of the entities and laws of the next. A more modest analysis, beginning with the more complex phenomena and breaking them down in scale, might however be possible and profitable. So "biology has certainly taken on a whole new aspect from the reduction of genetics to biochemistry and biophysics."[102] Whereas Schrödinger had anticipated that fundamental science would need to be reconsidered in light of new scientific knowledge of biology, Anderson's point was instead that the different properties of chemical and, here particularly, organic phenomena required reconsideration of what the idea of a fundamental science was.

This reconsideration focused attention, in turn, on what was involved in any imagined movement down the ladder of physical argument through analysis or up it through synthesis. As he had explained toward the start of his remarks, his own presentation, in terms of the recent developments of theories of "broken symmetries," was intended to clarify the "breakdown of the constructionist converse of reductionism."[103] Whereas Engels had criticized the naïve materialism that simply attempted to reduce all nature to underlying physical laws, Anderson critiqued an apparently naïve, synthetic foundationalism.[104]

Foundational dogmas had not been overcome, neither in physics nor in the chemical and life sciences, but Anderson's argument, as well as Dobzhansky's, demonstrated the extent to which prominent scientists found the idea of disciplinary organization on the basis of a foundational synthesis uninviting, elusive, and spurious. At the same time, the reductive impulse voiced in these dogmas could instead be discerned in alternatively conceived synthetic genres. Even Tompkins's fabulaic dreams were predicated on a notion of fundamentality or the foundational in the narrow sense of physics determining basic features of the world, where in turn the relative size of fundamental constants was linked to different arenas of a possible experience. There is a potential affinity here then with fabulaic worlds and Teilhard's different spheres or orders of nature and with Anderson's emergent physics laws pertaining to each order suggesting at least the possibility of reduction in his sense. But any synthesis across worlds, spheres, or levels for Tompkins or Teilhard or Anderson is not built upon a mathematico-logical structure connecting the different sciences, from more to less fundamental, from more basic to more complex, through a chain of deductions.

Nevertheless, foundationalist dogmas could be abstracted from scientific syntheses predicated primarily on other logics and figures, and were forceful as scientific creeds, motivating counterarguments such as those of Eiseley, Dobzhansky, and Anderson. The dispute over Teilhard's writings likewise demonstrated the balance of status, license, knowledge, and presumption required in the task of authoring any historical synthesis addressing a scientific audience. But the fact of alternative syntheses and other unifying dogmas highlighted the contingencies of any such belief or effort and the cultural resonance and attractions that the more popular of these had. Multiple reflections on an organic evolution of consciousness and on a scientifically revealed sublime nature thus were themselves a condition for a wider reception to a new scientific myth, though not necessarily for other possible syntheses judged too narrow or too insubstantial to be read in positive myth terms.

PICARESQUE FABLES AND SCALAR ZOOMS

Even with the old examples of the "bad Comte" or the "mystic Wallace," a cosmic religion or evolutionary spirituality seemed to linger at the fringes of scientific syntheses. As Dobzhansky saw it, foundational beliefs that imagined science could capture some set of fundamental laws governing human life were deeply misleading. But part of Dobzhansky's resistance to foundationalist dogmas and the hierarchical ordering of the sciences that they implied suggested a contrast with the holistic, mythic possibilities of a historical synthesis. And Dobzhansky's position was not an isolated one. If the research and writing of scientists such as Dobzhansky and Shapley, Bronowski, Eiseley, and Gamow are a fair measure, the foundational, perhaps the most discussed unifying dogma and potential synthetic genre in historical and philosophical scholarship, played a more marginal role in efforts by physical and biological scientists to produce a scientific synthesis up through the 1960s and 1970s. It functioned neither as an active research program nor unproblematically or extensively as an apology for the sciences as a whole. Harlow Shapley's work across the century offers testimony to the resistance or simply indifference to the foundational—not just Shapley's later work, in the conception of a psychozoic kingdom, but also his earlier work, which was more thoroughly dismissive of human-inhabited spheres. The bulk of his writings suggest a lack of interest in any thoroughgoing form of foundational ambition. Gamow's example also indicates that part of the reason for the comparative marginality of foundational theories among physical scientists (as opposed to dogmas) was the belief in the futility of the search for fundamental laws.[105] And while Dobzhansky identified and combatted a foundationalist dogma upheld by many of his fellow biologists, he did not find in it any active, synthesizing genre, arguing instead that actual research practice persistently contradicted foundational views. Neither for Gamow in the physical sciences nor for Dobzhansky in the life sciences did final laws or reductive disciplinary hierarchies play a constructive role.

Instead, by the 1960s and early 1970s, the question of a foundational synthesis—or indeed, any synthesis—was open to overt contestation from a variety of perspectives and disciplines challenging the visions of unification underlying them. The philosopher, sociologist, and theorist of urbanization Henri Lefebvre surveyed synthetically minded strategies, targeting the plurality of ways in which the "fragmentary sciences" were then conceived in relation to each other. Among those unifying dogmas he addressed were: convergence, "the hope and myth expressed at interdisciplinary conferences"; integration, "But with what? With some intellectual discipline that has been made dominant?"; pragmatism, "the use of information supplied here and there by someone or other"; operationalism, a "variant of pragmatism" in which concepts have "some *administrative* capacity"; and experimentalism, in which "analysts provisionally dissect 'abstract' objects; they study with the help of different descriptions, temporarily considered auxiliaries." A last such mode is hierarchization, which he challenged in a spirit similar to that of his natural scientific counterparts: "Who is going to establish valuations? Who is going to claim that sociology is worth more than geography or demography?"[106]

The continuing projects of Gamow, Eiseley, and Bronowski highlight other beliefs and efforts to make the fragmentary sciences whole in tension with or simply orthogonal to foundationalism and hierarchization. But two further narrative forms, the picaresque excursion and the scalar tour of the universe, from beyond the galactic to the subatomic, via the biological, suggest something of the other synthetic strategies, themselves entangled in narratives within final theories or positing final stories. Meanwhile, Eiseley established a sketch toward a historical synthesis underpinning ethical and political arguments against what he regarded as attempts to abandon humanity's intellectual, spiritual, and physical homes.

Gamow's and Shapley's notions of scales or appeals to scientific materialism, for example, related one discipline to another insofar as they established juxtapositions, distances, or pathways between them, over and above the projection of any potential disciplinary unions or reductions upon which a foundationalism could work. The adventures of Tompkins built a path bridging the sciences, with bridges that a foundationalist view would conceive of as ladders. The ways in which these were represented, the potential synthesizing narrative genres of a scalar traverse of the material universe and a picaresque traverse of the different worlds revealed by the scientific disciplines, did not produce uniformly resonant representational and organizational structures in the middle of the century, up through the time of Dobzhansky's critique of "Comtian hierarchies" and foundationalism.[107] Nevertheless, the form of these genres was in play in scientific ambitions for rendering final theories as final stories (final theories generally understood as nonnarrative) and for positing historical final stories as themselves finalizing theories (the constraints of narrative form often unnoticed or naturalized as nontheoretical). Hence these genres were part of concrete literary forms, and helped shape the expressions of ambitions by scientific thinkers to chart new paths for disciplines such as molecular biology, at least from the viewpoint of some who signaled distress with new trends. These genres based their appeal on discursive structures in the form of a macroscopic/microscopic change in focus in differing scales of objects throughout the universe, on the one hand, and of a road through worlds, on the other.

Each of these syntheses also met filmic or televisual adaptations, the former in the film adaptation of later Tompkins adventures, the latter in the collaborations of Judith Bronowski with the filmmakers Ray and Charles Eames.

By the 1960s, Gamow's Tompkins stories were published in book form and republished many times. Early in the decade, Stan Brakhage based a film, *Mr Tompkins Inside Himself*, on Tompkins's biological adventures. By that time, Gamow and his first wife, Rho, had divorced, and Gamow had married Barbara Perkins, whom he had first met as the publicity manager at Cambridge University Press. She counted as friends several prominent literary, artistic, and intellectual figures of the time, including e. e. cummings, Edmund Wilson, and the up-and-coming Brakhage. It is likely through her that Gamow made the connection to Brakhage.

Loosely based on Gamow's *Mr. Tompkins Learns the Facts of Life*, the film is an early work, bearing little immediate resemblance to Brakhage's more tender, lyrical pieces. Though Gamow's correspondence refers to a number of attempts to translate the Tompkins books into film (including as a cartoon), Brakhage's early attempt seems to be the most complete and perhaps the only such to have been made. Gamow appears at the start of the film, introducing the character of Tompkins to the audience. Though it was not widely distributed, it testified to the cultural weight of the lighthearted works. As in the case of the filmic adaptation *Of Stars and Men*, a measure of the impact of universal historical productions was the extent to which they inspired other cultural actors, whether those others encountered the scientists through personal connections (as Brakhage likely had Gamow) or were drawn to the work through more independent readings (as the Hubleys had come to Shapley).

The final installment of the Tompkins series was the 1967 *Mr Tompkins Inside Himself: Adventures in the New Biology*, published a year before Gamow's death, two years after an experiment that was soon taken to confirm Gamow's cosmogony. For this text, Gamow collaborated for the first time on a Tompkins work, with Martynas Yčas, the Lithuanian-American microbiologist who had formed part of the RNA-Tie Club. In December 1966, with the book soon to appear more widely, Gamow mentioned to Yčas that "Beatrice [Rosenfeld, Gamow's editor at Viking] insisted not to mention Prof.'s [Tompkins' father-in-law] death. When Mr. T met him in 1940 he was white-bearded old man (see illustration of Mr. T. in Wonderland) probably about 70 years old and had the daughter of maryable [*sic*] age."[108] Gamow concluded that the professor, if alive, would be approaching his one hundredth birthday, which "brings to my mind another discrepancy: Mr Tompkins do not show the signs of aging. May be it would be a good idea to put some white hair in his mustiçh."

However, it was not the synthesis of the sciences offered through Tompkins's accounts that captured the filmic/televisual imagination of the coming decades, and perhaps in this sense, Tompkins did show signs of aging. The picaresque was given no apparent weight by those deliberating over concrete organizational schema. In the picaresque narrative, the succession of worlds necessary to know this one might only proliferate. Or homecoming might come too soon. By contrast, although the narrative created by zooming across different scales might lack the broader cultural resonance of the literary genre of the picaresque, the picaresque lacked the sense of completeness and weightiness provided by the representational strategy of the zoom.

This popularization of a concise and provocative form of scalar synthesis was carried out through the work of Judith Bronowski, the youngest daughter of Jacob, a representation rooted in pedagogy, text, and image. Shapley's synthetic vision had been adapted to film by the Hubleys, one fabulaic voyage of Tompkins was adapted by Brakhage; Bronowski took a more thoroughly scalar vision mapping the universe through astronomy, physics, biology, and chemistry and produced an early version of the 1977 short film *Powers of Ten*. By 1968, she was working in the studios of designers and filmmakers Ray and Charles Eames. According to art historian Michael J. Golec, the Eameses set their new researchers the task of producing a film based on Kees Boeke's 1957 *Cosmic View: The Universe in 40 Jumps*. Bronowski at times had referred to what became her 1968 *Rough Sketch for a Proposed Film Dealing with the Powers of Ten and the Relative Size of the Universe* as, appropriately enough, *Cosmic View*.[109]

Boeke's book had itself been published a decade earlier, in 1957, with a preface by the physicist Arthur Compton. It was composed of a series of forty numbered frames separated from each other by scales of ten, where each number referred to the power of ten setting the scale. The first frame pictured a student embracing a cat, seated in front of the school Boeke founded in Bilthoven, Netherlands, where the scale is one centimeter in frame to ten centimeters in the area the frame represents: 1 to 10^1. By repeated factors of 10, the book leaps outward, increasing the scale within the frame, so that one centimeter represents exponentially more area, frame to frame. Frame 14, $1:10^{14}$, captures the whole solar system, frame 22 the galaxy, frame 24, the local group of galaxies, on to Shapley's vision of metagalaxies. In the negative numbers, it pushes through the skin down to bacteria ($1:10^{-3}$ or 1,000:1) and bacteriophages (10^{-4}), down eventually to molecules, and atoms (10^{-6}

to 10^{-8}), through to the nucleus of a sodium atom (10^{-13}). As Compton observed, "Fifty years ago our cosmic view would have been much more limited. Nothing could have been drawn with confidence in pictures 20 to 26 or in pictures –8 to –14 [*sic*]. There is reason to question whether we shall ever be able to draw what would be in the next pictures, 27 or –14."[110] The scales suggested the limits not merely of the known, but of the knowable.

The book developed from a school exercise, the original pictures drawn by Boeke's students in an effort to cull out a "sense of scale." Such a sense provided a universalizing orientation in a manner similar to what Shapley had argued decades before: "We tend to forget how vast are the ranges of existing reality ... and our attitudes may become narrow and provincial. We need to develop a wider outlook." It was an outlook that would also give orientation to knowledge in full: "At school we are introduced to many different spheres of existence, but they are often not connected with each other, so that we are in danger of collecting a large number of images without realizing that they all join together into one great whole. It is therefore important in our education to find the means of developing a wider and more connected view of our world and a truly cosmic view of the universe and our place in it."[111] But these images also offered the exotic visions of a Tompkins, since "they portray a wonderland as full of marvels as that which Alice saw in her dreams."[112] For Boeke, his class, and his book, these are static images, leaps in space, but not in time. Boeke makes this point himself, in frame 16, positioning the reader "5 million million kilometers" above the schoolgirl who occupies the first frame in their timeless trip: "As we have imagined all along that we are making our trip without spending time, this means that it would have taken the light rays which we now see more than six months to cover the enormous distance from the earth."[113]

Judith Bronowski added speed, taking Eames's exercise in the narrative of scalar zooming to a fuller realization. Golec describes the extent to which the film became her problem: "She wrote the script, drew all the storyboards, calculated the tracking moves (with the help of her father), and edited the sound and the final film. She even costarred in the film. Bronowski is the woman sitting at the man's feet in the opening picnic sequence. The only component of the film that Bronowski was not responsible for is the picnic scene."[114] This comprehensive labor produced a filmic flipbook of Boeke's playful text, revising and recontextualizing Boeke's images. Instead of a girl with a cat in the Netherlands, it begins with two people on a picnic blanket on a Miami golf course, the choice of Ray Eames.[115] Explained by Bronowski's narration, the *Rough Sketch* then zooms out to the farthest reaches of the known universe, taking note of time-dilation effects, and then rushes back to the picnickers, diving into the skin of the napping man, penetrating into skin layers, down through chromosomes and genes and through to the nucleus of a carbon atom within his hand.

The dark spaces at different scales are the unknown and the discontinuous; in the context of Boeke's book, these are what leaps must always efface.[116] Setting the scale leaves certain elements out of the frame, and others unresolved. As with any synthetic genre, the structure of the zoom produces a finality that is always ready to be upset or further refined. But in sweeping out beyond galaxies or penetrating to biological, chemical, or nuclear scales, the film does not assert that any one power of ten is fundamental or more valuable than any other. The overall perspective, propelling the viewer through the pages of an updated, relativized *Cosmic View*, emphasizes the necessity of every scale to an appreciation of what is known and knowable of the whole.

Popular receptiveness and resistance to the power of science to supply meaning for that whole grew together, weighing the possibility of an orienting scientific myth adequate to modern life. Accounts of universal history thus began to enter popular culture via media other than the essay penned or lecture given by a prominent scientist or through popularizing books such as Gamow's universal historical trilogy or Tompkins's adventures. Newer orientations and representations of the history of everything were produced through, for, and with television, a medium now playing an increasingly important role in the construction and dissemination of a final story and of representations of science and scientists as meaning makers for a new age. Arguments continued to be staged over who if anyone was authorized to give that universal story its shape, over whether it was a story of an entire species or of one small group among many.

PART III

CHAPTER NINE

Scientific Tribes and Totalizing Myths

Natural scientific syntheses of different forms had been repeatedly essayed and repeatedly constructed as a problematic. Following Herschel's definition of science as "the knowledge of many, orderly and methodically digested and arranged, so as to become attainable by one," Humboldt's cosmography, Somerville's connections, Comtean and Spencerian systems, Chambers's gravitational-developmental epics, T. H. Huxley's cosmic evolution and Julian Huxley's evolutionary humanism, Helmholtz's, Thomson's, and Eddington's heat deaths and Lemaître's fireworks birth, Shapley's renewed cosmography, J. D. Bernal's play, Gamow's multivolume universal history, Kees Boeke's and Judith Bronowski's cosmic views, to name a selection: these deliberated over the possibility, sense, legitimacy, and desirability of surveys and final theories of the universe, of final stories about it. The broader contexts in which these deliberations on synthesis as frame and unity as values were developed and in which they were imbricated varied, embracing questions of expertise, intellectual and social ambition, civic unity and fragmentation, secularism, and political authority. The modes in which candidate syntheses were sketched, more fully expressed, or tacitly presumed also varied, structured in different measures according to genres founded on historical, foundational, scalar, and fabulaic imaginations.

However, the latter part of the twentieth century saw a heightened self-consciousness over synthesis broadly construed—of any kind of totalizing account—with a pronounced rejection of many modes of natural scientific and wider synthesis emerging with and through debates over the meaning of peoples and nations. These scholarly and wider critiques of universalism have been traced to numerous causes encompassing American civil rights movements, the perceived failure of the American "creed," European-tinged critical theory, the acknowledgment of the failures and violence of colonial experiments, the exposed provincialism of past universalizing efforts, persistent cultural and subcultural conflict, the professional interests of scholars and cultural actors defending the rising tides of postmodernism, and multiculturalism.[1]

This contestation of varieties of universalism affected and was affected by science studies, as well. In the 1990s, reflecting on the sway of multiculturalism, the historian David Hollinger noted that "the science we believe and practice derives its warrant, we are sometimes told, not from its presumed capacity for verification across the lines of all the world's cultures, but from the authority of . . . distinctive social entities."[2] Hollinger tied this view of scientific legitimacy to the influence of Thomas Kuhn's work and its emphasis on the process of enculturation into professional communities. The rejection of the virtues of scientific community grew more pronounced, as did the rejection of the presumptuousness of speaking for the whole species: "The obligations we owe to one another and the rights we claim for ourselves derive not from our common membership in a species; our new moral philosophers tell us that these obligations and rights derive from the ordinance of the traditions of our singular tribe. . . . The very impulse to produce 'meta-narratives' and to develop 'totalizing' perspectives is, in itself, to be overcome."[3]

Toward the middle of the 1970s, the journalist and political scientist Harold Isaacs penned a characterization of a fragmentation of the world by ethnicities. This "retribalization," as he referred to it, was redolent with the language of cosmology: "What we are experiencing, then, is not the shaping of new coherences but the world breaking into its bits and pieces, bursting like big and little stars from exploding galaxies, each one spinning off in its own centrifugal whirl, each one straining to hold its own small separate pieces from spinning off in their turn."[4] Hollinger dated the mounting resistance to universalism to the 1970s, finding in Isaacs's "retribalization" an early perception of these trends.

In contemporaneous contrast to such a rise, Hollinger adduced the elder Bronowski: "At almost the same moment, a resolute defender of the older universalism produced an ambitious popular book in relation to a thirteen-part series of programs for public television: Jacob Bronowski's vigorously and explicitly species-centered *Ascent of Man*. . . . Soon, the book and the series were all but forgotten. The unself-conscious masculinity of Bronowski's sense of the species and its history renders *The Ascent of Man* an especially poignant example of a mode of thought placed sharply on the defensive immediately after Bronowski wrote."[5]

Yet, it is in the strength of scientific universalism that science itself begins to be structured along the lines of a new ethnos. The identity construction of the nineteenth-century "man of science" (as with Huxley or Tyndall) could be seen in opposition to the epistemological and institutional base of an older scholarly, largely humanism-minded elite; a century later, the "professional scientist" might be held up as cultural chauvinist, an opponent to the political weight given various contestants in pluralist, multicultural societies. The generational ties of the Huxleys and Haldanes to later scientists and popularizers, the emphasis given to great men by the Bronowskis, all represented science as the province of the few and the specifically Western, even as they promoted scientific truth as a contribution to world culture. This scientific personhood, as both elective and intergenerationally communal, as decidedly situated in a developed West in which its clearest oppositions were drawn, could nevertheless conspire to reduce the status of its truth as one among many—another ethnos making universal claims not requiring the corroboration of cultural others.

In the same period, reinforcing the idea of scientists as an almost-ethnos, some analysts undertook the study of scientific "tribes," as anthropologically minded work at times referred to observational studies of scientists in the 1970s and after. If the sociology of science had helped construct scientists as a "community," the anthropology and ethnology of science flirted with a new ethnogenesis, encountering scientists in the laboratory as a tribe calling for ethnographic examination.[6] Or an anthropologist's methodological emphasis on practice could represent scientists as typical of any members of an institution, however broad or thin such an institution might be.

So in a view shared to a degree by those commentators baffled by the rise of multiculturalist discourse, Hollinger observed that the early 1970s "is the chronological point at which the contrast between an older species-consciousness and newer ethnic-consciousness can be seen most vividly."[7] If this is the case, it is not only because of the emergence of more pointed critiques of universalism and stauncher defenses of ethnic-centered claims, and it is not the result of any simple overshadowing of universalism, even as universalist voices themselves could appear tone-deaf, or insensitive, or dated by the lights of some contemporaneous social and critical theorists.

By the late 1970s, confident natural philosophical defenses of universalism also had emerged in strength. Jacob Bronowski's work was living on for at least a time and had helped establish a filmic narrative that would soon produce a still more ambitious projection of scientific and cultural universalism, sublating to a degree ethnic claims of right within it. In fact, totalizing narratives were being advanced throughout the 1970s and thereafter by a number of scientists and popularizers, Bronowski among many others. And these other, later voices pronounced more confidently on behalf of the entire species than even did Bronowski. These syntheses, examined here and in the coming chapters, adopted different positions with regard to prevailing humanistic scholarly trends. The scientific syntheses put forward by Carl Sagan from the point of view of the planetary sciences and exobiology, by Steven Weinberg from the point of view of theoretical physics, and by E. O. Wilson from the point of view of evolutionary biology and sociobiology, drew from a spectrum of synthetic genres, often overlapping, but with the historical still the most prominent and, in the period, the most culturally charged. From such a perspective, the broader scope of their claims, the still wider scalar and historical totality, was tempered by the assertion that humanity itself is a local phenomenon, provincializing the earth in the face of astronomical, cosmogonic, foundational, and adventuring universalization.

At a moment when cultural and epistemological syntheses were declared impossible, naïve, and reprehensible, these scientists moved out from textual essaying and conventional pedagogy, promoting what they and others pronounced instead as a new myth, shedding Eiseley's negative usage of the term and endorsing a broad universalism. Theirs were syntheses embedding, revising, and reconfiguring the older syntheses, debating the efforts of past and present actors in local and globalized specialist discourse. Their accounts communicated the new myths through an elective scientific practice, the authors of which were at times themselves received and studied as a new ethnos: a tribe of disciplinary consent and scholarly descent characterized by the defense of a universal language conceived as beyond any tribe and perhaps beyond any species.[8]

A COMMON MULTIPLE: HISTORY AND *THE ASCENT OF MAN*

The cultivation of a larger audience and the models for the most concise forms of synthetic histories were themselves directly indebted to the contemporaneous work of older figures active across different media—figures who did not themselves adopt mythic registers, even as they helped established the conditions for doing so. Ironically enough, Eiseley and Bronowski, while flustered by the disrespect shown by the young, were at the same time finding a wide audience and receptivity to their mythically laden truths.

In the late 1960s, the BBC sought to produce a series that would relate science to a wide audience. According to Bronowski and the naturalist filmmaker David Attenborough, who had been controller of BBC 2 at the time, the conceit for the series came from Aubrey Singer, director of series and features for the channel. BBC 2 had aired the thirteen-episode series *Civilisation* in 1969, which Attenborough recalled as having inspired Singer.[9] At the same time, Bronowski claimed that it took Singer, "who invented the massive theme," two years to persuade him and that it was only in the middle of 1969 that Bronowski composed the initial outline for the series.[10] In a 1977 interview, his widow, Rita, recalled how her husband had attended a screening of an episode of *Civilisation* with Singer and had railed at the end against the use of the term "civilization" to describe the series. It was at that point that Singer began urging Bronowski to take on a project concerning civilization and then, ultimately, the history of science.[11]

Bronowski was in many ways a natural choice. He had significant radio and television experience: since the 1940s, he had been a familiar voice on the radio. His invitation to discuss the first US postwar atomic bomb test, in which he described his visit to Japan in deliberate and evocative terms, had launched his career as a public intellectual. He had also appeared as a panel member for the informational discussion program (originally itself on radio) *The Brains Trust* before producing other programs with the BBC and with director and producer Adrian Malone. Bronowski worked for roughly three years (three-and-a-half, according to Rita Bronowski) to produce *The Ascent of Man*, subtitled *A Personal Essay*, involved in every aspect of the filmmaking, from the choice of shooting location to editing decisions. The effort involved traveling to several different countries and institutes to construct in an essayistic, personal, and not entirely linear fashion what Bronowski took to be the "ascent of man"—a structure the spine of which was nevertheless a historical synthesis. That structure, as Bronowski emphasized, involved an extended amount of sometimes surprising research finds.[12] The support that Jacob Bronowski gave to his youngest daughter, Judith, was reciprocated, with Judith playing a role in the research that helped shape *The Ascent of Man*.

The series was loosely structured according to a progressive history, where each episode focused on a long-historical revolution or a comparatively recent revolution in the history of science. The first episode, "Lower than Angels," defined this in more explicit terms of "cultural evolution." In

Bronowski's view, it is remaking the environment, rather than accepting it, that has distinguished humanity—though he would focus on other distinguishing marks in the series, whether, for example, the dexterity of the hand or the wielding of technology. So he argued that "that series of inventions, which man from age to age has remade the environment, is a different kind of evolution—not biological, but cultural evolution. I call that brilliant sequence of cultural peaks *The Ascent of Man*."[13]

"Ascent" signified a definite direction for Bronowski, though not a predetermined one. This was clarified more overtly in the text accompanying the series than in the episodes themselves: "He [man] makes plans, inventions, new discoveries, by putting different talents together; and his discoveries become more subtle and penetrating, as he learns to combine his talents in more complex and intimate ways. So the great discoveries of different ages and different cultures, in technique, in science, in the arts, express in their progression a richer and more intricate conjunction of human faculties, an ascending trellis of his gifts."[14] But there are moments when the technological interventions of humanity appear pregiven. In such a way he discusses an almost natural progression of rectangular beams versus arches, a discussion emphasizing how different cultures failed to realize the arch, distinguishing Roman from Greek, Old World from New.

To represent these ascents televisually, the filmmakers along with Bronowski employed rhapsodic images of nature, illustrating revolutions of agriculture and biology and reflecting the ways that humanity came to dominate its environment and alter its social structures. That ascent would leave some nomadic families behind. The series employed the transhumance way of life of the reindeer-dependent Lapps of north Scandinavia and mountain-traversing Bakhtiari of Iran as a touchpoint for life without progressive civilization or progressive time, making the point that "civilisation can never grow up on the move"—provocative characterizations in the light of ethnologically minded trends underscored by Hollinger and others. Throughout, technologies are identified through a juxtaposition of shots; in voiceover or on site, Bronowski reflects in an often apparently unrehearsed manner, whether from the Asian foothills, the campus of Humboldt University or Cambridge University, or below the Thingvellor, the Icelandic site of the Althing—now, a UNESCO World Heritage Site. Throughout the series, he emphasizes the social structure underlying science and technology. So in the third episode, "The Grain in the Stone," Bronowski attempts to demonstrate that the stratification of the natural and social world reflect each other, inserting humanity and its society into a scalar structure suggested to him by the surrounding geology.

The history of science that Bronowski narrated connected the dots through well-known figures, from Archimedes and Euclid through Galileo, Newton, Darwin, to the more recent figures of the physical sciences, Einstein, Bohr, Heisenberg, Fermi, and John von Neumann. The seventh episode, "Majestic Clockwork," in which Bronowski discusses Einstein's relativity, revisits the trams that Einstein would have taken, propelling them to near light speeds, a very Tompkins-like exercise that once again produces an "Alice in wonderland world of relativity." But though he emphasizes the importance of adventure, the series in general relies less on the fabulaic and more on other synthetic generic devices.

The resulting synthesis is both scalar and foundational, on the one hand, and historical, on the other. Bronowski's justification for the weight he gives the physical sciences is revealed in the tenth episode, "World within Worlds." As a result of its penetration of atomic structure, initiated through an understanding of the periodic table in terms of atomic numbers, "physics becomes in those years the greatest collective work of science. No, more than that. The great collective work of art of the twentieth century." This truth is so universal as to demand that qualifications be stripped away. It is a point underscored with a dramatic perspectival shift during a walk through the salt mine at Wieliczka, near Kraków, a site that has since become inactive and, within a few years of filming, was also pronounced a UNESCO World Heritage Site. Bronowski's was a species truth told at a cherished species site.[15] But, in a chronology consistent with Bronowski's other writings, the roots of this collective artwork are older, with physics already demonstrating its glory in the seventeenth century. Biology and the human sciences are comparatively less in evidence, presented as newer sciences, saved for the later and later-dated episodes of the work.

Despite the relatively narrow set of specific men he celebrates in this foundationalist strain, Bronowski asserts that science is a "social enterprise," an idea he dates to the Industrial Revolution. As a result, the ascent of man is represented as a collective work, but with a special and essential role played by a few men of genius, stymied when their truths fail to disseminate broadly. It is also a picture largely consistent with his older writings, intersecting, for example, with his book *The Common Sense of Science* published two decades before. The older text and the newer series balance the assertion that scientific work is the property of all with depictions of it as the privilege of an elect. Its audience is universal, its great authors few.

The eighth episode, "The Drive for Power," suggests the degree to which the overt sympathies of the more mature Bronowski may have been departed from older socialist or Marxist models by this point, if not earlier still.[16] Though an apparently extensive MI-5 file had been compiled on him since the postwar years on the basis of his purported leftist views, the message he broadcast largely not only embraced the mechanisms of the Industrial Revolution, but also apologized for the conditions they required and the industrialists who gained by and promoted them. His portrayal of industrial leaders such as Josiah Wedgwood and the products for the poor that Wedgwood provided, set against an uneasy characterization of the state of workers in the factory, struck an incongruous note by comparison to that earlier leftism, his sympathies for which led Bronowski to be numbered among the long list of Cold War scientists investigated by their governments. In the series, he sought to defend these industrialists less in their capacity as factory owners and more as contributors to the ascent of man: "The men who made the Industrial Revolution are usually pictured as hardfaced businessmen with no other motive than self-interest. That's certainly wrong. For one thing, many were inventors who had come into the business that way."

Bronowski was not entirely consistent on such topics. He conceded that "the factory system was ghastly," and adding, "it was ghastly in the old, traditional way. Mines and workshops had been dank, crowded and tyrannical long before the Industrial Revolution. The factories simply carried on as village industry had always done, with a heartless contempt for those who worked in them." However, "the new evil that made the factory ghastly was different: it was the domination of men by the pace of the machines. The workers for the first time were driven by an inhuman clockwork—the power first of water and then of steam."[17] He admits here, almost poetically reinforcing the word "ghastly," that "the change in the scale of time in the factories was ghastly and destructive." Nevertheless "the change in the scale of power," which to his mind drove that change in the scale of time, "opened the future."

Across the episodes, the central periods of time in the camera's eye tended to be in forward march. The Big Bang was not in evidence. The account was largely prospective, and a fuller narrative of natural history was embedded in the series in the final episodes. In episode 9, "The Ladder of Creation," Bronowski depicted biographical accounts of Wallace, Darwin, and Pasteur. In the episode's final set of sequences, he emphasized the scientific work that Eiseley decried as myth: Stanley Miller's simulations of the potential first emergence of life and the very different cold environments theorized and experimented on by Leslie Orgel, a former member of the RNA-Tie Club and, like Bronowski, first-generation Salk Institute faculty member. "Generation upon Generation," episode 12, Bronowski devoted to Mendel and to Watson and Crick's discoveries, endorsing a common understanding, that biology in the twentieth century was becoming ever more deeply a science.

But in the tenth episode, Bronowski pulled this origin account further back in time. "World within World" depicted the contributions of, among others, Mendeleev, J. J. Thomson, Rutherford, Moseley, Chadwick, Fermi, and Bohr—turning back ultimately to Rudolph Clausius and Ludwig Boltzmann. Giving particular emphasis to this line of scientists, culminating in Hans Bethe's explanation of solar power in the fusion of hydrogen into helium, Bronowski declared, "Matter itself *evolves*. The word comes from Darwin and biology, but it is the word that changed physics in my lifetime." His own picture of cosmic evolution did not invoke any of the ultimate states of matter then proposed. Indeed, in the book accompanying the text, he cited the well-known *Reviews of Modern Physics* article coauthored by Margaret and Geoffrey Burbridge, William Fowler, and Fred Hoyle.[18]

Invoking Boltzmann and his formulation of entropy, upending a simple picture of the universe "running down," Bronowski presented order as a statistical fluctuation: "Statistics allow order to be built up in some islands of the universe . . . while disorder takes over in others." Where order arises, he gives a picture of stepwise movements from atoms through to more complex forms, explicitly marrying a scalar organizational structure to a historical account: "Evolution is the climbing of a ladder from simple to complex by steps, each of which is stable in itself." He endorsed this as "very much my subject," "Stratified Stability:" "That is what has brought life, not only here but constantly up a ladder of increasing complexity—which is the central problem of evolution. And now we know that that's true not only of life but of matter." Here, his celebration of physics is at its height: "Physics in the twentieth century is an immortal work. The human imagination working communally has produced no monuments to equal it, not the pyramids, not the *Iliad*, not the ballads, not the cathedrals."

Given the language of ascents and the rough Comtean narrative movement from physical to human sciences, the series suggests that cultural ascents are part of the selfsame process of increasing complexity. Of the physical process, he declares in tones redolent of those such as Teilhard or Hegel—Hegel, whom he explained in the next episode "Knowledge or Certainty" he "specifically detests"—that "the ascent of man is a richer and richer synthesis. But each step is an effort of analysis, of deeper analysis." In the text accompanying the series, he added to these words the ringing phrase, "world within world," here inflected by his visual and textual treatments of atomic and subatomic realms.[19] The spoken words are voiced over the image of a diamond cleaver "feeling for the structure of the crystal," chipping way to a glistering diamond—itself dissolving into a scene of Bronowski standing on site in Oak Ridge invoking commentary on the sixth day of Creation and the power linked to Chadwick's discovery of the neutron.

This synthesis, and the nuclear power it soon provided, had to his mind an ethical charge. At several points throughout the series, Bronowski emphasizes the social responsibility of science. In the most famous sequence, in "Knowledge or Certainty," he explains how Heisenberg's principle of uncertainty should instead be known as the principle of tolerance—the tolerance allowing for the unknown. This translates to a moral principle that he takes to be defied by the architects of the concentration camps, a point he makes as he bears dramatic witness to the suffering there—a debt owed, he explains, to Leo Szilard and to his own lost family members—bending down into the pond of the Auschwitz prison camp, pulling up ash from it, as the film slows.

In the episode itself, primarily through the figure of Szilard, he exculpates scientists themselves, a now long-standing conviction on his part, especially with regard to Hiroshima and Nagasaki. The "community of scientists failed," but only in their attempt to prevent the use of the bomb. He informs viewers of the dogged efforts of Szilard to prevent an actual strike. After a sequence of shots beginning with a burning clock melting at 8:15 A.M., the time of first strike on Hiroshima, Bronowski reported Szilard's retort to the idea that Hiroshima and Nagasaki represented the "tragedy of scientists that their discoveries were used for destruction": "Szilard replied, as he more than anyone else had the right to reply, that it was not the tragedy of scientists: 'it is the tragedy of mankind.'"[20] In turn, with a transition to the rising, creaking gate at Auschwitz over which the words "Arbeit macht frei" are visible, Bronowski begins the concluding sequence, arguing that the reduction of humanity to numbers was not a consequence of science, but a rejection of its ethos, which sees its own constraints. There is no contradiction, on this representation, between the progressiveness he claimed for science in the early 1970s and the military efforts of the early 1940s.

The topics of the final episode, "The Long Childhood," like the final event of other universal histories in circulation, are human justice and the mind, illustrated in part by an Eiseley-like encephalization of the hand. Set partially in his own California home, Bronowski reinvokes static and minority cultures such as the Bakhtiari, who "are as near to as any surviving, vanishing people can be to the nomad ways of ten thousand years ago," a period when children were expected to be iterations of their parents.

Here also he makes visible the recent concerns he had declared with Eiseley: "I am infinitely saddened to find myself suddenly surrounded in the west by a sense of terrible loss of nerve, a retreat from knowledge into—into what? Into Zen Buddhism; into falsely profound questions about, Are we not really just animals at bottom; into extra-sensory perception and mystery."[21] Having earlier tied the utopian literature of Thomas More "to the students riots of the day," he asked what had resulted in More's death and claimed that his execution was the consequence of his being "what every strong intellect wants to be," a "guardian of integrity." This retreat from knowledge amounted to retreat from Western universalisms, at least in the shape of scientific truths. His reflections also quietly recall his earlier consternation at youthful riot, reactions to youthful accusations that he regarded as working to deprive his generation of that guardianship of integrity.

Though he saw the intellectual leadership of the day as emerging through science, Bronowski once again rejected the idea of a world run by specialists, the "aristocracy of intellect" represented to him by John von Neumann. The rebellious young, the severing of scientific and humanistic cultures, and the privative intelligentsia threatened the further ascent of man at a moment when a nineteenth-century natural scientific promise was to be consummated, when there was "at last the bringing together of all that we have learned, in physics and in biology, towards an understanding of where we have come: what man is"—a full species-wide understanding, endangered by division and by non-Western conceptions.

Despite his fears, Bronowski's opinions were ready to be seen, heard, and read, at least in the short term. Geoffrey Wansell of the *Times* of London titled his review of the series "Bronowski, Messiah of Culture," concluding with the claim that "Dr. Bronowski admits to being messianic in his approach: 'It is part of my nature, I suppose, to save the human race. We are at a peculiarly important time in the development of society.'"[22] The text of *The Ascent of Man* made the *Sunday Times* bestseller lists, and the series was aired on public television in the United States in 1975.[23] By that point, it had already been screened at the Smithsonian in Washington, DC, to sold-out audiences. As reported in the *Washington Post* at the start of 1975, the series "comes to American TV with an additional impetus: The University of Virginia, the University of Maryland and American University are all allowing students to view the show as a full credited college course." The *New York Times* added to this list the State University of New York and Penn State, as well as the Universities of Alaska, Hawaii, Michigan, and New Mexico, among others.[24] A reviewer from *The Washington Post* included his own opinion that each individual episode "constitutes, in fact, exactly what an idealized college lecture would be: stimulating, poetic, eye-opening, intelligent, witty, wise, demonstrative—the more so because the film medium allows these segments to far exceed the potential of any behind-the-podium lecture."[25] Some professional historians were cautious, however, concerning its use in the classroom, finding Bronowski's historiography problematic.[26]

Nevertheless, *The Ascent of Man* was being used for college credit in the United States in several hundred colleges—according to some reports, as many as five hundred, and for twenty-five thousand students—by the time of the second broadcast of the series in the fall of 1975 (the first beginning in January of the same year). Miami-Dade Community College in Miami, Florida, and the University of California, San Diego, in La Jolla, California (among other schools) had constructed learning packages to go along with the series broadcast, packages tailored to more general or to scientific curricula, respectively, and available for broad university use in time for the first US screening. According to the *New York Times*, Dr. Robert McCabe of Miami-Dade had earlier organized a "widely used extension course employing television, 'Man and Environment.'"[27] McCabe is quoted as observing that "with a national program, we all share in what would have been economically prohibitive for any school to have done alone. We're all interested in getting high quality educational materials, and Dr. Bronowski's television series lends itself better than anything before it."

Bronowski encouraged and participated in such pedagogical efforts before his death in August 1974. According to Marjorie Hoachlander, who had been commissioned by the Corporation for Public Broadcasting to judge the success of the experiment, the publishing company, Little Brown; the sponsors, the Arthur Vining Foundation and Mobil Oil; and PBS itself were involved in efforts that ultimately led to mass mailings of brochures and the production of educational materials, including administrative and student guides.[28]

According to Hoachlander's study, the number of colleges that used *The Ascent of Man* from the fall of 1975 to the spring of 1976 was possibly closer to three hundred and fifty than to five hundred; the 252 survey responses she received constituted approximately half of the total number of schools contacted.[29] About half of the 182 administrative responses to Hoachlander's study (with fewer faculty respondents, 118) reported *The Ascent of Man* as their first academic use of television—understanding the series as an academic work of significantly better quality than other televisual materials. Roughly 10 percent of the institutions offered *The Ascent of Man* on the graduate level, and it was also offered as its own course, a replacement for a prior course, or as part of an already existent course. It was found in a variety of departmental homes, including history, anthropology, and, presumably, the natural sciences.[30] Underscoring the multiplicity of the uses to which the series could be put in a variety of environments, at different disciplinary levels (belying simple distinctions between popular and scholarly), and to different-sized audiences, Hoachlander referred to it as a "common multiple."[31]

The synthesizing arguments at stake in *The Ascent of Man*, partly developed earlier through prior essays, were becoming part of intergenerational educated public opinion. In classrooms and in wider intellectual exchange, the series acted as the basis for discussion, reading Bronowski's visual and textual statement as a more firmly rooted representation of science and scholarship. While it was less forlorn in tone and in argumentation than some of his personal essays might suggest, however, Bronowski's voice did not assume a mythic register, even if it flirted with the messianic, projecting a mythically scaled, historicized, and totalizing knowledge. If it did not last years in the future, that might have been as much or more the result of the success of the template it provided than a failure to resonate with what appeared to be an increasingly multiculturally minded audience. The next generation of public scientists overshadowed Bronowski's reception. Perhaps the most prominent among them were Carl Sagan, Steven Weinberg, and E. O. Wilson, with Sagan relying most clearly and directly on Bronowski's model.

MYTH, SCIENCE, AND NARRATIVE IN THE MID-1970S

From his undergraduate years in the mid-1950s, Carl Sagan was already connected to prominent scientists in different fields. He had worked in the laboratory of the genetics researcher Hermann Muller—a scientist numbered among those who linked political reform to the need to reform society scientifically (as with eugenics). Over the same years, Sagan had also worked with Harold Urey, a past political confrere of Shapley, Stanley Miller's advisor, and the 1934 Nobel Prize winner in chemistry for the discovery of deuterium. Urey was an important participant of the Manhattan Project and, along with work on isotope chemistry and separation, a researcher on cosmochemistry and the origins of life. While still an undergraduate, Sagan attended early defenses Miller gave on his results, work in which Sagan would remain strongly interested. So from his undergraduate years, he had been exposed to a number of scientific disciplines among scientists engaged by questions of origins theorized from different disciplinary points of view and relating to different environmental contexts.

As a graduate student, Sagan studied astronomy and astrophysics at the University of Chicago's Yerkes Observatory. His dissertation advisor, Gerald Kuiper, was regarded as one of the founders of modern planetary science, along with Urey, with whom from the late 1940s on Kuiper waged a fierce and extended debate on surface features of the moon, a debate still raging when Sagan took up these studies.[32] During these years, Sagan also met Joshua Lederberg, another in the collection of past or future Nobel Prize winners who took an early interest in Sagan. In the National Academy of Science, Lederberg had himself become the head of the recently formed Space Science Board's subpanel on extraterrestrial life—a connection of no small importance to Sagan.[33] Indeed, of the four papers Sagan submitted in pursuance of his 1960 doctorate in astronomy and astrophysics, the second, "Biological Contamination of the Moon," analyzed four different possible situations in which contamination might be possible and ultimately recommended that "all future lunar probes be scrupulously decontaminated."[34] Partly as a result of a 1957 conversation with Haldane, Lederberg had became concerned with the possibility of contaminating the moon and other planets in the course of space exploration and by potential politically motivated demonstrations of space supremacy. Through Lederberg's intervention, Sagan played an early role in NASA's exobiological efforts. By the early 1960s, Sagan was writing to Haldane about criticisms of George Gaylord Simpson on the probability of life elsewhere; it was criticism such as this that prompted NASA, according to Sagan, to determine whether planetary exploration was worthwhile.[35] Sagan's consultation and his wider advocacy were to help tie together the search for life elsewhere and the origins of life in the public imagination.[36]

But already back in 1957, Sagan had organized a public lecture series on the creation of life in the universe. He was able to entice George Gamow to participate, then still at the height of his fame. The idea for such a series would not have been new to Gamow; it might have reminded him of his co-organization of the Washington Conferences on Theoretical Physics two decades earlier, bringing together prominent scientists of the era addressing similarly far-reaching themes.[37] The precocious Sagan had spent the summer of that year working with Gamow in Colorado, where Gamow had taken up his own professorship only the year before. As a boy, Sagan had read

Gamow's works, and the chance to forge this connection with the senior scientist might have been a kind of consummation for him—all the more likely, given Gamow's work on the origins of the material world and of life and his popular speculations on space exploration. Gamow was far from Sagan's only scientific and literary model, however. According to his own recollections, he knew of Harlow Shapley from early in childhood, and later on, in his collaboration with his third wife, Ann Druyan, Loren Eiseley was noted as a particular influence.[38]

Sagan was in turn not the only reader of midcentury universal histories who would achieve scientific celebrity or even establish his own career as a successful popularizer. The physicist and future Nobel Prize winner Steven Weinberg was in his youth another of Gamow's admirers—and at least indirectly gained from Eiseley. Weinberg would later recall his excitement and inspiration when reading Gamow: "Theoretical physics, at least as Gamow portrayed it, had an element of the paradoxical, the counter-intuitive, to it; I felt that if I could understand theoretical physics I could understand anything."[39] In 1957, Weinberg was himself a young ambitious scientist, having just completed his own graduate studies at Princeton. The popular accounts of Sagan and Weinberg, along with the next generation of boosters of cosmology and theories of universal history, were not written until these younger scientists themselves rose to prominence.

During his graduate work, Sagan, however, was already achieving a certain notoriety, not just for organizing lecture series. In addition to consulting for NASA and the National Academy of Sciences on the question of planetary exploration, he was attracting press coverage (at times misleading) and at times irking faculty members.[40] Sagan's doctoral research was more broadly on the question of organic molecules on the moon and other planets, making an early attempt to produce a computational model of greenhouse gas effects on Venus—"The Radiation Balance of Venus," not yet published at the time of his submission. He advanced the picture that Venus is a hot planet, a view later accepted. The subtext of this doctoral work soon matured into what became known as "exobiology," the study of the possibility of life on other worlds.

On the basis of such work, Sagan took up a professorship at Harvard. And over the same years, Weinberg began to make a name for himself, as well, working in a variety of subjects at the forefront of theoretical physics.[41] In the early 1960s, while based at Berkeley, he began developing a research interest in astrophysics, one that persisted beyond his move from Berkeley to Harvard in 1966. If Weinberg's scientific reputation was to be the firmer of the two, neither in those years nor thereafter did he attract the attention Sagan did.

But it was not only from astronomy and physics that the next generation of popularizers emerged. In the 1960s, biologist E. O. Wilson was also working at Harvard. He had gone there as a graduate student in the early 1950s, having studied at the University of Alabama and then the University of Tennessee. While still a graduate student, Wilson recalled having been visited by Julian Huxley to discuss ideas Wilson had proposed concerning the demography of ant castes, which he (and apparently Huxley) had seen as following on the latter's research. These ideas Wilson in turn saw as important to his later work on sociobiology.[42] Wilson documented his own fascination with synthetic thinkers such as Ernst Mayr and Julian Schrödinger from an earlier period still, from his undergraduate years at the University of Alabama.[43]

By the early 1960s, Wilson was known as a prominent authority on ant species, having spent part of the previous decade in field studies in the South Pacific and Australia, among other sites. The early 1960s also saw collaboration with the ecologist Robert MacArthur on biogeography, a study charting and examining the conditions for species diversity in different geographical regions. This work led to further experimental field research in the Florida Keys in the mid-1960s to test the models of dynamic equilibria between immigration and extinction that MacArthur and he had proposed. Wilson's research interests were in many ways distant from his sometime Harvard colleagues Sagan and Weinberg, but questions relating to the organization of life and world, often conceived as distant from humanity, began to link them all in their treatments of broad, unifying themes.[44]

As with others, Wilson credited Haldane and the evolutionary biologist William Hamilton with some of the initial views relating to sociobiology, those on the "genetic evolution of social behavior," but not on "functional sociobiology," relating to the evolution of social structures.[45] Be that as it may, Wilson did not articulate the principles of sociobiology as a separate discipline until the 1970s, his theories meeting with strong support and severe criticism from the time of his publication of his 1975 text *Sociobiology: The New Synthesis*.[46] The more important point, for the moment, is that he drew on the work of those associated with the evolutionary synthesis across various parts of his different labors. Wilson's own appeal to sociobiology and universal history and the dilemmas they raised eventually reconfigured the ethical, political, and synthesizing concerns of a number of members of the evolutionary synthesis on both sides of the Atlantic, the Huxleys and

Haldane, Mayr and Dobzhansky, theorists, experimenters, and field researchers, among others. At the same time, Wilson's reflections on scientific myth were in dialogue with the contemporaneous presentations of those such as Weinberg.

SAGAN

In the 1970s, these three authors began to find a very wide public for works at times conceived and elicited by their editors and publishers. Carl Sagan was approached by the book producer Jerome Agel to write a volume on his professional efforts and enthusiasms, resulting in *The Cosmic Connection: An Extraterrestrial Perspective* in 1973. The book traces Sagan's extended engagements in three interrelated, overlapping parts: first, conveying a "cosmic perspective" historically rooted in scientific theory and in points of view advanced by Shapley, Haldane, and others; second, discussing aspects of the exploration of the solar system, to which his involvement with NASA and his planetary research was central; and third, looking at then-recent efforts in the search for extraterrestrial intelligence (SETI) and communication with other forms of life while disputing the claims of UFO enthusiasts. Overlap across the book is in evidence when, for example, in the final part, Sagan identifies the interest in life forms on other planets with the search for extraterrestrial intelligence, claiming that "in the period just after the Second World War, there was—in all of the United States—only one astronomer doing serious physical investigations of the planets, G. P. Kuiper, then of the University of Chicago. Not only had astronomers been turned off extraterrestrial life, they had been turned off planetary studies in general."[47] But he found that "new measuring instruments (a by-product of World War II)" had renewed interest in planetary astronomy, so that "young scientists have again been attracted to planetary studies, not only astronomers, but also geologists, chemists, physicists, and biologists. The discipline needs them all."[48] The understanding of the physical environments of other planets and the potential for life was once again tied to a synthetic disciplinary possibility.

The first few pages of *The Cosmic Connection* relate the history of five billion years, from the sun "turning on" through to formation of life and then intelligence. But this history also quickly moves from early human tribalism to a universalist humanism: "Many visionary leaders have imagined a time when the allegiance of an individual human being is not to his particular nation-state, religion, race, or economic group, but to mankind as a whole;

when the benefit to a human being of another sex, race, religion, or political persuasion ten thousand miles away is as precious to us as to our neighbor or our brother." Though it was "agonizingly slow," this was the dominant trend and necessary for survival: "There is a serious question whether such a global self-identification of mankind can be achieved before we destroy ourselves with the technological forces our intelligence has unleashed."[49] Sagan extended this species consciousness to interspecies respect: "The time has come for a respect, a reverence, not just for all human beings, but for all life forms—as we would have respect for a masterpiece of sculpture or an exquisitely tooled machine."[50] And this respect he observed as shared by some religious perspectives, such as Jainism, finding resonances with this species and science-centered narrative in specific cultural traditions.

Sagan advocated the perpetuation of what might be called a protomulti-culturalism for the advantage of the species as such. But achieving any representation of a universalist synthesis was difficult, even with regard to the representation of the human. The gender dynamics and racial overtones of the images produced for the identical 6-inch-by-9-inch gold-anodized aluminum plaques for the *Pioneer 10* and *11* spacecrafts had been the subject of debate and critique, for example. The external female genitalia were absent, the man, rather than the woman raised an arm in greeting, and to the eyes of many, the figures depicted a couple of European descent. This was despite "a conscious attempt to have the man and woman panracial." The idea of "pan-raciality" was a combination: "The woman was given epicanthian folds and in other ways a partially Asian appearance. The man was given a broad nose, thick lips, and a short 'Afro' haircut. Caucasian features were also present in both. We had hoped to represent at least three of the major races of mankind." However, the final engravings were not as Sagan had hoped: "The epicanthian folds, the lips, and the nose have survived into the final engraving. But because the woman's hair is drawn only in outline, it appears to many viewers as blond, thereby destroying the possibility of a significant contribution from an Asian gene pool. Also, somewhere in the transcription from the original sketch drawing to the final engraving the Afro was transmuted into a very non-African Mediterranean-curly haircut." Despite these lapses, Sagan still held that "Nevertheless, the man and woman on the plaque are, to a significant degree, representative of the sexes and races of mankind."[51]

For other elements on the plaque, relating to the hyperfine structure of the hydrogen atom, basic numbering notation, and representation of the initial vehicle trajectories, however, Sagan claimed: "We think we have written the message—except for the man and woman—in a universal language. The extraterrestrials cannot possibly understand English or Russian or Chinese or Esperanto, but they must share with us common mathematics and physics and astronomy. I believe that they will understand, with no very great effort, this message written in the galactic language: 'Scientific.'"[52]

Sagan's views seemed to project some multiculturalist concerns out to a cosmic level, to an ever-widening circle of who "we" might be, allowing for radically different lives and projecting back to the varied cultures on earth in a view asserting a species/human truth while provincializing at least some human concerns. On earth, though, the homogenizing emulation of the West that Sagan found in "non-Western, nontechnological societies" constituted a problem. "There must be some way to preserve the adaptive elements of our societies—painfully worked out through some thousands of years of sociological evolution—while at the same time coming to grips with modern technology. The principal immediate problem is to spread technological achievements while maintaining cultural diversity." On earth, diversity mattered because "Experiments in Utopias" were required, since the social systems could (still) not yet be captured through present theorization: "What is clearly needed are experimental societies," and abandonment of "many ancient traditions, world-views, and ways of life" might prevent the achievement of a more utopian condition.[53]

In *The Cosmic Connection*, this concern with diversity is writ large: "We read that 'life as we know it' is impossible on this planet or that. But what is life as we know it? It depends entirely on who the 'we' is." The primary context Sagan had in mind for this discussion was the evolution of biochemistry elsewhere in the universe, and the varieties of "chauvinism" that might mask the possibilities of life elsewhere, from oxygen, temperature, and carbon chauvinisms (to which Sagan admitted his own guilt) to sun-like star and planetary chauvinism.

Space travel itself thus held the key to an ever more expansive "we," but that claim to expansiveness could be dismissed as a falsifying and dangerous myth. In his 1971 book *The Invisible Pyramid*, Loren Eiseley had criticized the space race using the figure of an "invisible pyramid" as the structure of Western knowledge and its object, linking it to the disillusionment of youth that he also criticized.[54] The invisible pyramid was a figure for "a biological urge to complete what is actually uncompletable," to liberate humanity from its "cosmic prison" constituted of its language, culture, body, sensorium, evolutionary history, environment—"the cosmic prison which many men, in the excitement of the first moon landing, believed we had escaped still extends immeasurably beyond us."[55]

The fact of such critique was not new, on Eiseley's part and others. Over a decade earlier, Hannah Arendt began the prologue to *The Human Condition* with the flight of Sputnik, an event "second in importance to no other, not even to the splitting of the atom," revealing a new conception standing behind Eiseley's ambivalent metaphor, that of "the earth as a prison for men's bodies," a prison to be escaped.[56] In 1963, in an essay drawing on the works of Bronowski, Eddington, Einstein, Gamow, Heisenberg, Planck, and Schrödinger, among others, she reinforced these critiques arguing that even the vast expenditure on the space race was beside the point, given that scientists would do whatever they thought possible, regardless of consequences: "Man, insofar as he is a scientist, does not care about his own stature in the universe or about his position on the evolutionary ladder of animal life; this 'carelessness' is his pride and his glory. The simple fact that physicists split the atom without any hesitations the very moment they knew how to do it, although they realized full well the enormous destructive potentialities of their operation, demonstrates that the scientist qua scientist does not even care about the survival of the human race on earth or, for that matter, about the survival of the planet itself."[57] As a class or community, the fact that they collapsed or ignored the distinction between their actions as scientists and their actions as political subjects demanded that they be answered with an immanent rather than a consequentialist critique. The outlines of this critique were visible, she argued, in the inherent subjectivity of humanity's objective science, as quantum mechanics (voiced by Heisenberg) suggests in principle. An appreciation of this would reveal the inescapability of humanity's anthropomorphic and geocentric views, any meaningful success in escaping from which—in moving toward scientific viewpoints increasingly untethered from the world itself placed under a scientific lens—could only signify such a change in human nature, "some kind of mutation of the human race," as to equate to self-destruction. Here the danger instead was in a failure to appreciate the limits of a species voice to express such approximating universality intelligibly and in reference to actual experience, a danger entailing a collapse of meaningful language into the "meaningless formalism of mathematical signs."[58]

And likewise, Eiseley, who did not stop at making clear his indignation at the costs of the space race,[59] feared a species-suicide in human exploration of space. His 1953 article "Is Man Alone in Space?" had concluded with a report of a migration of "lowland butterflies" through the Himalayas: "High in that desperately cold and thinning air, the delicate-winged insects, strung out over a great distance in a long, flickering line, were moving upward! The tattered columns wavered; stragglers dropped frozen in the snow." Though their ascent meant death, they persisted "as though the march might have been boldly outward toward the moon. They were a living manifestation of discontent; they were life going about its immense business of changing worlds or perishing in the attempt."[60]

Despite such critiques, Sagan celebrated space travel as a universal "human enterprise," a claim that he elaborated across three chapters of *The Cosmic Connection*. In the third of these, "Space Exploration as a Human Enterprise: Historical Interest," Sagan reinvokes pyramid building. Urey had "perceptively referred to the space program as a kind of contemporary pyramid-building," and Sagan elaborated on the analogy: "the pyramids were an attempt to deal with problems of cosmology and immortality. In the long historical perspective, this is precisely what the space program is about. The footprints left by astronauts on the Moon will survive a million years, and the miscellaneous instruments and packing cases left there may last as long as the Sun." But he found disanalogies in the ancient Egyptian focus on the single figure of the pharaoh. A stronger analogy, to his mind, was with the Babylonian and Sumerian ziggurats, "the places where the gods came down to Earth and the population as a whole transcended everyday life."[61] Whether Sagan embraced what Arendt and Eiseley warned against depended on the extent to which that transcendence lost sight of or repudiated everyday life.

The penultimate chapters of the book relate a "kind of scientific fable," also of three parts, "Starfolk." It begins with a retelling of the Big Bang account in the idiom of the Book of Genesis: "Once upon a time, about ten or fifteen billion years ago, the universe was without form. There were no galaxies. There were no stars. There were no planets. And there was no life. Darkness was upon the face of the deep. The universe was hydrogen and helium. The explosion of the Big Bang had passed, and the fires of that titanic event—either the creation of the universe or the ashes of a previous incarnation of the universe—were rumbling feebly down the corridors of space."[62] The first chapter of that account, along with the Gamowian-inflected accounts of the Big Bang, went on to relate the stellar nucleosynthesis associated with Hoyle and his network, the formation of generations of stars, and the formation of planets via a briefly sketched nebular hypothesis, on to ever more complex formulations.

Sagan concludes the fable on the vision of science as a quest for self-knowledge in terms reminiscent of Friedrich Schlegel's aphorism that "Man is Nature creatively looking back at itself": "And then one day there came to be a creature whose genetic material was in no major way different from the self-replicating molecular collectives of any of the other organisms on his planet, which he called Earth. But he was able to ponder the mystery of his origins, the strange and tortuous path by which he had emerged from star-stuff. He was the matter of the cosmos, contemplating itself. He considered the problematical and enigmatic question of his future. He called himself Man. He was one of the starfolk. And he longed to return to the stars."[63]

Sagan claimed that as history, the fable had a firm scientific basis. It "is more or less what many modern scientists believe on the basis of available evidence. It is the outline of the emergence of man, a process wending through billions of years of time and driven by gravitation and nuclear physics, by organic chemistry and natural selection."[64] Drawn so quickly, the outline bears a thumbnail relationship to the evolutionary epics of the preceding century. For example, in comparison with Robert Chambers's account in *Vestiges of the Natural History of Creation* of 1844, Chamber's "one final comprehensive law," or the inorganic "GRAVITATION," is replaced by "gravitation and nuclear physics," and his "other great department of mundane things," which "rests in like manner on one law," "DEVELOPMENT," by "organic chemistry and natural selection." Among other differences however, Sagan related this as a more empirically minded and empirically constructed account at a time when the conservation of energy and the increase in entropy (not yet articulated with the first editions of *Vestiges*) were almost truisms and from a position of scientific epistemological and technological credibility in the Cold War period. The reformulation of Genesis did not only or even primarily act to present Sagan as a new biblical author. Instead, it partly effaced the modernity of the account, overstated to a degree perhaps the consensus, and cast the fable as a truth belonging to a very particular population—scientists bearing a privileged relationship to truth on the basis of unique practices that could be adopted universally.

The final words of the book end on a consummation of that vision, transforming the synthesizing work and account of the fragmentary sciences into a more celebratory, physical union: "The deaths of massive stars may provide the means for transcending the present boundaries of space and time, making all of the universe accessible to life, and—in the last deep sense—unifying the cosmos."[65] In the following year, *The Cosmic Connection* was one of the recommended texts for an introductory humanities-oriented course in astronomy at Cornell, initially cotaught with Frank Drake.[66] Sagan's work of unification was taking place across pedagogical and popular lines.

WEINBERG

In 1973, the same year that *The Ascent of Man* was first aired in the United Kingdom, Weinberg chose "The Early Universe" for the title of his dedicatory lecture for the new Harvard Science Center Building. At that point, he still felt it necessary to argue that "there was an early universe" as such at a time when "conditions were grossly different from what they are today."[67] The lecture, pitched for a general audience, was the first of a symposium dedicated to new advances enabled by observational space science. Weinberg set out to explain the primary evidence for the emerging consensus that the universe has a beginning and the ways in which this beginning is connected to fundamental physics.

Weinberg was mindful of what increasing numbers of cosmologists and physicists came to see and would come to see as confirmation of the theory, the so-called cosmic background radiation. The Gamow set had predicted that the contents of the very early universe would constantly be absorbing or scattering radiation—it would be "opaque" to radiation. But after new atoms had begun to form and with expansion the universe became less dense and hot, light would be able to travel much farther without scattering or absorption—the universe would be "transparent" to most radiation. As a result, this radiation would be released and should remain detectable in the present. It was this background radiation that was deemed to be discovered in 1967 by the physicist Arno A. Penzias and the astronomer Robert W. Wilson. The work was not primarily linked to Gamow or Alpher or Herman, the ideas for it at times credited to the Princeton physicists P. J. E. Peebles and Robert H. Dicke. (Dicke was then directing a small-scale search for the background radiation.) For those willing to embrace the results of the experiment—and some, such as Fred Hoyle, were not—it succeeded more fully in making cosmology an empirical science.

The central history that Weinberg relates is captured by what he repeatedly refers to as "a movie": a series of half a dozen slides tracing the history of the first three minutes at intervals determined by the ambient temperature in the universe dropping in successive factors of three. The freshness of the subject seemed evident in the apparent unintended laughter occasioned by Weinberg's claim that his first slide characterized the universe a millisecond after its creation.[68]

After showing the final slide of his film ("182 seconds into the history of the universe"), the universe shortly after the three-minute point of the universal clock, Weinberg makes another telling, now intentionally humorous remark: "Nothing of any importance has happened since that time." The "stop-action" film conceit resonated with the filmic and other documentary aids to the articulation of previous histories and syntheses of the world, those of Eddington, Lemaître, Shapley, Gamow, Hoyle, and Judith Bronowski. The quip in particular resonated with a claim made by Gamow that after the first hour following the Big Squeeze, "nothing of particular interest happened for the next 30 million years."[69] Gamow had first written this some two decades before in *The Creation of the Universe*, and let it stand in the second edition a decade later. His quip allowed for interesting events within the first *hour* and conceded that *something* new of interest had emerged some 30 million years later. The laughter Weinberg's facetious remark elicited pointed to something long-standing in Weinberg's sensibility and popular work: the extent to which for Weinberg, the crucial history of the universe took place long before the arrival of humanity on the stage.[70] By these lights, fundamental science treats the earliest things, and the earliest things are the most fundamental. The foundational and the moorings of the historical syntheses were thus bound together. For Weinberg, the crucial matter was not an all-embracing, epiphanic view of all history. Little other than the bare-bones sketch of a movie was necessary to capture the central moments of the history of the world.

At the invitation and urging of the president and publisher of Basic Books (an acquaintance of whom had attended the address), Weinberg eventually reworked the lecture, leaving the structure by and large preserved, producing his 1977 book *The First Three Minutes: A Modern View of the Origin of the Universe*. In many ways, it was *The Creation of the Universe* of the 1970s. In a retrospective assessment, Ralph A. Alpher and Robert Herman went so far as to claim, "In our view, the discussion of these [Big Bang] concepts by Steven Weinberg in his book *The First Three Minutes* . . . made a significant contribution to the widespread acceptance of the Big Bang model by many scientists, despite its targeting a general audience."[71] A blurb on the jacket of a first-issue copy of *The First Three Minutes* included an endorsement by Dicke: "This is an exciting book and I recommend it to both laymen and scientists. It is comprehensible to the former, significant to the latter and interesting to both." In this sense, too, it was like a third, more stoic edition of Gamow's generalist *Creation of the Universe*, targeting the scientist and layperson alike—and perhaps with success, if there was any truth in Alpher and Herman's belief. The book preserved Weinberg's little film as its structuring device.[72] Weinberg's jest was not written into the text, but its sensibility was maintained, relegating humanity to a chance, potentially farcical production of the essential principles of the universe, principles themselves firmly established within minutes of the Big Bang.

Weinberg begins the body of the text of *The First Three Minutes* with an Icelandic origin myth as found in the medieval *Younger Edda*, an origin story of primeval frost and fires, of giants and cows, which he dismisses as a "not very satisfying picture." But Weinberg would not go so far as to dismiss it entirely: "We are not able merely to smile at the *Edda*, and forswear all cosmogonical speculation—the urge to trace the history of the universe back to its beginnings is irresistible."[73] If Weinberg did not see the coevolutionary (cosmic to human) model as mythic in the same way that the Norse *Edda* was, he nevertheless saw both accounts as part of a wider cultural category: stories that could make serious claims on relating the origin, the history, and the essential elements of the world, claims that an entire society adopts or must seriously consider. His account was myth, then, in no dismissive sense, but a story based on empirically tested theories grounded in the authority of science.

Near the end of the book, Weinberg explicitly returns to mythology, comparing the possible future of the universe to the bloody ending of Nordic myth. Out of this discussion, he concludes despondently that "the more the universe seems comprehensible, the more it also seems pointless." Nevertheless, Weinberg suggests that there is some consolation in the fact of the research itself, closing with the only slightly less well-known claim that "the effort to understand the universe is one of the very few things that lifts human life a little above the level of farce and gives it some of the grace of tragedy."[74] The phrase recalls the first line of *The Eighteenth Brumaire of Louis Bonaparte*. According to Marx, "Hegel remarks somewhere that all facts and personages of great importance in world history occur, as it were, twice. He forgot to add: the first time as tragedy, the second as farce."[75] Weinberg's universal history reverses Marx's chronology—a world naturally farcical could be partly redeemed as tragic through the mission of science.

The same year that *The First Three Minutes* was published, 1977, also saw the release of Charles and Ray Eames's *Powers of Ten: A Film Dealing with the Relative Size of Things in the Universe*. Another prominent exobiologist and friend of Sagan, Philip Morrison, provided the narration in the place of Judith Bronowski. The departure point for the scalar quest was now Chicago, and the scales ranged across forty orders of magnitude, from 10^{24} down to 10^{-16}, penetrating deeper than Compton's -14 exponent, if well below 27. At the same time, in the spring semesters of 1977 and 1978, for example, the schedule for Sagan's introductory Astronomy 102/104 course listed lecture titles resonant with *Powers of Ten* and the same forty orders of magnitude, while anticipating the calendric, spatializing devices for universal time at play in the later *Cosmos*.[76] Along with Sagan's own popular work, the course attempted to take the scalar and temporal series as found in the cosmic perspective and scales advocated by those such as Boeke, Judith Bronowski, the

Eameses, Morrison, and Shapley and stitch them together with Gamowian and Huxleyan universal history—a version of which Weinberg had just begun to advocate himself.

WILSON

Weinberg was not the only scientist in the same period to appeal to universal scales or to the mythic or to link the scientific story to a perceived crisis of value. Eiseley was still a recent example—1977 was the year of his death. E. O. Wilson was also engaged in the production of a new synthesis of a different kind. Wilson's 1978 *On Human Nature*[77] "could not be a textbook or a conventional synthesis of the scientific literature," he wrote, but "its core is a speculative essay about the profound consequences that will follow as social theory at long last meets that part of the natural sciences most relevant to it."[78] Like Weinberg's book, Wilson's was based in part on previous lectures and articles, acknowledging connections to many other Harvard intellectuals, including Gerald Holton, W. V. Quine, and B. F. Skinner while paraphrasing Weinberg on the second page in the claim that "physical reality remains so mysterious even to physicists because of the extreme improbability that it was constructed to be understood by the human mind."[79] (According to the published record of the talk, Weinberg's was in turn an explicit paraphrase of a much repeated half-century-old statement of Haldane's that, "My own suspicion is that the world is not only queerer than we suppose, but queerer than we *can* suppose"—a claim Weinberg invokes partly to suppose otherwise, to conclude that it would be the "queerest thing of all" if this suspicion were wrong.[80]) This observation was in a sense central to Wilson's arguments, in which myth played the role often allotted to mathematical law in coming to grips with that reality.

In *On Human Nature*, Wilson claimed that contemporary scientific materialism produces a particular sort of story, one whose "narrative form is the epic," and declares that "the evolutionary epic is mythology in the sense that the laws it adduces here and now are believed but can never be definitely proved to form a cause-and-effect continuum from physics to the social sciences, from this world to all other worlds in the visible universe, and backward through time to the beginning of the universe."[81] For Wilson, "the evolutionary epic is probably the best myth we will ever have."[82]

What is the nature of this claim, its motivations, and how is that probability judged? For Wilson, "epic" and "mythology" aren't chance words. The specific thesis at stake is that there is a sociobiological explanation for the emergence of religion, rooted in its benefits in the construction of human identity. But, in turn, mythology, for Wilson, was one of the central, inevitable mechanisms sustaining both religion and the construction of identity throughout human history and into the current day. At the time of his writing in the late 1970s, there were three contemporary, large, world-framing myths that Wilson believed to be exercising political and intellectual conflicts: apart from "traditional religion" composed of ancestral myths and scientific materialism forming the scientific epic, Marxism for Wilson was a "sociobiology without biology," a mythology he saw as faltering and failing.[83] Though Wilson saw the evolutionary epic and traditional religion as having "zones of conflict" in which the scientific story proved its greater force, he dismissed the possibility that religion could be overcome, precisely because of the evolutionary forces that produced it. Marxism, on the contrary, had in his view dimmer prospects. The ultimate hope and plea of the book was that the evolutionary epic could be framed with a spiritual/religious force, a sentiment reinforcing the now long-standing call of Haldane and the Huxleys. If so, this could allow civilization to overcome the moral disorientation and valuelessness that Wilson saw as then endemic, resulting from the recognition that there was nothing that stood outside humanity to guide it, the first of the "spiritual dilemmas" that structured his book-length argument, dilemmas that readdressed and reformulated the questions of ethics, evolution, social planning, and untapped literary, axiological energy among the synthesizers preceding him.

The second of these dilemmas was related to the first—that ethics, though constrained by human biology, allows conscious choices to be made among our innate propensities. Here he addressed human nature as "a hodgepodge of special genetic adaptations to an environment largely vanished, the world of the Ice-Age hunter-gatherer." For him, modern society hid its origins and was "nevertheless only a mosaic of cultural hypertrophies of the archaic behavioral adaptations." Given the outdatedness of modernity, Wilson offered a more concentrated form of dilemmas surrounding the evolution of ethics in the Huxley line. "At the center of the second dilemma is found a circularity: we are forced to choose among the elements of human nature by reference to value systems which these same elements created in an evolutionary age now long vanished."[84]

Religion was one such value system for Wilson. To resolve this dilemma, he suggested that "the principal task of human biology is to identify and to measure the constraints that influence the decisions of ethical philosophers and to infer their significance through neurophysiological and phylogenetic reconstructions of the mind." This he believed would "alter the foundation of the social sciences" and produce a "biology of ethics."[85] Science could take up the nineteenth-century dream of a more satisfying science of society and right. By the light of a new ethics, it might be that religion and the "dogmatic secular ideologies" (Marxism) "can be systematically analyzed and explained as a product of the brain's evolution, its power as an external source of morality will be gone forever and the solution of the second dilemma will have become a practical necessity."[86] This practical necessity, Wilson believed, would be served by the evolutionary epic.

The third and final dilemma, which Wilson touched on only in the conclusion, was that as the capacity to change human biology grows through human genetics and other sciences, so will the question/dilemma arise of which changes humanity should pursue. "Will it remain the same, teetering on a jerrybuilt foundation of partly obsolete Ice-Age adaptations? Or will it press on toward still higher intelligence and creativity, accompanied by greater—or lesser—capacity for emotional response?" Here Wilson inserted an almost dramatic pause in the voyage that he envisioned for humanity: "But we are talking here about the very essence of humanity. Perhaps there is something already present in our nature that will prevent us from ever making such changes. In any case, and fortunately, this third dilemma belongs to later generations."[87]

This third dilemma reveals a crucial tension in Wilson's account, returning to the claim for the timelessness of the evolutionary epic: that though the future changes in humanity are unpredictable, nevertheless, the epic, Wilson believed, is the best account humanity will have of itself. On this view, what it means to be human might very well change, but humanity's narrative would not. The story would still be shaped by its earlier form, but Wilson claimed that the argument that humanity has evolved past the utility of religion probably would not also apply to the scientific epic.

Wilson held the scientific epic to be the best myth possible on the apparent basis that it yielded the greatest survival advantages. In this regard, he shared a spirit with mythographers of the day such as Hans Blumenberg, who detected in myth the desire to subordinate the world, to make it hospitable to humanity, and to establish a framework beyond questioning. But in being so keenly aware that the authorship of this myth was humanity's and that humanity could in his view evolve past myths tied to its sociobiological condition—as it was in the course of doing in religion—Wilson was flirting with, if ultimately sidestepping, a deeper dilemma. It was a dilemma akin to and also suggested by Weinberg's (and Haldane's) observation that the universe is likely not meant for human understanding. What form of myth or truth was humanity assured not to evolve past? What form could be assuredly true, regardless of the species that produced it?[88] And "best" by whose lights, at which particular cultural, historical, evolutionary moment or position?

What was quietly both raised and contained in *On Human Nature* was the question of whether there is any guarantee that the universe is given in such a way as to be able to be captured by human narration. In turn, it became unclear how timelessly secure were the truths linked both to humanity's subjective condition and to its evolutionary history. In the form of the question at play in Wilson's account, the issue turned on how evolution could produce the best of stories, if what amounts to the best is always dependent on an evolutionary struggle that always produces change.

Wilson (and Weinberg) did not raise these questions explicitly or directly, in Weinberg's case perhaps because he accepted the possibility of humanity's ability to appreciate transcendent truths even if from a nontranscendent position. And in Wilson's case, it was not the prime motivation of his work, which was to defend the new epic. The question of the future of the myth that he himself was helping to write was perhaps too far linked to the nature of the third dilemma, of how humanity would choose its own changes, the dilemma he left to later generations. In so doing, he also left the question of whether humanity could evolve away from the evolutionary theory that it itself had posited. In both announcing and disengaging from the questions of that future, Wilson revealed more about the curious epistemological condition of the contemporary myth, self-conscious of having its author as its hero and its hero as its author.

Directly after this, in the concluding paragraphs, Wilson invokes the "true Promethean spirit of science" that gives humanity "knowledge and some dominion over the physical environment." He concludes stridently: "But at another level, and in a new age, it also constructs the mythology of scientific materialism, guided by the corrective devices of the scientific method, addressed with precise and deliberatively affective appeal to the deepest needs of human nature, and kept strong by the blind hopes that the journey on which we are now embarked will be farther and better than the one just completed."[89] Though Weinberg did not endorse the view that the evolutionary story was either an epic or mythology in Wilson's sense, both Wilson and Weinberg revealed a shared register of the mythic, a shared anxiety concerning the valuelessness of life, and a shared belief that the emergence of and the answer to this feeling was strongly linked to the scientific.

MYTH AND SCIENCE

The increasingly iconic Sagan, by then based at Cornell rather than Harvard, continued to emphasize universalizing, scientific accounts. As did Weinberg and Wilson, he compiled addresses and essays written across the 1970s into popular books. One of these Sagan delivered as the first Jacob Bronowski Memorial Lecture at the University of Toronto. He expanded on this lecture to produce his 1977 book *The Dragons of Eden: Speculations on the Evolution of Human Intelligence,* a wide-ranging effort the major theme of which is the evolution of intelligence. Within it, Sagan also describes science in terms of myth, once more using the term positively, approvingly citing the ancient Roman historian Salustius: "Myths . . . are things which never happened but always are." Sagan clarifies: "I am not here employing the word 'myth' in its present popular meaning of something widely believed and contrary to fact, but rather in its earliest sense, as a metaphor of some subtlety on a subject difficult to describe in any other way."[90] Myths for Sagan were what "in former times [formed] the richest, most intricate and most profound" among the "insights obtained from human introspection."[91] Myths were thus both introspective insights and contours of knowledge, starting and ending points of figurative epistemic quests.

The consonance in spirit between Wilson's and Sagan's conceptions of the relation between myth and science was brought out still more evocatively in Sagan's 1978 *Broca's Brain: Reflections on the Romance of Science.* This book brought together and expanded on a number of lectures and addresses from the previous four years, including ones delivered to the American Psychiatric Association and the National Space Club, and a sermon before the Sage Chapel Convocation at Cornell University. In the final section of the book, "Ultimate Questions," Sagan explicitly articulates inquiries relating to the history of the universe. In the essay "Gott and the Turtles," like Weinberg, he contrasts contemporary experimental cosmogony with earlier myths, Vedic, Greek, and Asian, attributing to the latter a cosmology that had the world sitting on the back of a "world turtle," itself standing on another turtle and so on "all the way down."[92] In the final essay of *Broca's Brain,* the "Amniotic Universe," Sagan, like Wilson, suggests a naturalistic (if more physiological and individualistic) origin of religion, but also a potential physiological, "amniotic" origin to the possible choices of mythic/scientific universal histories. Sagan sees suggestive parallels between phases of human birth and recent theories for the origins of the universe: a steady-state, eternalist, stable universe might ultimately have drawn on the stable experience of being in the womb; an oscillating universe of perpetual collapses and contractions perhaps drew on the experience of perinatal uterine contractions and the Big Bang on the experience of a violent emergence from the womb, followed by the relative calm in the world immediately thereafter.[93]

The thought these ideas produced, like Wilson's, reflected on the confidence with which such universalizing accounts could be drawn. Sagan launched into a series of questions, almost restatements of each other: "Are we incapable of constructing a cosmology that is not some mathematical encrypting of our own personal origins? ... Can it really be that every possible mode of origin and evolution of the universe corresponds to a human perinatal experience? Is our ability to know the universe hopelessly ensnared and enmired in the experiences of birth and infancy? Are we doomed to recapitulate our origins in a pretense of understanding the universe? Or might the emerging observational evidence gradually force us into an accommodation with and an understanding of that vast and awesome universe in which we float, lost and brave and questing?"[94] This exploration of the biological origins of myth and the potential mythic character of scientific work was not unlike that of Wilson's "biology of ethics" and sociobiologically grounded epic questing narrative, if still more self-conscious of the potential limits of such an account.

Likewise, calling (knowingly or unknowingly) on a long-standing universal historical tradition, Wilson and Weinberg asked how humanity resolves troubling questions about its own self-conception. For Wilson, if science is mythology, what justifies its choice above other explanations/mythologies of the world? For Weinberg, if science promotes absurdist conceptions of history, how could individual meaning be redeemed? The answer that Wilson and Weinberg gave was the same: efforts in the production of the scientific epic itself worked to resolve these problems, demonstrating the evolutionary/survival advantages of scientific materialism, on the one hand, and revealing the dignifying tragic depth of the human situation, on the other.[95]

At the same time, as their more popular works revealed, these scientists were writing against the backdrop of an extended questioning of myth that included many of the theorists to whom their works briefly alluded or that they quickly paraphrased as a basis for speculation. Such speculations at times paid little heed to ethnocentrically minded sensibilities, sensibilities as little receptive to overtly universalizing claims as Dorothea Brooke became to Edward Casaubon's unfinished *Key to All Mythologies*. If myth and epic were to be drawn together in a finalizing account of the nature of everything, the potentially problematic nature of both concepts would have to be confronted in more depth. And both terms came laden with meanings not entirely commensurate with how, so far as their scientific authors were concerned, they needed now to be employed. Shortly after his sally tying together history and farce in the *Eighteenth Brumaire,* Marx had gone on to observe that "men make their own history, but they do not make it just as they please; they do not make it under circumstances chosen by themselves, but under circumstances directly encountered, given and transmitted from the past." With different emphases and interpretations, the new generations of universalizing historians—Sagan, Wilson, Weinberg, and others—were in different ways engaged in deliberations over Marx's further claim that "the tradition of all the dead generations weighs like a nightmare on the brain of the living."[96]

CHAPTER TEN

Cosmos and the Structure of "Epic Myth"

And so, in mythological terms, what is to happen now?
All of our old gods are dead, and the new have not yet been born.

JOSEPH CAMPBELL, *The Inner Reaches of Outer Space: Metaphor as Myth and as Religion*, 2002

Mythology is recreated by the poets in each generation, while science goes its own way.

NORTHROP FRYE, *Creation and Recreation*, 1980

By the late 1970s, prominent scientists located in significantly different disciplinary positions were paying attention to and advancing the idea that a new totalizing myth was at hand. In one sense, their claims were merely another iteration of those made by the several generations preceding them, announcing a synthesis dawning across the fragmented sciences—claims made from Shapley and Gamow, Haldane and Julian Huxley, into the present—or, still earlier, from T. H. Huxley, Spencer, and Helmholtz. These newer scientists simultaneously commanded both scientific and popular audiences. This attention at times provoked debate, given mounting suspicion of certain universalizing claims.

Nevertheless, for them and the audience that received and helped shape the expectation, what constituted such a myth and what such a myth might signify was uncertain. Prior to the 1970s, there was no *one* candidate mythology, no one scientific history that could command the consensus and attention of the scientific community however construed. At the same time, these synthesizers were writing and involved in producing television series in a broader cultural context that was not necessarily receptive to any universalizing gesture. Moreover, their understanding of mythology was not uniform, even taken individually across their own works. They explicitly, if loosely, drew on a cluster of theorists (Lévi-Strauss, Jung, and Freud, in particular, for Carl Sagan and E. O. Wilson) and texts (scriptural, antique, traditionally mythological) and endorsed a view that a profound and universal story was at stake in their work.[1]

Thus, while scientific authors in the latter third of the century persistently connected universal history and myth, they constructed this connection on the basis of varying conceptions of myth and its relationship to notions of an epic cosmic history. Their putative final stories exhibited conspicuous consonances at times also obscuring these differences in understandings of both myth and history. Though Steven Weinberg represented the scientific account of origins as a cosmogony sharing a strong enough relationship between past and present accounts that it could invoke and replace them, he did not imagine that a contemporary universal history could be either reduced to past myths or be dismissed as a falsehood. Sagan and Wilson argued to similar ends: the scientific account was greater than those that preceded or contended with it. Even when he explicitly speculated, as he did in *The Dragons of Eden*, that there might be inherent limits in the human ability to achieve a valid cosmogony, Sagan did not conclude that a scientific cosmogony is no more true than other accounts; he did not treat modern scientific history as on the same veridical footing as past myths. Instead, the greatest flattery he paid to past truths was a demonstration of the continuity of their line with present-day accounts or of the resemblance of their truths to science in the present.

Except through implicit comparisons or in characterizing the account given in *The First Three Minutes* as confronting "the problem of Genesis,"[2] Weinberg did not however generally describe contemporary universal history as myth or theorize the idea of "myth" as such. Sagan and Wilson did both these things to different degrees, despite the fact that "myth" was not the least controversial of terms. Both were aware of the ways in which such a term might mislead, at least insofar as each explained the specificities of their own usage.

Wilson's explicit definition of myth, for example, returned to the language of "tribe": "the narratives by which the tribe's special place in the world is explained in rational terms consistent with the listener's understanding of the physical world."[3] Though it was thereafter that he referred to the "scientific epic" as "probably the best myth we will ever have,"[4] he did not otherwise portray scientists as a special tribe, much as his definition also invited this identification. Nor did he suggest this epic was merely one among other tribal myths or scientists one among other tribes.

Other prominent authors of scientific universal history avoided mythic formulations and ambiguities, while presenting scientific universal history as an epic form. In courses at Harvard and thereafter in popular and more technical works, the astrophysicist and educator Eric Chaisson revised Shapley's "cosmography" into what he would come to term the "epic of evolution."[5] Chaisson had taught with George Field, the astrophysicist and director of the Harvard College Observatory, who himself had lectured on cosmic evolution in courses succeeding from and inspired by Shapley's long-standing interdisciplinary and curricular efforts.[6] Resonances of the terms of Chaisson's new cosmography were apparent in emerging origin-based initiatives, and were appealed to in attempts to incorporate scientific universal history within academic history departments. But Chaisson generally avoided mythic language, or used it to characterize a scientific account in a more moderate register to indicate the necessity of simplification.[7]

The increasingly confident projection of this scientific epic amplified and made more customary the understandings of "myth" and "epic" as the products of science rather than its obstacles, thereby reconstituting and further multiplying the terms' meanings. In this context, Sagan's 1980 TV series *Cosmos* merits particular attention, partly for its own place in relation to the popular and scholarly culture of the day and the ways in which it structured and spread the account that Sagan, as early as seven years before, had expressed as the scientific consensus. At the same time, it was only the most prominent example of efforts to produce and relate such universal history as consensus among students, scientists, and a wider public. In stitching together "epic" and "myth," Sagan and his coproducers broadcast an alliance of the terms with each other and with science, where the broad fact of that association was more culturally resonant than the specific conceptions making that association possible. Looking to contemporaneous thinkers who studied myth and epic brings out further the motivations for establishing the epic-mythic science of *Cosmos* and beyond it, clarifying the structure of the accounts that resulted, and the resemblances and resistances "myth" and "epic" bear not only to "science" but to each other.

COSMOS: A PERSONAL JOURNEY

From the 1970s and in force in the 1980s, the genre of universal history defined and established a widespread cosmological language and imagery. And, even with scalar syntheses partly at stake in complexity theories and the foundational syntheses soon to find forceful advocacy in the work of Weinberg, universal history rose to a decisive prominence. The most prominent site of this scientific epic at the start of the new decade, with its dilemmas and resolutions, was ultimately not textual and did not emerge until the new decade had begun. It was Sagan's thirteen-part television series, *Cosmos: A Personal Journey*.

Sagan was becoming increasingly famous through his popular science works, his NASA advocacy and consulting, and his television appearances. Already by 1976, he had been approached by a former NASA Viking mission colleague, Gentry Lee, with the ambition, as Lee recalled, of making a production company, "the Walt Disney of science and technology."[8] Shortly thereafter, KCET-TV in Los Angeles approached what Lee had coined Carl Sagan Productions for the purpose of making "13 one-hour presentations entitled 'The Heavens,' along the lines of Bronowski's 'Ascent of Man.'"[9] Production of the series, soon redolently titled *Man and the Cosmos*, began in 1978 and it was completed with a format whose structure, Ann Druyan has observed, in no small measure was adapted from Sagan's introductory courses at Cornell.[10] Originally designed and taught with Frank Drake, the course was intended for humanities students, modifying a basic course the Astronomy Department thought too challenging for them and not challenging enough for their science students.[11] The course did indeed canvass much of the subject matter that would appear in *Cosmos*, as it did Sagan's earlier works, particularly *The Cosmic Connection*, itself a recommended text for his astronomy course.[12] As with Shapley, the unanticipated laboratory for popularization was the classroom.

The television series was the product of an evolving, extended collaboration between Sagan, his wife, Druyan, one of his first graduate students, Steven Soter, and a full production team.[13] The promotional packet prepared for the series prior to its broadcast announced that "audiences will travel on the unique space journey" through the cosmos, the journey setting "astronomy and space exploration in the broadest possible human context."[14]

"For the intergalactic voyage on COSMOS, viewers will join Dr. Sagan aboard the 'spaceship of the imagination,' and it is through the spaceship's gigantic window that Dr. Sagan and the television audience will watch the journey through space." It was a spaceship that, in the conceit of one sequence in the film, transformed from a dandelion to what appeared to be a traveling starlike dandelion floret.[15]

The first airing of the series was on Sunday, September 28, 1980. It aired on successive Sundays thereafter, each episode one hour long, and the series as a whole was screened repeatedly throughout the decade. It was through *Cosmos*, itself a kind of weekend ritual for the wide audience watching it on first broadcast, that the idea of the new scientific myth was given its clearest and most evocative form to date. A shared, extravagant impossibility was constructed by the series: witnessing the origins, history, and exploration of the universe and, at same time, witnessing the work of those who had constructed that account of universal history.

Bronowski's earlier series (itself, again, a format explicitly modeled on Kenneth Clark's 1969 series *Civilisation*) and Sagan's bore a strong relationship to each other from the conception of the latter. Like Bronowski's series, *Cosmos* had a thirteen-sequence format, extended attention and comparisons to other times and cultures, filming on location, a classical (and in the case of *Cosmos*, often electronic) musical score, the use of computer simulation, and voice-over and narration by one scientific and inspirationally intoned speaker. Each, as well, produced a book along with the film, each book thirteen chapters long, with each chapter corresponding in subject and title to successive episodes of the film. Each, as well, had pedagogical spin-offs. The film producer and director Adrian Malone, who co-conceived *The Ascent of Man* project, acting as its director and co-producer, helped to craft the treatment of *Cosmos*, ultimately serving as a director and producer for the Sagan series, too.

Early in correspondence concerning the new project, Gentry Lee and Malone confirmed the connection explicitly, Malone affirming the synthetic function and novelty of such documentary series.[16] *Cosmos* thus was constructed in a sense as the American version of *The Ascent of Man*, updated by the decade of research in between and by the shift in documentary sensibilities, production values, cultural/national, political, and scientific contexts. And like *The Ascent of Man*, it was to establish a synthesis, but based more insistently on "cosmic evolution" or the "scientific epic" than was Bronowski's. In it, a species voice was sustained, although with an effort at least to address different cultures in the account, if not always genealogically, then structurally, drawing parallels between Western scientific and older non-Western cosmological beliefs.

There were significant differences in regard to the synthetic projects of Bronowski and Sagan. *Cosmos* focused much more strongly on astronomy and the idea of exploration—whether considering the past history of civilization, the present of space travel, or the future of humanity. Whereas the ultimate origin alluded to by Bronowski was stellar nucleosynthesis, Sagan rhapsodically projected the Big Bang cosmology. As one reviewer in the *Chicago Tribune* put it, *Cosmos* was "very much his [Sagan's] Big Bang baby."[17] The story, inflected by the theories of the Gamow set, was in quiet contrast to the more exclusive story of the Hoyle network told in *The Ascent of Man*. Likewise, to Bronowski's generally more lyrical and humanistic reflections, the Sagan series projected more totalizing reverie and rhapsody. And though Bronowski's *Ascent* was more directly apologetic for what he regarded as specifically Western commitments to the value of natural scientific truth, Sagan's was still less qualified in its embrace of a species and extraspecies totality while attempting to show sensitivity to non-Western truths.

A survey of the episodes, together with elements of their production process and certain telling sequences, gives a sense of the scope and roots of the series. The first, "The Shores of the Cosmic Ocean," launched three narrative principles that operate throughout *Cosmos*: Sagan on a spaceship questing through the cosmos; the narration, simulation, and invocation of the history of knowledge/science (in the first episode, through a simulated recreation of the Library of Alexander); and the "Cosmic Calendar," Sagan's calendar year of the universe. The latter condenses billions of years to one, a dioramic construction in which Sagan is able to "walk" through scenes of the days and months of the universal year itself. Extending these constructions, the following two episodes, "One Voice in the Cosmic Fugue" and "The Harmony of the Worlds," connect human evolution to the possibility of life elsewhere, invoking both the history of science—particularly Kepler's achievements—and fiction: Kepler's dreaming travels in his early seventeenth-century *Somnium*, what Karl Guthke has referred to as the first science fiction novel.[18] The second episode includes a sequence of sketches of different life forms evolving into later forms, ending in human form, repeated in the different episodes.[19] "Heaven and Hell," the fourth in the series, connects this organic evolution to cosmic evolution, examining the development of the planets, their births and potential deaths.

The question of the Mars canals was a more particular research interest of Sagan's, a problem he dealt with in collaboration with his student James Pollack. Through H. G. Well's *War of the Worlds* and Percival Lowell's explanation of the Mars canals, the next episode, "Blues for a Red Planet," contrasts history and science fiction to explore past scientific error and the current state of research. And consonant themes are explored in the following episode, "Travelers Tales," connecting past voyages of exploration to current voyages outward to space.

Sagan and his coauthors, Druyan and Soter, linked Sagan's personal boyhood interest in astronomy and tales of space to his mature studies. A representation of his personal journey was itself woven into the representation of an evolving universe and a journeying humanity, relating ancient myths about stars, relativistic considerations of space, time, and gravity, the Big Bang and other "creation myths." These are the subject matter of the next four episodes, "The Backbone of Night," "Travels in Space and Time," "The Lives of Stars," and "The Edge of Forever." In the concluding sequence of "The Backbone of Night," the seventh episode, Sagan teaches basic astronomy to a class in his own former elementary school. Here, in the classroom, he mentions to the students Shapley's work on globular clusters, invoking Shapley's favored trope of the suburban position of humanity and the earth.

In a quiet personal testimony, in "Travels in Space and Time," Sagan presents the relativistic demonstrations of the Tompkins series. The documentary first explains the basic principles of special relativity, specifically, the special nature of the speed of light. In the episode, as he enters a market square in Leonardo's home of Vinci, Sagan ties cultural relativity to special relativity by way of an emphasis on inertial frames: "Before Einstein, physicists thought that there were privileged frames of reference, some special places and times against which everything else had to be measured. Einstein encountered a similar notion in human affairs. The idea that the customs of a particular nation—his native Germany or Italy or anywhere—are the standard against which all other societies must be measured. But Einstein rejected the strident nationalism of his time. He believed every culture had its own validity. And also in physics, he understood that there are no privileged frames of reference. Every observer, in any place, time or motion must deduce the same laws of nature." The language of the accompanying text is still more definitive: "Europeans around the turn of the century generally believed in privileged frames of reference: that German, or French, or British culture and political organization were better than those of other countries; that Europeans were superior to other peoples who were fortunate enough to be colonized. . . . The young Einstein rebelled against the notion of privileged frames of reference in physics as much as he did in politics."[20] Modern science, in this account, is not in opposition to all forms of multiculturalism and does not endorse the correlates of all political universalist pretensions, imagined or actual.

In *Cosmos*, both text and film, Sagan thus espouses a particular multiculturalist sensibility that those such as Bronowski or Wilson and commentators thereafter saw as more inimical to science. At the same time, Sagan warmly speaks in a species-wide voice, with what could appear to be an ultimately uniformizing view of human wishes and wellsprings, a voice that clashes with the beliefs of those more suspicious of any science-minded universalism. In *Cosmos*, new cultures, new worlds, new cultural truths also come with new or variant physical laws, even if that physics is itself also universalizing in its ambitions. With two brothers, one on a scooter, the other awaiting him on a bench in Vinci, the film enacted a version of the "twin paradox," examining how two siblings age differently if one takes an outbound journey traveling near the speed of light and returns while the other stays home. To make this effect apparent during a short earthbound trip, Sagan instructs his readers in the accompanying text, "Following George Gamow, let us imagine a place where the speed of light is not its true value of 300,000 kilometers per second, but something very modest."[21] Examining the twin paradox in view instead of potential experience made actual by future journeying into space, Hannah Arendt observed, "It as though Einstein's imagined 'observer poised in free space'—surely the creation of the human mind and its power of abstraction—is being followed by a bodily observer who must behave as though he were a mere child of abstraction and imagination. It is at this point that all the theoretical perplexities of the new physical world view intrude as realities upon man's everyday world and throw out of gear his 'natural,' that is, earthbound, common sense."[22] Sagan, as with Gamow before him, attempted to find forms through which to reconcile common sense not only with uncommon knowledge but an embodied if still projected common experience of a modified world.

Here Sagan conceives a historiography that could be fundamentally altered, altered in the large, only by a careful selection of events: "History consists for the most part of a complex multitude of deeply interwoven threads, biological, economic and social forces, that are not so easily unraveled. . . . Random minor events generally have no long-range consequences, but some, which occur at critical junctures may alter the weave of history." Sagan speculates on the extraordinary example of the inevitability of the European discovery of America around 1500, given the economic and technological imperatives of the historical moment, noting that even if Columbus had not set sail, "the big picture would have turned out more or less the same." History and the possibility of altering it in the "big picture" focuses on such "critical junctures." Apart from them and in contrast to views of history suggested by rising scalar-minded complexity theories, history (both natural and human) is not presented as prone to ramifying complexity, despite the scale.

To this picture the story adds pivotal persons: "It is a lovely fantasy, to explore those other worlds that never were. If you had H. G. Wells's time machine maybe you could understand how history really works. If an apparently pivotal person had never lived—Paul the Apostle or Peter the Great or Pythagoras—how different would the world really be?" Borrowing the model from the 1960 filmic adaption of Wells's famous story, Sagan sits in the moving time machine, reflecting on the possible flourishing of the Ionian tradition and counterfactual histories stemming from it. But instead of moving through simulated worlds arising from the absence of such pivotal figures, Sagan imagines "a really ambitious time traveler" who would return to a seminal natural historical period—the origins of the solar system. He narrates a contemporaneous nebular hypothesis for the monogenetic origins of the sun and the planets. Alternative possibilities for the formation of the solar system, pictured through related simulations, suggested that physical evolution, like human history, is a contingent process, though how far that contingency might extend is unclear. Refracting narrative possibilities radiate out of this, evolution on earth and potentially elsewhere, the other worlds that the earth could have been and the evolution that brought humanity to its present point. Here, the series projects on to a cosmic backdrop its sequential simulation of organic evolution. The series as a whole builds toward the projection of a fuller historical and synthetic arc, from the Big Bang to the human present.

"The Edge of Forever" allows Sagan to expand further on shared universal inquiry across variant cultural frames. The episode introduces the language of myth in relation to contemporary cosmology, claiming that "the Big Bang is our modern, scientific creation myth. It comes from the same human need to solve the cosmological riddle." Sagan makes this claim over pastoral and quotidian images of India, setting modern cosmology in the context of more ancient beliefs, establishing cosmological inquiry as a profound human universal, "a hallmark of our species." While the series was in production, Sagan had written to Malone in a spirit that appeared open to multiculturalist discourses, looking to include non-Western developments.[23]

Perhaps in an attempt to avoid too narrow a frame, the documentary emphasizes the monuments, influences, and beliefs of countries from Egypt to India. Against the backdrop of Indian landscapes and religious artifacts, Sagan notes that "the Hindu religion is the only one of the world's great faiths dedicated to the idea that the Cosmos itself undergoes an immense, indeed an infinite, number of deaths and rebirths. It is the only religion in which the time scales correspond, no doubt by accident, to those of modern scientific cosmology. Its cycles run from our ordinary day and night to a day and night of Brahma, 8.64 billion years long, longer than the age of the Earth or the Sun and about half the time of the Big Bang."[24]

Cosmos attempts to balance cultural relativism and universal truth on the fulcrum of scientific inquiry, the spirit of which could persistently be traced to earlier belief and investigation. The next episodes develop these points through the more explicit themes of biological, cultural, and alien evolutions. Sharing a view with Lederberg and Medawar, Sagan claims that "the units of biological evolution are genes. The units of cultural evolution are ideas." Progress, whatever the culture, requires resisting destructive or authoritarian tendencies rooted in the lower brain. By contrast, Sagan stands before a model/graphic of a giant brain, explaining that "the cerebral cortex is, in a way, a liberation. We need no longer be trapped in the genetically inherited behavior patterns of lizards and baboons: territoriality and aggression and dominance hierarchies." This opens onto the possibilities of human transformation that had so much exercised Huxley, Haldane, and, in a different mode, Lederberg: "We are, each of us largely responsible for what gets put into our brains for what, as adults, we wind up caring for and knowing about. No longer at the mercy of the reptile brain we can change ourselves. Think of the possibilities." In this episode, those possibilities extend beyond the individual to a global technological culture and planetary intelligence.

The possibility of intercultural connection is made still more explicit in "Encyclopedia Galactica" in the transition from the successes and travels of Champollion to potentially more distant civilizations. Initiating a sequence describing the search for extraterrestrial intelligence (SETI) and the Drake equation guiding estimates of the number of alien civilizations for which communication is possible, Sagan sits at first at the base of the columns of the Temple of Karnac, noting that "today we also are seeking messages from an ancient and exotic civilization, a civilization hidden from us not in time but in space. Today we are searching for a message from the stars. . . . Extraterrestrial beings will have a different biology, a different culture, a different language. How could we possibly understand their messages? Is there in any sense a cosmic Rosetta stone? I believe there is. All the technical civilizations in the cosmos, no matter how different they are, must have one language in common, the language called science." As with *The Cosmic Connection*, science is depicted as the true universal language, a language so fully universal that it extends beyond human culture and embraces the cosmos.

In the final episode, Sagan clarifies what he took to be the possibilities of the end of history, at least human history. Contrasting the peaceful exchange between the historic scientific expedition of La Pérouse and the Tlinget people with the technologically empowered violent colonization of the conquistadors, Sagan concludes: "We revile the conquistadors for their cruelty and short-sightedness, for choosing death. We praise La Pérouse and the Tlinget for their courage and wisdom, for choosing life. The choice is with us still, but the civilization now in jeopardy is all humanity." The background of an expanding mushroom cloud clarifies the nature of the danger. In a reflective walk through a garden, sitting underneath a tree that by now the viewer is invited to associate with the Tree of Life, Sagan observes with gravity that all the destructive force of the Second World War, some two megatons of TNT, was now concentrated in one nuclear bomb, of which there were thousands.[25]

The episode speaks in these terms to the question of life, of whether a comedy (a conciliatory, redemptive account) or tragedy (an alienating account) is humanity's fate and its narrative form.[26] The primary discussion in this context concerns tragedy. But even here, there is a hint of what the comic, happy ending would be: choosing life, arriving at ultimate reconciliation. Throughout the series, that comic end is linked more specifically to the quest on which humanity is embarked, the increasing freedom to explore forever and together all of the universe, a quest that makes everything "ours," the universe as a whole enfolded in the embrace of humanity.

This is made particularly clear in the final rhapsodic sequence of the series, a crescendo of music and images patched together, synthesized, from earlier scenes, concluding on Sagan's words on redemptive possibility: "We humans have set foot on another world in a place called the Sea of Tranquility, an astonishing achievement for creatures such as we, whose earliest footsteps three and one-half million years old are preserved in the volcanic ash of east Africa. We have walked far." The next statement clarifies who this journeying "we" is:

> These are some of the things that hydrogen atoms do given fifteen billion years of cosmic evolution. It has the sound of epic myth, but it is simply a description of the evolution of the cosmos as revealed by science in our time. And we, we who embody the local eyes and ears and thoughts and feelings of the cosmos, we have begun at least to wonder about our origins—star stuff contemplating the stars, organized collections of ten billion billion billion atoms, contemplating the evolution of matter, tracing that long path by which it arrived at consciousness here on the planet earth, and perhaps throughout the cosmos.

Humanity as a whole constitutes this "we." That humanity is a local emanation of the cosmos, a process of evolution into form and knowledge redolent with Spencerian and Tyndallian proclamations of core mystery in the fact of lawful development. And with that species-wide self-recognition and voice also came a species-wide ethic: "Our loyalties are to the species and to the planet. We speak for earth. Our obligation to survive and flourish is owed not just to ourselves but also to that cosmos ancient and vast from which we spring!"

In this new myth, journeying heroes quest for the authorship of the new myth itself. In one sense, the hero is humanity in the context of the natural world, the species that speaks for the earth. On the level of the structure of *Cosmos*, its own narration across past scientific discovery and discoverers, the voice of the earth throughout history is heard from those writing this history, contemplating the stars to determine where humanity has come from. The voice of the earth is the voice of scientists in a certain sense, transhistorically understood, with the ancient cosmogonists and myth makers preparing the way for the myth of the scientific tribe of the present day.[27] This end glosses over precisely the set of dilemmas that Wilson attempted to provoke: the fact of authorship, of humanity as making its own meaning and deciding its own fate and purpose. Sagan embraces the resolution that both Wilson and Weinberg suggest—the value of the quest for meaning itself, putting aside for this final moment the need to distance science from the technologies that might prove humanity's story as self-authored tragedy and perhaps even farce.

RECEPTION AND CRITICISM

Though it was represented in part as a personal journey, Sagan narrates *Cosmos* as a confident and universal declaration of historical truth and destiny, a way of connecting Western science to other views of universal history. He repeatedly emphasizes what Gerald Holton had referred to as the "Ionian Enchantment," a phrase also adopted by Wilson. Across the episodes of *Cosmos*, Sagan relates a fuller evolutionary tale, from the Big Bang to the present, from ancient Greece to present-day Ithaca, New York, with humanity now distinguished by its ability to speak for the entire world, to author its own history, to quest outward into that world, and in so doing to speak that tale more fully.

Much of this was planned, partly on the basis of Sagan's resonance with audiences prior to *Cosmos*. In 1977, in the cover letter to Malone's introduction to the series—an introduction meant to attract funders—a member of Malone's staff explained that Malone expected to transmit that ability to establish such a resonance on the basis of Sagan's enthusiasm for science.[28] In the introduction, Malone characterizes Sagan in a way hearkening back to characterizations of and by Eiseley as Eiseley was himself celebrated in text and documentary, in prophetic and Homeric terms. Like Eiseley, Sagan would be an American seer, but one instead inspired by the space race, constructing the series around it: Sagan would travel through the cosmos in a filmic vehicle, itself designed to provoke an association with cathedral spaces.[29]

Sagan's saga took his viewers through different, overlapping stages and interpolated elements of the tale from week to week. A characterization of the initial *Cosmos* screenings as a Sunday ritual may threaten to mask the diversity of experiences in screening such a series to a wide audience. Scholarship in the phenomenology of film testifies to the importance of paying close attention to individual and diverse encounters with screenings. In her reading of Jean-Pierre Meunier and Maurice Merleau-Ponty, Vivian Sobchack points to the contingencies and specificities of the experience of the documentary—elements in what she terms the "documentary consciousness."[30] For Meunier and Sobchack, the classification and reception of a film—whether as *film-souvenir* (for example, a home movie), documentary, or fiction—turns on the question of the viewer's relative acquaintance with and attitude toward the film's subject.

Such phenomenological analysis necessarily depends on the uniformities of reception—of widely shared response—as often dictated by the larger context in which the documentary is screened. In Sobchack's terms, a film is marketed and received as a documentary on the basis of successful assumptions about its ability to provoke a documentary consciousness, which itself hinges on the extent to which the filmmakers can expect their theatrical or televisual audience to share a specific set of concerns and acknowledge the film as in some manner expositional. In the case of *Cosmos*, this points out the extent to which the evolutionary epic was taken as a matter *not* known to the wide audience for which it was designed—just as in *The Ascent of Man*, Bronowski regarded the project of science as only dimly understood by his audience.[31] The majority of the public, to whom the evolutionary epic was new, lacked a personal involvement with the crafting of that epic—and that lack of involvement acted to harmonize authorial (and production) intention and audience response, much of the public viewing the film as the intended expositional documentary. Likewise, direct pedagogical moments in the series—whether Sagan teaching astronomy to a classroom of children or, more directly still, walking the audience through an explanation of the Drake equation—participated in widespread pedagogical experiences, whether by way of elementary school lessons or past instructional programming. And, as must be considered below, there were still other pedagogical experiments deriving from the series, an important element in its reception.

The majority of *Cosmos* screenings in 1980 were at home, at the time of broadcast, making any sense of a typical screening both easier and harder to affirm and track. In details, the private viewings reflected an unrecorded diversity of experience. At the same time, however, the television format and airtime, the moment and extent of broadcast, were a strong shared constraint. This spoke to the differences not only in the typical watching of television versus cinema, but also to the different forms of circulation that the television narrative occupied. Nevertheless, in many ways, *Cosmos* expected, demanded, and, if critical response is a fair measure, received the attention of an audience resembling filmgoers, in the minimal sense that it promoted sustained viewing and scrutiny (even if itself critical and diverse) through much of its broadcast.[32]

The reviews at the time suggest that few, even among Sagan's critics, adopted a dismissive attitude toward *Cosmos*. Some commentators hinted at or complained explicitly of the vanity of the production or its cursory treatment of its content, particularly when it treated historical or religious materials.[33] The more negative criticisms saw in Sagan's popularizing voice and charm a more disturbing cultural weight. Five years later, with a wide audience and readership (including claims for *Cosmos* as the best-selling English-language science text to date), one journalist could refer to Sagan as "that charismatic tour guide to the cosmos" without significant fear of contradiction.[34] Nevertheless, there was variance in responses to Sagan, even among largely positive reactions.

Central trends in these reviews were the emphasis on costs/special effects/production values; the figure of Sagan himself (how effective a science advocate, popularizer, presence, and so on); and the amount, use, and quality of scientific and historical content packed into the series (the scope of the theories, the scientific details discussed, the specific discussions of exobiology), along with some limited mention of BBC precedents. A lighthearted, largely positive review in the *Washington Post*, "Cosmos—Public TV's Big Bang," referring to Sagan as "the Barry Manilow of Science," reemphasized the connection to modern cosmology: "It could be the biggest home-grown public TV hit (though produced in conjunction with the BBC) ever. Or it could make more of a big whimper than a big bang."[35] Looking ahead to the final episode and the synthesizing consummation it promised, the reviewer, Tom Shales, noted that in it, "Sagan will look into the future, such as it may be, of this particular planet in 'Who Speaks for Earth?' and 'Cosmos' special effects will compress the last 15 billion years, from Big Bang to the present, and recap the series." In including questions posed to Sagan, Shales also underscored the activist sensibility at play in the series: "In a sense, 'Cosmos' is not just science, but practical philosophy, a lesson in self-preservation, isn't it, Doc? Aren't you saying that this is such a wonderful old world, it would be grossly impolite for the human race to pollute it into a coma or blow it to kingdom come?" And though he endorsed the cosmic scope of the project, Shales also found moments that promoted meaninglessness; in reference to how small a space on the Cosmic Calendar the human race occupies, Shales concluded: "How very very impressive. How very very depressing."

The *New York Times*'s John J. O'Connor, by contrast, was more measured. Noting that Malone's previous efforts at the BBC included his work on "the superb 'The Ascent of Man' with Jacob Bronowski and the disastrous 'The Age of Uncertainty' with John Kenneth Galbraith," O'Connor presented *Cosmos* as neither the achievement of the former nor the disaster of the latter. He emphasized the variation in quality across the first three episodes, the material upon which his initial review was based.[36] "Grandly on display are some of the more spectacular special effects designed to reduce immensity to comprehensible visual concepts." Among these, he praised the Cosmic Calendar especially, but found that elements of the use of the Spaceship of the Imagination "border on the ludicrous." He noted that, "For a science illiterate such as myself, whose college exposure to science in the pre-Sputnik late 1950s was perfunctory and unimaginative, the series has many moments of indisputable value." Having already observed that "some of the material receives superficial treatment," he noted with apparent ambivalence that "an awful lot of information is packed into two [*sic*] small a space. Each hour could easily spawn its own 13-hour series."

The *Los Angeles Times* review of September 30, which focused more on Adrian Malone than it did on Sagan, noted that *Cosmos* was "the biggest, the costliest and the most ambitious program ever attempted by public television."[37] The review reported Malone's claim that "it was the special effects that almost frightened me out of my wits. At the BBC, the only special effects we do are with bits of string and sealing wax or maybe we blow up an old car in the parking lot."

Among the more caustic reviews was Richard A. Baer Jr.'s in the *Wall Street Journal*. Baer was then near to Sagan, at least institutionally, teaching at the New York State College of Agriculture and Life Sciences at Cornell. Sagan's cursory claims regarding religion and his treatment of history nettled Baer, all the more since Baer found in Sagan's own self-presentation a strong religious inflection—a point that ran enduringly through commentary on Sagan. As with others, Baer saw in the spaceship the religious, "cathedral-like" structure anticipated by Malone, observing that apart from the view from the ship's window, "the scene is like that of a priest presiding at Mass."[38] To Baer's mind, the scientists in *Cosmos* were represented less as a tribe than as the priest class, "the high priests of humanity." And *Cosmos* itself was Sagan's "religious testimony—a blend of nature mysticism, materialism and scientism." That testimony was the more problematic for its own flawed representations of past confessions: "Not only does Mr. Sagan ignore the role of the church in the rise of the university in the West, but when he claims that science in Kepler's day 'still lacked the slightest notion of physical laws underlying nature' he betrays his ignorance of the vital science discussions that were part of the life of these institutions already in the 13th Century." But Baer went further, not only to discredit Sagan's historiography, but to characterize his science as suspect, repeating and concluding with the charge of scientism: "But perhaps such a truncated and superficial historical perspective should not be too surprising to the thoughtful viewer, for Mr. Sagan seems less concerned to interpret history and culture sympathetically than to discredit rivals to his own scientism." Sagan as scientific priest or scientific prophet were leitmotifs of a kind, as "prophet" had been with Eiseley, and as culture messiah and sage had been applied to Bronowski and Bernal.[39] Such descriptions were variously invoked to indicate mockery, praise, and academic assessment.

The question of scientism hung over Sagan in relation to the latter episodes, in particular in his hopeful discussions of SETI and life on other worlds. One such review by Peter Gormer in the *Chicago Tribune* focused on the accompanying text to *Cosmos*, finding too much of the personal in Sagan's "personal journey" in the ardor and centrality of his support for exobiology.[40] Nevertheless, the same reviewer observed that the series and text "are providing the biggest classroom for his [Sagan's] theories that he could ever have hoped, backed by college courses, viewer guides, planetarium shows, etc." Likewise, at the start of the following year, Cecil Smith in the *Los Angeles Times*, naming *Cosmos* one of the ten best television programs of the previous year, noted that in the "controversial series," Sagan "approached the audience as if it were a classroom of pimply freshman," but that it was "compelling in its exploration of the vast sea of the universe in which ours is a most insignificant star."[41] A lesson once associated with Shapley was now regarded as a Saganism.

As *Cosmos* came to a close, O'Connor in the *New York Times* returned to the series, which already had "established itself as a phenomenon demanding, in varying degrees, admiration, substantial reservations and serious objections."[42] O'Connor noted again the production costs, but that these were already being partly defrayed by sales of the book *Cosmos*, which over the last weeks had "shot to the top of the best-seller lists"—testifying in turn to the *Cosmos* phenomenon itself. But once again, O'Connor compared Sagan unfavorably with Bronowski: "Unfortunately, Dr. Sagan is not quite as telegenically effective as Jacob Bronowski. Dr. Sagan tends to play too forcefully the role of 'merchant of awe.'" By Malone's own benchmark, the critical response to *Cosmos* was at times wanting, though the figure of Bronowski did not appear to persist as a strong association with Sagan or with *Cosmos*. More substantively, O'Connor seconded Baer's criticism in the *Wall Street Journal* of Sagan's historiography, particularly with regard to religion, portrayed in the series as "opposed to enlightened progress." To this, O'Connor added that "science is not as single-mindedly enlightened as Dr. Sagan would pretend." O'Connor resisted the view he saw in *Cosmos* that "science is approached as a monolithic repository of truth," finding disturbing Sagan's "grand pronouncements" and attesting opposing reader responses as evidence of the need for recognition of dissenting views. As with Gormer's piece in the *Chicago Tribune*, O'Connor held as particularly suspect Sagan's attraction to exobiology, indicting him for having "progressed blithely from his personal assumptions to grand speculations": "Dr. Sagan dabbles, imaginatively but dangerously, in pop science fiction," the kind that O'Connor thought sold to "pulp fiction and big-budget Hollywood movies." His final assessment was devoted to the last episode. In contrast to the views countering E. O. Wilson's science to non-Western or feminist-centered critique, O'Connor indicted Sagan for "dropping his mantle of the 'objective scientist'" as he "pleads for the causes of nuclear controls and, using the example of Hypatia in Alexandria, feminism."

Sagan's presentation of science could as well be accused of universalization as of overpersonalization, politicization, or, perhaps, its own variety of cultural relativism. These accusations were mobile—accusations of Sagan, but also of opposing parties in different scientific controversies: Wilson was accused of representing as scientific and universal views that were subjective, just as Wilson or his defenders could return the accusation that his accusers were essentially hostile to universalist claims, characterizing this skepticism as dogmatically antagonistic to natural science. What was understood as a universal claim at times depended on whether that term was itself taken to apply the social or natural world and of whether it appealed to one flag or other in ambitious cultural disputes.

The deliberation over the *Cosmos* "phenomenon" did not prevent its figurative classroom from being rendered literal in ways that recalled the pedagogical efforts using Bronowski's series. One early case of this was a "special seven-session" seminar given at the Chicago Museum of Science and Industry during the weeks of the initial *Cosmos* screening. Marilynn Preston in the *Chicago Tribune*, providing an early enthusiastic review of the series, cautioned potential readers that "your curiosity and willingness to explore new worlds is much more important to your enjoyment than the number of astronomy courses you've taken. The universe is not a Mickey Mouse course." In that regard, she made special mention of the museum seminar course, one available to the public or that could be taken for undergraduate or graduate credit with Governors State University that used the study guide prepared for the series. The seminar was led by Michael Turner, a University of Chicago physicist whose research work included contributions to theoretical astrophysics and cosmology.[43]

And like *The Ascent of Man*, *Cosmos* was integrated into still more creative pedagogical experiments. The following year, the Public Broadcasting Service offered a "television learning system," targeting students who could not attend university full time, and involving over one hundred stations ("at least 111 of 159 non-commercial TV licensees," according to the *Los Angeles Times*), and, by some reports, about five hundred colleges and universities, with up to twenty thousand students already enrolled by the fall of 1981. The program offered nine courses, one "the origin of the universe," following Sagan's *Cosmos*.[44] By 1984, the press was reporting that the program itself had expanded: "245 public television stations work with more than 600 colleges and universities to offer television courses distributed by PBS for credit." Over the two years preceding, "more than 130,000 people received college credit," with estimates that "millions more have watched for their own educational enrichment."[45]

A particularly interesting pedagogical use of *Cosmos* took place at the Center for Teaching and Learning in the University of North Dakota in the early 1980s. The center had been created in response to an invitation by the state to revise and reform public education. It hired teacher-educators together with mathematicians and scientists to refine and teach problem-centered, integrated pedagogical methods. A particular desire was to find a way to help elementary school teachers feel confident and enthusiastic enough in the sciences to engender enthusiasm in their students.

As an experiment, a group of faculty members focused on Sagan's *Cosmos*. In their report on the results of the project, the authors endorsed the reflections of Charles Silberman, who emphasized the idea of flexibility in education by appeal to a "project" the topic of which "cuts across the boundaries of subjects," ensuring that "artificial barriers do not fragment the learning experience."[46] In broad terms, there was a consonance between the theories of pedagogy informing this work and the integrated curriculum that Julian Huxley had advocated decades earlier.

The authors/experimenters chose *Cosmos* as "the vehicle with which we launched our experiment with integration." As they explained: "Here was a series of programs that dealt not only with astronomy and other scientific areas, but also with history, mathematics, geography, literature, human ambition, and human frailty."[47] To their minds, the third episode, "The Harmony of Worlds," was the "most memorable" of the series, because it "tied together the whole fabric of crucial events within a historical period" relating to Kepler's discoveries as an integrated account of culture and science. Drawing on the episode and further research stemming from or inspired by it, student-teacher groups prepared presentations meant to help structure potential elementary school exercises. Group presentations included dramatic performances of religiously persecuted scientists; taped "street-corner" interviews focusing on the question of life on other worlds; and students dressed "with long, pointed noses and straggly hair"[48] targeting the stereotypical portrayals of witches. According to Sherry Walton, one of the teacher-educators and the lead author for the publication of the results of the experiment: "Sagan's work was a way to try to give them entry into one area of science—how could the questions arising out of the series engender fascination for science for the student-teachers and the fire for their future students?"[49] She and her coauthors reported that "judging from the student responses and from our own perspective, the experience was clearly a success." To their minds, "*Cosmos* has served effectively to focus faculty thinking on teacher preparation and its impact on children," an impact that they hoped in turn would inspire their students "to feel more competent and committed to providing integrated, holistic learning experiences for the children with whom they will work."[50] Hopes similar to those Huxley had entertained for his scientific or evolutionary humanism thus were active in the reception of Sagan's epic myth.

MIRROR AND WINDOW

Wilson's characterization of myth as a cosmological and naturalizing explanatory tribal narrative or Sagan's as an almost inevitable explanatory

metaphor shared in theorizations of myth current at the time they articulated their ideas. Theories of myth and ideas of scientific universal history
evolved in relation to each other, whether with affirmations that they be-
longed to the same category or assertions that they had little to do with
one another.[51] What was implicitly asserted (with Weinberg) or explicitly asserted (with Wilson), that the construction of the new cosmology was doing
work continuous with myth, could nevertheless instead be denied as based
on a strong misunderstanding of myth itself.

Thinkers as varied in situation, disciplinary background, and approach
as Joseph Campbell, Roland Barthes, Hans Blumenberg, and Northrop Frye
discussed or referred to ideas of myth and modern cosmology, even as that
cosmology was hardening into a different shape than they at times were in
a position to appreciate. Campbell's own "natural history" of myth saw in
a rich and varied array of scholarship, from "comparative symbolism, religion, mythology and philosophy," a "new image of the fundamental unity of
the spiritual history of mankind," the "*membra disjuncta* of a unitary mythological science."[52] What he had repeatedly attempted to confront over the
preceding decades was "a natural history of the gods and heroes." Campbell's negotiation between archetypes or monads that were said to form universal mythic constants across different cultures and the unique developments of specific myths in different traditions paralleled a Darwinian view
of species requiring a negotiation between some stability and incremental
change. The analogy to scientific, post-Herschelian, post-Darwinian natural
historical synthesis was overt: "For, as in the visible world of the vegetable
and animal kingdoms, so also in the visionary world of the gods: there has
been a history, an evolution, a series of mutations, governed by laws; and to
show forth such laws is the proper aim of science."[53]

Quoting Yeats's "The Second Coming," affirming its observation that
"surely some revelation is at hand," by the early 1980s, Campbell asserted
and asked a question recurrent in different forms, but in reference to modes
of natural knowledge already apparent in the time of Comte: "The old gods
are dead or dying and people everywhere are searching, asking: What is the
new mythology to be, the mythology of this unified earth as of one harmonious being?"[54] For Campbell, Yeats was announcing the coming of a new
myth; Sagan and his fellow authors were in the midst of both announcing
and enacting it.

Campbell articulated two "services" of myth. The first and what Campbell held to be "most essential service" is "opening the mind and heart to
the utter wonder of all being." The second is cosmological, "representing

the universe and whole spectacle of nature, both as known to the mind and as beheld by the eye." To Campbell's mind, the new mythology he saw as emerging, the new "image of the universe," though it pertained to the lesser mythological function, would move beyond past cosmological thought in order to produce a new myth composed in relation to "(as of today, at least) an inconceivable immensity of galaxies, clusters of galaxies, and clusters of clusters (superclusters) of galaxies, speeding apart into expanding distance, with humanity as a kind of recently developed scurf on the epidermis of one of the lesser satellites of a minor star in the outer arm of an average galaxy, amidst one of the lesser clusters among the thousands, catapulting apart, which took form some fifteen billion years ago as a consequence of an inconceivable preternatural event."[55] If mythology is written by poets, for Campbell, it was clear that these would have to be poets with a grasp of the new cosmology.

However, different approaches to myth in this period at times emphasized starkly different aspects of the new myths being propounded than those to which Campbell pointed. In a semiotic approach that revealed myth in the everyday images of life, Barthes set myth and history—the particular history of the production of a myth, rather than the putative history such a myth relates—against one another in ways challenged and made consonant by this account. Though such a contrast was hardly new, he diagnosed the production of what he saw as bourgeois myths and the manner in which they circulated in the society around him.[56] At the same time, and by contrast, philosophers and intellectual historians such as Blumenberg saw in the essence of myth an attempt to render safe a threatening world. According to such investigation, modern physics was carrying out aspects of the work done by older mythologies, rendering an uncanny and unsettling world familiar and habitable. The contrast between that later physics and the early cosmological myths clarified for him the earlier working of myth.[57] Indeed, in a manner in aspects consonant with Campbell's species views, for Blumenberg, the idea of a *new* myth would involve an essential misunderstanding of what myth is.[58] Part of what determines what constitutes myth is a long-lived proofing, "a process of selection that has been at work for a long time": "No imagination could have invented what ethnology and cultural anthropology have collected in the way of regulations of existence, world interpretations, forms of life, classifications, ornaments, and insignia."[59]

Among work critically teasing out the functions, dispositions, and character of myth and narrative in this period, the criticism of literary theorist Northrop Frye is particularly apposite for revealing the narrative structures at play in the scientific epic articulated by scientists such as Sagan and Wilson,[60] providing historically consonant terms with which to describe and understand how the universal historical narrative is sustained across and inhabits the different media in which it is found: filmic, textual, programmatic, testimonial, or lecture-related.[61] And as Fredric Jameson suggests, the connection of Frye's characterizations to the social and to conceptions of power are immediate and transparent, helping to bring out more starkly the political and social dilemmas they encode and confront.[62]

Toward the start of the 1980s, Frye delivered a set of lectures on the subject of "Creation and Recreation," dwelling on the meaning of "creation" in relation to different discourses in which it operates. Frye described being haunted by the image of "the window of a lit-up railway carriage at night." At night, such a window would often be more mirror than window. "As a mirror, it fills us with the sense that the world is something which exists primarily in reference to us: it was created for us; we are the centre of it and the whole point of its existence." The mirror suggests that humanity is the axiological and meaning-making center. But when the reflection becomes partial enough to be transparent, "the mirror turns into a real window, through which we can see only the vision of an indifferent nature that got along for untold aeons of time without us, seems to have produced us only by accident, and, if it were conscious, could only regret having done so. This vision propels us instantly into the opposite pole of paranoia, where we seem to be victims of a huge conspiracy."[63] Humanity, from the view at the paranoid pole, is at the margins of first nature.

Frye's mirror-window figure inverts the medieval picture as summed up by C. S. Lewis, in which humanity is at the physical center, but at the spiritual perimeter. By contrast, for Frye in the present, humanity is at the physical margins, but at the spiritual or axiological center. There is an oscillation between poles, of mirror and of window, of nature without humanity and humanity in nature, the latter emphasizing man's centrality, the former depriving him of it: "The cultural aura, or whatever it is, that insulates us from nature consists among other things of words, and the verbal part of it is what I call a mythology, or the total structure of human creation conveyed by words, with literature at its centre. Such a mythology belongs to the mirror, not the window."[64]

This demonstrated to Frye how readings such as those of Weinberg in relation to the Norse eddas, or even of Genesis, were not germane to myth. Mythology "is designed to draw a circumference around human society and reflect its concerns, not to look directly at the nature outside." Reflecting and reflecting on one conventional understanding of the evolution of science, Frye claimed: "When man finally gets around to doing that"—to looking through the glass and his partial reflection out to nature—"he has to develop the special language of science," and "when it [science] does begin it does not descend from or grow out of mythology directly. Mythological statements about nature are merely grotesque or silly if they are thought of as pre-scientific explanations of it." Hence, as explanations of the workings of nature, such myths will always appear "not very satisfying," as Weinberg described one example. But in contrast to such views of myth as found in James Frazer's *Golden Bough* or Max Müller's *Comparative Mythology*, which he diagnosed as "mainly a by-product of a European ideology designed to rationalize the nineteenth-century treatment of non-European peoples," Frye asserted that "mythology is the embryo of literature and the arts, not of science, and no form of art has anything to do with making direct statements about nature, mistaken or correct. Similarly, as science does not grow out of mythology, so it can never replace mythology."[65] Mythology, he argued, is the continual province of poets, not scientists. But even in Frye's argument here, the space of mythology is negotiated in relation to the space of science. What is partial reflection and what is fuller transparency is determined in this argument by the scientific view of nature itself.

"WHO SPEAKS FOR EARTH?"
POWER AND FATE IN COSMOS

In the attempt to produce a schema of narrative as such, Frye emphasized a different aspect of myth than those that to his mind distinguished myth and science. Epic myths center on heroes and one of the distinguishing characteristics of the heroes suggested by these universal histories, whoever they were, is that they show little or no interiority, no mental landscape hidden from the outward world. Humanity and human intelligence in Sagan's "epic myth" are engaged in life-and-death struggles with truth and nature, not with personal, psychological questions. Even Wilson's dilemmas were framed as collective, not personal issues. This lack of interiority is one of the central reasons for turning to Frye's work here, because of the focus it offers on the exterior markers of character in attempting to describe the shape or genre of an account.

By the late 1950s, Frye had established a typology of narrative that mapped myth and other literary works according to the nonpsychological axes of the ontological status and power of the hero vis-à-vis the natural and social world, on the one hand, and the fate of those heroes, on the other.[66] Their fates are determined by whether the heroes are ultimately reconciled with or alienated from the rest of the world.[67]

The interplay of power and fate in the epic story of a hero in *Cosmos* can be seen in a five-minute animated sequence from the final, thirteenth episode of the series, "Who Speaks for Earth?" combining the previous separate strands of stellar evolution and organic evolution. The techniques by which the basic themes of power and fate are articulated in Sagan's universal history are illustrated most vividly via the special effects, to which a number of the reviews drew attention. These project the new natural universal history through the orchestration of animation, musical score, dioramic historical reconstructions, images of the sublime depths of space, and the abundance and variety of life, elements that acted in concert across the series to imbue it with a kind of moral authority. This was an authority that its critics deprecated, but that its wider audience often celebrated, with Sagan increasingly cast as a kind of prophet of, for, and through science.[68]

The visual images in this sequence progress melodically from a sparkling Big Bang through to the expanding early phases of the universe, the coagulating, emerging, vaporous cradles of stars, galaxies, to the formation of the earth, seas, through self-replicating life, helical DNA, the accelerated shifting forms of evolution in the "human evolution" sequence, up to the emergence of humanity in the figure of a sketched nude woman in profile, in midstride, standing before a figurative and naturalistically sketched tree. Sagan narrates this sequence in a reverentially toned voiceover, accompanied by first hushed and then the increasingly exultant electronic music. The sequence is both a summary of previous elements of the series—the animated sequence itself fused from the animations of previous episodes—and an attempt at presenting a fairly thorough account of the contemporary natural history of the world. The sequence culminates in the claim that "we," all humanity, are united in the universal inquiry into common origins.[69]

There are four prominent traditional elements of myth in such simulated/projected scientific accounts or stories: protagonists, conclusions, narrators, and audiences. First, there are the heroes or protagonists, those people or things to whom or to which things are happening, those whose fate is narrated in the myth. Second, the arc or plot of the story leads to that fate—to a definite conclusion, whether it ends well or unhappily. Third, there is the teller of the story and fourth, the myth is addressed to and constructs a discernible or implied audience. An additional element is of course the story's beginning. These elements relate to simple, prominent expectations:

a story relates a set of happenings that occur to a set of heroes for whom things go badly or not. It is told by someone or someones and to someone or someones.[70]

In reference to the simulated natural history in the "Who Speaks for Earth?" episode of *Cosmos*, the beginning is, in *some sense* (the ambiguity is in the materials), the beginning of all things; the end is the present day. Who, however, is the hero of the account? Who is the narrator, and who is the audience?

Largely speaking, according to Frye's typology, the heroes in natural historical synthetic narrative projected through *On Human Nature* or *Cosmos* (and in future series, as will be seen in later chapters), can be understood as romantic heroes in ways consonant with traditional, generic literary understandings, and this conventionality grounds the use of scientific terms in the synthetic histories developed by these authors. In Frye's schema, just below the domain of the mythic, where the heroes are gods—those who can make both the natural and social world bend to their wills—is the romantic, where the heroes are lordly humans, above their social world and in special communion with the natural world.[71] In the case of *Cosmos*, the hero of the series, viewed as a particularized identity, is the scientist as representative of humanity itself, a questing humanity that seeks to uncover the secrets of the origins of the world and life. It is science—and scientists—that here give humanity tremendous power over nature, life, and fate, the fate that humanity produces and confronts, just as Wilson had suggested. In this narrative, scientists are the delegates of humanity, its essence and its hope. Their science and technology are what give them social authority and arm them in their quest. The tremendous power of scientists, who are often represented as virtuous or even ascetic inquirers uncovering the secrets of the world, is romantic in this sense.

But viewed in the consonant but less particularized and characterological terms of what grounds and gives meaning to the scientific quest, of what birth and what fate is represented as of the most consequence (and so, straining in the first instance a straightforward application of Frye's typology), the hero of *Cosmos* is not the scientists as representative of humanity, its best and brightest, but life itself, a claim shared with the earlier accounts or drafts of Bronowski, Huxley, and Shapley, if not with those of Gamow or Weinberg. It is life that is represented as what is important and potent, the fuller flower or potency of which the audience waits to see emerge from the cosmos, and that is the focus of the tale. Over the course of the series, this hero nevertheless takes shape as more particularly human life and intelligence in the period of modern science, *now*. This follows the model of earlier popularizations, where human intelligence or consciousness is the hero, both for Sagan who made it the main theme of his book-length investigation in *The Dragons of Eden*, as well as for Wilson, who in *On Human Nature* called for it more in the voice of a kind of manifesto: "the human is subject to the network of causal explanation. Every epic needs a hero: the mind will do. Even astronomers, accustomed to thinking about ten billion galaxies and distances just short of infinity, must agree that the human brain is the most complex device that we know and the crossroads of investigation by every major natural science."[72]

The narrator of *Cosmos* is of course most patently Carl Sagan. But what allows him to carry out the functions of the narrator, as the audience would understand it, is his position as an authority. Indeed, the location of power in myth can be understood in relation to the narrator.[73] The narrator of myth can be understood as providing stories through which a society constructs visions of itself, its structure, relations, and even purpose, a claim common to the theorists of myth alluded to above and central to mythic work. Indeed, the poet who acts as a spokesperson for society, the epic poet, performs that role according to theorists as diverse as Frye, Mikhail Bakhtin, Theodor Adorno and Max Horkheimer, and Georg Lukács.

In this sense, the narrator—given voice in *Cosmos* by Sagan—could also indeed be called a mythic epic hero,[74] the scientist speaking for humanity. And that narrator is conceived as, in principle, not just speaking *for* all humanity but, ultimately, in intent, *to* all humanity as his audience. The history being related is universal not only in the sense of a history of the material world, but a history for everyone in that world. It is doubly universal, and so, it appears, doubly ambitious. This is not a universalism that Sagan hid: he celebrated it constantly, enjoining "us," his viewers, to celebrate ourselves as citizens of the cosmos. It is this universalism that he tied to the quest for our common origins, which at same time was a political exhortation against nuclear armament and, more still, against nuclear war.

It is with the element of fate, however, that a central aporia in all such accounts begins to arise. Humanity, the implicit audience represented in *Cosmos*, is not in a position to see whether its own story ends happily, but so far as the emergence of life is concerned, the long present is portrayed as a positive, if provisional moment. However, strongly at play in these materials and embedded as well in the historical moment is precisely the provisionality of the conclusion: How does the quest end—badly or well, tragically or comically, as Frye would term it?[75] The tragic possibility in the *Cosmos* series, for humanity and perhaps for all life on the planet, is that of technological and more specifically nuclear annihilation—not the end of the universe, but the end to the hero of the tale. Or the possibility that humanity's descendants will curse their ancestors as a result of technological choices speaks to a kind of alienation from future children, the future society of descendants, a danger assumed in Wilson's dilemmas—an alienation that in Arendt's possibility of the disappearance of humanity in mutation can emphasize instead the rejection of the present-day from the vantage point of that future horizon. An imbalance in the harmony of those descendants with their world, or their ascendants in the present suffused by an inhospitable future, could well be the fault of the technological legacy of today—in those terms, a properly tragic conclusion. Here, if the hero of this quest narrative is understood as science or scientists, ultimately, that hero might indeed become alienated from the humanity to which it owes its creation. In Sagan's time, this potential alienation was marked by the vilification of science for its role in creating the potential for nuclear destruction, an association already so dramatically different from that of Gamow's time, though perhaps evident as a possibility in Bronowski's apologetics or in Arendt's reflection on the readiness with which scientists make possibility actual. By contrast, the comic resolution that Sagan urged and promoted in *Cosmos* and in his NASA consulting and activism is a human reconciliation made possible by collectively searching the stars.

But while that specific simulation concludes with the emergence of humanity, it is a humanity that the series argues is facing the possibility of universal catastrophe: "choosing death." The history of the world that it relates focuses on the fates of the human authors of that history. By the light of the mythic claims circulating among these scientific authors and more generally, the specificity of this modern myth thus partly reveals itself as being open to the future in two central ways. First, the conclusion of this final story is, again, divided between happy or unhappy choices. It is open only in a highly delimited way in terms of its own account, once that account becomes stable. Only life or death are its stark options.

But, second, the stability of this account is put in question by the light of its own arguments. Scientists such as Wilson were aware of the historical character of their own production, at times in the thick sense of their personal biographical accounts, at times in the strict formal, epic sense of inscribing themselves into that epic. They saw and represented themselves in a tradition of other past cosmographers, mythographers, and scientists, symbols of the myths they promoted.

Theirs was a myth for which they, as short-lived authors, were looking to write a timelessly valid account submitted to an evolutionary process that could render their truths invalid in time, should humanity evolve past a belief in its own evolutionary story. This was generally not an explicit point, but one that animated the dilemmas they explicitly did see at work in their histories. Scientific authors opted for narrative strategies that had to produce closure to a story that remained in these senses open, testifying repeatedly to its inability to be concluded in any determinate way—as a matter of both its material scope and its historical ambition—even as the scientific authors were in the process of projecting its conclusion. Furthermore, the scientific epic was being posited, then and since then, as a site for the proposed resolution of wider social, political, and cultural dilemmas. The solution it proposed was its own composition, where that composition turned on an unprecedented reduction of empty chronology to meaningful time, shaping the story to the events it shaped, where the mythmakers were ultimately written into the myth being written, showing the marks of that composition. And through these marks, and in this position of authorship, the myth was vulnerable just as it was being assembled—but also just as might have been required by an idea of a new and of a secular myth.

EARTHRISE: NO CENTER, ONE FRAME

In 1976, the German newspaper *Der Spiegel* published a 1966 interview with Martin Heidegger, which they had agreed to publish only after the latter's death. After attempting to address his actions as rector of the University of Freiburg in 1933-1934, the interview shifted ground to a discussion of the technological inflection of the present. Heidegger's apprehensions had been particularly stirred up by the first picture of the earth taken from the perspective of the moon, a photograph transmitted by Lunar Orbiter I in August of that year.[76] Resonating with the concerns of his past student Arendt, and those in the starkly different idiom of Loren Eiseley, he declared: "We don't need any atom bomb. The uprooting of man has already taken place."[77] To this uprooting, and the larger disturbing planetary movements at play, he declared that "only a god can save us," a god who could not be summoned by philosophy or any other human effort: "we can at most awaken the readiness of expectation."[78] A new revelation was possible, and even necessary, but Heidegger did not see it as being at hand in the work of natural scientists of the day.

The relativity of the center, and the readiness for a new revelation, were captured for Joseph Campbell by another picture taken two years later: "The idea, it seems to me, is in a most appropriate way illustrated in that stunning photograph ... taken from the moon, and now frequently reproduced, of an earthrise, the earth rising as a radiant celestial orb, strewing light over a lunar landscape. Is the center the earth? Is the center the moon? The center is anywhere you like. Moreover, in that photograph from its own satellite, the rising earth shows none of those divisive territorial lines that on our maps are so conspicuous and important. The chosen center may be anywhere. The Holy Land is no special place. It is every place that has ever been recognized and mythologized by any people as home."[79] Campbell appeared to accept the mythological terms of Sagan's cosmic perspective in awareness and affirmation of the impact of space travel. And consonant with Sagan's appeal to Einstein, almost in dialogue with it, Campbell noted that "this understanding of the ubiquity of the metaphysical center perfectly matches the lesson of the galaxies and of the Michelson-Morley finding that was epitomized in Einstein's representation of the utter impossibility of establishing absolute rest. It is the essence of relativity. And, when translated from the heavens to this earth, it implies that moral judgments depend likewise upon the relation of the frame of reference to the person or act being measured."

Hans Blumenberg also found an affirmation in the earthrise photograph and the scientific understandings from which it emerged, but an affirmation that emphasized the earth as the cosmic exception, as the special frame that Eiseley too had stressed in his critique of the exploration of space. As Benjamin Lazier has pointed out, Blumenberg recognized in the context of space travel and the earthrise photograph the possibility of embracing rather than repudiating the earth. "Astronoetics"—Blumenberg's term for an intellectual exploration of and deliberation over space given the experience of space travel—"he might say, helps us see how in looking back at the Earth we escape some entanglements, but with a newfound sense for others we would like to affirm. Even if Earthrise was picture first, experience second, it could still prompt a transformation: a return *to* Earth by way of the rise *of* Earth in the pictorial imagination."[80] Thus the very relativity of science and the actualization of that relativity through space travel that troubled Arendt, was rendered as a condition for the possibility of celebrating the earth however noncentral its place or any place.

It is unlikely whether any such scientific or scientifically inflected synthetic account having the status of myth on such understandings was possible prior to or in the absence of a collective belief that humanity could rapidly and radically destroy itself and the planet or could transcend its physical environment as had been so recently experienced. The specter of that fate, the belief in that power, and the stark options in the application of that power were what were captured in the evolutionary epic's romantic conception of the scientist and in the question of whether humanity's story would conclude tragically or comically. Without belief in and evidence of that potential power, understood perhaps most evocatively at the time in terms of nuclear devastation and the exploration of sublimely represented distances and regions of space, a mythic account with humanity as its hero, as represented by the figure of the scientist, would indeed have seemed farcical. And given a conviction in that force, a myth in the absence of the heroizing of humanity would likely have seemed quixotic. If scientists were announcing their scientific epic as a new myth, a mixed genre of epic-myth, some mythographers, too, were willing to give that claim a hearing. They might at least acquiesce that a new myth was at hand, made necessary by the dissolution of old cosmologies, by the recent rising consensus over new cosmogonies, voiced in different emphases, scopes, and idioms by those such as Wilson, Weinberg, and Sagan.

However there were unique constraints on what some of these authors referred to as a new myth, shaping the poetics of this account. Its secularity had to be accepted, as an apparent function of the success of science itself and its immense scope. Science was either proving older stories false if read as straightforward historical accounts (a tone-deaf reading, as far as Frye was concerned) or demonstrating that they had no monopoly on natural history. Yet as the myth it was coming to be, the narrative had to be transcendent (beyond the specific situation of those writing it) and transcendental (participating in a realm of truth that might lie beyond the human). Human authorship was emphasized, and yet the objectivity or finality of the story was stressed. These constraints produced dilemmas on the level of its claims to knowledge, the difficulties of producing a story that was timeless even as it was regarded by Sagan and Wilson (if less perhaps by Weinberg) as connected to human organization and biological capacity.

It was a final story self-consciously produced through an evolutionary process at a particular position in time and space—at a particular time, at the final, minute area on the epic's cosmic calendar, at the initiation of a broad, new moment of a new human truth. And that position suggested to scientists such as Wilson that humanity could perhaps evolve past the evolutionary epic and/or have little choice but to accept it—as Sagan's amniotic universe and Wilson's sociobiological truths, for example—even if these were possibilities treated as marginal or unlikely.

As these authors understood it, this new myth had to be historically and metaphysically adequate to both the contemporary scientific world and the social world while acknowledging in a minor key that it might forever be inadequate to either. Weinberg's ambivalent invocation of Haldane's suspicion, that the human mind is perhaps unable to grasp the laws of the world, but nevertheless the source of an ambition to do so; Wilson's unlikely chance that humanity would produce a better story; Sagan's disquieting observation that our candidate theories of the origins of things were projections of our own births onto the birth of the universe: none entailed the falsehood of the scientific myth, much less its identification with earlier mythic or religious truths. But all noted that the subject position of the author of scientific myth rendered the potentially final story open to question.

Insofar as that narrative was open to uncertainty in its possible future or conclusions, and insofar as it still demanded refinement and understanding of its central events, it proclaimed this narrative uncertainty and made it central to the promise of meaning offered by the narrative itself. That promise required a delimited form of incompleteness in order to have a finality to perfect. Insofar as it was an inevitable account whose broad features and central processes and events were known, the narrative was already final. As acceptance and awareness grew, and the story was accepted by the scientific and broader public, the claim for myth broadened and deepened.

But myths are generally not new. Rather, as for Blumenberg attributing a "Darwinism of words" to the successful survival of mythic stories, myths are *essentially* old. Nevertheless, insofar as the historical actors we consider here embraced this term ever increasingly, insofar as their stories did the work of myth in reconciling dilemmas and organizing the world, and insofar as they did this according to the constraints of a contemporary period that demanded a secular account, the scientific epic began to define what a contemporary myth could and had to be. Its secularity, its circulation in secular spaces, did reveal essential differences from its mythic predecessors. Given what its practitioners believed science had done to the metaphysics underwriting and embedded in older myth or religion, those older myths were scientifically false, if poetically true. The contemporary dilemma was instead this: Given what human truth meant, its provisionality, its connection to its evolutionary history, its ever more provincial position in an ever wider world, its status more as sought after than received, how could a new myth be established if its authors could only be human?

OF AUTHORS AND HEROES

In the scientific epic in the form expressed by the authors and scientific institutions considered here, the universal story being produced and the story of those who produce it—those who carry out the story as a task—cannot be kept apart. The heroes of the universal story are the authors at work writing it, both because their own emergence is represented as a pivotal event in that universal tale and because of the power that they represent themselves as having in authoring and relating it. It is the power to tell a story of and for all—and by all, in the use of the collective subject "we"—but also the power to serve as its protagonist, not just its author, finding, forming, and voicing the knowledge that it relates, knowledge that holds open the possibility of great rewards and dangers, winning conquests or deserving curses: a modern universal history in the varied dimensions of the word "universal."

This entanglement of author and hero complicates any genre analysis taking the measure of the narrative through the power of hero or the conclusion of the tale. The power of the modern universal historical hero—whether of the scientist or of the human mind—is revealed in the sometimes self-conscious ability to author the story of everything and everyone, including that of the hero. This self-consciousness, however, reveals that humanity is the author of the story that in some sense is meant to stand outside itself.

The story does not generally emphasize the fact of humanity telling it, but simply goes about being told, even, paradoxically enough, as it shows humanity and its representative scientists in the act of its composition. As with most histories, this turns on the notion that there is a history of facts prior to there being a history of words: that natural history is there to discover facts and principles that are in some sense independent of its authors. But this act and relative self-consciousness of authorship involves the move from Wilson's first dilemma to his second, from the disorientation of an isolated humanity in modernity to the necessity to choose its own myth. These are moral dilemmas that Wilson saw as having been occasioned by a history of the universe that may suggest no particular place for humanity. As such, the creation of the earth, of human consciousness (and any possible consciousness), may not matter or, at any rate, could be written very differently, where humanity and intelligence are far from pivotal and could very well be excluded from the universe's history. And yet, as a structural matter, this is impossible, because the narrative cannot easily be sustained as compelling if the authorship is irrelevant. To dismiss the author threatens to dismiss the tale. Though a final story might thematize the irrelevance of humanity, and a final theory might make claims for the centrality of humanity according to an anthropic principle, the latter genre was not generally understood as a defense of human centrality, just as the former was not generally read as a calculated dismissal of it.

The conflict on this level can be understood as one between, as they might be termed, authorial and heroic codes: the author and the hero are in some sense meant themselves to participate in different universes, just as the frames of the subject/experimenter and the object of natural scientific study are to be held as far apart as nature allows. In a completed narrative work, the fates of the heroes of the story are at best tenuously tied to the life of the author, all the more so given the different lifetimes of text and author. Yet the scientific epic is on some level fashioned as a perpetual work in progress, even if its outlines and major points are clearly defined. And so the authoring of the tale brings into crisper outlines (from the perspective of the narrative itself) the nature of the human hero, which in turn clarifies the power of author and hero both. To know the hero better is to know the author better. That a story meant to be *about* "us"—where the species "we" are its heroes—is written and rewritten *by* us, where we read it to find out who we are, necessarily entangles the fate of the hero and the author, an entanglement that could be described as animating all three of Wilson's dilemmas. His third and final dilemma is that in mastering the scientific epic we might very well learn the ways in which we can modify ourselves, and so potentially modify our story.

And yet this is precisely the possibility Wilson inveighed against: "The evolutionary epic is probably the best myth we will ever have," he insisted. There is no evolving past it, a point in tension with Wilson's claims that we evolve past stories that are linked to older forms of our sociobiological condition. Wilson's account then demands asking what happens if and when humanity's organization changes drastically from what it is today. Can it preserve an epic that will seem as distant to it as religion seems to Wilson? Will humanity's descendants want to write a different myth about themselves and their world? In a secular grammar or secularly construed world, humanity is forced to know that it writes it own myths. The resonant solutions are variations on the same central theme: that whatever the future condition of species being and knowing, the universal human story must be maintained in the generic form of an epic.

CHAPTER ELEVEN

Political Cosmologies

It is not possible to judge the intrinsic worthiness of important experiments in different fields of science. It is fruitless, for example, to compare the importance of finding out more about the origin of the universe with that of finding out more about the origin of life.

T. H. GEBALLE AND J. M. ROWELL, "Funding the SSC," *Science,*
February 26, 1993

The fact of the all too human composition of varieties of mythic synthesis was more evident once such a synthesis guided the overt justification and organizational arrangement for widely discussed and debated collective scientific projects. The increasing importance of the modern universal history, the cultural and scientific commitment it attracted—its emergence as a cultural object and as an organizational scheme for science—affected and was affected by the rise of origins-minded, large-scale scientific projects taking clear shape in the 1980s and early 1990s. The blunt political fact that such large-scale projects drew on public funds marked only the most conspicuous ways that universal syntheses, their mythic expressions and genres, also constituted and depended on political forms. These projects invoked different synthetic genres, foundational and historical, relying on varying origins—biological and physical, logical and historical—as summoned in the expression of hopes to map the human genome and its diversity or to find more fully and finally the evidence of a universal material and organic evolution, and of necessity, they did so in deliberations in the public sphere.

Among these syntheses, the scientific epic as a historical synthetic genre had a singular resonance. This resonance made itself felt in the overt ways the epic represented and framed the activity of one notable scientific institution, NASA, in its emergent, overarching scientific mission. And in addition, NASA appealed to the epic in response both to its own perceived failures as a publicly funded institution and to those of other larger scientific projects.

NASA was facing and countering existential crises. As a number of historians have documented, the years following the broadcasting of *Cosmos* were among the most difficult for the space administration.[1] The *Challenger* explosion of January 1986 and the investigation that followed it had contributed to undermining public confidence in NASA. The catastrophe also set back the building of the Hubble Space Telescope (HST), intended to be the first to provide images free of atmospheric distortions, offering clearer evidence of the history of the universe.[2] And soon after its April 1990 launch, the HST project was itself in crisis. NASA engineers discovered a problem with the HST's light-gathering mirror, a defect that blurred its images. By the end of June, that problem was made public. These blows, with exposure peaking in a series of grim media reports that summer, undercut the reputation of NASA still further. With this "credibility problem," NASA was perceived as in crisis by critics within and without, and the institution looked to reinvent itself, not for the first time.[3]

Also looming in the early 1990s was the extensive debate over and eventual cancellation of the much-contested Superconducting Supercollider (SSC) project, a project the foundationalist conceit of which Weinberg most fiercely and publicly defended. The tendentious nature of that conceit, that other natural scientific phenomena are the more or less distant "mathematical consequences" of a set of basic laws, was revealed in professional resistance to the SSC, expressed in P. W. Anderson's critiques and in those of other physicists who refused to accept high-energy physics as the fundamental field in any significant sense.[4] Already on record dismissing such foundationalist views, Anderson was central among several prominent outspoken scientific opponents of the SSC.[5]

The contrast between a core, self-defining NASA narrative, inclusive in its historical emphasis on origins, and Weinberg's narrative, divisive, despite efforts to placate the wider community, appears with particular clarity in the narrative inhabiting the emerging NASA Origins Project. Emerging in the mid-1990s, the NASA Origins Project became an umbrella for different projects, many of which were already in play. The project outlined a twenty-year period (2005 to 2025) for some sixteen different investigations at varying levels of progress, from conception alone to data analysis, across seven different research areas, ranging from investigations of "black holes and structure in the early universe" to investigations of "optimal biosignatures for life on other worlds."[6] These investigations were laid out in succession in the series of *Roadmaps* composed by the subcommittee of NASA's Space Science Advisory Committee in collaboration with NASA field centers and academic astronomers from several institutions.[7] The latest edition of the *Roadmap* was published in 2003.[8]

The analysis of the political discourse and scientific narratives at play in these varied, but related projects will also help illustrate the distinction in these accounts between a final theory and a final story not just in terms of apparent language,[9] but also in terms of the different forms of organizational and conceptual synthesis toward which they tend, the different sensibilities and politics they address, and the different visions of finality, comprehensiveness, and futurity they put forward. Though the foundationalist and historical syntheses were not overtly or necessarily opposed, their differences (on one level, differences of *logos* versus *mythos*, of theory versus history) nevertheless established contrasting terms for defining the organizational principles of large-scale scientific projects in the post–Cold War era and offered alternate forms of justification.

ORIGINS AND EXOTICA: THE GENEALOGY AND STRUCTURE OF A NASA NARRATIVE

Near the turn of the new century, the appeal to origins remained a large part of the stated motivation for research in observational astronomy. In this regard, NASA was spearheading efforts to observe the universe not just in the visible range of light, but in the ultraviolet spectra (with wavelengths shorter than visible light) and infrared spectra (with wavelengths longer than visible light), as well. Hubble is primarily an optical telescope with observational capability in the visible and infrared regions.[10] In September 1993, the Association of Universities for Research in Astronomy (AURA), a consortium of universities founded in the 1950s with an interest in space telescopes, appointed the cumbersomely named "HST & Beyond" Committee. The committee was charged to produce a "consensus vision" for the future of ultraviolet optical infrared (UVOIR) astronomy for the next phase of research after the HST. AURA operated the Space Telescope Science Institute under contract for NASA, an institute that it helped found along with NASA in order to manage the research, information, advocacy, and outreach associated with the HST.[11]

The committee was appointed while the fate of the HST was still very much undecided. The *Endeavour* mission to service and repair the HST took place only that December, three years after the discovery of the flaw in the HST's light-gathering mirror. The success of that mission and the publicity that followed was a boost not only to the scientists who had long worked toward the construction of the telescope,[12] but to NASA itself. The success and the evocative images produced from the telescope's pictures no doubt encouraged the

"HST & Beyond" Committee in their meetings, the first of which convened the following April. In this context, in subsequent committee meetings and correspondence over the next two years and in line with other consonant space science efforts, a larger NASA narrative began to emerge in broad strokes.[13]

The committee, whose eighteen members were experienced with UVOIR projects including the HST, published their final recommendations in their report May 1996 report, "Exploration and the Search for Origins: A Vision for Ultraviolet Optical Infrared Astronomy." The document identified two major goals for future UVOIR research, worthy of attention: "(1) the detailed study of the birth and evolution of normal galaxies such as the Milky Way, and (2) the detection of Earth-like planets around other stars and the search for evidence of life on them."[14] These two large questions were presented as quests for "origins" and "exotica." Contextualizing these goals, the report characterized the present as "a golden age of astronomical research" in which "scientists have been able to explain major parts of the sequence of events leading to life on Earth."[15] NASA quickly embraced its findings, soon to be integrated into its Origins Project and its *Roadmaps* for future research.[16]

The extent to which the NASA Origins Project was constructed on narrative lines is better appreciated in light of this genealogy, which reveals much more explicitly the drive to produce a story bringing together and re-inflecting different research contributions. This was a narrative synthesis that could be used to justify additional expenditure for research programs already in progress, a circumstance of increasing and explicitly voiced concern to government-funded scientists in the post–Cold War political context. Thus, in the language of the "Exploration and the Search for Origins" report: "It may be decades before the discoveries of our time are absorbed by the majority of Earth's people, but many are following our journey closely. Many more will join in the years to come. What inspires them, as it inspires so many scientists, is the possibility of retelling the story of our origins, as we now read it from the sky, a story told in the relatively new language of science."[17]

The historical narrative thus was taken to be an essential part of the appeal of origins research to scientists and to the wider public alike, and the story-telling and historical nature of the research was invoked repeatedly throughout the report. Alan Dressler, the chair of the Origins Subcommittee, and Edward Weiler, chief scientist for the Hubble Space Telescope and then associate administrator of the NASA Science Missions Directorate, were two of its more prominent authors. The authors on the whole demonstrated a marked historical sensibility, invoking the history of both the cosmic and the scientific past, as well as speculating on the future of both science and the universe. In language very close to the earlier words of Sagan (and the later vision of the *Roadmaps*), the report declared: "Many astronomers consciously identify their investigations as part of a larger endeavor, one that we acknowledge as a fundamental human ambition. We seek, as did our ancestors, to tell the story of our origins, to describe our place in the universe and to understand how events transpired to lead to it."[18]

The appeal to the perennial created a sense of continuity across human history, characterizing humanity through such universal questioning/questing. In a consonant vein was the Origins Project's appeal to popular narratives of space exploration: "Many have not found it easy to follow our scientific findings, but very few have missed the point. This realization is perhaps why we find so many modern tales linked to space: a trip to the bookstore, or a glance at a movie listing or television schedule, is enough to convince any of us that, increasingly, great themes of human existence are being projected into space."[19]

The AURA documents reveal an intriguing self-consciousness on the part of practitioners as to the sensibility of their proposals. Linked to its title as well as to its two key goals, "Exploration and the Search for Origins" explicitly stressed the importance of both this search for humanity's origins as well as the lure of the exotic: "As the HST enters its prime years of scientific productivity, it can already lay claim to exciting discoveries that have captured the public imagination, in arenas of both 'origins' and 'exotica.'"[20] At times, these twin themes were promoted out of an unapologetic pragmatism: "This great appeal of astronomy may have several sensible explanations. We have found it particularly useful to consider two overall themes in research that have broad appeal to the public: (1) the quest for the exotic, and (2) the quest for place and origins."[21]

SETI

The strength of any case for undertaking a quest for the exotic and for origins as appealing or pragmatic was not entirely evident, however. As stitched into the HST report, accounts of the origins of life intersected with the search for life elsewhere. But NASA's funding for the search for extraterrestrial intelligence (SETI) came to an abrupt end the same year that the "HST & Beyond" Committee was established, in 1993, even though SETI's overarching universal historical narrative also bore many Saganesque markings.

In reference to the emergent field then known as "exobiology," Steven Dick and James Strick have observed that "from the start many academic biologists criticized the putative discipline, saying that, because there is no known life on other worlds, its creation amounted to establishing a field of science that has no subject matter."[22] Stephen Garber has argued that the reasons for the failure of SETI to sustain government support had less to do with the perceived value of the scientific experiments it proposed than it did with political timing. Garber proposes a number of other central factors, including the small scale of the project (which meant it lacked support from either contractors or lobbyists), the stringent federal funding mood, the political capital otherwise spent on getting through the Hubble Space Telescope problems, the ease with which the subject could be called into ridicule, the difficulty in aligning it with a specific subdiscipline (translating to shifts in divisional homes in NASA itself and to lukewarm support on the part of astronomers and life scientists), and the at least apparent inability to produce consensus.[23]

In this instance, a history of public debates might not have served SETI well, but such debates did occur. From the early 1960s, G. G. Simpson had criticized the idea of exobiology, explicitly targeting its funding. His argument was partly a disciplinary one—he believed biologists were being given short shrift in the exobiology research. In making his own assessments, based explicitly in part on Shapley's estimates of the number of earthlike planets in *Of Stars and Men*, Simpson adopted an interdisciplinary strategy, giving due weight to evolutionary lessons, arguing against any likelihood of contact with humanoid intelligence. Wolf Vishniac and Sydney Fox debated him in the pages of *Science* by disputing his scientific arguments, his understanding of exobiology, and his claims that any significant budgets were involved. Two decades later, Ernst Mayr undertook very similar criticisms of SETI, ultimately underlining the connection between his own critiques and the earlier arguments of Simpson. The scientific dissensus would continue to cast SETI as speculative science, at best, despite the project having won the endorsement of many well-established scientists.

Importantly, however, as far back as Simpson's critique, a possible exception to the implausibility of extraterrestrial contact that exobiology opponents were at least willing to deliberate over was the discovery of fossil, microbial life. Apart from the possibility that seasonal changes in Martian coloring were caused by vegetation—a conjecture that Sagan with James Pollack were soon to dispute in favor of the effect of windblown dust—Simpson had claimed that "the only other direct evidence for extraterrestrial life worthy of serious consideration is derived from meteorites. It has been claimed that some of these contain hydrocarbons of organic origin and even actual fossils of microorganisms."[24] Simpson was appealing to the recent analysis on and debate over the Orgueil meteorite, so called for the location in which it fell in France in May 1864. A 1961 study had concluded that "mass spectrometric analyses reveal that hydrocarbons in the Orgueil meteorite resemble in many important aspects the hydrocarbons in the products of living things and sediments on earth," evincing "biogenic activity."[25] In light of such potentials, Simpson added that "if confirmed, these observations would indicate that life (now extinct) had occurred on a planet of our system that has since been disrupted."[26] Though Simpson did not stress the point here, the possibility of such biogenic activity notably did not require a belief in a humanlike intelligence or any intelligence whatsoever.[27]

Overall, Simpson concluded that "in the present or any foreseeable state of our technology, the only way we could learn of other humanoids would be by their sending us a message or actually visiting us." Living or dead, this communication would to his mind require miraculous timing on the part of both senders or visitors, receivers or hosts. Identifying exobiology with the space program, he adopted a register consonant with Eiseley's during the same years, critiquing the misuse of "badly needed engineering and scientific manpower on space programs," costing billions.[28] When SETI was conceived and eventually funded, and doubts over the worth of the project were prominently aired by Mayr in terms revisiting Simpson's objections, Mayr stressed still more the fact that different disciplinary outlooks affect any estimate of likelihood of SETI's success. What Simpson had described as a "gamble at the most adverse odds in history" was realized as an actual gamble for Mayr, a small bet of five dollars between him and the astronomer Donald Menzel as to whether the NASA missions to Mars would discover life. Despite the fact that "determinism is no longer the fashion in modern physics," Mayr noted that "in conversation with physical scientists I have discovered again and again how strongly they still think along deterministic lines." He portrayed evolutionary researchers, by contrast, as "impressed by the incredible improbability of intelligent life ever to have evolved, even on earth."[29] Thus, in the mid-1980s, at a time when NASA was funding "definition studies" for SETI and when some political resistance had been overcome, Mayr concluded that "SETI is a deplorable waste of taxpayers' money, money that could be spent more usefully for other purposes."[30] His objections ultimately led to a debate with Sagan, which might in other circumstances have produced more interest, but by the time it took place in 1996 was well after the decision to rescind SETI's funding had been made.

The earlier responses to Sagan's *Cosmos* had suggested that newspapers and cultural critics who were otherwise receptive to Sagan's universal history were more resistant to his speculations on the possibility of life and civilizations elsewhere. The HST project made sure to emphasize origins together with exotica and to include in the latter a sense of questing that was comparatively more restrained in any specific speculations over what life might look like. But if, as Simpson had suggested, life were to arrive by itself and, as with the Orgueil meteorite, in a less presupposing form, a new origins program could launch from surer ground.

LIFE ON MARS

In May 1996, an ambitious and very Sagan-like story was taking shape in the context of NASA's HST research. That summer, an opportunity presented itself for NASA to seize upon and publicly endorse the themes of origins and exotica. In August, it announced the conclusion of a two-year investigation involving a number of scientists and scientific teams, including Johnson Space Center planetary scientists David McKay, Everett Gibson, and Kathie Thomas-Keprta of Lockheed-Martin, a Stanford team headed by the chemist Richard Zare, and "six other NASA and university research partners."[31] The team believed that they had found the "first organic molecules thought to be of Martian origin; several mineral features characteristic of biological activity; and possible microscopic fossils of primitive, bacteria-like organisms inside of an ancient Martian rock that fell to Earth as a meteorite." The disappointments of the earlier Viking missions in this regard could in some sense be redressed, but more importantly, the effort to find the signs of life elsewhere could refocus NASA's efforts. Though it was believed to have been on earth for some thirteen thousand years, a meteor with signs of life had in some sense crashed onto the public scene only in 1996.

The meteor in question, ALH84001, had been discovered in the Allan Hills area in Antarctica in 1984 by an expedition of the National Science Foundation's Antarctic Meteorite Program. The conventions of naming decided on "ALH84001," "ALH" for the Allan Hills ice field where it was found, 84 for the year of its discovery, and 001 for its sample number. The Johnson Space Center preserved it for study, but it was not until 1993 that scientists suspected Martian origins. Of the twelve other meteorites that matched the chemistry of Mars as measured by the 1976 Viking spacecraft, ALH84001 was the oldest. On August 16, *Science* published the team's results.[32] The paper began by invoking the earlier Viking mission: "A longstanding debate over the possibility of present-day life on Mars was addressed by the Viking Lander experiments in 1976." Nevertheless the authors observed, "Although the results were generally interpreted to be negative for life in the tested surface soils, the possibility of life at other locations on Mars could not be ruled out."[33] Appealing to a number of different observations of the rock,[34] the authors held, "Although there are alternative explanations for each of these phenomena taken individually, when they are considered collectively, particularly in view of their spatial association, we conclude that they are evidence for primitive life on early Mars."[35] (This is a claim now largely disputed; since then, still other meteors with other possible evidence have emerged as candidates for extraterrestrial life.)[36]

With NASA's announcement and publication in *Science* came a flurry of international attention, facilitated by dissemination over the Internet and through an accompanying proliferation of media resources. NASA defended a claim that some found difficult to believe, that there may once have been life on Mars. This brought renewed attention to Mars, to life elsewhere, and in turn to the origins of life, focusing NASA's efforts to overcome its own sense of crisis. In the extended wake of the discovery, Louis D. Friedman, executive director of the Planetary Society (a lobby cofounded by Carl Sagan, for which Sagan had served as president), entreated: "NASA can get back that can-do image." And: "Mars can return NASA to its origins—the idea of discovery, of surpassing limitations. These could be great years for NASA."[37] NASA's origins, the origins of life, and life elsewhere were all linked in this meteoric event of natural historical and institutional-biographical significance.

The discoveries also sparked wider political élan. On the same August day of NASA's meteor announcement, Bill Clinton reacted by proposing a space summit to be held to explore the implications of the potential discovery. The following month, on September 12, a congressional hearing was held on the question "Life on Mars?" Chairman F. James Sensenbrenner tied the news of possible life on Mars to the film *Independence Day*, a film in which the United States leads an international effort to save the world from alien invasion. Sensenbrenner also linked the discovery to the works of Steven Spielberg and to the celebration of the centennial of the serial publication of H. G. Wells's *War of the Worlds*—adducing all in all a "cultural fascination with life in outer space."[38] In what was not unlikely a veiled allusion to Arthur C. Clarke's *2001: A Space Odyssey*, in which more advanced life forms who had helped spur human evolution have left a monolith on the moon for humans later to discover, the associate administrator for space science at NASA, Wes

Huntress, remarked: "Now, the way I've come to look at this is about 13,000 years ago, a messenger arrived on this planet in the form of a meteorite. It landed in a very remote part of the planet, in Antarctica, and was buried in ice. And for 13 million years, it's waited patiently for the human species to get out of their caves and produce a civilized society and to develop the capability to go to such an inhospitable place and to find it."[39] His words were also redolent of the terms of Simpson's argument, of what skeptics such as Simpson thought might be necessary to make contact with life elsewhere. Microbial fossil life was not intelligent life or evidence of a more advanced civilization. In a romantic tone more rhythmically redolent of the literary Caesar, perhaps another veiled allusion, Huntress added: "We found it. We brought it back. We opened it up. And if we're reading the message correctly that it contains, it says, you are not alone."[40]

Here and elsewhere, he introduced connections to the Origins Project: "One of the things that I think is important to understand is just how the search for life on Mars fits into this new origins theme in space science and the agency. That program is directed towards asking some of the fundamental questions we could ask—where do galaxies, stars, planets and life come from? And second, are there worlds like the earth around the nearby stars? And if so, are they habitable or is life as we know it present there?" As much as a telescopic survey of the skies was an inquiry into the past, so was the search for life elsewhere: "And the search for evidence of life on Mars is as much about a search for origins as it is about Mars exploration."[41] Huntress concluded his testimony by announcing "an exciting new era of discovery and knowledge about the place in which we live," with the place in which we live understood in the broadest possible sense as everywhere, the entire universe.

Lieutenant General Thomas Stafford, a commander of Gemini and Apollo space missions, emphasized international cooperation, as did the other witnesses, their testimony falling within the still-recent shadow thrown by the Superconducting Supercollider debates over the nature and lack of such cooperation. But as with the leading role of the United States in the successful international effort that repels the aliens in *Independence Day*, Stafford ultimately tied exploration to national purpose: "However, I believe that before engaging in discussions with potential international partners, we should first establish the exploration of Mars as part of our national strategic plan." It was an exploration that would "rejuvenate our sense of challenge, our sense of competitiveness, and of national pride. As Americans, we must ask ourselves what our role will be in the human exploration of the solar system—to lead, to follow, or just step aside." Mindful of congressional budgeting, Stafford introduced costs into the high register of national meaning: "Let me emphasize that the safe and affordable exploration of Mars should not be our only goal, it should be our national purpose."[42]

A preliminary space science workshop was organized in late October 1996 (October 25) cochaired by Claude Canizares of MIT Physics, who served as director for its Center for Space Research, associate director for MIT of the Chandra X-ray Observatory Center, and (from 2000 to 2002) president of the American Astronomical Society. Apart from the Steering Group, the workshop consisted of four subgroups: the Universe, Planetary, Life, and Spacecraft Technology Subgroups staffed with prominent scientists in disciplines speaking to these large topics, each issuing series of reports sketching out loose chronologies. Connecting to the question of the pervasiveness of life, attention was given to extremophiles—creatures living in extreme conditions, conditions perhaps as extreme as those obtaining on Mars. Their example opened the door not only to adjusted expectations in encountering life elsewhere, but to the suggestion of humanity's origins potentially being still more foreign than anticipated, life still harder to define.

Sagan initially accepted an invitation to this meeting, but ultimately declined on the advice of his doctors. In his October 26 letter of regret addressed to Wesley Huntress and John H. Gibbons, assistant to the president for science and technology, Sagan reiterated the claim that 1996 had been "an extraordinary year for NASA," detailing a number of reasons for this fact: first and foremost the announcement of ancient life on Mars, but also including the discovery of planets orbiting ten "sun-like" stars, as well as the Hubble Telescope's images of the births of galaxies.[43] In this letter, Sagan noted:

> Many of these accomplishments, especially the first two [possibility of Martian life and planets orbiting sunlike stars], have garnered extraordinary public attention, and potentially give NASA a degree of public support it has not enjoyed in years. Most of these accomplishments are in space science and demonstrate the power of exploration to kindle the public's sense of wonder and excitement. In particular, the possible discovery of ancient life on Mars and the discovery of abundant planets around nearby stars imply a series of follow-on missions immediately apparent to the general public and extending into the far future.[44]

With these claims in mind, the workshop convened and set out a disciplinary agenda in a briefing book prepared for the vice president, *The Search for Origins: Findings of a Space Science Workshop*, which concluded with the summary:

The Hubble Space Telescope pictures of embryonic solar systems and the discovery of evidence for possible past life on Mars have created intense public interest in the ORIGINS of the Universe and its contents. These breakthroughs are astonishing returns being reaped from years of investment in many scientific disciplines, a major legacy of the 20th Century. Now is the time to leverage that investment into the 21st Century to create a profound new legacy of answers to the age-old question of ORIGINS.[45]

Sagan expressed his intent to attend the White House briefing, but his illness worsened, and he died the same month.

The briefing on December 11, held in the Indian Treaty Room of the Eisenhower Executive Office Building, brought together "a range of scientists, theologians, and other experts to discuss the implications of recent findings in space science and their significance for the future," as the White House press release termed it.[46] Among the actual attendees were Stephen Jay Gould, Jack Minogue (a Jesuit priest who was president of DePaul University and who had been a member of the NASA Advisory Council), Lynn Margulis (who, it might be recalled, was also Sagan's first wife), Bill Moyers, and NASA administrator Daniel Goldin.[47] Canizares gave the opening remarks, noting in regard to the October workshop that "this very diverse group, many of whom had never met before, came to the realization that their several scientific disciplines were all converging on a core theme, what we've called the quest for ORIGINS."[48] The briefing as a whole concluded that the origins theme should be a keynote unifying the sciences, a political effort endorsed by Al Gore.

The historical thus was formally endorsed as the synthetic genre of choice. In advancing this generic narrative, the space scientists were not only responding to their own past failures. They were offering a kind of belated retort to varied charges against the utility of at least some of their efforts, as had emerged, for example, in the context of the earlier SSC debate.[49] On this view, fundamental science did not need massive new expenditure or even sharply new efforts, and they could diffuse potential alarms they now understood better as a result of the controversies over those other origins-related projects. Canizares, for example, claimed in December that in regard to the questions of a new origins initiative suggested by the Mars meteor, "answering these questions will not require a major new commitment comparable to the canceled Superconducting Supercollider or the human genome project. . . . Rather, it requires pushing ahead with projects such as a new wave of robotic missions to Mars, ongoing and planned telescopic searches for planets around other stars, and continued efforts on Earth to explore the kinds of life found in the most extreme environments, such as volcanic hot springs, frozen lakes, and rock formations deep underground."[50]

AMBIVALENT FUTURES

The Origins Project took the origins of life and of the cosmos as what provided its thematic unity. This turned on the understanding that, as Sagan stressed, we are all stardust: "Only in the last few decades have we come to realize how closely bound our own existence is to the birth and death of these stars."[51] This rhetorical point, that humanity is constituted of the same stuff as the stars, acted to contract the universe to the human or to promote the human to the universal, in either case making all the universe accessible by casting scientists/the species "us" as romantic heroes, equal to nature. In this regard, "We now know that the universe was not born with these materials, but that the stars themselves are the sites of their manufacture. This discovery—that the heavy elements essential for a living being come directly from stars—ranks among the greatest human achievements in understanding the universe and our place in it."[52] Nothing then is other to humanity—we, in the greatest circle the pronoun entails, as everything else, are Sagan's star stuff. Conceptually, this might or should suggest conversely that humanity is not other to anything, that even physical materials are part of the circle of existence. Nevertheless, in these materials, humanity is given a special place. As with Sagan, the response to the question "Who speaks for the Earth?" is "humanity."

In a similarly spirited epilogue entitled "Origins Horizon" and specifically credited to Alan Dressler, the chairman of the Origins Sciences Subcommittee, the *Roadmap* asked: "Where did we come from? Are we alone? When the answers to these questions are known, our civilizations will evolve new visions of who we are and what our futures might be. Already we have learned enough to appreciate that the universe is enormous and ancient, but life tiny and transient is its precious jewel."[53] This peroration established the subject and object of its universal and historical quest again as this universalized "we," which itself takes its place in the history of life and more specifically of intelligent life. The language of history as quest is common throughout, establishing a vision of knowledge as historically and narratively constructed.

Throughout the *Roadmap*, as a conceptual matter, science is close to achieving a full narrative framework, a full final story: after all, even if in the distant future, the NASA Origins Project vision *will* lead us to the answers. More directly, although the greater detail of that large chronology is far from fully clarified, future discoveries/historical details are already nested in that framework. What it means *not* to know is also bound up in that same narrative, framing and inflecting questions as to when and how events took place. There are of course gaps and alternatives in the story, pockets of uncertainty, but structured as sites of future research.

The greatest of the alternatives lies in the future, with what consequences future answers bring. In this light, Dressler concludes the epilogue in questing and sublime registers: "Perhaps our descendants will praise us for our initiative, perhaps they will curse the relentless curiosity that propels humans into greater accomplishments and greater peril, but our part in this drama is preordained, its resolution beyond our time and imagination, barely within our dreams. We go on."[54] The present (circa 2000) then functions as a pivotal moment—and that as a result gives the present the resonance of something distant, of something beyond daily life, connected to an exalted chain of events. But this also brings vividly to attention the question of the conclusion of the quest, of the narrative.

The scientific cosmology here, just as in Sagan's *Cosmos*, hovers between ultimate tragedy and comedy—a future utopia of overcoming and reconciliation or one of tragic self-devastation. This account, then, is an open-ended one, its forecast for the future less decided, though the options are in broad terms sharply limited as the curse or praise of humanity's descendants, if they have voices to curse or praise. This openness might be said to mark a break with epic as it is generally construed, as an ossified generic form. But if it is a break, it isn't clearly a profound one, marking out, as it does with Sagan, two stark options alone.[55]

Taking the documents in their own terms, the stories they suggest emerged as part of a wider appeal that included and motivated the scientists themselves. The register of these quests for origins and exotica, which are conceptually twinned in the belief that humanity will find the conditions for the possibility of its existence in looking elsewhere for life, is again romantic. It emphasizes the power of the scientists uncovering the secrets of the world, gives them control through the manipulation of nature and holds them in a class apart from their society, but in forecasts for the future, considers the limits and potential misuse of that power. In terms of audience, the account is mythic, because it seeks to tell the historical truth for all of the world in the midst of an intellectual climate that, as Hollinger for one observed, often deprecated that ambition.

It is important to keep in mind that there are differences in stories of origins and stories of exploration, and each must be treated in light of that specificity. Nevertheless, from the time of the Mars meteor and the AURA reports through to the NASA Origins Project *Roadmaps*, these stories were explicitly connected in two ways. First, again, the exploration of the exotic was taken to reveal information about the conditions of the emergence of life as such, since observing the remote universe by the theory it explicates provides information about the universe's past.[56] Thus, the quest for the exotic was explicitly and theoretically grounded in the quest for origins. Even more directly, both origins and exotica were regarded as quests that "we" undertake—that is, the narrative of the scientific questing of the species as a whole. This is related to the origin narrative precisely in inscribing us (scientists or humanity) within it, in the possibility that we will bring the conclusion to our own story. And that fact, beside recalling the dangers aired in Arendt's critique or Eiseley's, reinvokes the acknowledged ambivalences at play in Wilson's dilemmas, in Weinberg's elevation of farce to tragedy, and Sagan's contrast in the choice of life or of death.

FINAL STORY, FINAL THEORY

Contrasts with other synthesizing narratives and genres not only put in relief the structure of NASA's synthesis but reveal the varying shapes of candidate scientific syntheses of the day. Such a contrast also clarifies further some of the historical connections tying different projects across different bureaus. Of these, the Superconducting Supercollider commanded the most attention at the time, both in primary and secondary literatures, at times smudging the line between them.[57] Only a summary of that history, focusing on salient moments and exchanges relevant to totalizing claims, needs relating here. It is a record of claims advanced in the genre of foundationalism and the critique and rejection of those claims from the point of view of historicist and scalar-emergentist syntheses, syntheses that we have seen at play in domains such as NASA and in debates over foundationalist biology. It thus is an instance of competing final stories and synthetic frames affording resistance to research on a foundational theory defended as a candidate final one, as government support for it was debated in the public realm.

In 1982, the Division of Particles and Fields of the American Physical Society sponsored a summer study in Snowmass, Colorado, dedicated to the discussion of the construction of a high-energy particle accelerator to operate at unprecedented energies. The idea had been circulating among high-energy particle physicists since at least the late 1970s, with workshops on the question of a "proton-proton collider" with energies of 20TeV (a million million electron volts).[58] On the basis of this meeting and further workshops at Cornell and Lawrence Berkeley National Laboratory, the following year, the High Energy Physics Advisory Panel (HEPAP) recommended the "immediate initiation of the project."[59] By the fall of that same year, feasibility studies were commissioned by the Department of Energy (DOE) and the directors of high-energy physics laboratories in the United States. These studies were completed the following April, and a Central Design Group was formed that summer.

As early as 1985, Steven Weinberg had alluded to the SSC as one of a series of "next-generation" elementary particle accelerators that would settle questions concerning the state of the universe at a very early age, when it was younger than a ten-billionth of a second. In honor of the initiation of the Instituto de Astrofisico de Canarias on Tenerife, Weinberg gave a lecture on the origins of the universe.[60] He presciently remarked, "Unless someone thinks of some ingenious new technique for accelerating particles, we are not going to be able to go much farther in this direction. The SSC accelerator would have a circumference of about 100 to 150 kilometers and would cost $3 billion. Perhaps the world economy will be able to afford even larger accelerators, but surely not much larger."[61]

Weinberg was forthright about the limits of the SSC and the need to look for other ways to answer what he regarded as the most important questions: "Much more is at stake here than just extending our understanding of the universe backward in time. For there is a scientific problem even more fundamental than the origin of the universe. We want to know the origin of the rules that have governed the universe and everything in it." The two sentences that followed, bear apparently formulaic but nevertheless important qualifications having weight in running and emergent debates: "Physicists, or at least some of us, believe that there is a simple set of laws of nature, of which all our complicated present physical and chemical laws are just mathematical consequences. We do not know these underlying laws, but, as an act of faith if you like, we expect that eventually we will."[62] That some physicists did not agree with Weinberg's general outlook and his demand for some kind of faith in the mission and outcome of high energy physics became increasingly evident.

In the lecture to the institute, Weinberg was emphasizing the laws behind the origins, the theory behind the history. For this kind of information, the only present-day hope was to find "relics" of those early times, hope offered perhaps through observational astronomy and the question of "dark matter"—the contention that given the structure of galaxies, there must be more mass than presently visible to astronomer's telescopes. The origins of the origins were not in the purview of the SSC.

Toward the conclusion, in a Saganesque analogy expressed in measured tones, Weinberg appealed to another element of origins research: "For one reason or another, the exploration of the universe plays a role for us today somewhat like that played by the exploration of the earth in the time of Columbus. In many nations throughout the world, these explorations awaken the imagination of the public, they gain the sometimes generous support of their governments, and they attract the most strenuous efforts of the men and women who undertake the explorations." This much was a trope of intellectual endeavor, individual or collective. But Weinberg added to this a universalist point: "One great difference is that, while the exploration of the world set the nations of Europe at each other's throats, the exploration of the universe has tended to bring them together." International centers based elsewhere were welcome examples: "In my own area of subatomic physics, we have the example of CERN, the greatest international scientific collaboration of history. In the Canary Islands we see the productive cooperation of Denmark, Ireland, Germany, the Netherlands, Spain, Sweden, and the United Kingdom, all working together, as on a happy ship to find new worlds."[63] In that ship, opposing multinational currents were overcome by strong intellectual winds, those same winds prevailing over past violence, launching a peaceful exploration of new worlds less menacing to the sailors themselves and, presumably, those catching sight of the sails.[64]

Weinberg made this a suggestion about the past and perhaps sustained the suggestion that the tendency for such peaceable collaboration was one that would persist into the future. But he held his greatest hopes open for the discernment of the answers to the universe itself: "We can trace the history of the present period of expansion back to its first million years, or its first three minutes, or its first ten billionth of a second, but we still do not know if time really began just a little before then, or if so, then what started the clock. It may be that we shall never know, just as we may never learn the ultimate laws of nature. But I wouldn't bet on it."[65] Over the next few years, what Weinberg would increasingly bet on was superstring theory.

The following year, for the 1986 Dirac Memorial Lectures on "Elementary Particles and the Laws of Physics" in Cambridge, UK, Weinberg expanded on his optimism for the future of science in a lecture entitled "Towards the Final Laws of Physics." His subject was "the clues in today's physics that we can find that can tell us about the shape of a final underlying theory that we will discover in the future."[66] The lecture ultimately advocated for the possibility of string theory as the final theory of nature. By the terms of this theory, a string is a "glitch in spacetime" vibrating in a spectrum of modes, each mode recognizable as a different type of particle. Weinberg felt that string theory had a "smell of inevitability." In this same lecture, he advocated for what he termed "reductionism," for the logical reduction to fundamentals. He claimed in this regard that "we are all reductionists today," a fact to face and embrace. He concluded in a measured apology for *theoretical* beauty, where this was understood as inevitability. All these were said to be signs and properties of a final theory.

Roughly at the time of Weinberg's lecture, SSC designs were being formed and chosen, and by January 1987, after a DOE review, the president approved the project. As the scholarship on the SSC episode has explored, the political debate over the necessity of the project, given its costs, and the question of international financial support were prominent factors throughout its history, from early congressional hearings to White House debate and support.[67] In November 1988, the competition for a site was decided in favor of Ellis County, Texas, while the acquisition of the land by the state, approximately sixteen thousand acres, did not begin until March 1990.

Across these years, SSC advocates made their case in a number of forums. The physicists Sheldon Glashow and Leon Lederman were early supporters and in 1985 coauthored an article for *Physics Today* rallying the physics community at large. There, Glashow and Lederman prominently quoted Gamow, from *One Two Three . . . Infinity*, hazarding the opinion that in the apparent indivisibility of subatomic particles then known (nucleons, electrons, and neutrinos), "it seems that we have actually hit the bottom in our search for the basic constituents from which matter is formed."[68] Despite Gamow's own qualifications that the history of science had itself suggested that past confidence in indivisibility (as of the atom itself) was misplaced and that there "is no way to predict the future development of the science of matter,"[69] Glashow and Lederman judged Gamow guilty of "hubris," which "has never survived for long."

Anticipating the need to persuade those such as Anderson, the authors asked their readers to engage in a thought experiment: "Consider Arthur, an intelligent alien from a distant planet, who arrives at Washington Square (New York City) and observes two old codgers playing chess. Curious Arthur gives himself two tasks: to learn the rules of the game, and to become a Grandmaster. Elementary-particle physics resembles the first task.

Condensed-matter physicists, knowing full well and with absolute certainty the rules of play, are confronted with the second task." The effort to play at a grandmaster level characterized "most of modern science," not only other subdisciplines of physics. However, "It is only in particle physics and cosmology that the rules are only partly known. Both kinds of endeavor are important—one more 'relevant,' the other more 'fundamental.'"[70]

These arguments, along with Glashow and Lederman's further appeal to national pride, to potential technological spin-offs, and to the duty to human curiosity—the imperative to knowledge—failed to persuade Anderson. Among other physicists and materials scientists, Anderson confronted these arguments most directly in congressional hearings and in the press. In April 1987, he submitted a letter in a hearing on the DOE's funding request for the SSC before the congressional Subcommittee on Energy Research and Development. In his testimony, Anderson rejected dominant foundationalist pictures in a manner consonant if not fully consistent with his writings of the previous decade.[71] He once again counterposed a scalar sensibility to a foundationalist one, now invoking the misuse of scales on the part of his opponents: "The first slide in many general talks given by my high-energy colleagues is a length scale spreading from the 'Planck length' (way below elementary particle size) to the size of the cosmos. They gesture deprecatingly towards the center of this scale (where we and our atoms and all of everyday life sit) and say, 'of course, we know everything there, and the only fundamental science is at extreme scales.'"[72] The chess-game dichotomy of Curious Arthur, the ambition to grand mastery supposedly characteristic of the middle regions, did not to Anderson's mind capture the play of research.

Years earlier, he had already appealed to aliens in order to defend his more pluralistic vision of science, and in the course of his exposition in "More Is Different," argued that large many-body systems can display very different symmetry characteristics than physical laws demand of their components—that in limit cases, "matter will undergo mathematically sharp, singular 'phase transitions' to states in which ... even the microscopic equations of motion, are in a sense violated." He suggested that aliens would encounter science very differently on the basis of a very different experience: "Actually, for a hypothetical gaseous but intelligent citizen of Jupiter or of a hydrogen cloud somewhere in the galactic center, the properties of ordinary crystals might well be more baffling and intriguing puzzle than those of superfluid helium."[73] If Arthur is such a Jovian, chess at the foundationalist extremes constituted by the rules might be absolutely straightforward, while basic strategies familiar to human players might be much more difficult to fathom.

Anderson was quite willing to go along with the high-energy physicists that theirs was a less relevant discipline. Its irrelevance to other scientific disciplines was evident in the fact of "emergent properties," that is, "that complicated systems have properties which are not implied by the elementary pieces of which they are made."[74] But their irrelevance also foreclosed the potential spin-offs emphasized by Glashow and Lederman, since, to Anderson, innovations require innovative technologies, and large-scale projects such as the SSC, if they hoped to succeed, should be innovation averse. Moreover, the emphasis on particle physics was already having adverse effects on physics education and employment, since training and popular work overrepresented the place of high-energy physics in active academic and industry research.

His indictment here was married to a defense of small-scale science in contrast to big physics, the latter of which attracted disproportionate congressional funds, partly as a matter of pork-barrel pragmatics. Anderson affirmed that "I too want to know what the ultimate structure of matter in the world is going to be, as well as how the Big Bang happened, and I want to see an accelerator going far beyond the present limits eventually built." But in words quoted thereafter in future congressional debate over the SSC, he rejected any present need to build such an accelerator: "I do not accept that doing so is as urgent nationally as many other scientific needs of the country, in space science, science education and above all in the rescue of our strong tradition of innovative, fundamental small science." If this left as motivation competition with other countries, Anderson responded with a more thoroughgoing nonnationalistic scientific universalism: "It disturbs me to see accelerator physics seen as a nationalistic, competitive race; science is too serious a matter for that."[75] Overall, the foundationalist narrative underpinning the arguments for the SSC was failing to persuade prominent voices within the physics community.

Other aspects of that narrative, particular to Weinberg, also produced rebuttals. In 1990, the year of the HST launch, *Origins: The Lives and Worlds of Modern Cosmologists*, an edited volume of interviews with prominent modern cosmologists, was published. The publication gave some measure of the circulation of Weinberg's *The First Three Minutes* over the intervening years. Weinberg's 1977 remark, that the answers the scientific quest brings to light tend to indicate the pointlessness resulting from finding no particular place for humanity, had, it seemed, provoked strong responses from scientific audiences. The remark itself was the basis of what was usually the final question in these interviews. There the scientists expressed various reactions to Weinberg's point: agreement, consternation, disputation of the terms of the question (often through the charge of anthropomorphism)— but most demonstrated a familiarity with Weinberg's approach and statement. Weinberg himself was also interviewed and asked to elaborate on his remark: "I've gotten more negative comments about that sentence than about anything else I've ever written," he observed, admitting that "I certainly meant *roughly* what I said, but it didn't come out *exactly* as I wanted it."[76] He expected to produce a more satisfying expression of his statement in the near future.[77]

Major construction on the Ellis County site began the following year, in 1991. Hopes were high that the accelerator would determine the question of the existence of the Higgs particle, confirming aspects of the so-called standard model of fundamental physics long in question. But at the same time, as it was emerging from the design to the construction stage, the SSC itself came under attack by congressional cost cutters for inappropriate and multiplying expenditures, exaggerated indirect benefits, and overall mismanagement.

So when Weinberg's promised response to his fellow cosmologists' complaints found expression in 1992, the major themes he had promoted from his "Origins" lecture were marshaled in advocacy both of string theory (itself no stranger to controversy) *and* the SSC. As criticisms sharpened against the SSC, Weinberg's statements and lobbying in its defense emphasized more what could be gained from the project than, as in 1985, what was beyond it or any accelerator's grasp. By 1992, he was prepared to argue for the SSC and string theory in a book-length treatment, *Dreams of a Final Theory: The Scientist's Search for the Ultimate Laws of Nature*. In many ways, the Dirac lecture was a blueprint for the philosophical and scientific text, combined now with an apologetics for the SSC project that became the main thrust of the work. The book waged a number of related attacks: for scientific reductivism and against social constructivism; for the idea of a final theory and the hopes of superstring theory; and for the fundamental nature of high-energy physics over and above other scientific research. All these were in concert arguing for the importance of building the SSC.

To be sure, Weinberg's book was "the single most sustained argument yet made (on natural philosophical grounds at least) for building the Superconducting Supercollider,"[78] as Peter Galison observed in a review at the time. The qualification, "natural philosophical grounds," matters here: as with Gamow before him, Weinberg linked his defense of the SSC to the fundamentality of his physics, a science narrow in scales of space and time. As Galison observed, Weinberg's argument/characterization was avowedly foundationalist in an thoroughgoing reductionist sense:

> Roughly speaking, Weinberg's argument goes like this. Every question about the physical world leads to another. Why is this chalk white? Because the atoms within it have no preferential absorption of one color or another in the visible spectrum. Why do atoms absorb light? Because the electrons can only orbit with certain energies around nuclei of protons and neutrons. What are the protons and neutrons made of, and what are the laws that govern their behavior? Quarks and gluons transmogrifying under the laws of quantum chromodynamics. From just about any question the nature of anything from electricity, transparency, or ductility to the colors of clouds or why kites fly, one falls, like Alice, down a hole, at the bottom of which lies the domain of elementary particle physics. This is the foundational view of particle physics.[79]

Falling down a hole, that is, digging the tunnel in Texas, was precisely what Weinberg was advocating in his defense of the SSC. Weinberg was explicit and unapologetic about the reductionism that is at the heart of this view, observing that "the reason we give the impression that we think that elementary particle physics is more fundamental than other branches of physics is because it is. I do not know how to defend the amounts being spent on particle physics without being frank about this. But by elementary particle physics being more fundamental I do not mean that it is more mathematically profound or that it is more needed for progress in other fields or anything else but only that it is closer to the point of convergence of all our arrows of explanation."[80]

Here Weinberg has in mind the foundational synthesis of one science logically underlying another. Such a position need not contradict Anderson's earliest reductivist/non-constructivist vision, if it at the same time conceded not only the in-principle impossibility of the constructivist climb out of the metaphysical foundation or of following the arrows back from the point of convergence toward grounds distantly referred back to that point. A foundationalist position positing new in-principle foundations for successive stages of that climb, or still other points of convergence for as full a set of arrows would square with if not nearly constitute Anderson's view, while contradicting the satisfaction Weinberg found in his more thoroughgoing reductionism. And it is Weinberg's narrative produced on the basis of a foundationalist genre of synthesis cued by his reductivist foundationalism that came to circulate in political spheres: that humanity should find the final law, a bedrock to and frame for all science. Here, untheorized story sought to legitimate seemingly non-narrative theory.

The intriguing choice of the term "arrows" of explanation points to an evocative and insistent metaphor structuring a text designed to defend digging a tunnel in the Texan soil, that of exploratory directions. In fact, exploratory or expeditionary metaphors dominate Weinberg's book, explaining his disagreement with social constructivism[81]; the possibility of an origin to the universe that has no precursor, logical[82] or temporal[83]; his confidence in the existence of a final theory[84]; and, as we will see directly, his speculation concerning humanity's reaction to the possession of a final theory. Which makes it all the more striking that, contrary to Sagan (and, as will be seen, the NASA scientists and administrators), Weinberg appeared to put little store in the scientific benefits of actual, embodied exploration. Contrasting earlier periods, when physical exploration brought knowledge, Weinberg once again adopted an elegiac tone: "Still, with the discovery of a final theory we may regret that nature has become more ordinary, less full of wonder and mystery." Such a loss of wonder was not, to his mind, unprecedented: "Throughout most of human history our maps of the earth have shown great unknown spaces, that the imagination could fill with dragons and golden cities and anthropophagi. The search for knowledge was largely a matter of geographical exploration." Medieval cartographies and cosmographies flourished in the space of the unexplored, "but today every acre of the earth's land surface has been mapped, and the dragons are all gone. With the discovery of the final laws, our daydreams will again contract. There will be endless scientific problems and a whole universe left to explore, but I suspect that scientists of the future may envy today's physicists a little, because we are still on our voyage to discover the final laws."[85]

Physical exploration may uncover new forms, but not new dragons, not new wonders or new physics, not what is essential to understanding the universe. Exploration for Weinberg was the determination of the fundamental, not the discovery of the new. On a logical level, exploration here means following the arrows to fundamental matters. This exploration may bring researchers to an understanding of ultimate physical origins, but it does not in principle have much to do with the actual historical sequences that flow from those origins, much less the human life produced from them. It is a narrative of quest based on the ideas of a final theory, but a quest more figurative than embodied. On the empirical level, that kind of exploration required boring a tunnel in Texas and not, as with Sagan, launching out into the universe.

HAWKS AND HOUNDS

For Weinberg, the question of life's meaning in relation to science turned on his perception of whether science was itself held responsible for the loss of God, that is, for the naturalization of the understanding of the world. That in turn Weinberg appeared to accept: "All our experience throughout the *history of science* has tended [. . .] toward a chilling impersonality in the laws of nature."[86] Such a claim was not however scientifically nonpartisan, nor to an unsympathetic eye might it have appeared to be consistent with all Weinberg's science. Anthropic principles in cosmology were a subject of increasing attention, and Weinberg himself had already offered papers and lectures that deliberated over and appealed to anthropic principles.[87] To some critics and proponents, anthropic principles did personalize the laws of the universe, rendering them species- or intelligence-centered; if so, a foundationalist scientific view could posit humanity as central.

But Weinberg deliberated over only a "moderate version of the anthropic principle," or, as he characterized it, "an explanation of which of the various possible eras or parts of the universe we inhabit, by calculating which eras or parts we *could* inhabit." He contrasted this "weak anthropic principle" to two others: first, a "very weak version," that "we are here as one more experimental datum," a version he found so unobjectionable as to be without scientific virtue.[88] The second "rather strong version" held that "the laws of nature . . . are completed by the requirement that conditions must allow intelligent life to arise"—a view he responded to aphoristically: "although science is clearly impossible without scientists, it is not clear that the universe is impossible without science."[89]

For Weinberg, therefore, moderate anthropic principles did not impact the impersonality of natural laws. The history of the universe he articulated and elaborated over time continued to answer the spiritual/philosophical dilemma that it itself articulated with a consistent answer: to seek out persistently the full comprehension of the universe even if it should mean the loss of wonder in the world and of a point to existence, to find meaning instead in and through the tragic inevitability of facing meaninglessness, the task of lifting tragedy out of farce.

In *Dreams of a Final Theory*, Weinberg drew a contrast between philosophizing and the sober tasks of physics: "In our hunt for the final theory, physicists are more like hounds than hawks; we have become good at sniffing around on the ground for traces of the beauty we expect in the laws of nature, but we do not seem to be able to see the path to the truth from the heights of philosophy."[90] The remark underscores the contrast between Sagan and NASA's Origin Project narratives, on the one hand, and Weinberg's, on the other, if less about their activities as scientists. Putting to one side the bellicose connotations of the term, Sagan's would immediately appear to be more the hawk's narrative—announced more from the heights, from space itself, and from a moral loftiness.[91] Weinberg's has something of the sensibility characterized by the hound, a narrative defending underappreciated and largely unseen tunnels, a disparaged "hole" or "pit" in public dialogues.[92]

Though the comparison is not one of kind, the distinction extends back to Weinberg's "movie," as it appears both in the *film-souvenir* of his "Early Universe" lecture and in the book *The First Three Minutes*, as well as in Sagan's filmed *Cosmos*. In these accounts, on the one hand, the emergence of fundamental physics is itself at stake and is even the hero of the tale, while on the other hand, humanity is pictured as central. For Weinberg, the Big Bang story functions in part as a platform for exploring high-energy particle physics, for the very beginnings and, at last, the very end of the world. Sagan's interest or the NASA Origins Project's commitments were almost the inverse of this, with the very beginning and very end dwarfed by the importance of everything in between—how life emerged, the possibility that life is elsewhere, possibly pervasive. The opposition, which is not over the scientific theory but the kind of story being narrated, extends as well to Weinberg taking a dim view of certain NASA projects. Indeed, as with several other scientific experts debating the merits of the SSC, Weinberg repeatedly used the example of the space station as a project funded by Congress having much less scientific merit by contrast.[93]

Nevertheless, even as opponents such as Anderson stressed the scientific validity of the SSC—and again as opposed to their own attitudes to monies spent on such efforts as space stations and, earlier still, Star Wars programs—they made public their criticisms broadly in the media. The response made clear how off-putting they judged the emphasis on the extremes of big science, on the places where those such as Weinberg believed that all the explanatory arrows converged. So early in 1993, in a letter to *Science* assessing the 1992 debate over whether to move ahead with the SSC, physicists T. H. Geballe and J. M. Rowell dismissed what they regarded as the central arguments for the SSC, several planks of which were now long in place, laid down for the community a decade earlier by Glashow and Lederman. But Geballe and Rowell found that some of material of this older advocacy had become untenable with age. They judged the argument for national competition, for example, to be anachronistically quaint in a post–Cold War era.

And as Daniel Kevles has particularly stressed, in the post–Cold War era, "missing at the national level was what had made physics, including its high-energy branch, so important since World War II—real or imagined service to national security."[94]

Geballe and Rowell also refused to accept that over the previous decade, the SSC boosters had won over the majority of the physics community: "the impression has been created that support from the scientific community for the SSC was 'overwhelming.' Despite the roughly 2000 signatures collected in support of the SSC, there was strong opposition to it across some parts of that community, particularly among the condensed matter scientists who make up the largest division of the American Physical Society."[95] The physicist G. H. Trilling, who challenged them point for point in the pages of *Science*, granted at least this latter point: "Unfortunately, this is true, and it almost seems surprising, given the fact that no serious scientific arguments ... have been made to suggest that the physics prospects of the SSC are not outstanding, or that there is a different, simpler technique for arriving at the answers to the same problems. There is no way that our present deep understanding of the subatomic world could have been achieved without the large accelerator facilities that were built in the 1960s and the 1970s. These projects were also criticized by small-science advocates."[96]

Geballe and Rowell's response was that narratives such as Weinberg's demanded a hierarchy of importance in which biological origins were less important than cosmological ones. They observed: "It is fruitless ... to compare the importance of finding out more about the origin of the universe with that of finding out more about the origin of life. The most important experiments in particle physics deserve support because they are likely to provide new understanding of the structure of the universe, but they are not necessarily the most important experiments in physics, chemistry, or biology."[97]

This comparison informed congressional testimony on the SSC during August 1993, in hearings that brought Weinberg and Anderson together to the Senate, standard-bearers of their opposing sides. The prospects for completing the SSC were now sharply in decline, the House having voted down the project 280 to 150 in June of that year. To commentators such as Deborah Shapley, the granddaughter of Harlow Shapley and a science writer and journalist, the Senate seemed unlikely to rescue it. In testimony on August 4, Weinberg's claims reiterated many of his points in *Dreams of a Final Theory* and Anderson his own arguments from "More Is Different" on.[98] Their opposition was not on points of the intrinsic scientific validity of the SSC or over the claims of waste, mismanagement, or poor planning that were circulating in the press and so much the focus of congressional attack. In defense of the budget increases, for example, Weinberg noted that what was real in the reported figures came as a matter of learning from example, that of the HST (just as the NASA Origins Project was soon to learn from the example of the SSC): "In fact, that decision to increase the cost at that point represented the laboratory's judgment that we did not want to go the way of the Hubbell [*sic*] Space Telescope. The Hubbell Space Telescope, in order to keep costs to an absolute level, sacrificed testing their mirrors, and as a result we have a marred instrument. Some of the increase from $5.9 billion to $8.249 billion went into increasing the aperture of the magnets to make sure that the magnetic field would have a high enough quality so that the machine would do what is promised."[99]

The focus for the expert protagonists instead remained on the question of the value of what the SSC set out to discover and the probability of what would come with that attempt, simply as a matter of the undertaking itself. Weinberg defended the possible relevance of the SSC, partly by invoking a recent op-ed piece by Deborah Shapley. Her piece was not without its criticism for high-energy physicists, the "HE crowd": "They're arrogant, even by the standards of high science, seeing themselves as a priestly elite trying to nail down a Grand Theory of Everything that explains both the four fundamental forces and the birth of the universe."[100] Shapley, Weinberg noted, emphasized the "bargain," a "compact between science and society," that had followed the postwar period and had resulted in the advance of science. Shapley had deliberated over this compact and its oversimplifications in an earlier coauthored book with Rustum Roy. Their book, *Lost at the Frontier: U.S. Science and Technology Policy Adrift*, was manifestly and structurally in dialogue with the famed document authored by her grandfather's sometime opponent Vannevar Bush, the younger Shapley and Roy seeking to bring out aspects of the original document that supported their recommendations for a renegotiation of that science-society compact.

Given the attacks on the SSC from "citizen's groups" and congressional representatives, Shapley declared that "the old compact is dead." To her mind, however, this was a call to action for those such as Weinberg: "The HE crowd needs to muster the courage of its convictions and argue forthrightly that the most basic forces of matter and the birth of the universe are things worth studying in themselves and worthy of support." Nevertheless, at the same time, her erstwhile coauthor Roy had become one of the most cited and aggressive critics of the project. To the same August 4 hearing, he submitted written testimony denouncing not only the project in highly

charged terms, but the entire subdisciplines of astronomy and particle physics for engaging in "speculative science," the speculations of which "can go
on *forever, without any conceivable impact on humans or society.*"[101] He accused
the scientists standing to gain from the SSC of trying to "pass themselves
off as real scientists" and of using "every trick in the book to hang on to
their entitlements."[102] Less hyperbolically, the physicist Laurence J. Campbell noted that "the energies to be produced by the SSC, 40 teravolts, would
have relevance only to the historical physical world during an infinitesimal
interval of time near the origin of the universe. Knowing more about this
early interval of time will merely raise questions about the preceding intervals, involving ever higher energies—and so on to absurdity."[103]

But to Weinberg, such accusations missed central facts about science—
that the quest for knowledge is not to be judged only as a result of immediate relevance and that the SSC was launched in order to ground theoretical
speculations. The central thrust of Weinberg's apologetics remained the possibility of discovery and finality: "Now, some of us think, although it is not
a universal view among scientists, that if we keep going in this way we are
going to eventually come to a final theory. We are going to come to a simple
set of principles which will be almost self-evident and that will be at the bottom of everything, that will be completely universal and govern everything
in the universe. We do not know that that final theory really exists. We hope
it does. I think it does. We certainly do not promise that the super collider will
reveal the final theory."[104] Glashow and Lederman had accused Gamow of
hubris for his belief that physicists may have come to the bottom of things in
the late 1940s, but Weinberg was not deterred by that accusation in holding
out the hope that a final theory, partly with the help of the SSC, was possibly
forthcoming.

But on these scores, Anderson granted little. He remained an advocate
for the relevance of small science and spoke (as he had repeatedly before)
for imaginative solutions to the high-energy impasse in which the particle
physicists found themselves. And like Geballe and Rowell, in very similar
terms, he looked for a bigger tent than he believed Weinberg's narrative allowed: "There are many other really exciting fundamental questions which
science can hope to answer and which people like myself are, on the whole, too
busy to write books about"—as opposed, he had made clear, to particle physicists whose failure to make progress opened up time for authorship. "There
are questions like how did life begin? Why is biological catalysis so efficient?
What is the origin of the human race? How does the brain work? What is
the theory of the immune system? Is there a science of economics?"[105] These

questions to his mind turned far more on the study of complexity than they did on the analysis of the basic elements of matter.

Following their prepared testimony, Weinberg addressed Anderson's objections directly, noting to the committee that "he and I agree on so many things that I do not want to get into a dog fight with him."[106] In the course of his rebuttals of Anderson's arguments, Weinberg washed his hands of the task of choosing which scientific goals were more worthy, at least in relation to funding: "It is very hard to prioritize between different communities that have different goals, because the goals are incommensurate. How do you compare the discover [*sic*] of the origin of life with the origin of mass? I do not know now [*sic*] to make the comparison. That is your job." As defense of the SSC, Weinberg dismissed the interrogation by Howe, Geballe, Rowe, and Anderson, the scientific questions of which the NASA Origins narrative embraced non-hierarchically. But importantly, Weinberg did not refuse the sense of their query; the need to weigh the importance of the origins of mass to the origins of life in the context of foundationalism measured the extent to which Weinberg's account remained divisive.

If this was a dog fight, opposing sides both attempted to promote a hound-like vision of science that did not appear to presume or to ask for a lofty philosophical perspective, while nevertheless making clear that they defended different conceptual grounds. Except in a formal sense—the perspective required to assert that the world at different scales is multiple—Anderson's own preferred narratives did not claim to survey the world from any one privileged path, nor that those paths traversed the same limited ground. While neither his emphasis on emergence nor Weinberg's on foundations could be entirely free of a philosophical overview, Anderson's own arguments for emergence curtailed the extent to which nature at different scales could be captured either by overarching, deductively minded philosophical flights or by following any one of the possible trails converging on the deepest ground through a huge conceptual rabbit hole in Texas. Though both Anderson and Weinberg went out of their way to assert their agreement on the Big Bang,[107] they could agree on little of substance with regard to potential syntheses that flowed from it.

NARRATIVE, FAILURE, AND SUCCESS

The attempt to achieve political support and funding for the Superconducting Supercollider was based more explicitly on stories at work in a putative final theory or derived from it—on what could be gained from such a theory or from the worldview on which that story depended and which it expounded—than on a successful conceptual or technical explanation of what that theory entailed. In the debates on the Senate and House floors over what the SSC was being built to discover, basic puzzlements emerged over whether there is a Higgs Boson, whether it exists or would come into existence, whether it would matter much if it did. The uncanny quality of the debate is captured in an exchange between Senator Dale Bumpers of Arkansas, a staunch opponent of the SSC project, and Senator Bennett Johnson of Louisiana, one of its most forthright allies, on September 29, 1993, each of them members of the Committee on Energy and Natural Resources (Johnson the chair) and each having listened to and questioned Weinberg and Anderson. In this respect, the exchange bears fuller citation:

MR. BUMPERS. . . . Dr. Steven Ahlen—I think he is head of the physics department; he is a professor of physics, and he is a particle physicist at Boston University—He told me himself—and this is not a letter, this is not in testimony—he came into my office; he said, "Senator, every scientist in this country is hurting for research money." That is exactly what all of these physicists are saying. You are so much more likely to get something beneficial that we can all enjoy the benefits of if you fish in many ponds, instead of looking for one big fish in one big pond. He went ahead to say, "Do you know what they are looking for?" I said, "Dr. Ahlen, I never even had high school chemistry. I know they are trying to find the origin of mass." He said, "Well, that is correct. What they are looking for is what physicists call the Higgs Boson. The Higgs Boson is a particle that physicists think existed for something like one-millionth or one-thirtieth of one-millionth of a second at some time billions of years ago.

MR. JOHNSTON. If the Senator will yield at that point, Mr. President, will the Senator not agree with me that scientists believe or theorize that Higgs Boson exists today, and that it is the field through which all energy and particles are propagated, and that it can be found only with the energies of the superconducting super collider? It is not that it existed way back then, it exists today, and they theorize that; will the Senator agree with that?

MR. BUMPERS. Not totally. The Senator is partially correct. I think Dr. Ahlen—I am reluctant to quote a physicist on a technical matter about which I know very little—but the point I was about to make when the Senator asked me to yield to him is I think Dr. Ahlen thinks the Higgs Boson existed for a second many billions of years ago. But his concluding remark to me was: "I do not believe the Higgs Boson exists."[108]

These remarks reveal a perceived ambiguity in the basic objective of the SSC. Senator Johnston, sensitive to this point, sought to face it directly: "Let me say a word about the science of this project. It is awfully difficult to talk about the science because it is in this esoteric area that is hard for people to understand. We have talked here about the Higgs Boson. Just the very words, 'the Higgs Boson'—people do not have any idea what a boson is or what the Higgs Boson is or, indeed, it is not a certainty of whether it exists or not."[109] Johnston names and quotes from Weinberg's *Dreams of a Final Theory*: "A Higgs particle could not have been seen in any experiment so far if its mass were greater than about 50 times the proton's mass, which it might well be. We need to experiment to tell us whether there actually is a Higgs particle or perhaps several Higgs particles and to supply us with their masses." And he continues in a somewhat murky summary of pro-SSC arguments:

> Mr. President, he [Weinberg] goes on to say that while we do not know whether there is the Higgs particle or the Higgs field, with the superconducting super collider we will discover what the mechanism is by which mass and forces are transmitted. We will open the door to this empty room where the answer lies. The theory is that it is the Higgs particle, but if it is something else, that is knowledge which will be at least as valuable as the Higgs particle, perhaps more so, because that would then destroy all the theories upon which we operate now, because the scientific community now—most of them—operates on the theory that there is something like a Higgs particle, the mass of which we do not now understand.[110]

Weinberg's narrative was threatening to become the farce he feared.

Already in the afterword to the 1993 edition of *Dreams of a Final Theory*, he had begun to resign himself to the loss of the project. There he reported his impressions watching the congressional debates on the SSC, having "the surreal experience of hearing senators on the floor of the Senate arguing about the existence of the Higgs boson and quoting this book as an authority."[111] The fact of the surreal experience, the extent to which the senators were arguing confusedly over the basic terms of the project, its allies failing to convey the urgency and necessity of its goals, demonstrated the extent to which Weinberg's narrative did not resonate deeply with the politics of the day. A political dilemma for which the narrative had projected a resolution was proving intractable. Congress officially cancelled the SSC in October of that year, rejecting for multiple and multiply contingent reasons the aims of a foundationalist narrative requiring a large investment in a single big science project and justified by theoretical claims divisive among scientists and difficult to explain, such reasons including resistance to the narrative itself. Three years later, as a consequence of its own configuration of causes including the cancellation of the SSC and the political and cultural resonance of its own narrative, a historical synthesis based on the value of different theoretical contributions focused on issues surrounding the origins of life projected a disciplinary organization and orientation amenable to the justification of additional expenditure for largely existing research programs.

The argument here is not that the SSC failed simply because its narrative was less powerful and that NASA Origins Project succeeded simply through the force of its own contrasting story. After all, the similarities of the NASA Origins story to SETI meant that a universal historical narrative was hardly a guarantee of practical success. Rather, the difference in outcomes underscores narrative differences linked to differing political dilemmas affecting the success of both projects and the articulation of the competing narratives—NASA's existential crisis and the opportunity to stress the topics of origins and exotica as objects of an epic quest, on the one hand, and the difficulty of explaining the need for the tunnel in Texas in terms of a similar quest, on the other. With regard to the success of the projects in narrow terms, however, the narratives did at least play some role in the large cast of causes that produced the eventual outcome and helped to establish the discourse through which the public and the government decided the merits of the scientific projects being debated.

The narratives differed, too, over the nature of exploration. Sagan, Canizares, Dressler, and the many NASA fellow-travelers projected the task of science and humanity together as questing the wide universe and speaking its story. Weinberg's exploration was lofty, in an intellectual sense, but its actual physical exploration was perhaps too rooted in a specific place, a tunnel the evocations of which were mocked more than celebrated, and a specific constituency appearing to win the anticipated economic benefits. The figure of that "hole" worked against the SSC's claims to a national, scientific universality, a universality that appeared to be contradicted in failures to secure envisaged international cooperation. The two narratives also had to contend with differing political contexts, problems, and goals. Advocates of the earlier SSC Project needed to justify the expenditure of funds for high-energy physics above other branches of science and in contrast to philosophies and policies of science such as those advocated by Anderson. By contrast, the political effort to promote the NASA Origins Project included an awareness of the reasons why the SSC had been canceled and the need to address NASA's own sense of mission disorientation and unsteady public support.

The epic that NASA promoted appeared capacious enough to harbor differ-ent scientific disciplines and intelligible enough to scientists and the public to provide clearer bearings, while the foundationalist synthetic genre that Weinberg advanced in arguing for the SSC failed to strike a wider cultural chord, despite the fact that the SSC was tied to an origins story. Human-ity—or at least Congress—did not want to accept Weinberg's version of a putative finality, to the extent to which it even understood its terms. Less remotely, and of more immediate consequence for the SSC, the narrative resonated little with many in the field who felt that high-energy physics held too high a status, a status only reinforced to their minds by the Clever Arthurs and the explanatory arrows leading to the extremes of the spatial scales and potentially to final answers. To Anderson and likeminded scien-tists, regardless of whether a synthetic narrative was of hawks or of hounds, every symmetry-breaking scale produced a new foundation, opening out into a few fields of experience that might recalibrate the metaphysical and explanatory center of the natural sciences.

By contrast, the discussion and disagreements of the NASA Origins Proj-ect were of a very different nature from those three years before with regard to the SSC project or the termination of SETI. There were many reasons for this, particularly when contrasting the SSC and the Origins Project: the different nature of their relationship to the institutions and the at least pro-cedural differences this entails (Origins as part of the NASA budget, the SSC under the jurisdiction of the US Senate Energy and Natural Resources Com-mittee and the Subcommittee on Energy and Water Development within the Senate Committee on Appropriations); the significantly different eco-nomic climate between the early and mid-1990s; the different political cli-mate between the early and later years of the Clinton administration and the increasing distance from Cold War scientific rhetoric; and the example that the SSC and SETI failures provided for NASA commentators. But an additional factor as well was the failure of one origin narrative to achieve a firm footing with congressional members and the public, in contrast to the success of the other. The confusion of even the advocates of the SSC on the congressional floor might be contrasted with the ease with which the gov-ernment and the public took to the Sagan-like story, stripped of its emphasis on SETI, and based on what appeared to be concrete evidence of past life elsewhere traced to a culturally and historically resonant Mars.

More broadly, the genre that Sagan referred to as "epic myth" allowed the scientific community to produce a unified vision of science, accepting its scrambling diversity and admitting no clear hierarchies among its disciplines.

Every chapter of this universal history mattered, and so did the varied disciplines authoring them. Its narrative was a site of a projected and attempted resolution of several political, social, and disciplinary dilemmas at once. In the practical lobbying of Congress, it provided a justification for the political funding of science in a more arid post–Cold War context. The successful effort of NASA scientists to obtain funds for the Origins Project appealed to a vision of nation that gave American scientists a leading role while at the same time representing the search for origins (on earth and elsewhere) as humanity's— and therefore the nation's—mission. In Sagan's earlier documentary series, that epic appealed to a latitudinarianism allowing for a diversity of religious and mythic beliefs. At the same time, however, it established science as a consummation of this diversity, the deeper truth that was inherent in all of them. In this way it promoted and inscribed within its story the quest itself as the answer to Wilson's overlapping spiritual dilemmas of ethics and evolution, or as the answer to the longstanding deliberations on which Wilson drew self-consciously or otherwise (including not only Julian Huxley's evolutionary humanism but in a very different vein, Arendt's "mutation of the human race")—answering Wilson's plea that "the mythopoetic requirements of the mind must somehow be met by scientific materialism so as to reinvest our superb energies."[112] Finally, it promoted the scientific quest, with its hope of finding endless wonders and cures, as the best way to avert the tragic conclusions that could arise from the use of nuclear and other technologies.

With the formation of large-scale programs that recommended and defended themselves through this coevolutionary narrative, widespread and varied organized activity was also understood by the lights of that account, from further pedagogical programs and media debates[113] to the recalibration of scientific efforts in narrow and broad terms. Attending this myth were interpretative debates, the recommendation of moral behavior (the assumption and endorsement of the scientific quest itself), and the ability to dictate values, rather than have them dictated by an externally conceived moral guide. The evolutionary epic was defended as what could dignify human life with meaning, directing scientific, technological, industrial, and even national efforts, providing an orienting synthesis and goal.

CHAPTER TWELVE

A New Version of Genesis

"The message of the DEEP survey, and all the other information that we're getting, is one beautiful story, a new version of Genesis, a new version of the cosmic myth, only this time it's scientifically based, from the Big Bang to now: Big Bang, formation of galaxies, formation of heavy elements in supernova, sun, Earth, life—one unbroken, great chain of being." These are the final words of the 2004 NOVA series *Origins: Fourteen Billion Years of Cosmic Evolution*, as stated by the astrophysicist Sandra Faber. "The DEEP survey" refers to NASA's galactic, telescopic census of the remote universe, enriching the contemporary view of the history of the world—"all the other information"—as advanced by the NASA Origins Project and other origin-based scientific research. Faber's words immediately recall the final words of *Cosmos*, "These are some of the things that hydrogen atoms do given fifteen billion years of cosmic evolution. It has the sound of epic myth, but it is simply a description of the evolution of the cosmos as revealed by science in our time."

Faber's conclusion shares with that of Sagan the characterization of myth, putting it still more assertively—not the "sound of epic myth," but "a new version of the cosmic myth." Hers is not "the problem of Genesis" prefacing Weinberg's *The First Three Minutes* but the declaration of a new version of it, the emphasis on a specific text and tradition pointing to the shift in context from the quarter of a century separating Weinberg's book and Sagan's series from NOVA *Origins*. Whereas Sagan concluded his series by laying stress on continuity with the past and with the universal search for origins, the culmination of this more contemporary questing emphasizes the new, the overwriting of past mythic histories with a new version. This change in emphasis reflects the ways in which the prominent cosmological narratives and new synthetic structures of science had circulated through and responded to the contemporaneous political and social environment.

By the late 1990s, a comprehensive contemporary narrative of origins had expanded outward from scientific papers, White House space summits, congressional debates, and NASA planning documents. The most conspicuous forum for this expansion was the television documentary, the documentary cosmological tradition continuing on from *Of Stars and Men*, *The Rough Sketch* and *The Powers of Ten* itself, *Ascent of Man*, the predecessors of *Cosmos* into the contemporary era, a development of the medium that refracted and reflected varied cultural responses to scientific narratives while shaping them into new forms.[1] As the film critic André Bazin suggested and as has been reemphasized in more recent documentary studies, among their other effects, these documentaries do the anticipated work of amplification, calling attention to their subject matter and mobilizing political efforts.[2] The contemporary television documentary rendering of the new origins initiative posed and constituted projected resolutions of wider political and axiological issues. Attention to the historical context and nature of these materials addresses the question of what genre of history the history of the universe attempts to be, its status as a political, social, and conceptual intervention.[3]

Examining television documentaries as stations for and of cultural movements positing politically opposed universal histories, I focus here on the production, conception, and circulation of three series, the 2001 WGBH and Clear Blue Sky (now Vulcan) coproduction *Evolution: A Journey into Where We're From and Where We're Going*, the 2004 NOVA series *Origins*, and *Cosmos: A Spacetime Odyssey*, which aired in 2014 on the Fox network. What follows is concerned with how each series attempts to channel and address its wider context. The connection between two of the series was drawn early on by the producers of later *Origins*, who joked that they were working on "*Evolution*—the prequel,"[4] while connections with the final series were structured partly through the relationship of *Origins* to NASA Origins and therefore to Sagan's narrative, as well as through the shared voice of Neil deGrasse Tyson.

The *Evolution* series clearly acknowledges the political context for all these series as the cultural resistance to syntheses based on evolutionary science and the rise of the intelligent-design movement in the United States, particularly in the overt theme of the final episode of the *Evolution* series,

"What about God?" (September 27, 2001). Analysis of *Origins* clarifies how the coevolutionary tale, cosmic and organic evolution together, was established in such a context, as well as the direct connections to NASA Origins and the preceding documentary television and scientific tradition—the ways in which the documentaries reflect and present a unified picture of active research. Finally, the production and reception of *Cosmos: A Spacetime Odyssey* illuminates the degree to which even in a variegated cultural environment vocal organs of which could be hostile to its reception and circulation, the mythology of science, with its synthetic account of universal history developed across previous series and out of past and present scientific writings and research, for all its problematic aspects, had become established as a recognized and authorized element of contemporary culture. This element, a scientifically posited universal story, or a universal myth, embracing consonant stories universal in historical scope and in audience extent, established a frame of reference "spatializing" time and domesticating sublime spaces, voicing species-wide truths in a manner open to the embrace and subject to the critique now characteristic of the public and scholarly reception of science and technology.

EVOLUTION, A CONTROVERSIAL PRODUCTION

Production of the *Evolution* series was initiated in 1999, when executive producer Richard Hutton secured the backing of Paul Allen's Clear Blue Sky Films, along with that of the WGBH/NOVA Science Unit. The last public television foray into a like-minded origins series was the three-part 1997 *In Search of Human Origins*, narrated by paleoanthropologist Don Johanson, the discoverer of the famed *Australopithecus* skeleton Lucy. As Johanson voiced it in the second of the series (June 10, 1997), "We uncovered a skeleton of our earliest known ancestor, nearly everything but the skull. Affectionately we named the creature Lucy," after the Beatles' "Lucy in the Sky with Diamonds."[5] "She caused quite a stir in anthropological circles. She offered convincing evidence of something quite unexpected. We could tell right away from the shape of her bones that this creature walked upright. Lucy was the starting point for the human lineage."[6] Here, too, in the context of Johanson's discovery, was the language of journeying, when he proclaimed that "over the last twenty years, I've been leading fossil hunting expeditions in this remote part of Africa on the trail of our earliest ancestors. The journey takes me and my team right down to the floor of the Great Rift. It takes two days driving dawn to dusk, if our vehicles don't break down.

But it's only in places like this where the fossils we're looking for can be found." As throughout the 1990s, the origins scientist was represented as an explorer or adventurer, and this representation was generally more optimistic the more that image was understood as actual (Sagan's hawks) instead of metaphoric (Weinberg's hounds).

After this series, NOVA had been looking to expand more overtly on the theme of evolution, to treat it more directly and in such a way as to address the contemporary political context.[7] The rise of intelligent design (ID) in the 1990s was accompanied by an intensity of media attention, in part driven by a series of literary, legal, and political battles in the later part of the decade and into the new century. Contemporary justifications for intelligent design had already been published, but Berkeley Law professor (now retired) Phillip E. Johnson, who galvanized the movement, was able to bring them renewed force. He believed that arguments against earlier forms of intelligent design and for evolution turned on little more than old rhetorical strategies and unfairly ruled out nonnaturalistic factors. On the basis of this belief, he began to publish arguments on behalf of ID and to seek legal redress to the prohibition of teaching varieties of creationism in the public school classroom.[8]

Johnson also began to build institutions, a "'big tent,' under which a broad range of antievolutionists—from young-earth creationists to progressive creationists—could gather," as Ronald Numbers describes it.[9] The "big tent" began with conferences, from which emerged the most active of the ID advocates: biochemist Michael Behe, philosopher of science Stephen C. Meyer (trained in the Cambridge History and Philosophy of Science Program), and biochemist Jonathan Wells among them. A series of events brought this group to the attention of Bruce Chapman, the founder of the Discovery Institute, which up until that time, in 1990, had emphasized less-controversial issues such as "improving public transportation and communication in the Northwest."[10] Chapman supported Meyer and the political scientist John G. West in establishing the Center for the Renewal of Science and Culture (CRSC), which by 1996 had attracted on the order of a million dollars worth of grants. Behe, along with Meyer and Wells, were among the first appointed researchers of the CRSC, and in the same year, Behe published his *Darwin's Black Box: The Biochemical Challenge to Evolution*, promoting his idea of irreducible complexity, a central tenet of ID: "By *irreducibly complex* I mean a single system composed of several well-matched, interacting parts that contribute to the basic function, wherein the removal of any one of the parts causes the system to effectively cease functioning."[11] Behe used the example of the mousetrap to demonstrate a simple case of such irreducible complexity and the obstacle he believed it to constitute to any evolutionary account of the origins of such a system. On the grounds of such irreducible complexity, Behe rejected the evolutionary account of the eye, a subject whose subtlety Darwin had attested.

The advance of the Discovery Institute and the attention it brought to ID loomed large in the media and political context in which *Evolution* (and the later *Origins*) emerged. And the production of the series was launched in the midst of much-covered state initiatives and court decisions examining and ultimately discounting the legality of equal treatment of evolution and forms of creationism or intelligent design in public schools. One case had particular prominence: "In those days it was Kansas that was coming up, right when we were doing the film."[12]

In 1999, the Kansas State Board of Education voted to remove teaching the Big Bang, geological ages, and evolution—main supports of the scientific epic—from the science standards of the state. As Numbers characterizes it, the decision made "Kansas the Tennessee of the 1990s," referring to the famous Scopes trial of 1926, which prohibited the teaching of evolution in the state public schools. Johnson and the Discovery Institute celebrated and supported the Kansas school board vote. Informed by these recent events, *Evolution*'s producers felt the need to assure religious viewers that evolution is no threat to religion, that it need not have sparked the controversies surrounding its inclusion in public school science curricula.

With this understanding in mind, *Evolution* was aired, a series of seven shows, a total of eight hours in length, originally broadcast over a four-day period in 2001, from September 24 to September 27. The first, two-hour keynote episode, "Darwin's Dangerous Idea," intersperses contemporary evolutionary research with a costume-drama recreation of Darwin's life, from his *Beagle* voyage through to his death and interment at Westminster Abbey. The other episodes each focus on one theme regarded as central either to the understanding of contemporary evolutionary research or to the wider reception of the theory. The second hour, "Great Transformations," gives the history of evolution in wide terms, the evolutionary movement from single-celled organisms to humanity. "Extinction!" (with exclamation point) examines the five mass extinctions since the beginning of life recorded in the fossil record, while the next hour/installment, "The Evolutionary Arms Race," focuses more on the question of disease, on the action of evolution in shorter time periods. In so doing, the episodes articulate the dangers that humanity poses for itself and for other species in the future, the possibility that the earth is on the brink of another mass extinction. "Why Sex?"

relates the evolutionary explanation of dimorphism—at times diffusing sensitive subject matter with pacifying animation—including a sociobiological dimension, which continues into the following episode. "The Mind's Big Bang" looks at the evolution of early human culture, primarily art, language, and technology, linking these potentially to neurological mutations in the brain and to cultural evolution.

Interwoven in the series are responses to the recent political and pedagogical contests. In the first episode, an extended sequence details zoologist Dan-Erik Nilsson's work demonstrating that the human eye could have evolved in less than half a million years. Nevertheless, it is the last of the series, "What about God?" that directly addresses the then-recent controversies that gave the producers of the series so much pause. With the camera staring upward at clear blue sky, turning toward a rocky peak through which the sunlight radiates, the narrator, Liam Neeson, begins: "The majesty of our earth. The beauty of life. Are they the result of a natural process called evolution or the work of a divine creator?" The shot dissolves and cuts to a classroom and then to a headline in the *Nation*, "Collision in the Classroom." Neeson continues: "This question is at the heart of a struggle that has threatened to tear our nation apart." What follows soon thereafter is a focus on a set of students at Wheaton, a Christian college, and students elsewhere, their struggle with the question of faith and science, put into the context of the American controversy over evolution, from Scopes to the present.

The pedagogical mission of the series in such a heated political and social context would certainly risk being provocative. Despite the conciliatory overall tenor of this final episode promoting the view that religion and science operate in different fields, address different questions, and are not essentially incompatible,[13] the series did provoke. The most prominent and entrenched of the criticisms was by the Discovery Institute itself. Prior to the broadcast, the institute commissioned a Zogby survey to determine how much of the public was receptive to teaching evidence against "Darwin's theory" in the classroom and found that 71 percent agreed that such evidence should be taught.[14]

In contrast to the *Washington Post*'s expository review of September 24, the *Washington Times* the day before, on the eve of the series broadcast, had published a criticism flatly denying the knowledge claims made by the series. Jonathan Wells, who wrote the article, dismissed the "interesting stories" put forward by evolutionary accounts, finding that the series "distorts scientific evidence" in support of its platform. He offered several examples: "A physician claims he sees HIV evolving into new species in a matter of hours—yet the claim is false. We are told that apelike creatures that lived a million years ago were our ancestors—yet Henry Gee, chief science writer for *Nature*, wrote in 1999 that this 'is not a scientific hypothesis that can be tested, but an assertion that carries the same validity as a bedtime story.' We are shown a mutant fruit fly with an extra pair of wings that is supposed to be evidence for the role of genes in evolution—yet (as the discerning viewer will see) the extra wings are immobile. The fly is actually a deformed cripple, an evolutionary dead end."[15] Apart from being a biochemist, Wells was both a minister and a scholar of the Unification Church, the founder of which, Sun Myung Moon, owned the *Washington Times*. Regardless of the suggestiveness of such connections, the political tenor of the paper in itself made the publication of a positive response to the series unlikely.

But the Discovery Institute didn't limit their response to criticism in friendly newspapers or appeals to surveys. The institute composed a book-length guide to the documentary series, *Getting the Facts Straight: A Viewer's Guide to PBS's* Evolution. It distributed the book gratis online and also sold hard copies through booksellers and Internet marketers. The guide argues four major points against *Evolution*: first, that the series does not show the shortcomings of the theory; second, that it ignores scientific dissensus over evolution; third, that it has an "excessive and biased focus on religion"; and fourth, that it is wrong for PBS to promote "a controversial political action agenda" through the series.[16] The authorship of the guide is credited to the Discovery Institute as a whole, and the language is suggestive of at least the particular contribution of Wells. Thus, in the same thread as Wells's *Washington Times* review: "*Evolution* also claims that all animals inherited the same set of body-forming genes from their common ancestor, and that this 'tiny handful of powerful genes' is now known to be the 'engine of evolution.' The principal evidence we are shown for this is a mutant fruit fly with legs growing out of its head. But the fly is obviously a hopeless cripple—not the forerunner of a new and better race of insects."[17]

Along with the series, *Evolution*'s production team issued a wide set of accompanying materials, including its own book. For a broad audience, science writer Carl Zimmer composed *Evolution: The Triumph of an Idea*, the foreword of which was prepared by Hutton and the preface by Stephen Jay Gould. The educational outreach was extensive, with a free teacher's guide "mailed to every high school biology teacher in the country."[18] For more specific pedagogical purposes (but potentially reaching a still wider audience) classroom lesson plans and primers for teachers on how to teach evolution were made available on the Web site. Speaking to the entire effort, Julie Benyo, former director of pedagogical outreach at WGBH, observed: "Teachers can't read the journals, it's too technical for them . . . and so we wanted to give them something that was specifically meant for them, for what they're teaching. So we did a scan of all the high schools, the primary high school textbooks too . . . and understood kind of what the curriculum was like at the high school level too and how we could integrate with that and give them more than they were getting through the primary textbooks."[19]

To each of the films, the Web site also provided a specific link, each with still punchier and more straightforward titles: "Darwin," "Change," "Extinction" (exclamation point removed), "Survival," "Sex," "Humans," and "Religion."[20] Apart from the episode previews these sites provided, they were interactive, with polls and slide shows of relevant chronologies. In the documentaries themselves, animated arrowheadlike cursors "click" on small graphics, evoking links inserted judiciously in each of the episodes that encourage viewers to visit PBS online in order to explore further materials related to the programs and pedagogy.[21] Since the original broadcast, the site has remained online (updated in 2007), and according to WGBH data, attracted on the order of several million visitors per year in the first several years.[22]

The pedagogical project therefore was more extensive than the broadcast series alone, with the series as centerpiece. The accompanying book begins with Hutton's diary-entry description of the on-site filming at Darwin's Down House in Kent. "I'm sitting in Down House, watching Charles Darwin teach his 8-year-old daughter, Anne, about barnacles."[23] As Hutton describes the objective of the project, "Our job is to report what is known, and not known, from a scientific point of view. And that means examining facts and hypotheses—the accumulation of evidence over the past 150 years; the exploration of ideas that are testable; the explanation of experiments that are repeatable—all to frame the story. Then we offer it to everyone who has a stake—everyone whose life is affected by it, everyone who is curious about it, everyone who is troubled by its implications for who we are and how we came to be—in other words, everyone."[24]

The declaration underscores the intent of the project to produce a certain story, and despite the opposing use of the term by Wells and others, "story" here was understood in no derogatory sense.[25] Moreover, the intended audience of that story, those affected, the curious, the troubled, was, as Hutton observed, "everyone." He emphasized the fact of this all-embracing story still further: "Our project, *Evolution*, is therefore a story of change over time, of who lives, who dies, and who grabs the opportunity to pass on traits to the next generation—and the next, and the next. It's the story of a simple mechanism discovered by Charles Darwin, and how that mechanism has altered the way we view ourselves. And it's the story of why evolution is so hard to accept for so many people. In sum, it is the story, the true story, of all life on Earth and how we are inextricably bound to one another"[26]—the true story of everyone for everyone. The species voice here extends further to the rejection of its presumption, now less in the context of the science wars or the multiculturalist discourses of the previous decade and more in relation to a reconstruction of a religion-science divide.

To viewers embracing the educational mission of the series, the story across the episodes constituted the historical account of evolution to which Hutton refers. This was the result not of the success of the initial broadcast, but of the more extensive pedagogical program, the Internet presence, without which the episodes themselves would have been less enduring or pervasive. The initial broadcast was overshadowed by the events of September 11, 2001, less than two weeks before.[27] For *Evolution*, the period of production and the period of broadcast and dissemination were therefore divided by a changed political climate, with the result that the implied or intended audience of the series was in certain ways out of joint with the broadcast's reception. The audience that viewed the series and digested its pedagogical program was now still more sensitive to issues of religious doctrine, difference, and right. The impact of the series was diffused over a longer time, in a period in which the importance of the broadcast itself was diminishing, the extensive educational and secondary school outreach program all the more important.

In this respect, science education polls in general, combined with the Web site's usage and the extensive counterresponse, suggest how much *Evolution*'s story was taken to be a canonical account. What gives body to that story, the structure and character of the films, their reliance on scientific advisors and scientific heroes, their pedagogical and media outreach, is much of what evolution means broadly, as a direct matter in its functioning as a primary conduit for the transmission of its evolutionary narrative and in filming a story still more widely told.

EVENTS IN TIME: THE DILEMMAS OF *EVOLUTION*'S STORY AND RESPONSES

The story that the eighth hour of *Evolution*, "What about God?," puts forward is of the confrontation between politicized conservative Christians, on the one hand, and science pedagogues, on the other. Though ID is at play in the history it relates and in the arguments it counters, *Evolution* does not discuss it explicitly.[28] And the proposed and overt resolution that the series offers is that the story's religious opponents recognize that the root of the conflict is groundless, is no true root. The majority of questioning Christians in the program ultimately lean toward the view that science and religion complement, rather than undercut, each other. Science, they argue, is the domain of empirical facts, religion and scripture the domain of values. And people are free to take evolution, it is lightly suggested, as God's chosen mechanism for producing life and the world.

However, a tension inherent from the outset of the series is that the resolution proposed is not on terms acceptable to many of those it hopes to assure. Protagonists in the series such as the energetic creationist Ken Ham argue for scripture as an empirical account. That is, they are arguing for a specific natural history, not in the sense of a secular history, one that excludes supernatural influences, but one that is nevertheless a true, empirical, experiential history of nature. This is not just a generational question or a matter of political, institutionalized action, at least not clearly so. The conflict runs more deeply, for example in an attempt of a group of high-school students in Lafayette, Indiana, to lobby their school board to mandate the teaching of creationism in their science courses. The students, it appeared, acted without supervision, on their own initiative.

Here the series anticipates the response of organizations such as the Discovery Institute and tries to represent potential criticism in order to respond and diffuse it in advance. Partly as a matter of the depiction of these tensions, partly as a matter of the biographical material on the scientists, the story at work in the series, as in preceding accounts of universal history, operates at two large levels: the natural history of the universe and the story about the production of scientific natural history. As a matter of filmmaking, this distinction arises in part from how much *Evolution* is a NOVA production, which embedded such a distinction in its mission.[29]

It is the second register, the story of the creative work of the scientists, that often includes the conflict over whether it is a valid natural history. This story of scientific production and conflict is in many ways part of the story in process, whatever is entailed in society's continuing composition of a new myth. The docudrama account of Darwin in the first two hours of the series makes this particularly clear. The Darwin of *Evolution* works toward his theory in the persistent anticipation of what he is ultimately represented as facing, the heavy-handed antagonism of the religious. In the series, this establishes a historical pattern of conflict that persists until the present.

The natural history itself, on the other hand, is the history of the world simulated in the documentaries, a final story of the world. In *Evolution*, it appears less as the tale of the coevolution of matter and life and more as the history of life alone. But even here, the previous chapters of that history—from the Big Bang to life—are connected in important ways to Darwinian evolution. This difference in registers, between the story in process and the final story or complete natural history, extends to differences in emphasis on the kind of footage used. Historical reconstruction and on-site footage usually point to the story in process, while computer simulation and expositional interview footage largely point to the projected final story, itself conceptually projecting an embedding of the story concerned with producing it. The difference also signals the different dilemmas that the series addresses.

The long tradition of conflict that *Evolution* represents as at stake, from Darwin to the present day, is in strong tension with the solution it proposes: the idea that should a correct understanding of the relevant domains of religion and science prevail, then peace will follow. The tension is signaled by the responses of the Discovery Institute, which, regardless of the absence of explicit attacks on ID, were sensitive to *Evolution*'s use of religion, judging that it "speaks to the religious realm from start to finish." The text of their viewer's guide observed that "Episode One is organized around a fictionalized account of Darwin's life, which begins with a scene pitting Charles Darwin, the enlightened scientist, against Captain Robert FitzRoy, the supposed religious fundamentalist. In fact, however, the two men shared similar views when Darwin sailed with FitzRoy aboard the HMS *Beagle*, because Darwin at that time in his life was more religious and FitzRoy was more scientific than this scene implies." The scene thus establishes the strategy of "casting everything in the stereotype of scientist versus religious fundamentalist."[30]

The guide concludes by rejecting the resolution the series puts forward, laying bare that resolution in uncharitable terms: "Although the producers of *Evolution* promised not to speak to the religious realm, they speak to it forcefully and repeatedly. The take-home lesson of the series is unmistakably clear: Religion that fully accepts Darwinian evolution is good. Religion that doesn't is bad. Now, the producers of *Evolution* are entitled to their opinion. In America, everyone is. But why is this opinion presented as science, on publicly supported television?"[31]

The Discovery Institute's response also tends toward a rejection of the idea that scientists have any special status, as with an emphasis on the similarities between Darwin and Fitzroy, rather than their differences. In the series overall, scientists are represented as in some way special, partly because as in prior representations, they are privy to universal truths through their engagement in a rare and elevated work of scientific authorship: Darwin had a difficult, high vocation.

In line with NOVA's mission, *Evolution*'s scientists are presented as having a life in science, a fate revealed in vocational calls. Viewers learn, for example, the story of biologist Mike Novacek, who as a child was inspired by his readings of the expeditions of Roy Chapman Andrews to the Gobi Desert and Andrews's discoveries of mammalian fossils. These biographical facts more generally include scientific discoveries where accidents of fate play a particularly strong role: a problem with his car ultimately leads Novacek and his party to unprecedented mammalian fossil discoveries of their own. As such, the scientists are heroes precisely in what they do to become authors and write their chapters of the history of the world, a task represented as working to complete a final story twining through their own lives.

But this conflict between antagonistic stories that the series attempts to reconcile finds itself embedded more deeply in *Evolution* than at the level of its overt religious allusions or themes or the representation of the status of the scientists. It is woven into the account through the nature of *Evolution*'s central events and the way in which it shapes its time scales. The simulations of the history of the world are the clearest representations of the final story, the universal history, that the film seeks to tell, and of how much that story is incommensurate with the story announced by creationists, and not just by Ken Ham. A sense of this scientific history is clearly portrayed in the displays of lost species roaming long-gone habitats, cross-sections of a particular time in the past, or in quickly evolving biological forms: one form transforming into another, into another, and so on as in a simulation of the fin-to-limb transition in "Great Transformations" or in the *Cosmos*-like simulated evolution title sequence following the prologue of each hour, which takes the viewer from the beginning of life to the current moment. This is a history of the world that has ambitions to completeness (forcing the question of what completeness means), an ambition that shapes the accounts of conflicts and of the production of the story and is taken to shape the work of the scientific disciplines that in sum work to write that history. It is an account that extends beyond the origins of the world, embracing the current moment in the evolutionary past, looking to these evolutionary processes as guides to the possible catastrophes and salvations of a future.

The final story projected through the series is open and incomplete to the extent that its own events—particularly the greater, pivotal events, rather than the inconsequential or ancillary events—are called into question.[32] For *Evolution*'s final story, there may be some uncertainty concerning the precise nature of a pivotal event, but not the existence of events of great consequence. In the series, pivotal events are global changes that affect all life, presented in the story either as catastrophic or as life engendering, or both. One clear example of a pivotal event in the third episode is the global extinction hypothesized to have been caused by the "K-T event,"[33] when an asteroid larger than Mount Everest collided with the earth at a speed greater than twenty-five thousand miles per hour. A simulation shows the asteroid falling toward earth, then cuts to a small, ratlike mammal running from a bright earth's surface into a dark hole, escaping flame behind it, reemerging later in a destitute world, uninhabitable for large animals.

This simulation, as the series explains, is only one of a set of possible alternatives not explored, a choice in the composition of the final story that nevertheless returns to the next point of certainty by the time of the next pivotal event. The final story sustains these uncertainties, even while tending very strongly to one alternative (the asteroid collision is probably how the K-T event happened, the film suggests) and giving a sense of what a complete story might mean. It would at minimum mean a story relating the true set of pivotal events, the complete framework of the story.

For example, the part of the final story developed through the episode "Extinction!" is a tale of five pivotal events, specifically global catastrophes, whatever their cause. Across the series as a whole, the pivots can be arranged from first to last, where the first event is the beginning of life, middle events are the catastrophes, a later event is the emergence of not only humanity, but a humanity able to relate these stories, and the final event is the potential fate of the human. That final event puts forward the now familiar comic and tragic possibilities—life continuing to flower or the next worldwide catastrophe. Neeson raises the central question of the episode: "What happens when a planetwide catastrophe kills off many species in a great mass extinction?" Directly following this, the skies darken, lighting strikes, lava bursts over a dark, mountainous horizon, and pits open in a blackened earth. The earth on screen is in upheaval just as the viewer hears that the postcatastrophe earth is a "level playing field," one that "made our very existence possible after a mass extinction 65 million years ago. Now it's we who may be causing a new one, but this time we may not be as lucky as we face evolution's severest test."

The danger is not framed as climate change, but as human overpopulation, the status of *Homo sapiens* as a "weed species": "The world is bursting at the seams with people. . . . There are now 6 billion of us on the planet. Even the dinosaurs would run for their lives. We have caused the rate of extinction to soar. . . . Many scientists worry that we are the new asteroid bringing about the sixth great mass extinction on earth." The potential forthcoming catastrophe, which portrays humanity as more fearsome than dinosaurs, "the next asteroid," is humanity's technologically enabled domination of the globe. But this domination does not always imply mastery: the primary cause of species extinction is given as direct and careless human habitat destruction, humanity's ability to take over lands just as weeds take over pastures. In comparison with this, the utopian possibilities are more muted. The fourth episode focuses on evolution and disease and puts forward the hope that "working with evolution instead of against it, we might eventually subdue even the deadliest microbes." The future promises catastrophic annihilations or extraordinary cures.

Point for point, event for event, the scientific history is a completely different story than the conservative religious ones. And it is not a story that many of *Evolution*'s religious-minded critics found meaningful. Its final story then raises the axiological questions announced by both E. O. Wilson and Steven Weinberg: Is this a vision of the world that the consumers of such an account can celebrate—in which they can find meaning? It is a question explicitly asked in the voice of Ken Ham in the final episode. Ham and a colleague reflect on the inclusion of dinosaurs in Noah's Ark over a diorama of the same. Neeson observes: "Ham and millions of other conservative Christians are convinced that it is the biblical story, not the evolutionary story that America's children need to hear, not just in Sunday school but in every school." What follows is more of the talking-head interview with Ham, who concludes: "Ultimately, if you're just a mixture of chemicals, what is life all about? Why this sense of hopelessness, this sense of purposelessness? And the reason is because they're given no purpose and meaning in life."

The implied wish of the producers is to answer just this question, to show how these simulations gratify, how that far-reaching, deep time can somehow be redeemed to meaning. The logic of the film accepts that the expanse of history must be made meaningful, despite its inhuman dimensions. The simulations that actualize the final story, that form it, attempt not only to satisfy a desire to experience this history as eyewitnesses, but to appreciate it as meaningful. To Ken Ham's strategic question, "Were you there?" the series appears to answer, "almost." By virtue of these simulations, all *Evolution* viewers are invited to position themselves as eyewitnesses to the workings of evolution and to the broad sweep of natural history. Viewers are prompted to inhabit lost habitats, to witness the explosive events that shaped the world to come. And they are in a sense invited to live through eons to watch changes of biological forms that transform them into impossibly long-lived witnesses.

To the question of meaning, the series enacts several responses. Perhaps this history, this view of the world, can be made meaningful through its exotic displays—the antique earth as an extraordinary bestiary. Or on the contrary, meaning might be found in locating viewers in their own place in a conciliatory, organically filled time, offering humanity a place on the tree of life, the whole of which is surveyable from that human vantage point. Overall, however, the two responses invited most clearly by the film are that humanity does have some special place in the world and that, despite many statements to the contrary on the part of *Evolution*'s producers, time isn't overwhelming after all. Both responses, both proposed resolutions, are revealing of the larger social and political issues in which the filmmakers were involved.

The first response, that humanity matters, is not simply a question of making reassuring statements. Which events are judged to be important is decided by viewing them through the lens of generalized human concerns, so much so that another way of characterizing an event as pivotal is whether or not it can be argued to have had decisive consequences for the fate of humanity, leading to either humanity's birth or its possible extinction. By comparison, the extinction of even a majority of species—nine out of eleven bird species on Guam as the indirect result of human travel, a destruction recounted in "Extinction!"—while ominous and dreadful, is not represented as tragic in itself. That extinction foreshadows what may happen to humanity, what is and may be the cost to its own species existence. The identity of the hero of this epic myth, life and even more specifically human life, is not decided as a matter of these species dying.

This claim also turns on evolution as story. At the end of the second episode, "Great Transformations," and toward the end of the fifth, "Why Sex?"—to take two examples—the ability of humanity to tell its own story is characterized as one of the distinguishing human features. Paleontologist Neil Shubin, at the conclusion of "Great Transformations," makes the strongest statement in this regard: "Does that [our own evolutionary history] mean we are not unique in many ways? Of course not. We're the ones telling this story. That evolution, that life has gotten to the point where it can tell this story." This invokes most evidently, intentionally or otherwise, Sagan's Schlegel-like claim that humanity is star stuff, evolutionary stuff, having come to consciousness at last, contemplating and telling its own evolutionary story.

Shubin's assertion characterizes humanity as distinctively human partly by being able to appreciate and relate its own natural evolutionary history. This is what counts as self-knowledge, offering a kind of species meaning to human lives: a story we live to tell, but one also that tells us who we are, what we may or may not accomplish, how we may or may not collectively live our lives.

But this natural history, which distinguishes and gives humanity meaning, may in turn be characterized as provided by nature, a story that the rocks tell. This begins to open into the second way in which the final story seeks to reassure its unsettled viewers, using film to condense or "spatialize" time. Neeson comments in "Extinction!": "Five times in the past 500 million years a mass extinction wiped out most of the species alive at the time. The earth itself tells the stunning story with its geological and fossil record stretching back in time." Indeed, similar phrases are used elsewhere, for example in a later sequence in the same episode, "The sediment tells the tale." The language is figurative, but asserts the universality of the story—the true, unique story of evolution is embedded in the fossil record. It can be called up from the earth. The story is not represented as subjective.

Yet at the same time, the earth is not conscious of this story, as far as is known, and cannot itself give voice to it. If it could or would, it might conceivably leave out humanity altogether, depending on what the earth took to be the central events of its story. Paradoxically enough, therefore, the fact of a narrator is set in relief precisely in implicitly being denied by the story being told. And by virtue of calling attention to the narration, to who is telling the history, this final story reveals itself to depend critically on who narrates. That is, the fact that humanity is characterized in part by who tells the history of the world highlights the fact that it involves choices of what to look for—of what counts as a pivotal event—and which events to relate in the narration. This structure can be seized on by the Discovery Institute, by Ken Ham and others, to argue that evolutionary history is mere rhetoric or storytelling, that the facts *do not* speak for themselves. This is not God's eyewitness account, God's voice speaking for the universe.

Directly after the pronouncement that the earth tells its story, Neeson continues: "Today, sheep roam the highlands of South Africa's Karoo desert, but 250 million years ago, the Karoo played host to creatures we can barely imagine. It was their world and then they were gone. Geologist Peter Ward is here to study the secrets of history's greatest mass extinction which swept those creatures away. It's a challenge that anyone would find daunting." Pursuing this daunting feat, the camera soon after follows Ward up a hilly trail as he narrates his own movements: "We geologists can climb through time. I'm going to climb about 50 feet up here. I'll go through two to five thousand years of time when I do it. This is the very last layer of the Permian. As soon as I climb above this, I'm now in the Triassic." Here time is etched onto a rock face, where a few feet capture and condense thousands of years.

This easy temporal gesture is itself indicated throughout the films and in the accompanying pedagogical works. Thus, in his preface to the accompanying book, Hutton observes: "Telling the story of evolution is almost as daunting as filming it. Cameras record only what is occurring right in front of them, and evolution doesn't happen in a few minutes. Yet we've taken it on, and we find ourselves exploring 3.8 billion years of life on Earth in 8 hours—which comes to about 132,000 years of life per second of television time."[34] This is an incisive statement about the poetics of the series, of its logic, and the countering of the human to the historical in telling ways.[35] But it also effaces the ways in which large scales of time are constantly manipulated with apparent ease in the series, making what the producers and scientists may indeed find "daunting" seem yielding, within grasp. This follows into the accompanying Web site in interactive materials that explain that humanity will probably never understand the "immensity of time," but then project important events along a timeline, showing geological and organic evolutionary changes together, with little attempt to evoke the expanse lying beyond the understanding developed through these timelines.[36]

These manipulations of the historical exposition, of the shape of time, are central both to the pedagogy of the story and to its other face, its structure. The distinction between what might be called "story time" and "narration or broadcast time" plays a role here.[37] At its heart, this is the simple distinction between how much time is understood to take place in the story itself versus how long it takes to relate or broadcast that story. Hutton's observations on the temporality of evolution and its recounting can then be regarded as demands on the ratio of story to narration time, averaged as over one hundred thousand years of story to one second of broadcast time. The prescription for the wedding together of story and broadcast time has broad consequences for understanding how these stories relate the truth claims they make.

Pivotal events in such an account have long-reaching consequences in story time. From that point of view, with so few events and so much time in human history, the story tends to be one of broad strokes, with relatively few pivotal events, but a strong sense of what holds them together. In turn, from the point of view of the lesson learned through the story—from story as a knowledge-making discourse—to know and teach natural history becomes to relate broad chronologies that are analogous to human genealogies, where humanity *is* emphasized. Following a catastrophe, small mammals spread and diversified, then came larger mammals, then the early monkeys and apes, then humans—those who tell this story.

Regardless, however much the blindness of evolution may be asserted, the story of series takes "us" to ourselves. It is *"for us."* As a program of knowledge, it bends the events and their status as pivotal or not to the conceived conditions of humanity's own emergence and to its endurance in the future. The earth tells *"our"* tale. The contractions of epic time, as well as the spatial stitching together of different regions of the earth, contract the history and extent of the earth to a human scale. As a kind of lesson, then—as a type of knowledge—this universal history is by default teleological, looking to a specific point in time, insofar as earlier events look ahead to humanity's emergence and reassure humanity (as the series thinks necessary) of its own importance.

Both ways of demonstrating the meaningfulness of the evolutionary vision—that humanity is the storyteller and that the storytelling can capture everything and make it familiar—together endeavor to resolve the dilemmas of meaning posed in different ways by Wilson and Ham. Together, they ultimately suggest, contra Wilson and Weinberg, that meaningfulness is not a problem that plagues a scientist, but rather is a problem only for a person with an incorrect understanding of science. And in contrast to Sagan (or Weinberg), it is less now the epic scientific quest that renders life meaningful, from which the final story emerges, but much more the poetry of that final vision itself.

But even in comparison with those proposed in the previous decade, this resolution was not universal. *Evolution* isn't Sagan's enjoinment to everyone to embrace exploration of the universe together. Its cosmology recognizes that the story it claims as of and for all has, decidedly, been rejected as universal over the course of the past decades by a large proportion of society—and from varied quarters, even if the series focuses only on related religious claims. Ham, the Discovery Institute, and their many sympathizers certainly did not accept that this history related in any non-oppositional way to them. Stories that had been written for everyone, as in the case of Sagan, had now been rescripted as arguments against many. The Discovery Institute's extended response demonstrates how much scientific natural history was rejected as applicable to them, and the anticipation of this kind of response tempered the universal ambitions of the series. The Institute's claim that, as represented by *Evolution*, good religion accepts evolution and bad religion doesn't, expressed a similar point, hyperbolically stated as a firm dichotomy: the series did accept that there would be viewers who rejected the evolutionary story, people who had an incorrect understanding of the relationship between science and religion and their separate work.

But universal acceptance was less crucial here than with Sagan. This story was not part of an effort to save the world from nuclear disaster, as it had been for him, linked within his series and more broadly to his scientifically minded universalism and activism. For *Evolution*, no specific and worldwide exploration is necessary to take on the task of authoring humanity's origins. But neither does *Evolution* express Weinberg's tough-minded foundationalist vision. Instead, the scientists appeal to the more aesthetic satisfaction in this picture of the world, to Darwin's much-repeated claim (on which the series, like *The Origin of Species*, concludes), "There is grandeur in this view of life." From that point of view, the question of how meaningful life is becomes relegated to how beautiful evolution is found to be.

BLANK SITES AND THE VOICELESS CHOIR: NOVA'S *ORIGINS*

Directly after the prologue of *Evolution*'s final hour "What about God?" a sequence begins with a shot of a small church billboard reading "Answers in Genesis." This is a church-hall lecture by Ken Ham and the name of the apologetics ministry he founded, Answers in Genesis or AiG, which under his guidance constructed the Creation Museum in Petersburg, Kentucky. "Today Biblical Literalism has no more forceful an advocate than Ken Ham. Five- to ten-thousand people visit his website everyday and his two-hundred and fifty lectures each year reach over a million people."[38] As people enter this particular lecture and the audience claps, viewers hear a voice singing "I don't believe in Evolution. I know creation's true." A man on a church pulpit sings into a microphone, a man to whom elsewhere in the film Ham refers as a dinosaur expert (his colleague in the discussion over the ark diorama). "I believe that God above created me and you. So praise his name for what he made, give credit where it's due. I know creation's true." Neeson speaks over the music, commenting on conservative Christian resistance to evolution. The singing then concludes: "I didn't crawl out of a pond or swing down from a tree. Adam is my ancestor and not a chimpanzee. God created everything, in six days he was through. So the Big Bang Theory's just a dud, and the million years are too." In his lecture, Ham implicitly affirms this connection between the Big Bang and evolution: "If you look at what the Bible says, if we start with the revealed word of God, and we build our thinking on the Bible, it tells us about the history of the universe, from beginning to end. It says that God made everything in six days."

This affirmation is made explicit in the next scene, in an interview with Ham: "The Bible says God created the earth covered with water, the sun, moon, and stars on day four. Well that's very different to the Big Bang. If the Big Bang's true, well the Bible got it wrong in astronomy. The Bible says there was a global flood but today we have a lot of people saying, no there wasn't. Well, if the Bible gets it wrong in geology.... And the Bible says God made distinct kinds of animals and plants to reproduce after their own kind. Well today Evolutionists would say no, one kind of animal changed into another over millions of years, so the Bible gets it wrong in Biology. Then why should I trust the Bible when it talks about morality and salvation?" Throughout this episode, both biblical literalists and science expositors, invoking in broad strokes the historical synthetic genre, connect organic evolution to material evolution, if on very different grounds, even while the narration, voiced by Neeson, stays quiet on this connection. The opponents of evolution here balk at the Big Bang as part of the same false story, while those who defend science standards in the classroom see the coevolutionary story as the expression of a story that is true.

Three years after the *Evolution* series was broadcast, NOVA's *Origins* series was aired, a series of four one-hour episodes on cosmic evolution. The bipartite evolutionary structure of cosmic-to-human evolution was evident to *Origins* series producers from the start—it was the prequel to *Evolution*, as they had quipped.[39] Production of the series began in 2001, the idea for the series conceived and the effort behind it begun several years earlier. It was both a prequel and sequel with a strong and immediate link to the NASA program from which it took its name.[40]

At the time of the emergence of the NASA Origins Project, science writer and documentary filmmaker Thomas Levenson, who would also be executive producer, a writer and a director for the NOVA series, met with Edward Weiler and Alan Dressler, two of the prominent crafters of the NASA Origins *Roadmaps*. Dressler later became one of the two science advisors for the series, with Sandra Faber. According to Levenson, the narrative of the NASA Origins Project sparked the interest in the documentary: "We talked to a lot of [to] Alan Dressler who was the lead guy on the research team and it was very clear what they were doing. I got started doing *Origins* when I'd done another film about the telescope, and I got along well with the people that I filmed with him [Weiler], talking about him after the fact and I wish I could do something with you guys again and I'm looking for the hook, and he says check out Alan's Origins thing. This great narrative hook. . . . . . . I called Alan . . . it was a great narrative hook. So that's all there." The narrative didn't merely act as a sequel of sorts to the *Evolution*, but amplified the NASA Origins story. "The Origins program as it was conceived by the team putting it together for NASA was one that said we can tell a comprehensive narrative that embeds origins of life, origins of the universe, the two stories are connected and we now know enough and it can direct the research program to do that. My film series was built on that statement."[41]

The extent to which this is true is suggested from the outset of the series:

A hellish, fiery wasteland. A molten planet hostile to life. Yet somehow, amazingly, this is where we got our start. How, how did the universe, our planet, how did we ourselves come to be? How did the first sparks of life take hold here? Are we alone in the cosmos? Where did all the stars and galaxies come from? These questions are as ancient as human curiosity itself. And on *Origins*, a four-part NOVA mini-series, we'll hunt for the answers. This search takes unexpected twists and turns. Imagine meteors delivering Earth's oceans from outer space. Descend into a toxic underworld where bizarre creatures hold clues to how life got its start. And picture the view when the newborn moon, 200,000 miles closer to Earth than today, loomed large in the night sky. This cosmic quest takes us back in time to within moments of the Big Bang itself and retraces the events that created us, this place we call home and perhaps life elsewhere in the cosmos.

These words are spoken by the narrator Tyson, current director of the Hayden Planetarium, who himself met Sagan in his adolescent years, the two developing a relationship thereafter in the course of Tyson's academic and wider career.[42] Tyson initiates each of the episodes with this invocation of the themes of origins and exotica from the NASA Origins initiative. The minute-long prologue calls on a palette of filmic imagery: lava flows, interstellar expanses, scientists in the field, telescope domes unfolding open, all headily intercut with computer simulations of the early earth and universe. Tyson is at times a disembodied voice, and at times he walks through the sky, actively inserted in the timeline he narrates.

In a NASA publication, Tyson himself affirmed the connections between NASA and NOVA *Origins*, commenting as well on the ways in which the Origins initiative functioned both as a funding model and as a unifying project: "If we go back to the beginning, we can trace the origin of 'Origins' to some efforts by NASA to create a new funding umbrella that didn't have much precedent in the portfolio of NASA projects. This 'Origins' umbrella would primarily bring together biologists and astrophysicists, but of course others would join in as time went on." This connection extended to disciplines that P. W. Anderson, who in his opposition to the Superconducting Supercollider and its privileging of high-energy physics, saw as set aside through claims that "the only fundamental science is at extreme scales." Here the connections began near apparent extremes, in labor to build a wider disciplinary tent: "Great work was already being done in cosmology, where we'd partnered up with particle physicists to understand the origin and evolution of the universe. But then it came time to understand the origins of other things, like the solar system and life, and NASA came to realize it might be uniquely positioned to make a research statement about that."[43] Levenson approached Tyson in the late 1990s, but the project was delayed by Tyson's work with the Hayden Planetarium and by Levenson's need to raise further funds. Tyson observed that over the intervening years, "there were certain key developments in cosmology and in other branches of origin science. Those developments gave the final product a level of enrichment that it could not have had if we'd completed it in the 1990s."[44]

It was partly as enrichment that Levenson saw his project in relation to the earlier *Cosmos*: "We very explicitly thought of it as a kind of. . . . not an homage to *Cosmos*, but we wanted to retell that same basic story or relook at the material 25–35 years on. . . . When Carl presented *Cosmos* there was a clear understanding of [that] the universe as a whole was a doable object of study, but in many many ways that concept was known to be true but what it meant in detail were the kind of things that you could find out and how you could build the picture of evolution from beginning to end—and yet there were blank sites."[45]

Along with *Cosmos* as an acknowledged point of reference, there were more direct relations and dialogue to and dialogue with the makers of the earlier series: "*Cosmos* was very much in our minds, even though we weren't going to redo it in terms of style or presentation. . . . Our host Neil Tyson was a friend of Carl's and continues to be a friend of Ann Druyan." Levenson himself knew Sagan, if somewhat more distantly: "I knew Carl near the end of his life, admired him, spoke to him, met at a couple of times and spoke to Ann Druyan a lot as developing *Origins*." And Levenson saw in Tyson a Sagan-like presence, one affirmed in the popular media: "Neil was the first person that we came into who had that same quality of charisma and joy and ability to popularize science like Carl had."[46]

The series addressed new developments, filling in blank sites with Tyson's enrichments in a style very different from the *Cosmos* series. The first hour, "Origins: Earth is Born," which aired September 28, 2004, portrays the research work of a few different scientists set against a backdrop of scientific consensus concerning the emergence of large-scale features of the earth. It also introduced a twenty-four-hour clock to put the major events of the universe on a human scale: "Well, it turns out, Earth became a habitable planet only after a series of devastating disasters in its early years. And to see how this happened, let's imagine all of Earth's four-and-a-half-billion-year history condensed into a single day, just 24 hours on an ordinary clock or watch like this." Tyson, standing in Times Square, shows his watch face, an animated, abstracted version of which is projected across an image of the earth from space. "If we start right now, then the first humans walked the Earth only 30 seconds ago." The second hand moves backward to this and each event that follows: "Dinosaurs began roaming the planet just before 11 P.M. The first multicelled animals evolved at 9:05. Before that, mostly single-celled organisms existed, and we think the first of those appeared around 4 o'clock on the morning. Earth was born at midnight on this 24-hour clock, 4.5 billion years ago, but its violent history began well before that, when huge ancient stars that had reached the ends of their lives exploded."

Central consensus elements of this story, the markings on the clock, are the earth's formation from the coagulation of stardust; its subsequent molten state, causing the repositioning of its elements in the period termed the Iron Catastrophe; and the formation of the moon. The episode is starkly different from Bronowski's and Sagan's "personal" narrations, whether as essays or voyages, however different their own notions of the personal were. The narrative expression of earlier generations, the unitary voice of the series or text, whether Shapley's, Gamow's, Bronowski's, Sagan's, or others', was now multiple, a feature the series shared with *Evolution*. Tyson was not represented as giving a personal view, but the consensus, aggregated view, as expressed not only through, but by the community of scientists.

Combined with the previous hour, the second episode, "How Life Began" (September 28, 2004), and the third, "Where Are the Aliens?" (September 29, 2004), demonstrated and amplified to a wider audience the intricate, historically resonant relationship between the origins and exotica themes of NASA and NOVA. "How Life Began" looked at the period largely before the history of life treated in *Evolution*, examining hypotheses of how life may have sparked itself on earth, how comets may have seeded the world via the organic chemistry they carried with them, a contemporary panspermia theory. Here, as with the first of the series, the early history of the earth is dramatically and overtly characterized as a series of catastrophes.[47] But here the event that caused the beginning of life is less certain. There are tantalizing possibilities, even plausibilities, but not quite certainty. As with the pivotal events of *Evolution*, the certainty is sometimes more in the fact of a pivotal event and its consequences than in the precise nature of its causes. And the beginning of life is clearly a pivotal event for this story—a story of a world that, constructed as a series of origins, would without the immanent emergence of life have meant no life, a story then paradoxically it would seem without life as its hero and without humanity to author it.[48] Again, it suggests that pivotal events become known as certain less as a result of their causes and more by their effects: something happened, and life began. Something happened, and the dinosaurs disappeared. And it is in turn conveyed that scientists are almost sure what did happen.

In "Where Are the Aliens?" the overall history of earth itself is characterized as relatively peaceful, a contrast with characterizations of calamitous events in the other three hours[49]—peaceful enough to give evolution enough time to produce intelligence. Despite the tensions this might produce in the narration of the final story (the earth as peaceful from one point of view, violent from another), comparison of the different episodes gives some further sense of how the assemblage of these varied inquiries in different sequences or episodes is presented as coherent, harmonious. This episode deals with the question of the likelihood of life and intelligence elsewhere in the universe, examining life's adaptability and the general question of what is necessary for its emergence, which in turn invokes the question of extremophiles so prevalent in the beginnings of the NASA Origins initiative and in the NASA *Roadmaps*. "In some ways, we owe our existence to serendipity, and some argue that this makes the evolution of intelligence far less likely. Our brains evolved through many stages: the little rodents, the early primates, and later on we branched from the apes."

What immediately follows is the connection of origins to exotica, sharing in the spirit of Loren Eiseley's or G. G. Simpson's midcentury criticisms, now, however, constructed through questions that both the Origins Program and the series took on as part of their narrative and object, tying the variation of life on earth to the possibility of life elsewhere: "This worked for us, but is it the only route to intelligence? Would an alien species have to go through the same steps? There's no way to know for sure, but on our planet, lots of animals have remarkable brains and behavior, including some that are very distant from us on the evolutionary tree. Among them are the cephalopods, including octopus, squid and cuttlefish." The cephalopods are domestic aliens who stand here for the possibility of an alternative evolution on earth.

The fourth hour of the series, "Back to the Beginning" (September 29, 2004), is the story of the accidental discovery of microwave background radiation and NASA's ongoing work (through surveys made by the Cosmic Background Explorer satellite, COBE, and the Wilkinson Microwave Anisotropy Probe, WMAP) to analyze it. These experimental efforts are shown in concert with the emergence and refinement of the theory of the Big Bang. This radiation is understood as the cosmogonic record, the radiation freed once the universe cooled enough to become transparent to it, a radiation predicted by the Gamow set.[50] Whereas the first two hours/episodes employed the twenty-four-hour watch to capture all of time, there is no clock device here. Levenson resisted this choice in the last two episodes.[51]

Unlike Sagan, then, in the series, Tyson functions more as a scientific impresario, inviting the audience to watch the performance that the history of universe had become. This structure, a kind of direct advertisement for that scientific history and the work done to produce it—and NASA's work, in particular—seems less concerned with a lack of confidence in the scientific content of the history or with the form that the account should ultimately take. Nor does it in any overt way address the fraught political and ethical position of the powerful, knowledge-producing scientist. Rather, the structure in part argues for the inherent drama of the history of the universe while, through this argument, revealing a concern that cosmology would not be found universally interesting. This concern itself may have been linked to fears throughout the previous decade that scientific education was failing in the United States, a fear for which popularizers such as Gould and Sagan were prominent voices—and a fear with its own extended history. But as with *Evolution*, the pedagogical context or subtext was strongly related to the political-legal question of teaching some form of creationism in public schools. It is this question that is invoked often and explicitly elsewhere by NOVA.

In NOVA *Origins*, however, the religious is not given an explicit voice. Instead, religion is pointedly ignored, unnamed as the historical account that is effaced by the new mythology. As with the producers of *Evolution*, the debates over religion and science were very much in the minds of the producers of the series. Levenson, in particular, made a conscious choice to address the religious by effacing rather than engaging them: "What I did in *Origins* was I refused to engage in that discussion. I appropriated the language and some of the symbolism or affect of that language to a certain and equivalent of authority, not an equivalent of truth."[52] Hence the choice to end on a new "version of Genesis." "I used deliberately the swelling celestial chorus that sounded kind of like church music that I could get away with to introduce the final moment where the guy says, we have a story that connects with very beginning stuff, right here, right now, with this enormous choir that's bursting out of the galaxies."[53] Levenson characterized this "religious sound in voiceless choir" as a "lay sort of historical and emotional claim to the grandeur, the marvelous beauty of some of the great religious liturgies." For him, it explicitly suggested that meaningful beauty of evolution: "I was trying to steal that affect for the story. As Darwin said, 'there is grandeur in this view of life.' There's grandeur in this view of the universe folks."[54] In this sense, Levenson intended to rebuild and transform Sagan's cathedral.

Origins, like *Evolution*, here reveals a sensitivity to a shifting political climate that, following the rise of creationist and intelligent design efforts, brought the political consciousness of natural science in significantly greater tension with politically active religious movements. It marks a shift from the uneasy political and cultural balances between the scientific and religious in works of earlier decades where—in popular scientific works—by and large the religious was represented as an earlier truth now consummated by science and where there was less apparent concern in the scientific epic about treading on religious sensibilities.

In part, however, this decision to ignore religion in *Origins* and on Levenson's part was informed by the failure of public discussion or of interventions like those of the earlier *Evolution* series to achieve a less contentious understanding in the public domain of the relationship between religion and science. "One of the problems in dealing with those who reserve a special status for revelation is that you cannot in fact have a symmetrical argument with them," Levenson observes, "because when the argument strays onto something that they, the other side, interprets as a claim of, not of empirical fact or argument, what have you, but as a statement of revealed truth, all you can do is say, 'Yes, but I don't agree.' There's no further exchange possible. So the conventional strategies of both sorts, competition of narratives even, I tell this story, you tell your story, or formal argument no longer work."[55]

Such politicized differences are reflected in the polarization of reasoning and story-telling (as a disassociation of logos and mythos) dismissing opposed synthetic constructions—the opposing histories—from which these poles cannot easily be separated. Creationists such as Jason Lisle, a University of Colorado–trained astrophysicist, indict the immanent logos of the scientific epic and also judge it (as Levenson suggests) according to an alternative mythos—according to a correlate form of reading that has its own logic, distinct from other religious modes of reading. Levenson's observation that the grounds for narrative comparison are undermined suggests in turn that the modes of arguments stitched into the different accounts are incommensurate. But Levenson's decision to leave the reception of the account by the religious out of the account is itself a reminder of the exclusivity involved in the celebration of a new cosmological myth.

As with *Evolution*, popular reviews of *Origins* were largely positive, with an emphasis on the high production values and, indeed, the Sagan-like appeal of Tyson.[56] In contrast to the *Evolution* series, the first broadcast was not overawed by recent events, culling, according to Levenson, a Nielson rating of 2.7 in its first screening, a viewing of 2.7 million households, without considering its further use in other venues, from classrooms to Internet viewing.[57]

However, *Origins* did provoke a hostility similar to what *Evolution* encountered, and from familiar quarters, although no book-length retort. One institutionalized critique came through Ham's AiG in the voice of Lisle. A creationist counterpart to Tyson, Lisle was formerly the program director of the Creation Museum's planetarium and a celebrity on creationism lecture circuits. Lisle's critiques began with the assumption of the truth of scripture as a form of eye-witnessing: *Origins* "stands in stark contrast to the historical account of origins recorded in Scripture. Consider the many differences between the biblical account of history, and the evolutionary story promoted by the 'Origins' series. The Bible teaches that the world was not created through natural processes; it was created directly by God (Hebrews 11:3). Scripture indicates that the origin of the earth (as well as the entire universe) was thousands of years ago, not billions. According to Scripture, the earth had water right from the start (Genesis 1:2); it was never a molten blob nor did it require comets to deliver the water as the 'Origins' broadcast teaches."[58]

Here the contrast is not between scientific and sacred history, but, in Lisle's hands, between "recorded history and the secular story." And here again "story" is used as something of snide jab: "Although it was an interesting story, 'Origins' did not adequately answer any of the questions it raised in the prologue," Tyson's introductory preamble cited above. And "although the special effects are very impressive, the story is merely fiction presented as truth. It stands in stark contrast to the historical account of origins recorded in Scripture." "Evolutionary storytelling," as criticism recurs in Lisle's rhetoric.[59]

As with the Discovery Institute's critique of *Evolution,* Lisle indicts the science that he takes to inform the story: "there are many scientific difficulties with the ideas of the formation of Earth, the moon and life promoted by the PBS series . . . there are problems getting the dust grains to stick together to form planets; the decay of Earth's magnetic field is not consistent with a multi-billion year age; the rotation rate of the sun is far slower than would be expected from these evolutionary scenarios; and the recent discoveries of massive, Jupiter-sized extra-solar planets orbiting very close to their star goes against the predictions of the evolutionary model of solar system formation."[60] Lisle thus attempts to stage both an externalist and an immanent critique: the science fails not only by the lights of literalist scriptural reading, but also is counterindicated by scientific data. He attempts to sever the empirical moorings of the coevolutionary narrative, to demonstrate that the narrative is already groundless.

Lisle also gives special attention to the series clockwork, affirming an alternative, rigid chronology: "Both parts of the series," the first and second hours, "employ an interesting analogy: the supposed 4.6 billion years of Earth's past are compared to a 24-hour clock. Using this scale, we are told that human beings come on the scene only in the last thirty seconds of history. Yet, Jesus said that human beings have been around from the beginning of creation (Mark 10:6). If we compress the approximately 6000 years of true history as recorded in the Bible down to 24 hours, then humans existed within the first fraction of a second of history—right at the beginning as our Lord has said."[61] The ratios—one day to the history of the world, scales of broadcast time to universal time—constitute markers of belief, political and religious, and of the heroes/authorities that the histories celebrate. The different heroes of the different histories are understood as the authors of the widely disparate epics they relate.

The extent to which NOVA can be said to be playing a dominant role in establishing a new cosmology, a new version of the cosmic myth, a new choir, can be judged to a degree by the reception of the original broadcast audience, but also by the extent and success of the pedagogical mission. Here, indications of success are only indirect, but give a sense of its scope.

First, the film is pedagogical by its very nature, linking directly and explicitly to then active research at NASA. The series Web site was designed with links to more in-depth interviews with many of the scientists involved (Tyson, Andrew Knoll, Sandra Faber, Peter Ward). The site further included interactive media canvassing the research at stake in the films, teaching for example, the decoding of cosmic spectra, how to calculate probabilities of

other life in the galaxy, and the processing of Hubble Space Telescope images. As with *Evolution*, the slide shows also included interactive timelines
of the history of life ("A Brief History of Life") and on still more sublime
scales, the history of the universe ("The History of the Universe") from the
Big Bang to the future Dark Era, when "for all intents and purposes, the
universe as we know it will have come to an end."[62]

But through interactive media, NOVA (and NASA) also has pedagogical
initiatives that take the series, research activities, and scientists directly into
classrooms and museums. NOVA generally makes an extensive effort to produce documentaries amenable to coursework, including forming lesson plans
satisfying standards determined and set by the National Research Council's
National Science Education Standards or the Principles and Standards for
School Mathematics of the National Council of Mathematics.[63] In the context
of NOVA *Origins*, the overt pedagogy focused on a set of classroom-directed
research activities in conjunction with each hour of programming. These
range from minimeteorite searches, to dioramic exercises in determining the
characteristics of life, to model spectral analyses in the hunt for alien life,
to question-and-answer sessions recommending that teachers pose questions
such as "How are humans and stars connected?"[64] So far as the classrooms
are concerned, NOVA *Origins* was a part of a science lesson. The exercises call
attention to specific aspects of the final story of the series and also to the more
human aspects of the life of the scientists, with suggestions that, for example,
students debate the question of the value of competition among scientists.

The connections between series episodes and lessons are clearest in exercises directing prescreening and postscreening discussions. For the first
two episodes, prior to screenings, students are asked to construct their own
timelines of universal events, such as the formation of the earth. Postscreening, they are directed to revise those timelines on the basis of what they've
learned from the episode. As a result, these exercises also tend to emphasize
the simplicity of capturing the expanses of time, the basic but ambitious
work captured in the drawing of a timeline. From the first of the "Before
Watching" activities: "The universe is about 14 billion years old, while Earth
is estimated to be about 4.6 billion years old. Homo Sapiens evolved about
600,000 years ago. Have students calculate the length of a timeline that
would show all these events if 500,000 years were represented by one centimeter."[65] The exercises accompanying the second episode have a similar
time-line exercise, here based on the clock analogy.

From this point of view, one of the major pedagogical thrusts of the
documentary series reinforces its demonstration not of the impossibility of

conceiving of such vast expanses of time and space, but of techniques for condensing and capturing it. This, too, turns on establishing and focusing the strict chronology of events, so far as is possible and up to a level of accuracy immanently given. Classrooms could thus represent timelines of events recording the history of the universe laid out against or alongside timelines of civic histories. This form of exercise and of graphing underscores a teaching philosophy that sees chronology as a first step toward understanding a broader historical narrative. It gives some sense of why AiG's response would pay attention to the clock-based analogy, asserting a different ratio of human to universal history, a different way for humanity to appear, a different story.

ABSOLUTE MYTHS AND CONTEMPORARY EPIPHANIES

The ideas of myth and epic that were at work in the accounts of previous generations—whether in Frye's powerful heroes and epic authors or Sagan's inevitable metaphors—are still at work in these documentaries. But that scientific history became still more confident, thorough, and concise while involving still more authors and narrators. At the same time, its most vocal opponents (AiG, the Discovery Institute, and so on) increasingly fought its narratives as incompatible with their own histories of the world. The claims of myth in these documentaries and pedagogical enterprises went further than did their predecessors in the assertion of an account that left no space for the truth of alternatives. For both Levenson and Lisle, the scientific epic of NASA *Origins* and the history related in the biblical Genesis, could not both be true. The validity of one excludes the truth of the other. In that light, a measure of the status of the final story told in *Origins* as a new myth was how much it succeeded in drawing out responses such as those of AiG. Perversely enough, then, the extent to which the final story put forward by *Evolution* and *Origins* failed to convince its opponents that it wasn't contradicting religion was the extent by which measure could be taken of its status as the new myth it claimed to be.

Regardless of their overt stances, both series shared in the absoluteness of their truth claims. *Evolution* may indeed have attempted to effect reconciliation by arguing for the complementarity of religion and science, but only by denying the role of science in the field of human value or by denying the status of the Bible as a literal history of nature. The series made both claims and denied making both claims, reflecting the paradoxes of those positions in the culture and the political moment in which the series attempted to intervene.

The first claim, that science deals in facts, but religion in values, broke with the many attempts in *Evolution* to present evolutionary history as aesthetically satisfying and as leaving a special, meaningful role for humanity. The series made the second claim, that the Bible is not a literal chronology or a literal history, in a backhanded way: by showing how evolution comes into conflict only with the most literal-minded religious people. But however conciliatory the gesture, this was not a compromise on truth: scripture must be interpreted in order to produce the scientific account, while the scientific myth must not bend in order to redeem the religious account. It is religious scripture that needs all the padding of interpretation to make it true.[66]

The specifics of the different accounts alone were almost certain to produce conflict: to fail to include Moses in one account while emphasizing the importance of (for example) an asteroid was to deny the completeness of the religious framework. A history may not be far wrong if the dates are askew, but not if the central dates differ by scales of billions of years with readers/ viewers in the audience who balked at differences of thousands—and not if the heroes and their feats diverge entirely from one account to another. The complete literalist or conservative Biblical story would have to be the story of another world from the point of view of the contemporary scientific history, just as for people like Ham and Lisle, the scientific history could be no more than the historical account of a fictional world. And given their political-cultural context, other worlds must be false worlds.

The scientific and religious accounts also face very different problems when dealing with the scales of time involved for each. The difficulty at stake in the young-earth hypotheses of AiG and other like-minded institutions lay in finding naturalistic ways of explaining phenomena that could no longer easily be made to fit the "generational" time of the six to ten thousand years they posited. As Lisle discusses in his video introduction to the Creation Museum Stargazer's Planetarium, "We can talk about hot topics in the creation astronomy field, like 'how did God get the light from those distant galaxies to earth in just a few thousand years?' We can present some of the different models there."[67] *Evolution* and *Origins* faced the different problem of capturing an amount of time that they accepted as overwhelming and nonhuman within a broadcast period. In contracting the age of the earth to one day, the filmmakers collapsed over one and a half trillion days down to one, a kind of scalar contraction that they repeatedly emphasized when describing their work.

These contractions were of course not unique to these series. Among the many other examples over the years were Carl Sagan's Cosmic Calendar,

Stephen Weinberg's textual use of film reels, science popularizer Tim Ferris's use of 4.5 kilometers of highway for the 4.5 billion year history of the earth in his 1998 film *Life beyond Earth,* and the many illustrative animations on different NASA Web sites.[68] All of these "domesticated" the expanses of time and space that in interviews and anecdotes the authors and series producers repeatedly referred to as inconceivable. The contractions, all these different "spatializations" of time, smoothed out the raw conceptual questions and physical phenomena into familiar filmic and scientific imagery. As repre-sented in the series, conceiving of the enormity of the universe, its size and its age, did not require much imaginative energy. With audiences long since weaned on *Cosmos* and with the century-old manipulations of film, what it meant to apprehend the whole of the universe and its history could be little more than imagining a filmed history, reels running impossibly fast, even in a digital era. Or it meant rescaling time against an ordinary spatial mea-sure—a strip of road, the markers on a clock face. Each of these filmic and textual devices, even when attempting to underscore the inconceivability of the spans in question, swept the viewer across time with a few quick gestures.

Given the evident political context and the criticisms of these series as mere theory or mere story, it was more important for the documentary se-ries to stress how seductive these histories were, rather than stress how im-possible they were to understand. Levenson might have resisted the stale neatness with which the clock analogy packaged these sublime scales of time and space, but his series also gave in to the advantages of these devices. The dilemma was that *Origins* as well as *Evolution* had to do both at once: stress the exotic nature of a history on the order of billions of years, but also show how intuitive and nonthreatening a history on that scale was. The dan-ger was also to a degree implicit in the construction of the devices as such, in the appearance of having captured a whole of time the capturing of which in any substantive sense was declared impossible, their use in pointing to limits of understanding undercut by the totality the devices embraced. And, as with the NASA Origins Project, the question of the exotic was ultimately subordinated to the question of origins, with the universe presented on a human scale, even as such an act of representation was claimed to be impos-sible or in key ways misleading.

It is in part the ease of conception and reception enabled by these de-vices, how simple the viewer found it to feel at home in the vastness of time and space, that made these series, and the NOVA documentaries, in particu-lar, strong pedagogical tools and myth-making technologies. Many of these features were by this point familiar to an audience receptive to scientific accounts, if at times the features were hazy in their representations: that the universe probably started in some kind of explosion; that it has been cooling as is it has been expanding, an expansion that creates its own frontiers; that supernovas produced heavier matter; that from coagulating dust emerged larger bodies; that eventually the conditions for life emerged in a cocktail of the right chemical reactions in the right environmental conditions; and that once life emerged, natural selection eventually produced a rich variety of forms, if not the variability itself; and finally intelligence appeared, however understood.

This development in part appears easy to grasp because of the familiarity of the devices employed in the documentary tradition that explored it, at least in a contemporary Westernized context: bursts of light; sound tracks associated with the hollows of distant space; a disembodied voice explaining that matter is cooling and coagulating out of hazy nebulae; the molten early earth; jiggly single-celled organisms, pregnant with potential for future life; fossilized remains; images of early man in the wilderness, stooped over in an African, Edenic pastoral, Eden often an overt allusion. To claim to know the history of the world or that it might in fact be easy to visualize the evolution of the universe and of life became much the same as claiming that it is easy to recall the ways filmmakers have chosen to depict the theories at work in that history.

The compression to a human scale structuring these depictions might be an alternative mode for presenting science as a humanistic endeavor, with all the political and cultural controversies attending that visual and narrative technique. The universe is repeatedly, perhaps inevitably, represented by the human: by its clocks, its roads, the sizes of its screens. Still more, the universe is directly pictured *as* human, conceived of as a maturing human life. Among the documentaries, the most vivid if somewhat whimsical case appears in the first hour of *Origins*, when the filmmakers use Tyson's own pictures as a child (and an animation that "zaps" him into a baby) to draw the analogy explicitly: "But the early Earth bore little resemblance to the planet we're all familiar with. And today, working out exactly what Earth was like as a newborn planet is no easy task. It's sort of like looking at me as an adult, and trying to figure out exactly what I was like as a baby: When was I born? How much did I weigh?" Here the analogy reveals how self-conscious such constructions can be, so often disguised by the inconspicuousness of phrases such as "the infant universe."[69] And such constructions more deeply connected the history of the universe to the lives of the scientists and human actors.

The scientists themselves on an individual level are portrayed as having little more to their lives than their vocational calls and scientific discoveries. And they find so much in those discoveries that they take the meaning of their lives and the meaning of life as such to be at stake in their work, a representation consistent with the axiological stances taken by Weinberg and Wilson. In this sense, the consonant scientific biographies as portrayed in the documentaries are representative of the quest to write a history of the world that in turn allows them to find themselves—to find themselves at what they proclaim to be *the* moment in human history. As physicist Anthony Redhead states in the final episode of *Origins*: "This is a wonderful time in science. This is actually the best time of science, because we have the satisfaction of—through these observations and these discoveries—having confirmed certain predictions. We are actually on the brink of a revolution of unimaginable proportions."

Among scientists, this quest is most often represented as one they have undertaken and are undertaking together. This is best seen in how the independent research initiatives are represented in both *Evolution* and NOVA *Origins* as interconnected.[70] The role of Tyson as scientific interpreter or impresario and not as a Saganesque prophet speaking his own personal truth reflects this interconnectedness, as does the third-person narration of Leeson. Their work of narration appears as one that simply draws out the connections in research programs, tracing timelines in the history of the world.

Significantly, the research elements are not presented as potentially conflicting or as deeply competing paradigms, but as mutually reinforcing and consonant accounts. Even if competing among each other, the scientists are on the same quest.[71] When they do describe their discoveries as having been controversial, the controversy is presented as settled, as only in the past, to do with superseded paradigms not at stake in the final history related in the documentaries. Repeatedly, the scientists assert that only now is the time for a full account to be written—the current, relevant theories can largely stick together. And as with the promotion of the NASA Origins Project, this appears to be what scientists such as Canizares, Weiler, Dressler, and others took to be the case, from their preparations at the Washington space summits, to their congressional testimony, to their composition of the *Roadmaps*, to their preparations for TV documentaries.[72] Canizares saw the topic of origins as "encompassing really almost all of NASA's science program," "the theme that seems to run through the whole story from beginning of time to the present," and for different research investigations, "it provided the catalyst [to] synthesize a whole bunch of things that were already there."[73] This consonance between all the stories, like the consonance between all the different research programs under the NASA Origins Project, reaffirms the idea that this new myth involves a tendency toward a bundling together.

The human story of the series is this kind of cluster of different stories, the stories of the authors of the myth, of different research groups. By and large, it does not give an account of the historical ways in which the different research initiatives might be or are linked. As far as the viewer can know from the series, the initiatives do not take their cues from one another. This is not to suggest that the series asserts the opposite, that there are no interpersonal or professional relations between the research programs. Instead, in not assembling the historical discoveries as so many elements in *their* own chain, an alternative image of a kind of happy coincidence and consonance emerges, a Whewellian consilience on the level of narratively constructed synthesis. Paleontologists at work in discovering geological processes develop theories that fit with the theories of evolutionary life, with speculations of alien life, and so on. And as in *Evolution*, measurements of the extinction of animals in one part of the world are linked to the spread of tuberculosis and attempt to cure tubercular prisoners in Russia, to combating the spread of weeds in North Dakota, to the emergence of sexuality, to the battle of religion and science, and so on, all of which across both series can be made sense of and folded into the embrace of the theory of evolution and the Big Bang. This conveys effectively and seamlessly that the cosmology is there to be had, independent of the questing.

Here, the choir registers not only the new myth, but the epiphany of that current moment as represented in the series. In effect, the documentaries place the scientists both inside and outside time, as living their lives at a particular time, but with the epic authorship nearing completion, able to survey the entire sublimely scaled past and possibly the future.[74] The individual lives of the scientists are not explicitly included in the timelines that are established, thus poetically reinforcing their place outside of any clear time when they survey all of it and keeping the final story at a distance from the work that scientists do to write it. This fact is criticized in the form of Ken Ham's question, "Were you there?" By these lights, the series responds less with the apparent "almost" of historical simulation, and twice answers otherwise: that it hardly matters, being there or not, but also that in some sense, to know the history of the world requires the timelessness of surveying all of time.[75]

THE NEW *COSMOS*

The documentary tradition remains the most prominent platform for contemporary universal history and still goes by the name *Cosmos*. NOVA *Origins* has not been the only sequel to *Evolution*. For documentarians in this tradition, the defense of the role of science in education and the circulation of their own productions called for them to confront intelligent design directly, while conversely organizations such as the Discovery Institute continued to worry about the influence of PBS and of *Evolution* in the years following the broadcast of these series. In 2007, on the pro-ID blog *Evolution News and Views*, Jonathan Wells observed that "many public schools in the U.S. are still showing biology students the 2001 PBS *Evolution* series. This 8-hour propaganda extravaganza—like most modern biology textbooks—distorts and exaggerates the evidence to convince people that Darwinism is true."[76] In the same year, in preparation for concurrent release with NOVA's *Judgment Day: Intelligent Design on Trial* (November 13, 2007), the Web site of *Evolution* was updated, as well—"Evolution 2.0" as WGBH referred to it internally.[77] The Web sites for each production, *Judgment Day* and *Evolution*, featured conspicuous links to the other documentary production. And *Judgment Day* shared many of the major themes of the final episode of *Evolution*. Now, almost as final work to be done and a response to criticisms of *Evolution*, NOVA focused on the question of ID as promulgated by the Discovery Institute.

The specific trial alluded to in the religiously resonant title of the documentary *Judgment Day* was the court case *Tammy Kitzmiller, et al. v. Dover Area School District*. Kitzmiller and other parents of students in the area argued that a recent school board decision requiring students be told of the existence of alternative accounts to evolution was a violation of the establishment clause in the First Amendment, a challenge they prosecuted successfully. The Discovery Institute was involved in both the school board decision and in its defense. Members of the board were advised by the Discovery Institute prior to the requirement being instituted, and ultimately, much of the trial was argued on the merits of whether their version of ID was a scientific theory. This went to the question of how differently the ID theory was produced, was written, in comparison with the extended, intricate, empirical history that stood behind the coevolutionary account.

That account, as expressed in *Evolution* and *Origins*, ultimately emphasized less how the scientific arguments work and more their results, a choice that at least in the case of *Evolution* drew criticism. For one commentator, "the biggest frustration is the series' focus on the 'what' of evolution as opposed to the 'how.' My frustration is that there is so little emphasis on mechanism."[78] Though partly addressed through pedagogical efforts produced in tandem with the series, the criticism suggested a potential imbalance. The production choices, balancing thickly exhibiting "what" in relation to transparently simulating "how," constituted a shift from earlier series such as *Ascent of Man* and Sagan's *Cosmos*. The shift in cultural and political backdrop between earlier and later productions, the emergence of polarized opinions apparently segregating scientific from religious communities, added weight to this imbalance.

Such prior documentary choices and the shape of the contemporary cultural-political landscape helped define the production choices of the new *Cosmos*, a series called *Cosmos: A Spacetime Odyssey*, which aired in 2014. Even without the history detailed here, it would be impossible not to compare and contrast the old and new series. The latter repeatedly cites/invokes the former. Ann Druyan and Steven Soter authored the new *Cosmos*, as they coauthored the original. Neil deGrasse Tyson presented the new *Cosmos*, who since NOVA *Origins* has achieved something of Sagan's categorical status as the most famed US science popularizer of the day, reinforced by his appearances on popular media outlets such as the *Colbert Report*, just as Sagan's appearances on the Johnny Carson *Tonight* show had advanced his popular recognition.[79]

The continuities from the first to the second *Cosmos*, from personal voyage to space-time odyssey, extend into the overt subject matter detailing the impossible breadth of the cosmos through identical or similar devices used to survey it: the ship of the imagination; the cosmic calendar; the use in the later series of the animated sequence showing an evolution of forms leading to humans from the earlier series; the natural historical recreation based on the epic of evolution and the human historical recreations of the lives of past natural philosophical and scientific thinkers.

But the differences between the series are also manifest. Conspicuously, the possibility of and the search for life on other worlds is not a theme in the new *Cosmos*. The original series had provoked criticism on this score, with a variety of critics taking issue with what appeared to be an idiosyncratic interest of Sagan's.[80] In the new series, though there is the suggestion of the existence of other life in the universe, there is no encouragement to search for it. The difference speaks to the extent to which the new myth can treat life elsewhere as thematically secondary, even if it remains a conceptually and structurally necessary consideration.

More broadly, there is a comparatively greater emphasis on explanation in the former series, on the "how," a greater scope for naturalizing display in the second, on the "what." This is not only a function of differences in production values in a narrow sense and in filmic/televisual technologies. Differences in format between public and commercial television and between corresponding audience expectations foster the more expositional voice of the original public television series and the still greater emphasis on technological display of the latter, which was produced to be consistent with the demands of commercial television. The running time of the individual episodes is less in the second series, correlated with a more cursory use of theoretical detail and a denser visual landscape.[81]

This format shapes the content of the new *Cosmos* most obviously in sequence punctuation. Leads-in to the advertisements, themselves cutting further into the duration of the programs as compared with their public television counterparts (once the BBC-inherited 13 episode format and standard ranges of programming lengths are accepted), tend in the second series to be dramatized as moments of high tension or expectation. The series, much faster paced, is filled with event and drama, giving less time for the contemplative mood often sustained in the original series. Sagan's voice was at times exultant, and to some, spiritual or theatrical; Tyson's voice is more emphatic and filmic. Less room is made for milder illustrations. Instead, with computer-generated imagery at the forefront, the sensibility of the contemporary series is to nurture a sense of concreteness by opening up further the spaces where the ship of the imagination can go. So Tyson visits the slippery world of the cell or the violent event horizon of a black hole. As such, if the new series achieves "documentary consciousness," it does so by moving away from a sober, didactic presentation, leaning instead toward television as adventurous travelogue, as the more impersonal odyssey that its subtitle suggests. The historical genre of synthesis in this *Cosmos* is nearly overcome by a variety of the fabulaic.

With a certain care and conceptual risk, this difference might be captured by attention to standard uses of the terms "simulation" and "exhibition." Simulation has a number of connotations inherited from or invested in scientific or computer modeling or historical restaging.[82] As a matter of investigation of the natural world, it can often be understood as open to the future, as a process awaiting further refinement or revision. In this sense, it also requires a feeling for the theoretical tools necessary to stage it and/or the possibility of revising it. Exhibition, in the evidentiary sense, has a slightly more enclosed connotation, as exhibiting facts presumed or presented as decided upon, even if the exhibitors understand that not all accept such facts in this way. The forms of participation in dominant paradigmatic exhibition, depending on the weight given to different performative contexts, have tended to be sealed off enough that disruptive or interactive installations still have the power of surprise—the surprise of being free to touch an object or to circulate in a no longer rigidly demarcated space, or to contribute to the content of an exhibit.

By a comparison encouraged by the continuing life of the original *Cosmos* and the invocation of it by the later series, the treatment of science and history in Sagan's version appears to lean more on simulation. The viewer is given comparatively more time and programming space to reason and weigh the truth of contemporaneous scientific views (whether relativity or atomic theory), of the speculations of the *Cosmos* authors themselves (as with the probability of life on other worlds), and of the worlds they constructed as visualizable and explorable. Through the same filter, the historical episodes and scientifically disclosed worlds in the second series tend somewhat more toward (conventionally understood) exhibition. Viewers are still invited to reason along with the filmmakers over the emergence of scientific belief through, in particular, their treatment of historical subjects—especially in certain episodes, such as the representation of the geochemist Clair Patterson's work in the seventh episode, "The Clean Room," or the oceanographer Marie Tharp's work on continental drift in the ninth episode, "Lost Worlds of Planet Earth." But the pace and production values tend still more than in the original series toward evidentiary display, in the sense of "come and see." The two series turn on different balances between universal historical explanation/reconstruction and fabulaic exploration, balances propping up the different syntheses these series enact.

It is these different visual palettes that ultimately underscore differences in context, especially the extensive use of animation in the latter series. History in Sagan's *Cosmos* is constituted by costume drama, in the second, by animated depiction, from Giordano Bruno to Clair Patterson and the astronomer and astrophysicist Cecilia Payne-Gaposchkin. Early criticisms from the perspective of both the history of science and the history of religion, though less dismissive of the overall project of the new *Cosmos*, indicted the new series on the level of its historical portrayals. While holding out hope for the series and impressed by its effects, historians of science Robert Goulding of Notre Dame and Michael Crowe, emeritus of the same, critiqued an animated segment on Bruno in the first episode of the new *Cosmos*. They found historically inaccurate the representation of Bruno as a casualty of a conflict between dogmatic religious authority and scientific imagination.[83] Wherever the matrix of history of science, religion, and culture was represented as at play in the 1980 production and also in the 2014 production, analogous criticisms emerge: of inaccuracy and insensitivity to the sense of things of the past.

But even more than wonder-inspiring or tantalizing exhibition, the logic of the animation and the popular cultural associations with it tend to signal good-humored entertainment or youth-oriented pedagogy, even with the cultivation of adult audiences for animated productions. In turn, the strategies for promoting historical interest and for inviting intergenerational appreciation in the latter series turn less on quasi-realistic exhibit than did its more immediate predecessors, relying more on the friendliness of approachable cartoons.[84] In such a medium, the subject of history, even when depicting the burning of Bruno or the gender biases experienced by Tharp, isn't foreboding. The animation tends to leaven and lighten the "great person" historiography that focused on crusading or inspirational individuals, presenting them somewhat less as otherworldly geniuses or presenting true science less as the preserve of a distant elect. Such a science is accessible and adventurous, and though at times in conflict with political or commercial interests, it does not produce such essential conflict as to challenge advertisers who endorse the series vision or accept the financial and societal advantages of endorsing it, regardless of resistance to its narrative.[85]

The modes of computer-generated exhibition (cosmic space, evolutionary time) and cartoon-animated historicity (human time), however, can produce more immediately evident tensions, tensions in the definitions of science posited by the newer series, in its historiography, and in its politics. Initiating cosmic exploration, in the first episode Tyson speaks to the camera and testifies explicitly to a traditional view of the scientific method: "Test ideas by experiment and observation. Build on those ideas that pass the test. Reject the ones that fail. Follow the evidence wherever it leads and question everything." He adds, as commanding invitation: "Accept these terms, and the cosmos is yours. Now come with me." Bracketing the question of the validity of historical representation as raised by Goulding, Crow, and others, Bruno is not depicted as a scientist in the animated history the series projects—not by the lights of this definition of the scientific method launching the odyssey. This defiant Bruno is imaginative and soars in film-animated reveries beyond the world, but speculatively, without experiment or observation. Whether the Bruno of *Cosmos* is a kind of protoscientist or a scientific/natural philosophical fellow-traveler (as a freethinker) is unclear. But to include him in the lists of scientific thinkers—in the intergenerational "cooperative enterprise" embracing Sagan and Tyson as depicted at the end of the same episode—renders problematic the characterization of the progression of scientific thought that the series emphatically posits.

The shift in modes of viewership from the time of the first to second *Cosmos* underscores related elements sustained in tension, in what is no straightforward argument/advocacy for science. The old series produced simultaneous viewership over broadcast television, a collective conversation preparing viewers for that evening's showing, digesting it together during and after screening. The more flexible viewing possibilities in the present tend toward less ritualized, still more individualistic viewings of a less personal odyssey. The heated debates over universalism and multiculturalism, arguments over the validity of the species voice that those such as Sagan or Jacob Bronowski before him adopted, have cooled. But their lessons, already felt in Sagan's attempts to move away from a "provincial" view, as a matter not only of cosmic perspective, but of world culture (if perhaps to some minds reinforcing the very Western-centered universalism such critiques targeted), have been further internalized. The new series attempts to insist more on the scientific contributions of others than those included in disputational category of "dead white men," the ghosts haunting those intracultural and cross-cultural debates. This insistence has less the character of special pleading, less reliance on prior mythic truths in order to find continuities between Western science and a Near or Far Eastern mythos. This is in turn related to the more contemporary context seeing greater polarization between politicized putatively proscientific and proreligious camps. And so Bruno, if not the figure of the scientist as a matter of his method, becomes the figure of the scientific martyr as a matter of his opposition to authoritarian religion. The political/cultural atmosphere that the more recent series addresses takes this opposition as largely given, playing a role in resistance not only to historical truth, but to the recognition of the contemporary dangers of pollution and global warming.

Perhaps the most significant differences of reception already to be gleaned and speaking to the wider modes of reception of the contemporary myth are those related to ritualized viewing and documentary consciousness together. The "scientific epic" or "cosmic evolution" that the first series projected was less widely known in 1980. The work and collective reception of that series played a substantial role in familiarizing a television and reading public with its universal historical account. As we have seen, the continuities that the series shared with Sagan's earlier authorial and instructional efforts and in dialogue with works such as those of Wilson and Chaisson helped cultivate a sense of pedagogical mission—of a specific sort of documentary sensibility—establishing the symposia and other pedagogical experiments that grew out of the series, deepening its scientific outreach and cultural representation. The outreach of the later series amplified its account in a number of venues, popular and pedagogical. If a similar cultural resonance (and dissonance) is at play for the newer series, it thus is unlikely to be a matter of a collective disclosure of a new "epic myth" already disclosed—unless the series should succeed in finding/establishing new publics still unaware of the enveloping narrative of material and organic evolution. It might more easily play a structuring role in emergent forms of scientific exhibition, in the persuasive construction of fabulaic worlds, inviting new viewers to new vantage points overlooking a multiplicity of scientific worlds.

Regardless of the comparisons and contrasts with sacred histories, *Evolution* and *Origins* and the new *Cosmos* taught narrative lessons. Indeed, in such opposition, every argument sees truth as inherent in at least one narrative, whether sacred or scientific—or, for some, both, as in the case of the spiritually inflected treatment of the evolutionary epic offered by the biologist Ursula Goodenough.[86] What qualifies as well-informed opinion in and through these scientifically minded documentaries is the structure and sensibility of a universal history or final story: that projected narrative the scientists are depicted as working toward composing, and the story in process, what they undertake in order to finalize it and how far short they fall of completion by the measure of that projection. In none of these productions is there a great density of events, not judged against the timeline being established. There is little sense of the reader or viewer being lost as a matter of the intricacy of the central plot—though every moment, every event, can be opened to greater exploration, probed further, and represented through the cluster of connecting research stories emphasizing one part of the timeline above another. If time is judged by the number of events or, similarly, by the ease of temporal contraction, then comparatively little time has elapsed in the universe. If it instead is judged by a timeline that itself nests ever new potential stories and levels of detail, then how old the universe is remains to be decided and may forever be an open question.

In contrast to earlier generations, the more contemporary universality of these later series sees the limits of its embrace, even more clearly perhaps than when scientific universalism was understood as in opposition to multiculturalism. Over the past decades, the pragmatic concession to the reach of the scientific epic, a qualification of the universality of its audience, is in the acknowledgment of the evident and active rejection by some religious-minded audiences. Its story, as opposed to other epic stories at play in the period, establishes humanity as the hero and constructs that hero romantically, as being able to precipitate global change in the society and in the nature in which it is embedded and, in strong relation to this, as being able to tell its own sprawling, clustering tale. In so doing, the documentaries collapse the world onto a screen—and now, onto varied screens—and teach their audience how to do so—how to collapse geological ages into hours or into what can be seen on hillsides. They frame world life by species-wide human life, leaving open the question of whether this epic tale is tragic or redemptive.

As in previous generations, pride of place is given to the scientists in these works, the difficulty of their task emphasized repeatedly. Like Russian dolls, the universal is contracted into the world, the world is contracted into humanity, and humanity is contracted into the figure of the scientist. The scientists, as opposed to the filmmakers, are more prominently represented as the authors of the final story, the writers of a contemporary universal myth. But there is now greater room for generalists, for multiple authors, for many people to be speaking a story, and in so doing, endowing that story still more with the status of myth.

The construction of that myth requires the various feats and gifts prominently shown in the documentaries—travels across the world, adventures, and the kind of luck that suggests fate. There is an affinity between this story written for everyone and the camera's global appetite, skirting the world in order to recount the tale. Oriented by reference to a fraught cultural territory in which the later documentaries were filmed—so different from the nuclear theorizing of Gamow, the apparent cultural foment alienating Eiseley or Bronowski from the young, or the nuclear activism of Sagan—the documentaries argued that to have an educated understanding of the history of the world is to know it as history, to have a sense of what it means to complete it and who on earth is doing the work of completion.

Whatever the shifts in representational strategy and structure, in the different weights given to the historical and the fabulaic, where scales do not so much organize a synthesis as operate as sites of new worlds and where foundational structures are largely effaced, these documentaries have functioned as variations on contemporary cosmology and a record of active scientific work, as documentaries after all. Articulating a universal history, a final story, the contours of which extended beyond active research, they have established a framework that announces a wider meaning in many of the individual investigations they depicted, however the reception of that meaning by such investigators might have been or had been received.

Coda

Epic Humanisms

In 1920, the philosopher and literary critic Georg Lukács wrote of the novel that "it is the epic of a world that has been abandoned by God."[1] The word "epic" for him did not simply or uniquely relate either to a literary work or text that somehow could be understood independently from the historical world in which it emerged. Instead, he understood the epic as above all the expression of a meaningful totality of existence. To Lukács, the ancient world was closed and homogeneous, a world where everything had its proper place and no question was left unanswered. This was the world of the Homeric epic. The modern world, by contrast, the world abandoned by God, is fragmented and diverse, inharmonious and subject to dilemmas of meaning. In that sense, the novel is structured as the marker of something that could no longer be, the fallen epic. The novel is produced in the attempt to discover a totality, a whole, at a time when the existence of that whole can no longer be experienced. Art, which had suffused an integrated world, became only "one sphere among many."[2]

Decades thereafter, fragmentation and multiplicity were rendered the conditions of a fraught modernity and simultaneously the welcome opponents of false totalities, of quaint, if offensive universalisms. Lukács's idea of such a totality resonated with what E. O. Wilson saw at work in the evolutionary epic, the "everything" that this scientific epic was meant to capture. Wilson's words in this regard bear repeating: "When the scientists project physical processes backward to that moment," the moment of creation, "with the aid of mathematical models they are talking about everything— literally everything."[3] It is an everything "beyond the imaginings of earlier generations." Nevertheless, to Wilson's (and Lukács's) mind, in comparison with those earlier generations, that everything is less immediately meaningful, more problematic.

Wilson elaborated on the possibility of capturing everything as part of his explanation of the need to "cultivate more intensely" the relationship between science and the humanities. He approvingly quoted and endorsed J. B. S. Haldane's call to a new, scientifically informed literature and mythology.[4] The modern scientific epic, with its potential power to express everything, to assume the subject position of Job, but now able to answer God, was in this light both science and literary form. As a myth that was seen as able to enact its own mythic claims, if completed and accepted, it would be both a representation of the world and the essence of it. Reflecting back on Lukács, the idea of such a modern epic evokes the meaning-making and organizational powers implicit in narrative itself, powers that in the scientific epic of the modern era would be the product of at least partly self-conscious acts of collective authorship.

A meaningful totality was at stake not only in the genre of the final story, however, but also in a different sense in the final theory. In the context of explaining the ways in which science does not uphold the belief in an interested God, Steven Weinberg remarked: "If there were anything we could discover in nature that *would* give us some special insight into the handiwork of God, it would have to be the final laws of nature. Knowing these laws, we would have in our possession the book of rules that governs stars and stones and everything else."[5] Here, even more than in the context of discussions of the epic, there is a knitting together of representation, the universe described, and being, the essential principles constituting it. These laws allow us to know, in Stephen Hawking's phrase (also cited by Weinberg), "the mind of God," itself a trope. The confidence in the possibility of a final theory functions as a compelling dogmatic unity, which in turn can partly underwrite the finality of the scientific epic, insofar as the story relies on what it constructs as ahistorical scientific principles in accordance with putative final laws. As a synthetic genre, however, the final theory recommends a narrative and disciplinary organization in stark contrast to any final story composed primarily as a historical genre of synthesis.

Even if it were to capture fundamental things, there would potentially be much that would elude this final theory, according to Weinberg, above all the historical itself. There is "an important difference between biology and the physical sciences: the element of history."[6] Whereas physics concerns itself with universal laws, there are other sciences that need to concern themselves with historical accident: "Biology is not unique in involving this element of history. The same is true of many other sciences, such as geology and astronomy."[7] This was not a minor point for Weinberg: "The intrusion of historical accidents sets permanent limits on what we can ever hope to explain. Any explanation of the present forms of life on earth must take into account the extinction of the dinosaurs sixty-five million years ago, which is currently explained by the impact of a comet, but no one will ever be able to explain why a comet happened to hit the earth at just that time. The most extreme hope for science is that we will be able to trace the explanations of all natural phenomena to final laws *and* historical accidents."[8] However, Weinberg was not by any means pessimistic, expecting for decades that science would uncover the origins and the origins of the origins.[9] The qualification reveals as much about his *scientific* interest in historical accident as it does about the limits of any final theory. History itself is not essential. The realization of John Herschel's dictum that "science is the knowledge of many, orderly and methodically digested and arranged, so as to become attainable by one"[10] would not be temporal.

Still, Weinberg's stance often sounds more subdued and cautious than others with totalizing scientific ambition. The final story promoted by Wilson, Sagan, NASA, WGBH, and NOVA, and the many celebratory accounts that emerged along with them, generally embraced the historical accidents and the potentially provisional laws that existed at various levels of complexity and historical moments. They did not tarry as often with Weinberg's careful qualifications, though they would not dispute them. In seeing the scientific epic as projecting the immutable laws of physics and nature onto the historical emergence of different kinds and shapes of matter and life, the origin story, the new version of Genesis—all these synonyms for one another—needed less qualification. It attempted to capture the essential *and* the historical. And in so doing, in giving a naturalized account of the world dealing in timeless and temporal truths, this contemporary origin story was both immanent (as authored by people in the world) and transcendent (as capturing something apart from humanity). In a different spirit, this point nevertheless shares something with Lukács's characterization of the ancient epic: "the concept of totality for the epic is not a transcendental one, as it is in [classical] drama; it is not born out of the form itself, but is empirical and metaphysical, combining transcendence and immanence inseparably within itself."[11]

But to at least some of the advocates of cosmic evolution, such an immanence could not and should not assume the burden of meaningfulness and spiritual resonance reflected in the term "myth." Eric Chaisson, although opting for the language of epic, did not generally celebrate the epic as new mythology, the resonance of which remained ambivalent, despite the embrace by other scientific universal historians. Those authors of the scientific epic who celebrated it as myth or as doing the work of myth struggled to define a place for humanity in the story they were telling, from Weinberg's stoic assessments to Sagan's or Wilson's rhapsodic passages, differences not of kind, but of attitude toward the fact of that authorship. But all of these authors oscillated between doubts and hopes. For Sagan, the mood of wondrous hope was perhaps in the ascendency, even with the affirmation of science ("we have walked far"). The same seems true for much of the NASA literature, in which the embodied exploration humbles the explorers. With Weinberg and Wilson, what prevailed instead was the Promethean, the figure on which Wilson concluded. Weinberg cited Einstein on the "arrogance" and "Promethean element of the scientific experience." Here the wonder was more over the ability of humanity to grasp a universe sublime in its scales than over the sublimity of the universe itself. And should those final laws be discovered, it was a universe that would be "less full of wonder and mystery." The humanism at stake was one that might celebrate humanity's power to grasp everything or that might, through that grasping, render humanity and the universe together meaningless. In either case, the assumption was that humanity is central to this verdict. Science as the one enterprise that held open the possibility of making the world meaningful, after itself having undercut older modes of meaning making, was to these authors not merely one sphere among many.

At the same time, this status and mission were and are contrary to the more critical humanities, for which the figure of the human was more insistently subject and object and which have been differently exercised by the conundrums this subject-object relation has presented to theories of knowledge. The contemporary universal history was also contrary to that social theory and historiography that itself sat uncomfortably in the traditional humanities, resisting such origins accounts, rejecting straightforward endorsements of "Enlightenment" values, stances developed through Continental philosophy, postcolonial critiques, and other theoretical perspectives denying the figure of the human as more than a sociohistorical construction serving certain ends, an object of the special sciences that the sciences themselves have created. Moreover, the syntheses for which the sciences had apparently opted appealed to genres that in the contemporary moment critical theorists, from Lukács on, had found problematic—whether because this appeal smacks of wishful thinking, denied the historical conditions from which the accounts emerged and from which they worked to stand apart, or because of the more concrete political histories to which such universalizing efforts have been apparently tied. From these perspectives, new myth as category and instantiation was wrong-headed and wrong-footed.

There nevertheless have been attempts to find some theoretical accommodation on the basis of convictions in a common cause. So whereas Dipesh Chakrabarty defends "the hermeneutics of suspicion" and the critique of the category species (as "an effect of power") in postcolonial scholarship, and estimates that such hermeneutics is "an effective critical tool in dealing with national and global formations of domination," he nevertheless concludes that it is inadequate to address the Anthropocene present.[12] In this regard, "The task of placing, historically, the crisis of climate change thus requires us to bring together intellectual formations that are somewhat in tension with each other: the planetary and the global; deep and recorded histories; species thinking and critiques of capital."[13] Unwilling to surrender the "obvious value in our postcolonial suspicion of the universal," and in direct response to Wilson, Chakrabarty expands on the phenomenological nature of the problem, undercutting the broad "we" that David Hollinger and universalists looking instead to the problematic past of diversity seek to defend: "Who is the we? We humans never experience ourselves as a species. We can only intellectually comprehend or infer the existence of the human species but never experience it as such." Nonetheless, in his call for "a global approach to politics without the myth of a global identity," he finds it necessary to grapple with "an emergent, new universal history of humans that flashes up in the moment of the danger that is climate change." The universality of the species concept evoked by this historical illumination eludes any experientially grounded understanding while nevertheless intimating a condition common to illimitable and uncapturable individual experiences: "First, inchoate figures of us all and other imaginings of humanity invariably haunt our sense of the current crisis. . . . Second, the wall between human and natural history has been breached. We may not experience ourselves as a geological agent, but we appear to have become one at the level of the species."[14]

Embedded in this accommodation is a vision, built for Chakrabarty from the history of science he finds in those such as Paolo Rossi, that natural history and human history had long ago parted ways and, perhaps now, in order to address the dominance of humanity, its power to destroy itself and its world, must come together. This is a shared belief among a collection of scientists such as Goodenough and Wilson, but includes those historians establishing the correlate subfields or studies of "Big," "Deep," and "Evolutionary" histories.[15] As opposed to Chakrabarty, these more recent universal historians,[16] such as Daniel Lord Smail and David Christian, largely see their work as a recuperation of a secularizing and universalizing Enlightenment project. Cognizant of research within Western academia attempting to overcome a Eurocentric focus, and sympathetic with the scope and anti-provincializing spirit of world history movements, they invoke the category of the universal in giving voice to unrecognized historical actors, human and otherwise. Their histories address materials covering hundreds of thousands to several billion years, resisting the apparent divide separating the academic discipline of history with its focus on written records from the unscripted eras preceding it, those periods that are conventionally the province of the natural sciences.

But, the divergence of scientific and humanistic periods and themes, itself not strict—as figures from John Herschel, to Spencer to T. H. Huxley to Shapley to Sagan all attest—depended on narrative and synthetic devices repeatedly crisscrossing and reshaping natural and human histories: from the shared dependence on periodization, through to the repeated interrogation by humanist and scientific disciplines as to the nature of myth itself. The disciplinary narrative of a parting of ways is itself a disciplinary-specific view, one that entails ignoring those historical modalities that continually ran together human and natural history, personhood and species development.

When resistance on political-religious grounds became pointed over the past decades, questions were raised as to what is and isn't science, what is and isn't history, overshadowing the forms of these questions as posed by critical scholarship in the humanities. In this same period, modern intelligent design was set against evolution, creation science against the Big Bang. These disputes were often put forward as conflicting stories with conflicting forms of reasoning to support them and conflicting conceptions of value. "Story" at times had a dismissive tone in the context of this debate in a way not dissimilar to Eiseley's own negative usage of "myth," creationists and other opponents of the scientific epic dismissing it as so much "evolutionary storytelling." As far as they were concerned, the totality that the epic represented and enacted did not include them. Nevertheless, the idea of story was also embraced in the period, as scientific practitioners and popularizers explicitly asserted that they had written a new Genesis or new myth on very different and more valid grounds than the original. And in a period of increasing self-consciousness over the question of the impact of humanity itself, the entanglement of authorial and heroic narrative/mythic codes dramatizes the promotion of the human in the determination of the objective truth of things.

Since the aim to compose that new myth and to contend with negatively universalizing crises was intimately linked to the final form that a given synthesis was meant to establish, the logic or poetics of that form determined what finality would mean. From this perspective, the aesthetics and epistemological efforts were obverse sides of one another, constituting another synonym of this history and synthesis. In the context of the less culturally resonant final theory, the aesthetics were those of mathematical rigidity and inevitability, the pristine form of necessity derived from first principles and fundamental laws. Its completion entailed clarifying those laws and principles governing all the domains of being, from which any laws more specific to a given domain could be derived or at least to which they could be traced back or reduced. By contrast, the genres of synthesis most useful in articulating that scientific epic as a totalizing form—the scalar, the fabulaic, and particularly the historical as its ruling figure—appealed to a still more aggregative quality of myth and history, exploiting their openness to more substantive addition.

But the character of this synthesizing labor departed somewhat from the aggregative quality of older myths or universal histories. Those accounts were generally understood as given, as complete, and yet being open to addition, while newer accounts were understood as admitting of expansion on the details of one or another part of that myth. Yet in exploring the dilemmas present in establishing these modern myths, a conflict arose between the authorial and heroic codes—another language for the varying, related dilemmas of the Huxleys or Wilson, the disorientations of Weinberg, the anxieties of Sagan, the fears of NASA. The scientific authors who self-consciously crafted this new myth were aware of their own hands in the craftsmanship, aware of how far from complete by their own lights that work was and perhaps always would be. But because that universal history they crafted was also taken to decide humanity's role, it kept in suspense the questions of meaning and purpose. It could not, like older forms of myths, provide the absolute end, at least not yet and perhaps never, as Wilson believed and Sagan at times suspected, or treat its account as complete, even while tasked with similar purpose.

This point presses most on a distinction between epic and myth, at times used synonymously in the scientific work, at times distinguished. Myth is almost always understood for these authors as a broad, socially endorsed, world-organizing story. Myth speaks to questions of truth and being. The epic, on the other hand, was often understood as a particular kind of quest story, a particular narrative form.

In conveying the work of science as the quest for truth and frequently linking it to a political universalism, the work of the epic was seen to uncover and ground the basic principles on which a particular conception of nationhood might be founded and, toward the close of the twentieth century, particularly the American nation. NASA's lobbying of Congress took the form of defining the purpose of the nation through the scientific epic, a purpose that looked toward a future goal, but that nevertheless was understood as embedded in an older notion of the American people leading the world in its noble-minded quests.[17] Eiseley had decades earlier indicted NASA and the US commitment to it for just such a sense of resonant national purpose, but one that, in borrowing the words of Campbell, poured out another self-destructive blazing vision. At the same time, an absolute "epic distance" was still asserted, representing the proper sphere of the scientific authors as less in the embroiled political and social present and instead at a point in universal history in which consciousness, the category of "living beings," was finally able to relate its own story.

These questions of aesthetics and epistemology, in their links to the different synthetic programs, as with the historical and foundational in the context of congressional deliberations, bore directly on variant social and political battles and were suffused by them. How does science present and justify itself to a broader public? The epic suggested that it does so by answering fundamental perennial questions. But by the end of the twentieth century, the answers were also promising technological cures for both technological and ancient ills: new exotic wonders, new cures for old and new diseases, possible solutions for technologically enabled traumas, forecasts of the future, themselves discerning a choice between final condemnation or ongoing redemption.

In the later twentieth century, the epic was aligned with well-articulated but different political visions. With Sagan, the universal tale was allied with a political universalism that recalled Merton's arguments tying the universalism of the scientific ethos with a universalizing political spirit. In the hands of Wilson, the epic was instead a counter to other political histories, particularly to what he saw as the Communist myth or the multiculturalist fallacy. These oppositions held more firmly than the religious opposition, Wilson seemed to suggest, because (for himself as much as for Sagan) the scientific epic consummated the sacred stories. In the funding battles that followed the post–Cold War period, Weinberg's grimmer version of the tale promoted what he saw as the role of physics above and beyond other sciences, a physics ultimately allied to failed battles to save the Superconducting Supercollider. In one sense, this was a more narrow political battle, but in another, it was an argument over intellectual pursuit as such, over the reach of the imperative to knowledge, and over what were by those lights the most important scientific enterprises. Foundationalist views were contrasted with visions of scientific enterprise structured by a historicizing account of complexification drawing on scalar, foundational, and fabulaic logics. For NASA, meanwhile, the more Sagan-esque account was used to argue for funding in congressional hearings as well, but in that account, the vision of science was as a widely shared effort, without discrimination toward disciplines.[18] The study of origins here was a house with many rooms, where each science found its place, a vision finding its academic homes partly through the work of Chaisson, the outreach of public television, as well as through the efforts of universal/big historians to readdress the relationship between the academic discipline of history and this epic natural history. The first or last three minutes of the universe were less the emphasis for the contributors to such a synthesis. Instead, the story mattered for all the times and events in between.

The epic across its variants set the terms for the political and ethical in-terventions that science could make, and in so doing, opened an evocative space for deliberating over dilemmas in and of science. In speaking to scien-tific and nonscientific audiences alike, the story helped to define what it could mean for scientists to function as ethical and political actors *as* scientists. The stakes of the story, which repeatedly framed the future in terms of two stark possibilities, one catastrophic and the other redemptive, was the language in which scientists lobbied for funds, conceived of their broad mission, and oriented their students and each other. The story in its variations helped to provide terms for defining who scientists are in the world and how effective they might be in it. Apart from representing the basic features of the world and deciding its central events, the story also suggested who or what is good and bad (scientific or obscurantist, beneficent or dangerous), who or what are the protagonists, what is the essential organization and orientation of knowl-edge, and what is the contemporary state of things (in peril or not, and if in peril, what could be done about it?). It allowed the production of a narrative model of the world testing potential resolutions to scientifically inflected problems—of human meaning, of social value, of active politics—even while exposing the depths of those problems and the limits of the resolutions that science might provide.

The dilemmas of science framed in terms of the scientific epic have thus opened onto the question of whether humanity, having determined its origins and fundamental nature, should work to transform itself. The implicit ethic often endorses the imperative to knowledge, even while explicitly recognizing the possibility that the quest it champions might bring self-destruction, rather than new cures and wonders. The ethical position of the scientist has often been constructed, in effect, as one of a strange humanism defining what it means to be human through the destination of this scientific quest, even while raising the unsavory specter that being human means quite possibly to be engaged in an extended historical process of self-destruction. This, too, draws a contrast between epic and myth, resolving the epic quest of humanity in what resonates with a mythic apocalypse. But that strange quest has held open the possibility of a homecoming, even if that home could only be the exploration itself.

Acknowledgments

This book is indebted to discussions, workshops, reading groups, teaching, talks, and less formal conversations at several institutions and programs over as many years, including Harvard's History of Science Department and STS Circle of the Program on Science, Technology and Society, the Max Planck Institute for the History of Science, New York University Tandon School of Engineering, Lyman Briggs College at Michigan State University, Arizona State University's Center for the Study of Religion and Conflict, the Chemical Heritage Foundation, the Consortium for History of Science, Technology and Medicine, and the Department of Rhetoric at the University of California, Berkeley, and neighboring institutions. First among the many to whom I owe thanks are Peter Galison and Katharine Park. Katy and Peter have given abiding support and guidance, with intellectual and deeper generosity. The expertise, responses, and thought of Janet Browne, Peter Dear, Michael Gordin, Myles Jackson, Sheila Jasanoff, Rebecca Lemov, Erika Milam, Robb Moss, Afsaneh Najmabadi, Matthew Stanley, and Judith Surkis directly advanced the arguments of this book or contributed to its underpinnings. Several among them also helped make it possible to continue writing. Over that time, conversations with nearby colleagues and friends became infrequent but valuable exchanges of ideas, a discussion including Alex Csiszar, Megan Formato, Amber Jamilla Musser, Christopher Phillips, Chitra Ramalingam, Hanna Rose Shell, Hallam Stevens, David Unger, and Alex Wellerstein. In this list too are Carin Berkowitz, Daniela Helbig, Ben Hurlbut, and Ksenia Tatarchenko to whom I owe particular thanks, whether for extended specific manuscript readings and enthusiastic interchange, for rousing collaborative work that was also at times straightforwardly instructional, or for incisive comments, for consonance and productive dissonance of thought. The book furthermore owes an extensive and direct debt to Deborah Coen and Steven Dick, for their critical and consequential commentary on the manuscript, and the wealth of scholarship they brought to bear on it. There are in addition friends who supported this writing more distantly but significantly, including Ian Ayres, Ellen Bales, Katarina Dudas, Maya Jasanoff, Dániel Margócsy, Elizabeth Murphy, William Rankin, Marco Roth, Colin Rowat, Masha Salazkina, Elly Truitt, Matthew Underwood, Adelheid Voskuhl, Elizabeth Yale, and, particularly, Philippa Townsend; many early conversations with Philippa helped inform and inflect the questions addressed here. The hospitality and inspiration offered by a much missed periodic gathering with Michael Holquist and Ilya Kliger set horizons for this work, conveying a sense of the value and immediacy in what is beyond it. Ilya has been an inseparable part of the process of thinking, learning and writing; acknowledging him must entertain the risk of keeping a custom better honored in not being observed.

Further debts are owed to those who were welcoming and generous with their time, insights, and memories, particularly Julie Benyo, Claude Canizares, Eric Chaisson, Ann Druyan, Owen Gingerich, Thomas Levenson and Sherry Walton. Ellen Guarente and Jude Lajoie in turn gave material support with oral history interviews and their transcription. Archivists and staff at the American Institute of Physics, the Chemical Heritage Foundation, Harvard University, the Library of Congress, University of Pennsylvania, and the University of Louvain all helped with orientation with regard to relevant materials. Also the institutional support and dialogue of the CHF Matters and Materiality working group, and the specific ongoing exchange with Jody Roberts and Nick Shapiro, helped to cultivate a sense of other contexts for this work. The manuscript gained from the insights of Bud Bynack, from his thoroughgoing and structuring demand that the argument be unconcealed. It benefited further from Mary Corrado's thorough review. Karen Darling at the University of Chicago Press showed an early and enduring interest in this project, her support for which made its completion possible.

My remaining thanks are to those whose presence and consequence was manifold and ongoing: Ana Vollmar, who has engaged with this work in too many ways to trace, and family here, and gone, and still present.

Notes

INTRODUCTION

1. I will make use of several of these terms, partly as a matter of conceptual emphasis, partly as a matter of the preference of specific historical actors and commentators.

2. For extended discussions see, for example, Katy Börner, Chaomei Chen, and Kevin Boyack, "Visualizing Knowledge Domains," *Annual Review of Information Science & Technology* 37 (2003); Chaomei Chen, *Mapping Scientific Frontiers: The Quest for Knowledge Visualization* (London: Springer-Verlag, 2003/2013) and Sigmar-Olaf Tergan and Tanja Keller, eds., *Knowledge and Information Visualization: Searching for Synergies* (Berlin: Springer Verlag, 2005). For some of the historical roots of such projects, see Alex Csiszar, "Broken Pieces of Fact: The Scientific Periodical and the Politics of Search in Nineteenth-Century France and Britain" (PhD diss., Harvard University, 2010).

3. Such correlations and their disruption recall Hayden White's treatment of historiographic conceptions in the nineteenth century, where he examines the elective affinities between modes of argument, of historical emplotment and of ideological implication. See Hayden White, *Metahistory: The Historical Imagination in Nineteenth-Century Europe* (Baltimore: Johns Hopkins University Press, 1975). Following out the affinities between my own inquiry and White's, beyond the ways in which I am already indebted to his approach, would involve examining whether the correlations I describe here can be mapped to his framework. This is particularly suggestive in relation to affinities between historical and foundational syntheses on the one hand, and modes of emplotment and argument on the other. But though it might be worth drawing out the comparisons and contrasts elsewhere, in the context of my analysis here, the apparent similarities might ultimately be too misleading, and my own examination of scientific historiographic projects sits less well with the contrasts White draws between historical and scientific work.

4. These ratios moreover enable individual researchers to place themselves within the topography of a larger scheme, to be located within a representation that they could survey individually, even if not all aspects of it were readily visible. This is true even in imagining a multiplicity of worlds, for the very specific images it conjures have established models for those worlds as yet unknown, placing a given scientist in one world more than all others. Figuratively put, the terrain determines whether one is in the world of physics, of biology, of chemistry, and so on, disclosing its stable entities and characteristics. The implicit emphasis on individual understanding is yoked to the ability of self-emplacement, of representing the whole and one's place in it. Moreover, the emphasis on surveyability shows how much disciplinary self-understandings and goals can depend on the whole in which the research is imagined to participate, changing the nature of the discipline/part that whole was posited in order to embrace.

5. Literary theorist Mikhail Bakhtin discusses a more thoroughgoing notion of "epic distance." In relation to the scientific epics discussed here, this directs attention to the less quotidian registers the authors of this more modern epic at times assume, placing themselves at a pivotal and therefore fascinating point in the timeline of the universal history they were composing—elevating them to meaning bestowing, Olympian heights—even as it gave them the very pragmatic coordinates for their scientific research. For the discussion of epic distance, see M. M. Bakhtin, *The Dialogic Imagination: Four Essays by M. M. Bakhtin*, trans. Caryl Emerson and Michael Holquist (Austin: University of Texas Press, 1981), 17.

6. Theodor W. Adorno and Max Horkheimer, *Dialectic of Enlightenment: Philosophical Fragments*, ed. Gunzelin Schmid Noerr, trans. Edmund Jephcott (Stanford, CA: Stanford University Press, 2002), 9. The nature of "Enlightenment" is not straightforward in Adorno and Horkheimer; the partly Hegelian movement they detail itself demonstrates the difficulty in associating the term with a specific period or, alternatively, of simply diagnosing their use of it as transhistorical.

7. Outside the scope of my treatment are such earlier synthetic visions as Bacon's own "Division of all Human Learning" or Cartesian or Leibnizian "universal science." Also, another figure who, under an alternative organization of the first part of the book, would be included is C. S. Peirce.

8. "Myth," "epic," and related terms such as "romance" appear, for example, repeatedly in generalist or popular scientific work, adorn the epigraphs and allusions of more technical articles, and establish the terms of comparison for scientific syntheses in relation to other universal historical accounts. In scientific works, these terms can signal the endorsement of an account as of among the highest truths. This epic register alerts us to the moments when the idea of history is contrasted with the idea of myth, as truth is contrasted to lie, and when, as is generally the case in more recent universal historical syntheses, myth is regarded as the ultimate history, as a great truth that makes sense of all other truths.

9. In the attempt to show how these accounts themselves operate as and expose political, social, and mythic dilemmas of their day, I borrow intuitions from several theorists of the relations between society and narrative. The forms of romance, myth, and the epic have been examined by seminal critical theorists, such as Adorno and Horkheimer, Mikhail Bakhtin, and Georg Lukács, all in different ways, all moving toward the view that such stories have a political/national/populist dimension resonating with its moment, exposing dilemmas, but also establishing what functions as a nation's, as a people's history. Among these writers, the distinctions between generic forms are drawn in different ways, as are the characteristics of the forms themselves. Though these generic distinctions between forms and theorists of forms are relevant here, I emphasize instead the effort to track the scientific usage and scientific register of these words, which do not always honor any one set of generic characterizations. Generally, I appeal to these theorists in order to gain insights into the logic and functions of the scientific stories and myths, not in order to endorse their views of literature or mythology. In this regard, I also loosely engage with Fredric Jameson, who in the period in which the scientific epic begins to produce consensus, in turn rereads several earlier thinkers on myth—in particular Claude Levi-Strauss and Northrop Frye. Borrowing from Levi-Strauss, Jameson examines how different stories capture conflicts at work in the periods in which they are written and enact resolutions to those conflicts, an analysis that in turn defends aspects of Frye's genre analysis. These stories operate in immediate political terms, as with the narrow funding battles for different scientific programs; they operate in social terms, in the values and views science is taken to promote or undercut; and they operate as terms in a progressive narrative of history: the scientific epic is itself represented as the culmination of science, the discovery of truth enabling finer discovery still.

10. Confronting the structure of the story demonstrates how the story can itself function as an argument. The scientific epic constitutes one answer to the questions of whether and how it is possible to tell a history of billions of years that the scientific account demands. Through it, its authors

emphatically yet often implicitly argue for a set of core events, working to keep those events stable, as well as for what linked and bound the events together, how the sequence of events would be shaped into a beginning, middle, and end. (This latter borrows from the much discussed and problematized distinction of *fabula* and *sjuzhet*, a distinction scholars of narrative have made between the sequence of events at work in story, and the characteristic ways in which those sequences develop and produce meaning.) The effort to produce a scientific epic can therefore be understood as an extended effort to identify and sustain the pivotal events; the structural motifs give a sense of direction to the story; the ratios of how much time is described to how long it takes to relate the tale indicates the scope and the relative detail, and so on. The evolutionary formations of the solar system, of the earth, and of human beings mattered, as did their promised entropic or explosive dissolutions. In another form of the story, none of these need to matter; scientific synthetic thinkers of the twentieth century repeatedly quipped that nothing interesting happened after an early moment in the universe.

11. Steven Weinberg, *The First Three Minutes: A Modern View of the Origin of the Universe* (New York: Basic, 1977), 101.

12. Ibid.

13. Ibid., 112.

14. Ibid., 8.

15. Ibid., 112.

16. Ibid., 4.

17. Ibid., viii.

18. Edward O. Wilson, *On Human Nature* (Cambridge, MA: Harvard University Press, 1978), 202.

19. Ibid., 202. The invocation of Job in relation to technoscientific advance had precedents more than a century old. See Iwan Rhys Morus, "'The nervous system of Britain': Space, Time and the Electric Telegraph in the Victorian Age," *British Journal for the History of Science* 33, no. 4 (2000): 456.

20. Note that other translations of the Bible give "jackals" in place of "dragons."

21. Carl Sagan, *Broca's Brain: Reflections on the Romance of Science* (New York: Random House, 1979), 205. Adapted from the April 1977 banquet address at the annual meeting of the National Space Club.

22. Ibid., 206.

23. Ibid., 313.

24. The scientists' appeal to the category of "story" in turn repeatedly announced the ambitions and limits of science as such, by the lights of the authors of any particular account.

25. Such narratives inhabit different media and different genres. Seymour Chatman suggests ways of examining the many dimensions of the scientific narrative, from papers, to research work, to documentaries. Chatman distinguishes between particular kinds of events, those that have a pivotal role in the account being told, and those that do not. He also distinguishes between the time it takes for the story to be told (how long it takes to read it, watch it, or broadcast it) and how much time is being related in the world of the story itself. These distinctions clarify how the scientific stories are structured through central events, regardless of the medium, and how these help to determine what it means to relate a story that covers so much time. See Seymour Chatman, *Story and Discourse: Narrative Structure in Fiction and Film* (Ithaca, NY: Cornell University Press, 1978).

26. Pierre-Simon de Laplace, *Exposition du système du monde*, vol. 2 (Paris: De L'Imprimerie du Cercle Social, 1796), 303. "Quoiqu'il en soit de cette origine du système planétaire, que je présente avec la défiance que doit inspirer tout ce qui n'est point un résultat de l'observation ou du calcul; il est certain que ses élémens sont ordonnés de manière qu'il doit jouir de la plus grande stabilité, si des causes étrangères ne viennent point la troubler."

27. As Charles Gillispie has observed about Laplace's account of the birth of the solar system, "If we were to find a phrase that would characterize what Laplace had in mind about that event, it

would not be 'nebular hypothesis.' It would be 'atmospheric hypothesis.'" Charles Coulston Gillispie, *Pierre-Simon Laplace 1749–1827: A Life in Exact Science*, with the collaboration of Robert Fox and Ivor Grattan-Guinnes (Princeton: Princeton University Press, 1997), 173. The idea of such a nebular hypothesis is diffuse in its own right; John H. Brooke has offered an extended classification of different varieties of such hypotheses. John Hedley Brooke, "Natural Theology and the Plurality of Worlds," *Annals of Science* 34 (1977): 268–73. Paradigmatic forms address the varieties of nebulous matter in the skies and hypothesize that the solar system formed through gravitational attraction of more diffuse, nebulous matter or that even larger systems, such as galaxies, are undergoing a larger and endless process of such condensation, perhaps into stars. There are many sources that describe this history. Mary Bailey Ogilvie offers a concise formulation in relation to the later work of Robert Chambers. See Marilyn Bailey Ogilvie, "Robert Chambers and the Nebular Hypothesis," *British Journal for the History of Science* 8, no. 3 (November 1975): esp. 216.

28. For a more extended and intricate argument concerning their relationship, see Paolo Rossi, *The Dark Abyss of Time: The History of the Earth and the History of Nations from Hooke to Vico*, trans. Lydia G. Cochrane (Chicago: University of Chicago Press, 1984).

29. For the importance of both the political nature of universal history and its relation to "baroque orientalism," see Nicholas Dew, *Orientalism in Louis XIV's France* (Oxford: Oxford University Press, 2009) and see below. For a discussion of such sacred histories and other natural historical genres, see for example Stephen Jay Gould, *Time's Arrow, Time's Cycle: Myth and Metaphor in the Discovery of Geological Time* (Cambridge, MA: Harvard University Press, 1987), Martin J. Rudwick, *The Meaning of Fossils: Episodes in the History of Paleontology*, 2nd ed. (Chicago: University of Chicago Press, 1985), and Rudwick, "The Shape and Meaning of Earth History," in D. C. Lindberg and R. L. Numbers, eds., *God and Nature: Historical Essays on the Encounter between Christianity and Science* (Berkeley: University of California Press, 1986).

30. Such an imagined alliance predates by centuries the central history considered here: from, for example, medieval conceptualizations of human difference relying on naturalistic, but plastic/variant scriptural interpretation of the story of Noah's three sons, to voyages of exploration that brought European scholars in contact with recorded histories of other civilizations. See, for example, Hannah Franziska Augstein, "From the Land of the Bible to the Caucasus and Beyond: The Shifting Ideas of the Geographical Origin of Humankind," in *Race, Science and Medicine, 1700–1960*, ed. Waltraud Ernst and Bernard Harris (London: Routledge, 1999), and Benjamin Braude, "The Sons of Noah and the Construction of Ethnic and Geographical Identities in the Medieval and Early Modern Periods," *William and Mary Quarterly*, 3rd series, 54, no. 1 (January 1997).

31. For example David A. Hollinger, "The Enlightenment and the Genealogy of Cultural Conflict in the United States," in *What's Left of Enlightenment? A Postmodern Question*, ed. Keith Michael Baker and Peter Hanns Reill (Stanford: Stanford University Press, 2001), and David Sorkin, *The Religious Enlightenment: Protestants, Jews, and Catholics from London to Vienna* (Princeton: Princeton University Press, 2008).

32. See, for example, Ronald Calinger, "Kant and Newtonian Science: The Pre-Critical Period," *Isis* 70, no. 3 (September 1979). G. J. Whitrow notes that in Kant's 1763 *Der einzig mögliche Beweisgrund zu einer Demonstration des Daseins Gottes*, he summarized the earlier cosmogony and that in 1791, "an excerpt extending to nearly the end of the fifth chapter" was published together with "a translation into German of three memoirs by William Herschel on 'The Construction of the Heavens,'" suggesting how soon Kant and Herschel could be read alongside each other. Gerald J. Whitrow, introduction to *Kant's Cosmogony: As in His Essay on the Retardation of the Rotation of the Earth and his Natural History and Theory of the Heavens*, trans. W. Hastie, D.D. (New York: Johnson Reprint Corporation, 1970), xl.

33. For an analysis of Newton's cosmogonic views, see David Kubrin, "Newton and the Cyclical Cosmos: Providence and the Mechanical Philosophy," *Journal of the History of Ideas* 28, no. 3 (July-September 1967).

34. In the present, how are such Kantian state-of-nature, developmental arguments cued by cosmological evolutionary narratives and vice versa? Even given the conspicuous distances between Kant's early and later work, his interests in them are suggestive of at least the possibility of the continuing mutual implication of the universal historical and the natural historical. Kant clarified themes and structures that have in fact remain entangled, both in the structure of those natural historical genres and by the framing given to likeminded critiques of each.

35. For Kant, this position is explicit. A reading of his first critique is an investigation into the workings and possibility of what he terms synthetic a priori knowledge. This is knowledge that is valid "universally" (valid for whatever "potential experience") and that is possible through the connections of different concepts, which add something to a concept, rather than explicating it analytically. Mathematics supplies the chief example for him of synthetic a priori knowledge, as in simple arithmetic statements such as $5 + 7 = 12$.

36. John Swinton et al., *Universal History from the earliest Time to the Present, compiled from Original Authors, and illustrated with Maps, Cuts, Notes, Chronological and other Tables* (London: Printed for T. Osborne . . . , 1747–1768).

37. Kant, *Cosmogony*, 27. Whitrow notes that the italics follow Kant's usage of them. As he explains: "The Introduction quoted from deals with the Cosmogony or Creation of the World, and it gives an account of all the principal theories ancient and modern, as then understood . . . 'That philosopher' is Descartes; the 'others' are Burnet and Whiston.'" Ibid., 27.

38. Jacques-Bénigne Bossuet, *Discours sur l'histoire universelle à Monseigneur Le Dauphin: Pour expliquer la suite de la Religion et les changemens des empires*, vol. 1, *Depuis le commencement du monde jusqu'à l'empire de Charlemagne* (Paris: Sebastien Mabre-Cramoisy, 1681).

39. See Voltaire, *An Essay on Universal History, the Manners and the Spirit of Nations: from the Reign of Charlemaign to the Age of Lewis XIV*, trans. Mr. Nugent (London: J. Nourse, 1759), esp. 2–3 for Bossuet. See also Tamara Griggs, "Universal History from Counterreformation to Enlightenment," *Modern Intellectual History* 4, no. 2 (August 2007).

40. Stanley Jaki notes that Kant here was using the translation of Baumgarten. Immanuel Kant, *Universal Natural History and the Theory of the Heavens*, trans. Stanley L. Jaki (Edinburgh: Scottish Academic Press, 1981), 249. Baumgarten is one of the central figures discussed by David Sorkin in his work on the religious Enlightenment. See chapter 3, "Halle: Siegmund Jacob Baumgarten's 'Vital Knowledge,'" in *The Religious Enlightenment*, esp. 149–50 for his role in translating and commenting on the *Universal History*.

41. For example, at the start the eighth section of part II, "General Proof of the correctness of a mechanical doctrine of the arrangement of the world-edifice in general, and of the certainty of this present doctrine in particular," Kant claims: "One cannot view the world-edifice without recognizing the most excellent disposition in its arrangement and the certain signature of God's hand in the perfection of its correlation." The extent to which such claims are defensive themselves is impossible to determine and of no great significance here, though Sorkin's work cautions against being dismissive of them. See his related discussions in *The Religious Enlightenment*, "The Public Sphere," 16–18, and "The Enlightenment Spectrum," 19–21.

42. See, for example, Ronald Calinger, "Kant and Newtonian Science."

43. On Kant's choice of style in the first critique, see Ernst Cassirer, *Kant's Life and Thought*, trans. James Haden (New Haven: Yale University Press, 1981), esp. 139–44. Also, as Whitrow makes clear, there are at least aspects of his earlier cosmology that Kant was willing later to affirm and publish. See Whitrow, introduction to *Cosmogony*, xiii.

44. Immanuel Kant, *Critique of Pure Reason*, trans. Norman Kemp Smith (London: Macmillan, 1988), 436.

45. Ibid., 437.

46. Ibid., 448.

47. "Thus the antinomy of pure reason in its cosmological ideas vanishes when it is shown that it is merely dialectical, and that it is a conflict due to an illusion which arises from our applying to

appearances that exist only in our representations, and therefore, so far as they form a series, not otherwise than in a successive regress, that idea of absolute totality which holds only as a condition of things in themselves." Ibid.

48. Pierre-Simon Laplace, *The System of the World*, trans J. Pond, vol. 2 (London: Blackfriars, 1809), 373-75. The original French—*"De cruelles experiences ont prouvé dans tous les tems [sic], que ces lois sacrées ne sont jamais impunément enfreintes"*—might now be rendered more clearly in the translation with slightly different syntax, if "with impunity" appeared after "never." Laplace, *Exposition du système du monde*, 312.

49. From Kepler's *De Stella nova in pede Serpentari* as translated by Alexandre Koyré, *From the Closed World to the Infinite Universe* (New York: Harper Torchbooks, 1958), 61.

50. Alexander Koyré, *From the Closed World to the Infinite Universe* (New York: Harper Torchbooks, 1958), esp. chapter 3, "The New Astronomy and the New Metaphysics: Johannes Kepler's Rejection of Infinity," 58-87. I am indebted to James Voelkel for a helpful exchange on Kepler in reference to his response to Bruno.

51. C. S. Lewis, *The Discarded Image: An Introduction to Medieval and Renaissance Literature* (Cambridge: Cambridge University Press, 1994), esp. 63.

52. Quoted in Roger Hahn, *Pierre Simon Laplace: A Determined Scientist* (Cambridge, MA: Harvard University Press, 2005), 104-5. Hahn situates Laplace's move into a more public arena in the "vacuum" produced by the death of Condorcet and the new political disfavor of those scientists who had a sustained a voice during the Terror. Ibid., 108. See also Gillispie, *Laplace*, esp. chapter 19.

53. For recent scholarship in this regard, see Pauline Kleingeld, *Kant and Cosmopolitanism: The Philosophical Ideal of World Citizenship* (Cambridge: Cambridge University Press, 2012). Kleingeld situates Kant's cosmopolitanism in relation to other varieties of German cosmopolitanism in his day. In providing this extended explication of such cosmopolitanism, she largely brackets further contextualization. Nevertheless, she emphasizes the critical state of the "heterogeneous amalgam" that was then the "Holy Roman Empire of the German Nation." Ibid., 9. She and others have also emphasized the role of cosmopolitanism in other countries at the time, particularly France. Ibid., 11. See also Pauline Kleingeld, "Six Varieties of Cosmopolitanism in Late Eighteenth-Century Germany," *Journal of the History of Ideas* 60, no. 3 (July 1999). For the more specific case of German reception to "la Déclaration des droits de l'homme," particularly given prior German interest in related notions, see Jacques Godechot, "L'Expansion de la Declaration de Droit de l'Homme de 1789 dans le Monde," *Annales historiques de la Révolution française*, 50ᵉ Année, no. 232 (Avril-Juin 1978).

54. Immanuel Kant, "Idea for a Universal History with a Cosmopolitan Aim," trans. Allen Wood, in *Kant's Idea for a Universal History with a Cosmopolitan Aim*, ed. Amélie Oksenberg Rorty and James Schmidt (Cambridge: Cambridge University Press, 2009), 10-11.

55. Ibid., 11.

56. Ibid.

57. Ibid., 13. For Montaigne, see Immanuel Kant, *Toward Perpetual Peace and Other Writings on Politics, Peace and History*, ed. Pauline Kleingeld (New Haven: Yale University Press, 2006), 251-52.

58. Kant, "Idea for a Universal History with a Cosmopolitan Aim," 22. In the course of the brief universal history, Kant makes a further nonchalant but telling footnote to his sixth proposition, *"The problem* [the attainment of a civil society, universally and justly administered] *is at the same time the most difficult and the latest to be solved by the human species."* Ibid., 15. In a footnote to this section, he appeals in passing and without any apparent fear of contradiction to the histories of life on other planets, comparing the impossibility of achieving human destiny in an individual life with the potential for it in life otherwise. Ibid., 16. The ambition to address history, politics, and society in the idiom of the physical sciences is repeatedly invoked by those historical actors who examine the scientific universal historical genre thereafter. Sometimes at the margins of the

universal historical genre formation and sometimes directly infusing it, the question of the world as constructed by other beings according to other possibilities connects to this universal historical ambition.

59. The nature of this secular grammar is perhaps part of what is at stake in Kant's falling out with Herder. For the controversy with Herder, see for example Manfred Kuehn, *Kant: A Biography* (Cambridge: Cambridge University Press, 2001), esp. 295-301, and Ernst Cassirer, *Kant's Life and Thought* (New Haven: Yale University Press, 1981), esp. 228-31.

60. His initial results on the conformity of Jupiter and Saturn to Newtonian gravitation were announced in the middle 1780s.

61. Kant explicitly disclaimed mathematical demonstration. See for example Kant, *Universal Natural History*, 92 and 113-14. Stanley Jaki and Stephen Brush appear largely dismissive of Kant's cosmogonical reflections, and Whitrow (as Jaki observes in Kant, *Universal Natural History*, 19) of his critical writings (Kant, *Cosmogony*, xl). See also Stephen Brush, *Nebulous Earth: The Origin of the Solar System and the Core of the Earth from Laplace to Jeffreys* (Cambridge: Cambridge University Press, 1996), esp. 7. Regardless of estimations of either set of writings, the point here is rather to consider how they are historically and conceptually bound together in terms of the readership Kant received thereafter in the context of cosmic evolutionary narratives and in terms of debates over the limits of natural scientific work and representation.

62. Hahn, *Laplace*, 77; Gillispie, *Laplace*, 110.

63. For other immediate possible predecessors see Hahn, *Laplace*, 28-29 and 114-17.

64. Hahn, *Laplace*, 104.

65. Simon Schaffer, "Herschel in Bedlam: Natural History and Stellar Astronomy," *British Journal for the History of Science* 13, no. 3 (November 1980), and Schaffer, "'The Great Laboratories of the Universe': William Herschel on Matter Theory and Planetary Life," *Journal for the History of Astronomy* 11 (1980).

66. See Hahn, *Laplace*, 118, and Schaffer, "Herschel in Bedlam," 231-32.

67. See for example M. le Marquis de Laplace, *The System of the World*, trans. Rev. Henry H. Harte (Dublin: Printed at the University Press for Longmans, Rees, Orme, Brown, and Green, 1830), 336-37.

68. Hahn, *Laplace*, 175, 184-88.

69. In lectures compiled only after his death, Hegel put forward this much-read characterization of universal history, a preface to a more extended world history—what Leo Rauch characterizes as cultural history. G. W. F. Hegel, *Introduction to the Philosophy of History: with Selections from The Philosophy of Right*, trans. Leo Rauch (Indianapolis: Hackett, 1988), x. Hegel classified history into three types, the immediate, the reflective, and the philosophical. The first, referring to the work of historians such as Herodotus and Thucydides, gave what Hegel saw as a representation of the spirit of their age, privileging prominent state and collective events. The third sort, the philosophical, took as given the preeminence of reason and the fact of history as built on a rational plan. For Hegel, this presumption did not itself preclude an inquiry into facts that demonstrate the workings of reason in the world, and he took this latter itself to be involved in an attempt to come to a knowledge of God, rather than a complacent belief in a hazy idea of the action of providence. This view had implications for his second kind of history, reflective history, where the historian's reflections, a subjective position, was explicitly at work in the history. Reflective history had its own subcategories, of which two were the pragmatic or moralizing history and the critical history, which took as its objects the mistakes and omissions of prior historical work. It is as a kind of reflective history that Hegel introduces universal history.

70. For example, "The History of the world is none other than the progress of the consciousness of Freedom; a progress whose development according to the necessity of its nature, it is our business to investigate." G. W. F. Hegel, *Introduction to the Philosophy of History*, trans. J. Sibree (New York: Willey, 1900), 19.

71. Hegel's extended reflections on universal history, as capturing the spirit of nations, and through that, the spirit of humanity itself in its unfolding, also posited that history as simultaneously and identically the triumph of reason, of freedom, as the articulation of a providential plan, as a theodicy. In this movement, history was divided into times and spaces, characterized by the degrees of development in all these, but principally articulated through freedom. Unlike for Kant, Hegel found the universality of freedom in his own time, in his own culture, so that in a sense Kant's "last problem" was in this rendering perhaps already resolved. History ended, but only to be restarted by others who attempted to provide such totalizing accounts.

72. See, for example, Gerd Buchdahl, "Review: Hegel's Philosophy of Nature," *British Journal for the Philosophy of Science* 23, no. 3 (August 1972): 257. For a more extended and thorough discussion of Hegel coordinated in particular in relation to German thought in his day, see Charles Taylor, *Hegel* (Cambridge: Cambridge University Press, 1975).

73. Since spirit itself emerges in historical periods, the sciences themselves also emerge partly historically, with the oldest, as Auguste Comte would soon posit, the most certain. Here, too, in an understanding of nature through a development of stages that resisted a privative teleology, Hegel emphasized the connections between the living and nonliving while insisting on the importance of the former: "Nature remains, despite all the contingency of its existence, obedient to eternal laws; but surely this is also true of the realm of self-consciousness, a fact which can already be seen in the belief that providence governs human affairs. Or are the determinations of this providence in the field of human affairs only contingent and irrational? But if the contingency of spirit, the free will, leads to evil, is this not still infinitely higher than the regular behaviour of the stars, or the innocence of the plants?" From G. W. F. Hegel, *Encyclopedia of the Philosophical Sciences in Outline and Critical Writings*, ed. Ernst Behler, trans. Steven A. Taubeneck (New York: Continuum, 1990), §193.

74. This consonance has been remarked upon in the secondary literature. See chapter 1, n. 91.

75. Helmholtz, for example, targeted and protested against the influence of Hegelian logic. See chapter 3.

76. Kant, "Idea for a Universal History with a Cosmopolitan Aim," 19.

77. Ibid.

78. Hegel, *Introduction to the Philosophy of History*, 16. This is the Sibree translation. For an alternative translation and the use of "world history" in the place of "universal history," see the Rauch edition, 19.

79. Hegel, *Introduction to the Philosophy of History*, trans. Sibree, 16.

80. Ibid., 63.

81. Ibid., 63–64.

82. Ibid., 64.

83. Ibid.

84. Ibid., 23.

85. Ibid., 20–21.

86. Ibid., 21.

87. Ibid. Note that Erasmus Darwin had used a similar figure, "one great Slaughter-house the warring world," in reference instead to living creatures preying upon each other in *The Temple of Nature, or, The Origin of Society: A Poem, with Philosophical Notes* (London: Printed for J. Johnson by T. Bensley, 1803), 134.

88. Anthropologist of religion Talal Asad has advanced the view that the realms of the secular and religious are themselves grammars, as he terms them, with the distinction between them constructed and deployed in particular cultural and political contexts, resisting any simple picture of a process of extended secularization across the West. As such, secularization as an extended process must itself be rendered through such grammars. See Talal Asad, *Genealogies of Religion: Discipline and Reasons of Power in Christianity and Islam* (Baltimore: Johns Hopkins University Press,

1993) and *Formations of the Secular: Christianity, Islam, Modernity* (Stanford: Stanford University Press, 2003). By contrast, Charles Taylor has advanced a typology of secularisms emphasizing the modern construction of the self and the (loss of) the fullness of experience in *A Secular Age* (Cambridge, MA: Belknap Press of Harvard University Press, 2007). Recent "Anthropocene" histories have echoed this sense of a loss of fullness. Ironically, perhaps, at its most apparently secular, in Asad's sense of the term, histories of the universe shared in the shape and thematics of Kantian universal histories, bearing a complicated relationship to conceptions of providence and fate, militating toward or against environmentalist or pluralist politics.

89. To claim, as Georges Lemaître does (see chapter 5), that the whole story may not have been included in the first quantum, both says something of how that story may have been produced and implicitly lays claim to the idea that there is such a thing as a complete story. This is a question that has to be kept alive here: How is a story finalized? What completes a story? What is a complete or *final story*? I use the somewhat uncomfortable term "final story" here to contrast it with the much used term "final theory" prevalent in physics theorizing throughout the last century. This is not to claim that stories are not themselves theories or that theories are not suffused with narrative, but to call attention to the different if linked modes of finalization the terms suggest, and to underscore the collective construction of a finalizing account conceived in narrative terms. In many ways, just as confidence in reaching a final theory was fading at the end of the twentieth century, a renewed confidence in the final story emerged. A narrative structure (and indeed final theories and other genres of synthesis) can provide a working ideal or target vision that might not be possible to actualize, but that nevertheless serves as a guide to what the ultimate object of science might be. To see that final story as in itself a research target is not to suggest that the historical practitioners thought that reaching the target was possible only in the distant future. At times, they considered the completion of the new myth to be at hand and the promises and dangers it presented as imminent—just as a final theory might be declared as tantalizingly close to the scientific grasp.

90. Connections between some varieties of universal history and histories of the world/universe have become more conspicuous of late, with a consensus over a universalizing history tied in turn to shared global problems. The consensus has suggested to some scientific and humanist historians that the relevance of history now lies in the immediacy of its global lessons for living. See coda.

91. For a discussion of such trends, with indications of and arguments for a shift to longer duration studies, see Jo Guldi and David Armitage, *The History Manifesto* (Cambridge: Cambridge University Press, 2014).

92. This idea of myth providing a way to locate oneself as a scientist is apparent in seeing the scientific narrative as one scientific synthesis among others. An example in the contemporary organic evolutionary context is apparent in the much-repeated refrain that it is only through evolutionary theory that the contemporary biological framework makes sense, revealing its unity and clarity. The efforts of later synthesizers extend this refrain to cosmic evolution and the connections between cosmic and human evolution. These together establish a disciplinary view that ultimately has worked to locate scientists as authors of different periods in the history of the world. In speaking and evoking the universal histories, scientists were fashioning a vision of unity historically constructed, positioning their own efforts in ways that would inform institutional organization.

93. The relationship between the technical and the more popular is not a straightforward matter of a linear path from expert to popular knowledge simply understood. It is not a matter of vulgarization alone, or even primarily, of the move from research publication to popular announcement. Looking at the story being produced and understanding it as a story that no one can fully evaluate expertly, but that nevertheless functions as a compelling argument, underlines the appeal these accounts had to scientists and laypersons alike as collective accounts with privileged authors. In this regard see Mario Biagioli and Peter Galison, *Scientific Authorship: Credit and Intellectual Property*

in Science (New York: Routledge, 2003). By these lights, it is not surprising that the ambition to compose a history of the world was tied to the ambition to speak both to a lay and to a professional audience at once. So George Gamow wrote his 1952 *Creation of the Universe* in the hopes of reaching an audience of varied scientific training. And while Steven Weinberg wrote *The First Three Minutes* with the intent of convincing an intelligent layperson, the book was repeatedly endorsed by a scientific audience that to a certain extent also accepted its vocabulary, engaging and disputing its claims. Ralph Alpher and Robert Hermann, Gamow's two most enduring collaborators on the articulation of the Big Bang, went so far as to credit Weinberg's book with having made a significant contribution toward convincing the *scientific* community that the Big Bang theory is true. See chapter 9. Along with and after Gamow, Sagan, and Weinberg, many cosmologists and evolutionists wrote popular or semipopular works that transcended their specific expertise and provoked both public and more technical discussion, as well as criticism. They served to produce a shared language and a shared store of imagery between experts and the public. But with the self-conscious broadening of the story, in part that very conceptual distinction between public and experts became less easy to appeal to in the context of works such as these, even as that distinction remained largely unproblematic as a matter of the ways of authorizing such works as stabs at valid scientific popularization.

94. A distinction that seems more pertinent—what is demanded by the fact of the fragmentation of the disciplines—is what might be called *relative expertise* and *relative laity*. On the level of scientific training and research, *no one* was in a position to speak expertly to the entire chronology being traced in the full stories that were being produced, though many did put the full stories forward, as least as a kind of lightly sketched chronology, or even as an allusion just to the fact of that chronology. This distinction between expertise and a nonspecialist understanding was not only a matter of large disciplinary headings such as physics versus biology. The more refined the chronology scientists attempt to write, the greater the disciplinary divisions called upon when writing it. Which branch of physics speaks to this or that moment, which branch of chemistry, geology, biology? Thus, in a 1973 lecture discussed in chapter 9, Weinberg functioned as a relative layperson when he reminded his audience that he was not an expert on the Big Bang—whether he was speaking modestly or otherwise, and regardless of whether he was well on his way to becoming a cosmology expert. And his lecture was itself perhaps part of an effort transforming the proper expertise on the very first few moments of science to the domain of high-energy particle physics. This was a move to transfer the interrogation of the earliest, speculative moments of the universe to the tools appropriate to his relative expertise. (In his vision, the most constitutive had also become the earliest, and the earliest, the most constitutive.)

CHAPTER ONE

1. John F. W. Herschel, *A Preliminary Discourse on the Study of Natural Philosophy* (London: Printed for Longman, Rees, Orme, Brown, and Green and John Taylor, 1831), 18.

2. Ibid., 221.

3. Ibid.

4. "As it offers to us, in a confused and interwoven mass, *the elements of all our knowledge*, our business is to disentangle, to arrange, and to present them in a separate and distinct state; and to this end we are called upon to resolve the important but complicated problem,—Given the effect, or assemblage of effects, to find the causes." Ibid., 221–22, emphasis mine. Relaxing the demand of a particularly mechanical causality, an assumption Herschel would come to question, will alert us to still other forms of synthesis. Note also that Herschel defines natural *philosophy*, indebted to "our immortal countryman Bacon," as "a series of inductive generalizations, commencing with the most circumstantially stated particulars, and carried up to universal laws, or axioms, which comprehend

in their statements every subordinate degree of generality, and of a corresponding series of inverted reasoning from generals to particulars, by which these axioms are traced back into their remotest consequences, and all particular propositions deduced from them." Ibid., 104. Here, the work of natural history in the context of the physical sciences is central to natural philosophy, but the latter has a still broader, more generalizing compass.

5. For an assessment in the secondary literature, see Walter F. Cannon, "John Herschel and the Idea of Science," *Journal of the History of Ideas* 22, no. 2 (April–June 1961), 215 and passim; Peter M. Millan "The Herschel Dynasty—Part II: John Herschel, *Journal of the Royal Astronomical Society of Canada* 74 (1980). For further contemporaneous assessment, see below.

6. Cannon gives extended attention to John Herschel's "apotheosis." See Cannon, "John Herschel and the Idea of Science," 217. See also James A. Secord, *Victorian Sensation: The Extraordinary Publication, Reception, and Secret Authorship of* Vestiges of the Natural History of Creation (Chicago: University of Chicago Press, 2000), 404.

7. Robert Chambers, *Explanations: A Sequel,* first edition (1845), in Vestiges of the Natural History of Creation *and Other Evolutionary Writings* (Chicago: University of Chicago Press 1994), 178.

8. Iwan Rhys Morus observes the similarities in the "synthetic framework for the sciences" laid out by Herschel, Somerville, and Whewell in the early 1830s. See Iwan Rhys Morus, *When Physics Became King* (Chicago: University of Chicago Press, 2005), 76–77.

9. Mary Somerville, *On the Connexion of the Physical Sciences* (London: John Murray, 1834), 413.

10. [William Whewell], "On the Connexion of the Physical Sciences," *Quarterly Review* 51, no. 101 (March 1834), 59.

11. Ibid., 55.

12. See below for a more detailed discussion.

13. To name these threatens to be an exercise in the absurd. But among them, in a diverse and multinational body over a still more extended period, we might number Herder, Hegel, Wordsworth, Nietzsche, and in a different vein, as we will see, Whewell, among many others. So in the 1870s, in his critical examination of history as such, Nietzsche makes mention of the "mighty historical movement which, as is well known, has been in evidence among the Germans for the past two generations." Friedrich Nietzsche, "Advantages and Disadvantages of History for Life," in *Untimely Meditations,* trans. R. J. Hollingdale (Cambridge: Cambridge University Press, 1997), 59. Ernst Cassirer comments on the conventionality of this "historical consciousness" in *An Essay on Man:* "What we call 'historical consciousness' is a very late product of human civilization. . . . The concept of history first reaches maturity in the work of Vico and Herder." Ernst Cassirer, *An Essay on Man: An Introduction to a Philosophy of Human Culture* (New Haven: Yale University Press, 1944), 172–73. (Cassirer goes on here to distinguish mythical from historical time, a distinction to which we will have occasion to return.) For a more recent commentator, see Reinhart Koselleck, *Futures Past: On the Semantics of Historical Time,* trans. Keith Tribe (New York: Columbia University Press, 2004).

14. Richard Yeo, *Defining Science: William Whewell, Natural Knowledge and Public Debate in Early Victorian Britain* (Cambridge: Cambridge University Press, 1993), 149.

15. Ibid., 149. In this regard, Yeo draws on the scholarship of Philippa Levine, in particular Levine, *The Amateur and the Professional: Antiquarians, Historians, and Archaeologists in Victorian England, 1838–1886* (Cambridge: Cambridge University Press, 1986). In the rise of a Victorian historical consciousness, she attests as well to new conceptions of time linked to innovations of science and technology, including the structuring of daily life by clockwork and public time and a preoccupation with the past, among other factors. Ibid., 5–6. Note by contrast Alice Jenkins's problematization of space. See below.

16. Among texts that, together with other factors, observe the ways in which encounters with other cultures placed pressures on European natural philosophical or historical thought and timelines, see Frank E. Manuel, *Isaac Newton, Historian* (Cambridge: Belknap Press of Harvard University

Press, 1963), Francis Haber, *The Age of the World: Moses to Darwin* (Westport, CT: Greenwood, 1978), Paul Hazard, *La crise de la conscience européenne* (1935; Paris: Le Livre de Poche, 1994), M. J. S. Rudwick, "The Shape and Meaning of Earth History," in *God and Nature: Historical Essays on the Encounter between Christianity and Science*, ed. D. C. Lindberg and R. L. Numbers (Berkeley: University of California Press, 1986), and John D. North, "Chronology and the Age of the World," in *Cosmology, History and Theology*, ed. Wolfgang Yourgrau and Allen D. Breck (New York: Plenum, 1977).

17. See for example Alan Rauch, *Useful Knowledge: The Victorians, Morality, and the March of the Intellect* (Durham: Duke University Press, 2001), Alice Jenkins, *Space and the "March of the Mind": Literature and the Physical Sciences in Britain, 1815-1850* (Oxford: Oxford University Press, 2007), William Christie, "Going Public: Print Lords Byron and Brougham," *Studies in Romanticism* 38, no. 3 (Fall 1999), and Steven Ruskin, *John Herschel's Cape Voyage: Private Science, Public Imagination, and the Ambitions of Empire* (Aldershot, UK: Ashgate, 2004). Further relevant citations appear below.

18. The descriptions characterizing such figures cannot be straightforwardly or flatly made. In light of the still-hardening institutional boundaries of the sciences, the question of what popularization itself meant and who was attempting it, it is difficult to employ terms that do not mislead in relevant ways. For more extended discussions of these issues, see Bernard Lightman, *Victorian Popularizers of Science: Designing Nature for New Audiences* (Chicago: University of Chicago Press, 2007) and Secord, *Victorian Sensation*. It might also be noted that Lightman puts forward four historiographic approaches in treating the authorship of texts for popular Victorian audiences: the first focuses on the scientific elite (to praise or merely to analyze), the second on marginalized groups, the third on the institution of Victorian publishing, and the fourth on the different sites in which science for a general audience was disseminated and constructed. See Lightman, *Victorian Popularizers*, 13-17. In tending to focus on the dominant, authorizing narratives being debated and produced as themselves cultural artifacts and orienting bodies of knowledge, my analysis does not tend to emphasize the differences that emerge by focusing on the different approaches Lightman suggests. Nevertheless, Lightman observes that a "book remains to be written on the involvement of the scientific elite in writing and lecturing for the public." Ibid., 14. This is certainly not that book, but among the options in Lightman's scheme, it would align most closely to it.

19. For an overview of the extended historiography on reformations of the Cambridge curriculum in the first third of the nineteenth century with particular emphasis on the work of Robert Woodhouse in the generation preceding the Analytical Society, see Christopher Phillips, "Robert Woodhouse and the Evolution of Cambridge Mathematics," *History of Science* 44 (2006).

20. It is a concern evident throughout her recollections. See Martha Charters Somerville, *Personal Recollections, from Early Life to Old Age, of Mary Somerville with Selections from Her Correspondence by Her Daughter, Martha Somerville* (London: John Murray, 1873).

21. So much is true of the regard in which he was held by both Somerville and Whewell. See for example Whewell's review of the *Preliminary Discourse* for a contemporaneous assessment: [William Whewell], "A Preliminary Discourse on the Study of Natural Philosophy," *Quarterly Review* 45, no. 90 (1831). On the particularly important role of John Herschel in Somerville's life, see Kathryn A. Neeley, *Mary Somerville: Science, Illumination, and the Female Mind* (Cambridge: Cambridge University Press, 2001), 70-73. For her own descriptions of encounters with Herschel and correspondence between them, see Somerville, *Personal Recollections*.

22. As Silvan S. Schweber shows, after some estrangement from his father, this work became a more self-conscious act of consummation. See S. S. Schweber, "John Herschel and Charles Darwin: A Study in Parallel Lives," *Journal of the History of Biology* 22, no. 1 (Spring 1989): 14.

23. Simon Schaffer has particularly emphasized this point. See Simon Schaffer, "The Nebular Hypothesis and the Science of Progress," *History, Humanity and Evolution: Essays for John C. Greene*, ed. James Moore (Cambridge: Cambridge University Press 1989), esp. 137-38. In his 1833 *Treatise on Astronomy*, Herschel gives very few words to his father's nebular hypothesis, leaving the

judgment of its validity to the accumulation of future evidence. John Herschel, *A Treatise on Astronomy* (London: Printed for Longman, Rees, Orme, Brown, Green, & Longman, and John Taylor, 1833), e.g., 402.

24. See Thomas Keightley, *Outlines of History: From the Earliest Period to the Present Time* (Philadelphia: Carey & Lea, 1831). Note that some later American editions seem to credit Lardner with the authorship of the volume. For Keightley's attribution, see, for example, Morse Peckham, "Dr. Lardner's 'Cabinet Cyclopaedia,'" *Bibliographical Society of America, Papers* 45 (1951): 55.

25. Peckham, "Dr. Lardner's 'Cabinet Cyclopaedia,'" 37–58. Peckham emphasizes the importance of the utilitarian agenda in the "revolution in literacy" that forms the context for Lardner's efforts, as well as that of the Society for the Diffusion of Useful Knowledge and the University of London. Ibid., esp. 38–39.

26. Jenkins observes that the phrase "march of the mind" is often misattributed to Jeremy Bentham, but is coined instead by Brougham. See Jenkins, *Space and the "March of the Mind,"* 10.

27. For these points, see J. N. Hays, "The Rise and Fall of Dionysius Lardner," *Annals of Science* 38 (1981): esp. 535 and 532.

28. In assessing Lardner's writing style, Hays emphasizes its similarities to the styles of Herschel and Somerville over and above the "baroque excess of Nichol or the romantic lyricism of Humphrey Davy." Hays, "Rise and Fall," 530.

29. Herschel, *A Preliminary Discourse*, 103. See also 131–32, section 126.

30. Ibid., 219. Such sentiments weren't chance ones for Herschel. For example, in reference to Humboldt's argument for the importance of all the intellectual pursuits in the advance and preservation of culture and civilization, Herschel's only complaint was that Humboldt had not gone far enough: "Life, thought, and moral and social relations, are all equally *natural*—equally elements of the great scheme of the Kosmos with matter and magnetism. The only imaginable reason why the sciences growing out of these ideas are not regarded and handled, or have not hitherto effectually been so, as branches of natural science and inductive inquiry, is the great difficulty of arriving at true statements of facts." [John Herschel], "Kosmos: Entwurf Einer Physischen Weltheschreibung [*sic*]/Cosmos. Sketch of a Physical Description of the Universe, "*Edinburgh Review* 87, no. 175 (1848): 182. See also below.

31. Herschel, *A Preliminary Discourse*, 281.

32. James Hutton, *Theory of the Earth or an Investigation of the Laws Observable in the Composition, Dissolution, and Restoration of Land upon the Globe* (Transactions of the Royal Society of Edinburgh, 1788), 304.

33. Herschel, *A Preliminary Discourse*, 281.

34. Ibid., 281–82.

35. Note that in Herschel's review of Humboldt's *Cosmos* some eighteen years later, he noted himself: "As the distinction drawn in the Aristotelian Philosophy between celestial and terrestrial motions operated for ages to cut off the possibility of arriving at any just views of the Planetary System, so it is perfectly conceivable that, by gratuitous assumptions of another kind, we may willfully sever ourselves from the possible attainment of knowledge of a far higher order." Herschel, "Kosmos," 178. On the basis of the analysis here, there is a certain irony in this point, given the distinctions drawn between astronomy and geology.

36. See for example discussion in Stephen Jay Gould, *Time's Arrow, Time's Cycle: Myth and Metaphor in the Discovery of Geological Time* (Cambridge, MA: Harvard University Press, 1987), e.g., 78 and 97 for Lyell's contemporaneous appreciation of such a point.

37. John Playfair, *The Works of John Playfair, Esq. . . . with a Memoir of the Author* (Edinburgh: A. Constable & Co.; [etc.],1822), 496–97.

38. In this one regard at least, John Herschel had something in common with a Huttonian such as Playfair, who also dismissed origins. Playfair saw Hutton and Laplace as joined in barring the question of ultimate origins from the physical sciences, one through an examination of the earth,

the other of the skies. But if John Herschel was right, by 1830, only astronomy was the obstacle to a thoroughgoing historical vision. For Playfair, the sciences of earth and heavens had been free of historical vestiges and future visions.

39. Herschel, *A Preliminary Discourse*, 93–94.

40. Christie, "Going Public," 465.

41. Somerville later described Young's *Lectures* as "a mine of riches to me" and took particular delight in Wollaston having personally disclosed to her the secret of solar spectral lines, on a Fraunhofer prism he presented to her as a gift. Somerville, *Recollections*, 132, 134. See also Thomas Young, *A Course of Lectures on Natural Philosophy and the Mechanical Arts* (London: Printed for J. Johnson, 1807). Faraday, who became another acquaintance and correspondent, held perhaps the highest place in her scientific estimation; she regarded him as a latter-day Newton.

42. For further details of her life, see Elizabeth C. Patterson, "Mary Somerville," *British Journal for the History of Science* 4, no. 4 (December 1969).

43. Brewster makes this reference in his review "On the Connexion of the Physical Sciences," *Edinburgh Review* 59, no. 119 (April 1834): 171. The research in question was presented to the Royal Society in 1826. See Mary Somerville, "On the Magnetizing Power of the More Refrangible Solar Rays, *Philosophical Transactions of the Royal Society of London* 116, no. 1/3 (1826).

44. Lord Brougham, letter to Mary Somerville, Somerville, March 27, 1827, reproduced in Somerville, *Personal Recollections*, 161–62.

45. Murray thus forms something of a historical contrast with Brougham: their journals were competitors, and John Murray was also close to Lord Byron. Between Byron and Brougham there was overt and potentially dangerous antipathy.

46. See correspondence in Somerville, *Recollections*, 170–72.

47. "House divided against itself" though it may have been, to use John Brooke's characterization of natural theology. See John Hedley Brooke, "Natural Theology and the Plurality of Worlds," *Annals of Science* 34 (1977): 222.

48. See Lightman, *Victorian Popularizers*, 22.

49. Mary Somerville, *A Preliminary Dissertation on the Mechanism of the Heavens* (Philadelphia: Carey & Lea, 1832).

50. Miss Edgeworth, letter to Mrs. Somerville, May 31, 1832, reproduced in Somerville, *Recollections*, 203–6, esp. 204.

51. Mary Somerville, *Mechanism of the Heavens* (London: John Murray, 1831), xiv.

52. Ibid.

53. Ibid.

54. Somerville, *Mechanism of the Heavens*, lxvi.

55. Ibid.

56. Ibid.

57. She differed here to a degree with both Hutton and the younger Herschel. Geology disclosed information concerning the ultimate state of the world that astronomy could not yield.

58. Somerville, *Mechanism of the Heavens*, lxix–lxx.

59. Ibid., lxx.

60. And those interested as well, as we will see, in the debate over the plurality of worlds, itself an argument partly over the plurality of disciplines.

61. By this time, Herschel had published his treatise on astronomy with Lardner. There he alluded to his father's nebular hypothesis, taking a position similar to Somerville's, that more evidence was necessary before any conclusions could be drawn. See Herschel, *Treatise on Astronomy*, chapter 12, "Of Sidereal Astronomy," esp. 402 and 406–7.

62. James A. Secord, introduction to *On the Connexion of the Physical Sciences Collected in Works of Mary Somerville*, vol. 4 (Bristol: Thoemmes Continuum, 2004), xii.

63. Mary Somerville, preface to *On the Connexion of the Physical Sciences* (London: John Murray, 1834).

64. Ibid., 413.

65. This amounted to a consilience *of sorts*, as Whewell was soon to coin the term. For the disciplines to cohere, they had to share a conception of the phenomena they analyzed—their different approaches to the world had to share enough postulated facts to evince in common the truth of those facts. But this was not in reference to a robust theory that existed beyond that sharing. It was not therefore a consilience proper (itself emphasizing more a shared end point and divergent beginning points), except in the vaguest sense of adumbrating that unifying theory, to be evinced in manifold apparently unrelated ways. See below.

66. And even as in later editions she attested potential compensatory phenomena to avoid decay. See for example Mary Somerville, *On the Connection of the Physical Sciences* (New York: Harper & Brothers, 1871), 21.

67. See Yeo, *Defining Science*, 16–19, and M. J. S. Hodge, "The History of Earth, Life and Man: Whewell and Paletiological Science," in *William Whewell, A Composite Portrait*, ed. Menachem Fisch and Simon Schaffer (New York: Oxford University Press, 1991).

68. For an extensive study of the formation of Whewell's tidology, along with its relationship to his history and philosophy of science, see Michael S. Reidy, *Tides of History: Ocean Science and Her Majesty's Navy* (Chicago: University of Chicago Press, 2008), esp. chapter 4, "Tidology," and chapter 5, "The Tide Crusade."

69. William Whewell, "A Preliminary Discourse on the Study of Natural Philosophy," 378.

70. Whewell seems to have introduced a new nuance in the term itself in the review: "The step of which we speak is the marking and fixing our generalizations by imposing exact *terms*—a process which may be included in the word *terminology*." Ibid., 390. In the third volume of his 1837 *History of the Inductive Sciences*, he uses this term more explicitly in reference to natural history: "It must be recollected that I designate as *Terminology*, the system of *terms* employed in the *description* of objects of natural history, while by *Nomenclature*, I mean the collection of *names* of *species*." William Whewell, *History of the Inductive Sciences, from the Earliest to the Present Times*, vol. 3 (London: J. W. Parker, 1837), 307. For further comparison and distinction of Whewell's conception of induction over and above Herschel's and others', see Menachem Fisch, "A Philosopher's Coming of Age: A Study of Erotetic Intellectual History," in *William Whewell: A Composite Portrait*.

71. Whewell, "A Preliminary Discourse on the Study of Natural Philosophy," 388–89.

72. Whewell's early biographer Isaac Todhunter noted that John Gibson Lockhart, the editor at the *Quarterly Review* and the series editor for the Murray Family Library, turned to Whewell for this review because Herschel was unavailable and Brewster was reviewing *Connexions* for the *Edinburgh Review*. Isaac Todhunter, *William Whewell, D. D., Master of Trinity College, Cambridge: An Account of his Writings with Selections from his Literary and Scientific Correspondence* (London: Macmillan and Co., 1876), 92.

73. Whewell, "On the Connexion of the Physical Sciences," 59.

74. Ibid.

75. Ibid.

76. Ibid. As Sydney Ross observes, it was not until six years later that Whewell recommended the term more emphatically and earnestly, in the *Philosophy of the Inductive Science*. See Ross, "Scientist: The Story of a Word," *Annals of Science: A Quarterly Review of the History of Science and Technology since the Renaissance* 18, no. 2 (June 1962): 72.

77. Whewell, "On the Connexion of the Physical Sciences," 54.

78. Ibid., 55.

79. Ibid., 55–56.

80. Recalling Sydney Smith's quip concerning Whewell that "Science is his forte; omniscience is his foible." See Ross, "Scientist," 80.

81. Whewell, "On the Connexion of the Physical Sciences," 55. Compare the language here to Herschel's 1848 Edinburgh review of Humboldt's *Cosmos*. For analyses of passages such as these in reference to the "cult of the 'picturesque'" and to aesthetics more broadly, see W. H. Brock,

"Humboldt and the British: A Note on the Character of British Science," *Annals of Science* 50, no. 4 (1993); Simon Schaffer, "On Astronomical Drawing," in *Picturing Science, Producing Art*, ed. Caroline A. Jones and Peter Galison, with Amy Slaton (New York: Routledge, 1998); and Jenkins, *Space and the "March of the Mind."*

82. Whewell, "On the Connexion of the Physical Sciences," 55.

83. It was a form of synthesis that shared a spirit with the use of surveying or landscape metaphors at points in his work, e.g.: "By considering what is the real import of our acquisitions, where they are certain and definite, we may learn something respecting the difference between true knowledge and its precarious or illusory semblances ... by tracing the boundary-line between our knowledge and our ignorance, we may ascertain in some measure the extent of the powers of man's understanding." William Whewell, *Philosophy of the Inductive Sciences Founded Upon Their History*, vol. 1 (London: John W. Parker, 1840), 5-6. See also ibid., 12-13. We will examine other, later examples of surveying metaphors.

84. "If we apprehend her [Somerville's] purpose rightly, this is to be done by showing how detached branches have, in the history of science, united by the discovery of general principles." Whewell, "On the Connexion of the Physical Sciences," 60.

85. Ibid.

86. As Yeo has observed, for Whewell, "the unity of science consisted in epistemological analogies, not substantive natural laws. There was a uniformity in the way knowledge was achieved, not a unity of the laws of nature. Although Whewell believed that the history of science displayed a 'consilience' of inductions towards simplicity and unity, he also stressed the particularity and integrity of different disciplines." Yeo, *Defining Science*, 243. In this, Whewell differed as a matter at least of emphasis from those who saw themselves engaged in unifying or synthetic programs—Comte, Spencer, and Mill, among them.

87. William Whewell, *Astronomy and General Physics: Considered with Reference to Natural Theology* (Philadelphia: Carey, Lea & Blanchard, 1833), 145.

88. Ibid., 156.

89. Ibid., 159.

90. Ibid., 161-63.

91. For Herschel's allusion, see Herschel, *Preliminary Discourse*, 219-20. Though it was not yet entirely evident here, the Cambridge colleagues Herschel and Whewell ultimately came to define somewhat opposing positions in relation both to the nature of science and in their situation with regard to one another. Herschel advocated a more empirical/inductive vision, where conceptualization was comparatively downplayed, and Whewell a more dialectical one, in dialogue in part with his reading of Kant. See below. See also Fisch, "A Philosopher's Coming of Age," 63-66. For further speculation as to whether Whewell relied on Hegel in stressing the importance of the historical in understanding this process of emergence, despite Whewell's disdain for Hegel's critiques of Newton, see Michael Ruse, "William Whewell: Omniscientist," in *William Whewell: A Composite Portrait*, 89n3.

92. William Whewell, *History of the Inductive Sciences, from the Earliest to the Present Times*, vol. 1 (London: J. W. Parker, 1837), xii.

93. Geoffrey Cantor, "Between Rationalism and Romanticism: Whewell's Historiography of the Inductive Sciences," in *William Whewell: A Composite Portrait*, 68.

94. Whewell, *Philosophy*, vol. 1, xxxvi.

95. "We have repeatedly found ourselves upon the borders of inquiries of a psychological, or moral, or theological nature." Whewell, *Philosophy*, vol. 2, 279. In 1845, Whewell published his two-volume *Elements of Morality, Including Polity*.

96. Ibid., 278-79. There is a Kantian resonance to these deliberations.

97. Whewell's original formulation of such consilience was as follows: "*The Consilience of Inductions* takes place when an Induction, obtained from one class of facts, coincides with an Induction,

obtained from another different class. This Consilience is a test of the truth of the Theory in which it occurs." Whewell, *Philosophy*, vol. 1, xxxix. The term was soon taken up in by John Stuart Mill, in 1843, though not to Whewell's satisfaction. See John Stuart Mill, *System of Logic, Ratiocinative and Inductive, Being a Connected View of the Principles of Evidence, and the Methods of Scientific Investigation*, vol. 2 (London: John W. Parker, 1843), 563, 599, and for Whewell's response, William Whewell, *Of Induction, with especial Reference to Mr. J. Stuart Mill's System of Logic* (London: J. W. Parker, 1849), 61.

98. This is not to suggest that they were all quoted in equal measure or all read together. Cannon and Stephen G. Brush find that Kant's cosmogony had little or no impact among astronomers. See Stephen Brush, *Nebulous Earth: The Origin of the Solar System and the Core of the Earth from Laplace to Jeffreys* (Cambridge: Cambridge University Press, 1996), 7, and Cannon, "John Herschel and the Idea of Science," 229n45. Nevertheless, Kant's cosmogonic theories were not unknown to other natural historians and philosophers. In his *Cosmos*, Alexander von Humboldt refers to Kant's "celebrated essay *On the Theory and Structure of the Heavens* . . ." Alexander von Humboldt, *Cosmos: A Sketch of a Physical Description of the Universe*, vol. 1, trans. E. C. Otté (London: H. G. Bohn 1848), 31. He connects Wright, Kant, and Lambert, viewing their "purely speculative conclusions" as "confirmed by William Herschel." Ibid., 71. In the coming chapters, we address repeated invocations of Kant in universal histories written later in the century.

CHAPTER TWO

1. Among these, Silvan S. Schweber also lists Nicolas Carnot and Charles Fourier. Silvan S. Schweber, "Auguste Comte and the Nebular Hypothesis," in *In the Presence of the Past: Essays in Honor of Frank Manuel*, ed. Richard T. Bienvenu and Mordechai Feingold (Norwell, MA: Kluwer, 1991), 133. For further background on Comte's lectures and a broader overview of the movement of his thought in the context of "Paris between the Napoleons," see John Tresch, *The Romantic Machine: Utopian Science and Technology after Napoleon* (University of Chicago Press, 2012), and particularly chapter 9, "Comte's Calendar: From Infinite Universe to Closed World."

2. The *mémoire* is reprinted as "Premier Mémoire sur la Cosmogonie Primitive contenant verification mathématique de l'hypothèse proposée par Herschel et Laplace pour expliquer la formation de notre système solaire" in Auguste Comte, *Écrits de jeunesse 1816–1828 suivis du Mémoire sur la Cosmogonie de Laplace*, ed. Paulo E. de Berrêdo Carneiro and Pierre Arnaud (Paris: Archives Positivistes, 1970). See as well the twenty-seventh lesson of volume 2 of the *Cours*, "Considérations générales sur l'astronomie sidérale, et sur la cosmogonie positive." Auguste Comte, *Cours de Philosophie Positive* (Paris: Bachelier, 1830–1842).

3. Quoted in Schweber, "Auguste Comte and the Nebular Hypothesis," 158.

4. For a recent discussion of Saint-Simon and Saint-Simonianism, see chapter 8, "Saint-Simonian Engines: Love and Conversions" in Tresch, *The Romantic Machine*.

5. His organization of the fundamental character of the sciences is predicated on a division between organic and inorganic, regardless, he noted, of the nature of that difference, whether it is illusory or not. His ideas contrast with Whewell's fundamental ideas in much the same way as he saw himself as contrasted with d'Alembert.

6. Auguste Comte, *The Positive Philosophy of Auguste Comte*, trans. and condensed by Harriet Martineau (London: J. Chapman, 1853), 22–23. Note that here I use Martineau's nineteenth-century translation because of its ultimate currency in the Anglophone response to Comte. Frederick Ferré's more recent translation prefers to translate the distinction as the "Historical" and the "Theoretical." See Auguste Comte, *Introduction to Positive Philosophy*, ed. and trans. Frederick Ferré (Indianapolis: Hackett, 1988). In the original French, Comte contrasts "la marche *historique*," and "la marche *dogmatique*."

7. An arguably likeminded contrast was at work in Whewell too, in an 1832 text of Mechanics, in which he criticized what we might anachronistically refer to as the "textbook" style of presentation—a presentation proceeding from "ultimate axioms and most general principles" understood as "self-evident independently of experiment; or at most, manifest by reference to a few obvious facts." William Whewell, *First Principles of Mechanics with Historical and Practical Illustrations* (Cambridge: Printed for J. & J. J. Deighton, 1832), iv. The best counter to that style, to Whewell's mind, was to "give a sketch of the history of each of the leading principles of the science." Ibid., v. For more analysis of this work and its role in Whewell's growing interest in history of science, see Geoffrey Cantor, "Between Rationalism and Romanticism: Whewell's Historiography of the Inductive Sciences," in *William Whewell: A Composite Portrait*, ed. Menachem Fisch and Simon Schaffer (New York: Oxford University Press, 1991).

8. Comte, *Positive Philosophy*, 24.

9. And the vocabulary of Kant. See the Introduction.

10. I employ "dogmas *of unity*" because "unity" is an explicit, emphatic term of analysis across historical sources. It is a term of praise or critique in concerns over connections between the sciences, intimately entangled with, but distinct from, "synthesis" or even "unification."

11. Comte, *Positive Philosophy*, 18, 22. See what follows for Thomson making a similar point. In a later period, this inability to construct a consensus over an idea of unity (particularly in method) was a site of inspiration and conflict for the sociology of science, for which discipline Comte might stand as a foundational figure.

12. This dogma of unity might therefore be termed more specifically the "complicative dogma": how complex, how structured a science is in its dependence on the sciences preceding it historically and ideally, locates it in a general system of the sciences. Herbert Spencer challenged this aspect of Comte's law of increasing complexity. In a book review of 1854 entitled "The Genesis of Science," Spencer hazarded his own definition of science as "*an extension of the perceptions by means of reasoning*," disputing hard and fast separations between science and other forms of knowledge, cueing an "embryology" of the sciences. Herbert Spencer, "The Genesis of Science," *British Quarterly Review* 20, no. 39 (July 1854): 114: "Is not science a growth? And must not science, too, have its embryology? And must not the neglect of its embryology lead to a misunderstanding of the principles of its evolution and of its existing organization?" Spencer's review stands as a construction/recognition of past attempts at conceiving the arrangement of the sciences in relation to the origins of its branches. He critiques other systems of knowledge for appearing to take this history *less* into account: "There must be serious defects, if not a general untruth, in a philosophy of the sciences considered in their interdependence and development, which neglects the inquiry how they came to be distinct sciences, and how they were severally evolved out of the chaos of primitive ideas." Ibid., 113.

13. For a brief summary of this reception prior to and through Martineau's translation, as well as a characterization of the differences between French and English positivisms, see Susan Hoecker-Drysdale, "Harriet Martineau and the Positivism of Auguste Comte," in *Harriet Martineau: Theoretical and Methodological Perspectives*, ed. Michael R. Hill and Susan Hoecker-Drysdale (New York: Routledge, 2001).

14. In the lecture, Huxley discusses organic unity and origins hinted at by the chemical preconditions of living matter. These amounted to the protoplasm and its own chemical constituents. Huxley advocates the view that the living is distinguished from nonliving only by a differing organization of shared elements. T. H. Huxley, *On the Physical Basis of Life* (New Haven: The College Courant, 1869), 13, 17. Despite the review's brevity on Comte, Darwin appeared to recall its treatment as decisive. See Janet Brown, *Charles Darwin. Volume 2: The Power of Place* (Princeton: Princeton University Press, 2002), 354.

15. Silvan S. Schweber, "The Origin of the 'Origin' Revisited," *Journal of the History of Biology* 10, no. 2 (Autumn 1977): 250.

16. See David Brewster, "Cours de Philosophie Positive," *Edinburgh Review* 67, no. 136 (July 1838) and Schweber, "The Origin of the 'Origin' Revisited."

17. For the variety of Huxley's efforts, see Bernard Lightman, *Victorian Popularizers of Science: Designing Nature for New Audiences* (Chicago: University of Chicago Press, 2007), 358 and Adrian Desmond, *Huxley: From Devil's Disciple to Evolution's High Priest* (New York: Basic, 1997), esp. "The Season of Despair." Later in life, Huxley also articulated views of the sequential nature of science that, as Lightman has noted, bore some resemblance to Comte. See Lightman, *Victorian Popularizers,* 386. Huxley had granted some limited use to Comtean ideas, but he perhaps would have attributed his position more to the work of those such as Playfair than to Comte. See T. H. Huxley, "Science," *Westminster Review* 61, no. 119 (January 1854), esp. 257, and Huxley, "The Scientific Aspects of Positivism," in *Lay Sermons, Addresses and Reviews* (New York: D. Appleton, 1871), esp. 147-48.

18. Marian Evans (George Eliot), as assistant editor for the *Review,* in turn dismissed Huxley's powers as a reviewer. See Paul White, *Thomas Huxley: Making the Man of Science* (Cambridge: Cambridge University Press, 2003), 73. For an analysis of Huxley's review of Lewes, his efforts at the *Westminster Review,* and his attitudes toward popularization and practice, see ibid., esp. 69-75, and Lightman, *Victorian Popularizers,* esp. 359-60.

19. Huxley, "Science," 255.

20. Ibid., 257.

21. Leonard Huxley, *Life and letters of Thomas Huxley, by His Son, Leonard Huxley,* vol. 1 (London: Macmillan, 1908), 434. Comte's law of three states claimed that society and individual science progress through three stages, from the theological to the metaphysical to the positive. In the final line of the letter, Huxley made the disturbing comparison: "Comte in his later days was an apostate from his own creed; his 'nouveau grand Être suprême,' being as big a fetish as ever nigger first made and then worshipped." Ibid., 435. For the public version of this sentiment, see T. H. Huxley, "The Scientific Aspects of Positivism," *Fortnightly Review* 5, no. 30 (1869): 654. For the relevant letters to Darwin, see ibid., 432-33. By 1870, having seen confederates such as Spencer and Mill repudiate Comte in part or in whole and having come under attack for his apparently ill-informed characterizations in the "Physical Basis of Life," Huxley only enlarged on his critique. See Huxley, "The Scientific Aspects of Positivism."

22. Frank Egerton, "Refutation and Conjecture: Darwin's Response to Sedgwick's Attack on Chambers," *Studies in History and Philosophy of Science* 1 (1970): 176.

23. Whewell's review of Somerville's *Connexions* suggests a problematization of expert synthesis even in relation to the physical sciences alone. And Herschel in his review of Humboldt emphasizes the singular difficulty of the challenge. See below.

24. These would not likely be the terms with which Chambers would have seen his effort, at least not by the time he was to make Herschel an icon of overly restrained scientific labor. Whatever the case, his historical claim to originality in the effort is in certain ways difficult to evaluate. The seventeenth century saw efforts such as those of Thomas Burnet to rewrite sacred histories according to what were regarded as more immanent laws. Newton engaged in chronological work relying on a number of disciplines, though not of course in the nineteenth-century sense of scientific fields, and to different ends. See Frank E. Manuel, *Isaac Newton, Historian* (Cambridge, MA: Belknap Press of Harvard University Press, 1963). For a focus on relevant European sources, see the Introduction and Pietro Corsi, *The Age of Lamarck: Evolutionary Theories in France 1790-1830* (Berkeley: University of California Press, 1988) and Corsi, "Before Darwin: Transformist Concepts in European Natural History," *Journal of the History of Biology* 38, no. 1 (Spring 2005). Corsi intimates the extent to which Chambers, among others, relied on such sources, ibid., 81. Note that Chambers explicitly invokes Lamarck (e.g., first ed., 230-31) and Buffon (281), among others. For reference to further examples see Nicholas Jardine and Emma Spary, "The Natures of Cultural History," in *Cultures of Natural History,* ed. Nicholas Jardine, J. A. Secord, and Emma C. Spary (Cambridge: Cambridge University Press, 1996), 4-5. These earlier sources are often treated more narrowly as

cases in the history of geology or biology and at times as intrinsically nonscientific works. But such evaluations are unnecessarily delimiting in two different senses: the history of the earth often invoked the history of universe, particularly, but not exclusively, when the chronologies of earth and heavens were taken to coincide more nearly. And the "nonscientific" histories of the earth were nevertheless predicated on the state of natural philosophy at the time they were written.

25. Somerville was more receptive to Chambers. Writing in 1845 to her son from her first marriage, Woronzow Grieg, Somerville suggested Charles Babbage as the author of *Vestiges*. See Martha Charters Somerville, *Personal Recollections, from Early Life to Old Age, of Mary Somerville with Selections from Her Correspondence by Her Daughter, Martha Somerville* (London: John Murray, 1873), 278. As she also observed, a son of Babbage had been staying with them.

26. Whewell, *History*, vol. 3, 618-19.

27. But Whewell's overall position is more subtle. Whewell could not treat the nebular hypothesis as more than a worthy conjecture. For astronomy to have offered geology more thoroughgoing palaetiological analogies, in the context of his argument, would demand that Whewell treat the hypothesis as a certainty. But even granting the truth of the gradual development and decay of the solar system, Whewell saw astronomical phenomena as in effect stable and ahistorical, as his *Bridgewater Treatise* made clear. To treat astronomy as historical, even when granting that it had a history in a formal sense, would be to humanize and potentially distort the meaning of astronomical scales.

28. M. J. S. Hodge notes that these passages in Whewell's *History* were "almost certainly known to Chambers." See Hodge, "The Universal Gestation of Nature: Chambers's 'Vestiges' and 'Explanations,'" *Journal of the History of Biology* 5, no. 1 (Spring 1972): esp. 139-40.

29. For the central place of the nebular hypothesis in Nichol's progressive views and initiatives, see Simon Schaffer, "The Nebular Hypothesis and the Science of Progress," *History, Humanity and Evolution: Essays for John C. Greene*, ed. James Moore (Cambridge: Cambridge University Press, 1989).

30. See for example Immanuel Kant, *Universal Natural History and the Theory of the Heavens*, trans. Stanley L. Jaki (Edinburgh: Scottish Academic Press, 1971), 154-55. For treatment of Kant's views in the secondary literature, see, e.g., Simon Schaffer, "The Phoenix of Nature: Fire and Evolutionary Cosmology in Wright and Kant," *Journal of the History of Astronomy* 9 (1978): 187. For the relationship of renewable energies to the distinction between Newtonian and Leibnizian frameworks, see Alexander Koyré, *From the Closed World to the Infinite Universe* (Baltimore: Johns Hopkins University Press, 1991), 272. Nichol himself pointed to a more theistic Newton. See J. P. Nichol, *Views of the Architecture of the Heavens* (Edinburgh, 1838), esp. 194. In this current of thought from Newton through Nichol, the fact that the world seemed sustainable for only so long was itself a marker pointing to the undiscovered sources of ever-new changes to come.

31. See James A. Secord, *Victorian Sensation: The Extraordinary Publication, Reception, and Secret Authorship of* Vestiges of the Natural History of Creation (Chicago: University of Chicago Press, 2000), 68 and passim for a more exhaustive treatment.

32. Sondra Miley Cooney, "Publishers for the People: W. & R. Chambers—The Early Years, 1832—1850" (PhD diss., Ohio State University, 1970), 1. Cooney sees the emergence of the journal not only in the context of Reform Bill politics, but as an outgrowth of a specifically Scottish historical democratic sensibility. Ibid., 5, and passim. Secord further attests to the journal as "the basis for the firm's preeminent reputation as 'publishers for the people.'" Secord, *Victorian Sensation*, 69.

33. A number of historians have commented on his membership in scientific societies, his connection to scientific thinkers, and his publication of scientific papers. See, e.g., Frank Egerton, "Refutation and Conjecture: Darwin's Response to Sedgwick's Attack on Chambers," *Studies in History and Philosophy of Science* 1 (1970): esp. 176-77, and Hodge, "Universal Gestation," esp. 132-33.

34. For Nichol's relationship to and assessment of *Vestiges*, see Secord, *Victorian Sensation*, esp. 378, 466. Chambers had drawn from Nichol and was partly inspired by him and his writings to author *Vestiges*. Nichol was ultimately one of the few to whom Chambers disclosed his secret, and Chambers asked Nichol for corrections to the astronomical parts. Years earlier, the astronomer

had discussed a plan for his own synthetic cosmological account with Chambers, an idea that Nichol felt *Vestiges* to have scooped. See Secord, *Victorian Sensation*, esp. 466–67. Judging from Nichol's *Views of the Architecture of the Heavens*, his own history would have been less immanent in tone than *Vestiges*.

35. Mary Bailey Ogilvie's variorum-like work demonstrates the extent to which *Vestiges* was a more protracted endeavor, across which run different patterns of revision. See Ogilvie, "Robert Chambers and the Successive Revisions of the *Vestiges of the Natural History of Creation*" (PhD diss., University of Oklahoma, 1973), and Appendix A, "Revisions to Vestiges and Explanations," in Robert Chambers, *Vestiges of the Natural History of Creation and Other Evolutionary Writings* (Chicago: University of Chicago Press, 1994). Secord has extensively examined the variations in reception from a variety of reading-based historiographic perspectives.

36. The first chapters were heavily weighted by the generational contributions of the Herschels and of Laplace, but the presentation also followed a number of secondary sources. See Ogilvie, "Robert Chambers and the Successive Revisions of *Vestiges*," esp. 50–55. Commentators from the time of its publication to the present day have underscored the centrality of the nebular hypothesis, seeing the essential argument as an ever-increasing refinement of the conjecture or as a central explanatory figure. See Hodge, "Universal Gestation," 142.

37. In the first edition, Chambers notes in regard to the conspicuousness of skin color that the "opposition of two of these in particular, white and black, is so striking, that of them, at least, it seems almost necessary to suppose separate origins" and treats "the Negro" as the one exception to the common origin of languages and culture in northern India. Given their "inveterate black colour," they may constitute an exception to his developmental hypothesis explained instead by "a deteriorated offshoot of the general stock." Robert Chambers, *Vestiges of the Natural History of Creation* (London: John Churchill, 1844), 278, 295–96. As Ogilvie notes, Chambers vacillated on the importance of color, and in later editions (8th through 10th) treated it as more superficial, also not listing the varieties of man in these editions. See Ogilvie, "Robert Chambers and the Successive Revisions of *Vestiges*," 440–41. Ogilvie notes as well contemporary moral critiques of Chambers on the grounds of advancing views regarded as tending to support slavery (ibid., 447) and Chambers's more decisive switch to a polygenetic position in 5th and 6th editions (ibid., 448).

38. See Schaffer, "The Nebular Hypothesis and the Science of Progress," and Schaffer, "On Astronomical Drawing," in *Picturing Science, Producing Art*, ed. Caroline A. Jones and Peter Galison, with Amy Slaton (New York: Routledge, 1998).

39. This is perhaps why *Vestiges* is so often included in histories of organic evolution.

40. Chambers, *Vestiges of the Natural History of Creation* [1st ed.], 32–35, 27, 41–42, 65.

41. Ibid., 165–66.

42. Ibid., 176, 163–64.

43. Ibid., 36.

44. The possibility of its successive improvement was appreciated by other Whewell correspondents, such as the natural philosopher and explorer J. D. Forbes.

45. Quoted in Janet M. Douglas, *The Life and Selections from the Correspondence of William Whewell D.D.* (London: Kegan Paul, 1882), 316–17.

46. By the second edition of *Indications*, Whewell added a preface directly targeting the author of *Vestiges*. Chambers had by this time responded to his initial critics with his *Explanations*, giving considerable attention to Whewell's criticisms. Part of their exchange was on the nature of the "palaetiological," an exchange that clearly outraged Whewell.

47. Ibid., 317–18. See also Richard Yeo, "Science and Intellectual Authority in Mid-Nineteenth-Century Britain: Robert Chambers and 'Vestiges of the Natural History of Creation,'" *Victorian Studies* 28, no. 1 (Autumn 1984): 12, for the case of Darwin's own appreciation of the synthetic successes of *Vestiges*, of "the multiplicity of parts he brings together, though I do [not] agree with his conclusions at all."

48. Note that John Brooke has argued that Whewell's early sympathy to some versions of the nebular hypotheses became tempered possibly by his resistance to *Vestiges*, and possibly provoking Whewell's 1853 anonymous treatise, *Of the Plurality of Other Worlds*. John Hedley Brooke, "Natural Theology and the Plurality of Worlds," *Annals of Science* 34 (1977): e.g., 268. Brooke also observes that the anonymity was more formal than substantive—it was clear that Whewell was the author. Ibid., 230. The treatise, denying what in retrospect *might* appear the more broadminded position of the possibility of life elsewhere, appeared to constitute a change of opinion from Whewell's early *Bridgewater Treatise*. It engaged him in debate with the more religiously literalist Brewster, who supported (and found in scripture support for) a non-evolutionary view that other worlds must be populated. For an extended treatment of Whewell's views on the plurality of worlds, see part II of Michael J. Crowe, *The Extraterrestrial Life Debate 1750-1900: The Idea of Plurality of Worlds from Kant to Lowell* (Cambridge: Cambridge University Press, 1986).

49. So, for example, in Chambers's much-critiqued treatment of the fossil record as found in the stratified rock, *Vestiges* represented the progress of the narrator and readers through the geological record as advancing "to a new chapter in this marvelous history—the era of the Old Red Sandstone." Ibid., 66. Similar appeals were made in the text itself, as for example "the leaves of the *Stone Book*." Ibid., 57.

50. So, for example, at the transition from the primarily geological domain to the domain of the sciences of life, Chambers notes: "Thus concludes the wondrous chapter of the earth's history which is told by geology." Ibid., 145.

51. Chambers, *Vestiges*, 1st ed., 359-60. In the third edition, the final line is altered to "Nor may even these be after all twain, but only branches of one still more comprehensive law, the expression of a unity, flowing immediately from the One who is First and Last." Chambers, *Vestiges*, 3rd ed., 269.

52. A particularly relevant commentary on this attack came from Spencer. See [Herbert Spencer], "Recent Astronomy, and the Nebular Hypothesis," *Westminster Review*, new series 14 (July and October 1858): 185-86. Further discussion below.

53. Ogilvie observes: "Although Chambers insisted that the correctness of the nebular hypothesis was not essential to his theory, his behaviour suggested that he really thought otherwise. . . . He continued to defend the theory without offering a replacement throughout all of the editions, although he reduced the amount of space devoted to it." Marilyn Bailey Ogilvie, "Robert Chambers and the Nebular Hypothesis," *British Journal for the History of Science* 8, no. 3 (November 1975): 219.

54. From the third edition on, this treatment was largely abridged, and from the fourth edition onward, Chambers advocated and developed his genealogical classification. For a detailed treatment, see Ogilvie, "Robert Chambers and the Successive Revisions of *Vestiges*," esp. 370-439. For Secord's assessment and treatment, see *Victorian Sensation*, 388-89.

55. The particulars of the reception of *Vestiges* depended to an extent on who was imagined to be the author, by whom the imagining was being done, and how much the work was appreciated as a flawed, but successive and anticipatory, effort.

56. Secord has pointed out that the candidate initially given the most attention was the aristocratic Richard Vyvyan, whose status played no small role in the initial reception. See Secord, *Victorian Sensation*, 180-83.

57. For a case study of the drawing of these boundaries in the context of geological work in 1830s, see Martin J. S. Rudwick, "Charles Darwin in London: The Integration of Public and Private Science," *Isis* 73, no. 2 (June 1982). As Rudwick elsewhere observes, *Vestiges* put pressure on the "cognitive boundaries" stabilizing the disciplinary landscape, unsettling as well the sensitive accommodations between natural historical and theological concerns. Rudwick speaks to geology in particular, but the point extends to other disciplines. See Martin J. S. Rudwick, "The Shape and Meaning of Earth History," in *God and Nature: Historical Essays on the Encounter between Christianity and Science*, ed. David C. Lindberg and Ronald L. Numbers (Berkeley: University of California Press, 1986), 315.

58. One notable point of partial resonance and partial failure is Darwin's much-analyzed response. The treatment of Chambers as part of Darwin (and Wallace) historiography has more recently undergone extended critique, targeted as retrospective history masking the importance of Chambers's book in itself and its relation to its moment. The connection that both Chambers and Darwin drew between their works further suggests something more of the knotty boundary between the expert and the lay: the debates around Chambers's "amateur" work shaped work thereafter understood as scientific, Darwin and Wallace's together. The point is not that Chambers mattered to Darwin and Wallace and therefore matters, but rather that work regarded as scientific was crafted in no small measure in relation to work dismissed as unsound. It suggested lines of research (as, for example, Wallace seemed to read Chambers) and ways of delimiting inquiry. For this role of Chambers, see Egerton, "Refutation and Conjecture," 176. Darwin famously used Chambers as a point of contradistinction, establishing the boundary between his work and less respectable science. See Charles Darwin, *On the Origin of Species by Means of Natural Selection, or the Preservation of Favoured Races in the Struggle for Life*, 1st. ed. (London: John Murray, 1859), 3–4. And for Darwin, that delimiting produced in his work a much less overtly "cosmic" story, with a less explicitly narrative historical presentation. In contrast to Chambers, Darwin used Herschel and Whewell to adopt the methods and limits of scientific work, avoiding any attempt to provide as broad an induction as that of *Vestiges*. See the next chapter.

59. In the much-publicized address, Herschel distinguished sharply between the Laplacian nebular hypothesis and his father's. The former was pure speculation and was in no way demonstrated or linked to the latter, which was itself based on observation. Chambers, who had been present to hear the address, returned fire, privately and publicly. In *Explanations*, he derided Herschel's emphasis on the practical benefits of the sciences, and, with sentiments in line with Egerton's later thesis, claimed that the elite, having no orientation themselves, repressed the possibility of imaginative, ambitious scientific ideas. Chambers, *Explanations*, 179. But even given all this, Chambers seemed less in himself the very target of Herschel's outrage than Comte. Herschel was galled most of all by the arguments that Chambers and J. S. Mill accepted and promulgated as proof of Laplace's conjecture: "I really should consider some apology needed for even mentioning an argument of the kind to such a meeting, were it not that this very reasoning, so ostentatiously put forward and so utterly baseless, has been eagerly received among us as the revelation of a profound analysis." John Herschel, "Address of the President," in *Report of the Fifteenth Meeting of the British Association for the Advancement of Science; Held at Cambridge in June 1845* (London: John Murray, 1846), xxxix. In the printed report, he footnotes here Mill and *Vestiges*.

60. Even J. D. Forbes's mixed but largely critical review was more respectful of the figure of Humboldt. J. D. Forbes, "Review of the First Volume of Kosmos," *Quarterly Review* 77 (1845). In reference to this review, W. H. Brock writes: "Forbes could suspect him [Humboldt] of playing into the hands of sinners like the phrenologist George Combe or Robert Chambers . . . who were appealing to the need to obey the secular laws of nature rather than for mankind to seek for moral analogies within nature." Brock, "Humboldt and the British: A Note on the Character of British Science," *Annals of Science* (1993): 370–71. See also Jean Théodoridès, "Humboldt and England," *British Journal for the History of Science* 3, no. 1 (June 1966).

61. In his review of Humboldt's work, Herschel applauded Humboldt for not conflating those genealogies, even though the possibility was consistent with Humboldt's approach. Herschel took special note of what he referred to as the genetic and historical classifications available within it: the former involved how the rocks originated, emphasizing their efficient causes, the latter looked to their longer historical roots. [John Herschel], "KOSMOS, Entwurf einer Physischen Weltbeschreibung," *Edinburgh Review* 87, no. 175 (1848): 210.

62. Robert Chambers, *Explanations: A Sequel to "Vestiges of the Natural History of Creation" By the Author of That Work* (London: John Churchill, 1845), 179.

63. [Herschel], "Kosmos," 172.

64. See ibid., 182.

65. Reviewers such as Herschel and Forbes in fact felt that they had too vague a sense of the whole in Humboldt's work. Forbes, moreover, saw the work as outdated, as wasteful in its energies and scope, and as repetitive in a manner that a different technology of writing could have spared: "when he ought to have written a single work, or at most two, he wrote an encyclopedia." Forbes, "Review of Kosmos," 158.

66. [Herschel], "Kosmos," 182-83. Giving some broader context to the relationship between Herschel's conception of science and his own empirical methods, Elizabeth Greene Musselman has claimed in reference to Herschel's South African surveys that his "method represented in microcosmic form the macrocosmic progress of human society. Despite its presence at the peak of civilized culture, science had something to learn from the grosser errors of the less advanced empire." Musselman, "Swords into Ploughshares: John Herschel's Progressive View of Astronomical and Imperial Governance," *British Journal for the History of Science* 31, no. 4 (December 1998): 431. For a parallel claim in reference to Whewell and the definition of science he enunciates in his *Philosophy,* see Michael S. Reidy, *Tides of History: Ocean Science and Her Majesty's Navy* (Chicago: University of Chicago Press, 2008), 192-93. Note that Musselman and Reidy directly affirm the importance of Humboldt to Herschel and Whewell, respectively.

67. [Herschel], "Kosmos," 182.

68. One famous commentator who pointed to such a change was Thomas Kuhn. See the next chapter.

69. His classification or map of the sciences—published in his *Classification of the Sciences: To Which Are added Reasons for Dissenting from the Philosophy of M. Comte* (New York: D. Appleton, 1864)—could be regarded as presenting at least a *formal* synthesis of the knowledge of the world.

70. Spencer's own resistance to unity was expressed through the importance of similarity or resemblance to the process of classification, a process that simultaneously reveals differences or disunities across the categories of that classification. Applying this to the subject matters of the different sciences, he argued that "the broadest natural division among the Sciences, is the division between those which deal with the abstract relations under which phenomena are presented to us, and those which deal with the phenomena themselves." See Herbert Spencer, "Reasons for Dissenting from the Philosophy of M. Comte" in *The Classification of the Sciences.* This was a division that distinguished between those sciences that, like mathematics or logic, in effect study time and space itself and those that instead study objects given in time and space. The latter sciences further subdivide into concrete and abstract categories relating to objects understood through their components or to objects regarded in their totalities, a kind of analytical versus synthetic distinction. And so on to further subdivisions still. It should be noted that the nature and degree of the distance of Spencer's system from Comte's was a point taken up from the first, as for example in a debate between Spencer and Émile Littré. It so exercised Spencer that he devoted a significant portion of his classificatory system to distinguishing between Comte's and his own work.

71. [Herbert Spencer], "Recent Astronomy and the Nebular Hypothesis," *Westminster Review* 70, no. 137 (July 1858). Two further citations were François Arago's *Popular Astronomy* and, the most recent, *Recent Progress in Astronomy; Especially in the United States,* by Elias Loomis.

72. Ibid., 202.

73. It would not seem that Spencer saw in Whewell's work enough of a reflection on the interrelation of the sciences to demand a confrontation. There is a possibility that Spencer took what he *might* have seen as an oblique dig at Whewell—but what might equally have been an argument indebted to Whewell (or, for that matter, to Bacon). In his disputation of the serial arrangement of the sciences as put forward by Comte, he observes, "Where is our warrant for assuming that there is some *succession* in which they can be placed? There is no reason; no warrant. Whence then has arisen the supposition? To use M. Comte's own phraseology, we should say, it is a metaphysical conception. It adds one to the many cases constantly occurring, of the human mind being made the measure of Nature. We are obliged to think in sequence; it is the law of our minds that we

must consider subjects separately, one after another: *therefore* Nature must be serial—*therefore* the sciences must be classifiable in a succession." Spencer, "The Genesis of Science," 127-28. Despite Spencer's critique, Comte and Whewell did see the order of knowledge as linked to the (order of the) genesis of knowledge; they each understood this link differently, as we've seen, but nevertheless affirmed the fact of it.

74. Ibid., 133 and 145-146.

75. Ibid.

76. In countering Comte's views, Spencer draws from Brewster and, in the same passage, draws on Whewell's *History*, demonstrating ways in which the sciences were involved in one another in their development. Ibid., 153. And in what might have been a chapter-and-verse reading of Somerville, Spencer observed that "so complete in recent days has become this consensus among the sciences, caused either by the natural entanglement of their phenomena, or by the analogies in the relations of their phenomena, that scarcely any considerable discovery concerning one order of facts now takes place, without very shortly leading to discoveries concerning other orders." Ibid., 154. In his emphasis on nomenclature and classification and in his explicit reading of Whewell, Spencer could hardly not have been reacting to Somerville and to Herschel, as well.

77. [Herbert Spencer], "Progress: Its Law and Cause," *Westminster Review* 67, no. 132 (April 1857): 445.

78. Ibid., 446, 447.

79. Ibid., 465.

80. *"Every active force produces more than one change—every cause produces more than one effect."* Ibid., 466. In the latter part of the paper, he returned to his history to demonstrate the action of this law.

81. Ibid., 485. See also under "Propositions Which I hold" in Herbert Spencer, *The Classification of the Sciences*, 34-35.

82. William Whewell, "Comte and Positivism," *Macmillan's Magazine*, March 1866, 358. As a further parting shot, Whewell names Ampère as an earlier and better arranger, even as he reinforces again his curious claim to indifference to the whole exercise. Ibid.

83. Herschel, "Kosmos," 180.

84. Ibid. Also, it comes as no surprise in this context that some notion of unity is at stake, where perhaps Anson Rabinbach, among contemporary commentators, signals most clearly the importance of the doctrine of energy as supplying a kind of homogeneous fundamental, but many-faced element for the identity of all potential forces in the world, relating it to conceptions of social labor: "The discovery of energy as the quintessential element of all experience, both organic and inorganic, made society and nature virtually indistinguishable. The pioneers of energy conservation viewed the transformation of mechanical energy into heat, and subsequently, the transformation of all natural forces as manifestations of a single *Kraft*." Rabinbach, *The Human Motor: Energy, Fatigue, and the Origins of Modernity* (New York: Basic, 1990), 46. Energy in this guise is a particular dogma of unity, bearing a subtle relation to a notion of unification by forces in the more contemporary sense of the latter terms.

CHAPTER THREE

1. Darwin is not given a central place in my analysis. This is partly because, though deeply exercised by these debates and Thomson's critiques, Darwin engaged with them less actively than Thomson or Huxley and did not help establish as clearly or unapologetically as for example Helmholtz the contours of the synthetic narrative at stake. Additionally, Darwin's more direct role in establishing that narrative has been the subject of extended historical and philosophical treatment, which is examined only as is necessary for the historical analysis carried out here. Note also that Helmholtz was not "von Helmholtz" until the early 1880s.

2. There is a copious secondary literature upon which this analysis explicitly depends and leans. It is necessary to revisit this history in no small part because central protagonists established a set of expectations for a totalizing scientific narrative—the more narrow point of interest of this treatment.

3. Recent scholarship has disputed the claim that the articulation of conservation principles amounted to a case of international simultaneous discovery. They have emphasized differences between terms and concepts, between local contexts and motivations. Crosbie Smith argues that codified energy principles in the natural sciences were the contingent construction of a local, North British collective, committed to different political, physical, and metaphysical programs than other putative "discoverers." Morus, also sensitive to such differences, suggests how the arguments of a protagonist such as Grove were likewise simultaneously physical, metaphysical, political, and economically minded, staking out a very different vision of scientific community than the views of those such as Herschel and Whewell. See Iwan Morus, "Correlation and Control: William Robert Grove and the Construction of a New Philosophy of Scientific Reform," *Studies in History and Philosophy of Science* 22, no. 4. David Cahan, from a somewhat different standpoint, argues that the conservation of energy was "to no small extent an Anglo-German creation" for which Helmholtz was a kind of "bridge" between his own science and those he knew in the British context.

4. On the question of final theories in the late nineteenth century and thereafter, see Lawrence Badash, "The Completeness of Nineteenth-Century Science," *Isis* 63, no. 1 (March 1972). Badash reflects on the "next decimal place" trope, considering its relationship to the recurrently expressed sense of completeness in the sciences and the professional "pessimism" this entailed. On the basis of his treatment of Maxwell, Michelson, J. J. Thomson, and others he concludes that the feeling of completeness "was more a 'low-grade infection,' *but nevertheless very real.*" Ibid., 58. See also Peter Galison, "Re-reading the Past from the End of Physics: Maxwell's Equations in Retrospect," in *Functions and Uses of Disciplinary Histories*, ed. Loren Graham, Wolf Lepenies, and Peter Weingart (Dordrecht: D. Reidel, 1983). The specific phrase "final theory" in reference to scientific theories has been in use since at least the late nineteenth century. The term was not confined to natural philosophy and the physical sciences alone. For example, in 1879, John Venn, the logician of the famous eponymous diagrams, noted: "If Evolution in its present form be a final theory, then no doubt we may have got at something like an 'objective' connotation of some our class-terms, but only on this rather bold assumption." John Venn, "The Difficulties of Material Logic," *Mind* 4, no. 13 (January 1879): 46.

5. Though the latter two famously faulted his evolutionary writings, their estimation of Darwin's work was higher prior to the publication of *Origin of Species*. Whewell, for example, testified to the importance of Darwin's geological research and invited Darwin to the office of secretary of the Geological Society. Herschel, editing the 1849 *Manual of Scientific Inquiry*, requested that Darwin write its geological chapter. For reference to Herschel's request and its not entirely happy conclusion, see S. S. Schweber, "John Herschel and Charles Darwin: A Study in Parallel Lives," *Journal of the History of Biology* 22, no. 1 (Spring 1989): 63. The full citation of the work in question is *A Manual of Scientific Inquiry Prepared for the Use of Officers in Her Majesty's Navy and Travellers in General*, ed. John F. W. Herschel (London: John Murray, 1849). Darwin, reciprocally, found great importance in the writings of both Herschel and Whewell. They established a vision of science Darwin found inspiring and an understanding of scientific method through which he could ground his own research problematics. On a rhetorical level, Darwin appealed to Whewell's *Bridgewater Treatise* for his constant epigraph in the different editions of *Origins* and voiced superlative expressions of admiration for Herschel. See for example Martin J. S. Rudwick, "Charles Darwin in London: The Integration of Public and Private Science," *Isis* 73, no. 2 (June 1982): 197. Rudwick here endorses the importance of Herschel and particularly Whewell for the scientific methodology evident in Darwin's approach to geology, emphasizing *vera causa* and the consilience of inductions. Ruse, by

contrast, emphasizes their importance to Darwin's biological work. As Ruse, Rudwick, and Cannon have all argued with different emphases, Darwin endorsed and perhaps internalized their visions of scientific argument as such. He saw himself as addressing the problems of origins of species with their tools and in their frameworks. Both Ruse and Cannon occasionally put a "Herschel-Whewell" or "Whewell-Herschel" union to use in relation to Darwin. See, e.g., Michael Ruse, "Darwin's Debt to Philosophy: An Examination of the Influence of the Philosophical Ideas of John F. W. Herschel and William Whewell on the Development of Charles Darwin's Theory of Evolution," *Studies in History and Philosophy of Science* 6, no. 2: (1975): 167, and W. Faye Cannon, "The Whewell-Darwin Controversy," *Journal of the Geological Society* 132 (August 1976): 378. Darwin's structural reliance on Lyell has been discussed by many scholars, as has Humboldt's impact on British scholarship. For Darwin's more personal testimony to the inspiration of Humboldt, see for example Jean Théodorides, "Humboldt and England," *British Journal for the History of Science* 3, no. 1 (June 1966): esp. 49–53.

6. With regard to Huxley's relationship to Chambers, for example, Bernard Lightman observes that "The *Vestiges* and Lewes' book on Comte provided Huxley with models of scientific work that were unfit as vehicles for educating a popular audience." Lightman, *Victorian Popularizers of Science: Designing Nature for New Audiences* (Chicago: University of Chicago Press, 2007), 361. *Vestiges* was adduced as a model against which expert authorship could be fashioned. But also, Darwin, along with Somerville, saw more direct value in the contribution of Chambers. See Richard Yeo, "Science and Intellectual Authority in Mid-Nineteenth-Century Britain: Robert Chambers and 'Vestiges of the Natural History of Creation,'" *Victorian Studies* 28, no. 1 (Autumn 1984) and discussion in chapter 2.

7. So, too, Thomas S. Kuhn saw in her work a "new look" forming part of the conditions for the possibility of the articulation of the principle of the conservation of energy. Kuhn's observation and its own context points to related modes of connection and potentially unity that Somerville's work adumbrated and likely helped stimulate. Kuhn's essay on the principle of the conservation of energy and its implications for the nature of simultaneous discovery concerned what other critics thereafter have understood as a case of social construction. Kuhn himself named twelve possible contributors to the principle, independent or semi-independent, including, among them, Faraday, Joule, Carnot, Liebig, Helmholtz, and Grove, and emphasized three factors that the historiography thereafter has since debated: the "availability of conversion processes," the "concern with engines," and the "philosophy of nature." See Thomas Kuhn, "Energy Conservation as an Example of Simultaneous Discovery," in Kuhn, *The Essential Tension: Selected Studies in Scientific Tradition and Change* (Chicago: University of Chicago Press, 1977), 73. An extensive scholarship exists on the emergence of the principles of conservation of energy, showing its variegated roots, from the metabolic studies of animals to the rejection of the principle of perpetual motion. In the more contemporary form of this scholarship, Kuhn's is something of an urtext.

8. Iwan Morus, *When Physics Became King* (Chicago: University of Chicago Press, 2005), 56.

9. Iwan Rhys Morus, s.v. "Grove, Sir William Robert (1811–1896)," *Oxford Dictionary of National Biography*, Oxford University Press, 2004, available online at http://www.oxforddnb.com/view/article/11685.

10. Grove, like Mayer, has been a figure long associated with the development of energy principles, associations in each case defended and criticized from the nineteenth century until the present. For the changing response of Thomson's collaborator, P. G. Tait, to Grove's work, and correspondence with Thomson in this regard, see Crosbie Smith, *The Science of Energy: A Cultural History of Energy Physics in Victorian Britain* (London: Athlone, 1998), esp. 176–77. It is in this text that Smith roots his most extended argument that "energy physics . . . was the product of a 'North British' group concerned with the radical reform of physical science and with the rapid enhancement of its scientific credibility." Ibid., 2–3.

11. G. N. Cantor suggests that Grove's reading of Comte, as well as of Herschel, Berkeley, and Hume, helped Grove develop his critique. G. N. Cantor, "William Robert Grove, the Correlation of Forces, and the Conservation of Energy," *Centaurus* 19, no. 4 (December 1975): 275-78.

12. Herschel had discussed causal explanation in relation to the unification of electricity and magnetism when reviewing Whewell's *History* and *Philosophy* and alluded to critiques of causality in his 1845 BAAS presidential address. Cantor speculates that Grove had read Herschel's review. See ibid., 277. Note that Grove's critique of causality does not preclude his own reductionism, which, as Cantor observes, centered on "matter and motion." Ibid., 280.

13. William Grove, *The Correlation of Physical Forces* (London: Sam Highly, 1850), 13.

14. Ibid., 17. For Grove's explanation of his use of the term "correlation" see ibid., 93.

15. Ibid., 17.

16. In language of later chapters, it would be to resist the possibility of a foundational genre of synthesis.

17. See P. M. Heimann, "Mayer's Concept of 'Force': The Axis of a New Science of Physics," *Historical Studies in the Physical Sciences* 7 (1976). For an analysis of Mayer's conceptions of force particularly in relation to *Naturphilosophie*, see M. Norton Wise, "German Concepts of Force, Energy, and the Electromagnetic Ether: 1845-1880," in *Conceptions of Ether: Studies in the History of Ether Theories 1740-1900*, ed. G. N. Cantor and M. J. S. Hodge (Cambridge: Cambridge University Press, 1981), 271-75.

18. John Tyndall, "On Force," Royal Institution of Great Britain, *Philosophical Magazine*, 4th series, 24, no. 158 (1862): 64.

19. Ibid.

20. Ibid., 65.

21. Ibid., 64.

22. Ibid., 65.

23. See for example Crosbie Smith's review, "Robert Mayer and the Conservation of Energy by Kenneth L. Caneva," *British Journal for the History of Science* 29, no. 3 (September 1996): 372-73. Caneva provides a thoroughgoing contextual treatment of Mayer's work: Kenneth L. Caneva, *Robert Mayer and the Conservation of Energy* (Princeton: Princeton University Press, 1993).

24. Hermann von Helmholtz, *Vorträge und Reden*, vol. 1 (Braunschweig: Friedrich Vieweg und Sohn, 1884), 346.

25. His army service meanwhile continued until 1848, when Humboldt himself interceded on Helmholtz's behalf. Humboldt arranged an early release from the military so that Helmholtz could take up a teaching post in anatomy at the Akademie der Künste. See Arleen Tuchman, "Helmholtz and the German Medical Community," in *Hermann von Helmholtz and the Foundations of Nineteenth-Century Science*, ed. David Cahan (Berkeley: University of California Press, 1993), 28.

26. See Arleen Tuchman, "Helmholtz and the German Medical Community," for an extended discussion. Many of the biographical facts concerning Helmholtz presented here are from this essay and the others in this collection, as well as an early biography by Leo Koenigsberger. Leo Koenigsberger, *Hermann von Helmholtz*, trans. Frances A. Welby (New York: Dover, 1960).

27. Hermann von Helmholtz, *Selected Writings of Hermann von Helmholtz*, ed. Russell Kahl (Middletown: Wesleyan University Press, 1971), 11, 14.

28. W. R. Grove, "Presidential Address," *BAAS Report* 36 (1866): lv.

29. Ibid., lxxix.

30. For example, ibid., lxxix.

31. Ibid., lxxx.

32. Ibid., lxxx.

33. From the 1850s, Huxley too, as we will see in the next chapter, had weighed in on the question of science education and the curriculum, emphasizing science's ability to teach its students how to balance the forces of the natural world and its greater (though not unique) advantages in teaching the proper principles of scientific induction.

34. William Whewell, *Philosophy of the Inductive Sciences Founded Upon Their History*, vol. 2 (London: John W. Parker 1840), 505–21, and John Herschel, "History of the Inductive Sciences from the Earliest to the Present Times," *Quarterly Review* 68, no. 135 (June 1841): 236.

35. Silvanus P. Thompson, *The Life of William Thomson, Baron Kelvin of Largs*, vol. 1 (London: Macmillan, 1910), 239. Crosbie Smith excavating this history finds in Thomson a partial recapitulation of the theories of his father, James, and his Glasgow teachers Meikleham and J. P. Nichol. This text was apparently included in each subsequent introductory lecture in his course on natural philosophy in Glasgow College and suggests a verdict on the theorization of branches in Comte, Whewell, or even Hegel. However, Thompson describes this essay "as much revised in later years," indicating perhaps that the concerns therein were enduring ones for him, in need of continuing thought. Ibid., 190. Thomson follows this warning by articulating "in general terms the relation which Natural Philosophy bears to other branches of human inquiry, observation, science, and philosophy." Ibid., 239.

36. Ibid., 239–240.

37. They are also reminiscent of divisions that Huxley would appeal to in late popular work. See T. H. Huxley, *The Crayfish* (1880; New York: D. Appleton and Company, 1895), 2–4. It is in reference to this work that Lightman has observed that Huxley establishes a three-stage law of development imitating Comte's. See Lightman, *Victorian Popularizers*, 386.

38. Thompson, *Life of William Thomson*, vol. 1, 240.

39. Ibid., 251.

40. I borrow C. S. Lewis's term here, in his reference to humans as "creatures of the Margin." C. S. Lewis, *Discarded Image: An Introduction to Medieval and Renaissance Literature* (Cambridge: Cambridge University Press, 1964), 58.

41. But much as Herschel and Thomson had argued that science invites humility as it brings the scientist closer to the creator, nevertheless, future generations would often find in the scientifically infused argument for humanity's insignificance little humility and little solace. For Herschel, see John F. W. Herschel, *A Preliminary Discourse on the Study of Natural Philosophy* (London: Printed for Longman, Rees, Orme, Brown, and Green and John Taylor, 1831), 8. For the way that science nevertheless improves and promotes the state of the natural philosopher, see ibid., 16–17. See also Shapley's critiques in subsequent chapters.

42. Ursula DeYoung has noted that the lectures were particularly intended to provide resistance to the popular belief in "table turning," the movement of a séance table by spirits. Ursula DeYoung, *A Vision of Modern Science: John Tyndall and the Role of the Scientist in Victorian Culture* (New York: Palgrave Macmillan, 2011), 136. Faraday asked why a table turner wouldn't allow an experimental test: "Why should not one who can thus lift a table, proceed to verify and simplify his fact, and bring it into relation with the law of Newton?" Michael Faraday, "Observations on Mental Education," in *Lectures on Education, Delivered at the Royal Institution of Great Britain* (London: J. W. Parker and Son, 1854), 52. Faraday did, it seems, carry out experimental tests. See ibid., 60–61.

43. William Whewell, "On the Influence of History of Science upon Intellectual Education," in *Lectures on Education*, 15–16. For an in-depth discussion of Whewell's views concerning the place of the natural sciences in education, see Richard Yeo, chapter 8, "Science, Education and Society," in Yeo, *Defining Science: William Whewell, Natural Knowledge and Public Debate in Early Victorian Britain* (Cambridge: Cambridge University Press, 1993).

44. Whewell, "On the Influence of History of Science upon Intellectual Education," 29. Whewell also recommended the study of the history of science, because it "*may* do, and carefully studied, *must* do, much to promote that due apprehension and appreciation of inductive discovery." Ibid., 31.

45. See Joe D. Burchfield, "John Tyndall—A Biographical Sketch," in *John Tyndall: Essays on a Natural Philosopher*, ed. William Hodson Brock, Norman D. McMillan, and R. Charles Mollan (Dublin: Royal Dublin Society, 1981), 3.

46. John Tyndall, "On the Importance of the Study of Physics as a Branch of Education for All Classes," in *Lectures on Education*, 180.

47. Ibid., 180-81.

48. Ibid., 181.

49. Herbert Spencer's views on education, which likewise asserted the virtues of scientific education, confronting as well what Spencer took to be central activities of life and their relation to disciplines of knowledge, were also in circulation. See, e.g., Herbert Spencer, *Education: Intellectual, Moral and Physical* (New York: D. Appleton, 1860), the compilation of four of his review essays. For recent scholarship analyzing the global reach of Spencer's theories, of education and otherwise, see Bernard Lightman, ed., *Global Spencerism: The Communication and Appropriation of a British Evolutionist, Cultural Dynamics of Science,* vol. 1 (Leiden, NL: Brill, 2015).

50. For an assessment of the extent of Helmholtz's contributions, see David Cahan, "Introduction: Helmholtz at the Borders of Science," in *Hermann von Helmholtz and the Foundations of Nineteenth-Century Science,* 1.

51. See Helmholtz, "The Conservation of Force: A Physical Memoir," in *Selected Writings,* 3-55; Tuchman, "Helmholtz and the German Medical Community," esp. 28; and Bevilacqua, "Helmholtz's Ueber die Erhaltung der Kraft," in *Hermann von Helmholtz and the Foundations of Nineteenth-Century Science,* esp. 297-304.

52. Helmholtz, "The Relation of the Natural Sciences to Science in General," in *Selected Writings,* 124 and, for further points regarding Kant and Hegel, passim, esp. 124-27.

53. See Christa Jungnickel and Russell McCormmach, *The Intellectual Mastery of Nature: Theoretical Physics from Ohm to Einstein,* vol. 1, *The Torch of Mathematics, 1800-1870* (Chicago: University of Chicago Press, 1986), 3-8.

54. For the concern over any association of Helmholtz's natural philosophy with Hegel, see Koenigsberger, *Hermann von Helmholtz,* 140. It should be noted that Thomson apparently showed even less patience than Helmholtz for Hegel. Thompson, *Life of William Thomson,* vol. 2, 1122.

55. "Long may the German universities be preserved from such a fate! Then indeed would the connection among the different sciences finally be broken." Helmholtz "The Relation of the Natural Sciences to Science in General," 128.

56. Ibid.

57. Ibid., 141. As in the last chapter, the shifting meanings of "science" must be kept in mind, as well as the enduringly wider German sense of *Wissenschaft.*

58. Ibid., 129. For extended analysis of such efforts and their relationship to establishing scientific periodical publication as the conventional measure of scientific output, see Alex Csiszar, "Broken Pieces of Fact: The Scientific Periodical and the Politics of Search in Nineteenth-Century France and Britain" (PhD diss., Harvard University, 2010).

59. Helmholtz, "The Relation of the Natural Sciences to Science in General," 129-30.

60. Ibid., 141.

61. Ibid., 142.

62. Anson Rabinbach, *The Human Motor: Energy, Fatigue and the Origins of Modernity* (Berkeley: University of California Press 1992), 4.

63. Historians have seen in Whewell and Herschel conceptual, pedagogical, and cultural keys to the intellectual problematics of Darwin and Thomson in particular. See, for example, Michael Ruse, "Darwin's Debt to Philosophy," W. F. Cannon, "The Whewell-Darwin Controversy," and Martin J. S. Rudwick, "Charles Darwin in London." For Herschel and Whewell as constitutive of Thomson's intellectual backdrop and his responses to particularly compelling pedagogic and scientific principles, see Crosbie Smith and M. Norton Wise, *Energy and Empire: A Biographical Study of Lord Kelvin* (Cambridge: Cambridge University Press, 1989). Smith and M. Norton Wise observe that at Cambridge the man Thomson referred to as "the despotic Whewell" provided an educational backdrop of "moderate conservatism"—his mature, geometrically minded resistance to the analytic tradition—against which the more professionally minded "whigs in mathematics as

in politics" of Thomson's generation resisted but were also partly formed. Smith and Wise, *Energy and Empire*, 65.

64. In the execution of his scientific theoretical vision, he emphasized a Whewell-inflected conception of the importance of geometry. Likewise, Whewell's work on "labouring force" may have been formative for the Thomson brothers' understandings of mechanical effect, which was in turn crucial to advances in thermodynamics and its cosmological consequences. See, for example, Smith and Wise, *Energy and Empire*, 65. These are of course only some among the formative systems that the young Thomson encountered. Apart from the manifold traditions, cultural and political backdrop, and more specific aspects of his upbringing, Smith and Wise also emphasize the "Scottish tradition of natural philosophy" as important to gauging the character of Thomson's future work. Ibid., 88.

65. This dependence could hardly have been regarded as an innovation in itself. Laplace's solar nebular hypothesis made this argument, even while Laplace demonstrated the stability of that system once established. But in contrast to the celestial-minded accounts of many of the authors of different nebular hypotheses, Thomson's thermodynamic considerations were more insistently dependent on the identification of the sciences of the heavens and of the earth.

66. Smith and Wise, *Energy and Empire*, 502–4. Herschel suggested the possibility of an endless creation of heat through friction. Despite direct efforts, Thomson did not convert Herschel to the cause of energy. For discussion and reproduction of the correspondence between Herschel and Thomson on the subject of conservation of energy, see Frank A. J. L. James, "Between Two Scientific Generations: John Herschel's Rejection of the Conservation of Energy in His 1864 Correspondence with William Thomson," *Notes and Records of the Royal Society of London* 40, no. 1 (November 1985).

67. Crosbie Smith places Thomson's original and eschatological visions in a Voluntarist-theological line drawing in part on Whewell's *Bridgewater Treatise*. Crosbie Smith, "Natural Philosophy and Thermodynamics: William Thomson and 'The Dynamical Theory of Heat,'" esp. 303 and 314. Also see Herschel, *Treatise on Astronomy*, 211–12. Smith and Wise further speculate that Herschel's *Treatise* may have had direct influence on Mayer's 1848 work positing a meteoric theory of solar heat close to the theory of Thomson. See Smith and Wise, *Energy and Empire*, 503.

68. Note that still in 1866, Grove was remarking that "Indications of the connexion between cosmical studies and geological researches are dawning on us," giving as his example the belief that "we can trace many geological phenomena to our varying rotation round the sun; thus more than thirty years ago Sir J. Herschel proposed an explanation of the changes of climate on the earth's surface as evidenced by geological phenomena, founded on the changes of excentricity in the earth's orbit." (Spelling as in the original.) W. R. Grove, "Presidential Address," *BAAS Report* 36 (1866): lxviii.

69. See Joe D. Burchfield, *Lord Kelvin and the Age of the Earth* (Chicago: University of Chicago Press, 1990), 10–11.

70. Centuries prior, naturalistic estimates were made of the age of the earth. Nicolaus Steno analyzed sedimentary strata to determine the order of natural events. Based on the assumption that the earth was once covered with water, Benoît de Maillet obtained lower limits of the age of the earth as a function of apparent falling sea levels. Based on the alternative assumption that the earth was itself cooling from a hypothetical original molten state, the Comte de Buffon examined the cooling of iron balls of different sizes and extrapolated to the cooling of the full earth itself. The basic ideas motivating this conception are not altogether different in spirit from the thoughts animating Thomson's arguments concerning the secular cooling of the earth—see below. Work in the history of geology such as that of Claude Albritton and Martin Rudwick would suggest that these efforts were no less empirical than Thomson's. For various forms of such attempts, see Martin Rudwick, *The Meaning of Fossils: Episodes in The History of Palaeontology* (Chicago: University of Chicago Press,

1976); Claude C. Albritton Jr., "Geologic Time," *Journal of Geological Education* 32, no. 1 (January 1984); Stephen Jay Gould, *Time's Arrow, Time's Cycle: Myth and Metaphor in the Discovery of Geological Time* (Cambridge: Harvard University Press, 1987); and Francis Haber, *The Age of the World: Moses to Darwin* (Westport, CT: Greenwood, 1978). Meanwhile, though Hutton's geological work was not initially in the field, its more mechanistic aspects themselves gave empirical reasons for the perpetuity of the globe, to which exact figures would have been inessential. See chapter 3, "James Hutton's Theory of the Earth: A Machine without a History," in Gould, *Time's Arrow, Time's Cycle*.

71. Quantifying uncertainty indicates one of the strong analytical contrasts between Thomson's computational efforts and those of Darwin. Darwin notoriously and perhaps foolishly gave a precise, specific figure for a particular geological process in the first two editions of *On the Origin of Species*. He speculated that the denudation of the Weald Valley in southern England required some 306,662,400 years. But he stressed that he offered this age only to provide some "crude notion on the subject." Charles Darwin, *On The Origin of Species or The Preservation of the Favoured Races in the Struggle for Life* (London: John Murray, 1859), 286. It is clear that the disclaimer didn't resonate. It wasn't any ruse, however; Darwin believed that the earth was older, regardless.

72. Smith and Wise place theological commitments at the very heart of Thomson's articulation of the second law of thermodynamics, understood in his own view as a law about the necessary dissipation of work/energy. See Crosbie Smith and M. Norton Wise, *Energy and Empire: A Biographical Study of Lord Kelvin* (Cambridge: Cambridge University Press, 1989), 327-33.

73. It should be kept in mind that Thomson was resistant to specifically Darwinian evolution, but not to the possibility of evolution as such. So Thomson in his BAAS presidential address of 1871 entertained a vision laid out by Grove five years before, connecting his theory of panspermia to Grove's exposition of the "doctrine of continuity" where "all creatures now living on earth have proceeded by orderly evolution from some such origin." Thomson, "Presidential Address" in *Report of the 41st Meeting of BAAS held at Edinburgh in 1871* (London: John Murray 1872), cv. It was a lecture filled with encomium directed at the Herschels, John in particular, because he had passed away the May previous to the address. See also Michael J. Crowe, *The Extraterrestrial Life Debate 1750-1900: The Idea of Plurality of Worlds from Kant to Lowell* (Cambridge: Cambridge University Press, 1986), 402-4, and Thomson's remarks on "geological evolutionism" below.

74. See Thompson, *Life of William Thomson*, vol 1, 288-91 and below.

75. Note that Nichol had written to Thomson directly to complain of Chambers's *Vestiges* as a "foolish book." See Secord, *Victorian Sensation*, 466.

76. Thompson, *Life of William Thomson*, vol. 1, 42. See also "On the Linear Motion of Heat," part 1, 15, and "Note on Some Points in the Theory of Heat," 39-40, in William Thomson, *Mathematical and Physical Papers*, vol. 1 (Cambridge: Cambridge University Press, 1882).

77. Its theme, according to his biographer, was "the Age of the Earth and its Limitation as determined from the Distribution and Movement of Heat within it." Thompson, *Life of William Thomson*, vol. 1, 188. The dissertation, in Latin, Thompson notes, was "burned by Thomson the same day," so that "only one draft page exists." Ibid., 186.

78. Thomson, *Mathematical and Physical Papers*, vol. 1, 514.

79. Smith and Wise, *Energy and Empire*, 519.

80. And while the limits on time ranges remained uncertain enough, they still moved toward some consensus on the orders of magnitude. "After 1870, only a handful of investigators defended a terrestrial age that differed much from Kelvin's 100 million years. A few would reduce his limit to the order of tens of millions, and a few others would increase it to hundreds of millions, but almost no one thought it necessary to exceed his upper limit of 500 million years. Even within that narrow range, however, the opportunities for controversy were endless, and some degree of tension characterized almost every discussion of geological time." Burchfield, *Lord Kelvin and the Age*

of the Earth, 17. In its own day, it should be added, this would likely have been considered by many a wide, rather than a narrow window of time, a point Burchfield implicitly concedes elsewhere in his volume and that was testified to by Huxley in overt criticisms—see below.

81. For a more thorough presentation of these views, see Smith and Wise, *Energy and Empire*, 519-23 and 533-34. In 1850s and 1860s, Thomson detailed one possibility in which the universe began with nothing but gravitational potential, the fragmented materials of the current world not yet having fallen toward one another as they would over vast periods of time. But he did not find such a vision of a meteoric primeval universe congenial. Instead, he also considered, and found more appealing, the idea that the current state of things emerged slowly from a primordial homogeneous distribution of fundamental materials or ether throughout a preexistent space, coagulating over time under the action of gravity—a variety of nebular hypothesis. Ibid., 534. Smith and Wise observe that "Thomson's 'aer' [his ether] performed the same role as J. P. Nichol's nebular matter; likewise his views [as of 1862] on the unity of creation matched those of Nichol." Ibid.

82. On his own view of the crucial dependence of natural selection on uniformitarianism, see Darwin, *On the Origin of Species*, 282. As Burchfield observes, this strong sentiment is preserved in every future edition of *Origin*.

83. Perhaps chief among them were Thomson's friend and collaborator, the mathematical physicist P. G. Tait, and the versatile professor of engineering Fleeming Jenkin, who five years later, in a famed 1867 review of Darwin's *Origin*, extended critiques he had developed in conversation with Thomson. Fleeming Jenkin, "The Origin of Species," *North British Review* 46, no. 92 (June 1867).

84. Burchfield, *Lord Kelvin and the Age of the Earth*, 17.

85. Martin J. Rudwick, *The Meaning of Fossils*, 123.

86. William Thomson, "On the Age of the Sun's Heat," *Macmillan's Magazine*, March 1862. *Macmillan's* was in many ways a good fit. The magazine began its life through booksellers Daniel and Alexander Macmillan who, turning their hand to publishing, focused initially on mathematical and religious texts, numbering among them the *Cambridge Mathematical Journal* in 1846. Fiction did not become a part of the company's catalog until nearly a decade later. See George J. Worth, *Macmillan's Magazine, 1859-1907: "No Flippancy or Abuse Allowed"* (Aldershot, UK: Ashgate, 2003). Worth argues for the continuing influence of Alexander Macmillan over the magazine during the period of Thomson's publication. Ibid., 19.

87. Thomson observes in his own footnote to the same that he is referring here to Helmholtz's theory as articulated in the "Popular lecture delivered on the 7th February, 1854, at Königsberg, on the occasion of the Kant commemoration." William Thomson, "On the Age of the Sun's Heat," *Macmillan's Magazine*, March 1862, 393. Note that Smith and Wise observe that Helmholtz was in part inspired by Thomson's own early 1850s views, expanding on Thomson's ideas and attributing to him the idea of "heat death." Smith and Wise, *Energy and Empire*, 500-501. On the other hand, S. P. Thompson suggests at minimum that (as Smith and Wise would agree) Thomson had long earlier been extrapolating from his conceptions of heat to cosmology, as Helmholtz congratulates him for doing so well. Ibid., 500.

88. In remarking on this choice, Burchfield describes this "in the guise of a popular article." Burchfield, *Lord Kelvin and the Age of the Earth*, 27. It is an intriguing locution, but a guise in what sense? Burchfield is primarily referring here to the circumstances of its publishing. The ideas set forth in the popular article were first presented in the summer of 1861 to a meeting of the British Association, but for various reasons were ignored. See William Thomson, "Physical Considerations regarding the Possible Age of the Sun's Heat," in *Report of the Thirty-First Meeting of the BAAS* (London: John Murray, 1862), 27-28. Note that the article concludes with the same storehouse topos as the more popular piece. The *Macmillan's* article became the format for promoting those more technical ideas, but in a fashion that was rigorously reasoned and argumentatively focused. Apart, however, from questions it raises with regard to the content of the

article and the aptness of the publication, "guise" suggests a stable distinction that may not have existed so clearly at the time and arguably was always less stable than might appear. The distinction is a product of the very same disciplinary forces establishing a professional class of scientists. The opposite face of disciplinary formation is the construction of a layperson, a point to which we will return and which a widening scholarship has addressed. See subsequent n. 89 for examples.

89. For the public fascination with Victorian era science and the demand for popular accounts (with an emphasis on the work of popularizers who were not practitioners), see Lightman, *Victorian Popularizers*, and Aileen Fyfe, *Science and Salvation: Evangelical Popular Science Publishing in Victorian Britain* (Chicago: University of Chicago Press 2004). Fyfe argues for the greater importance of such authors over and above the more scientifically sanctioned popularizers discussed here. For the congeniality of such journals to more rigorous scientific argument, see Richard Yeo, "Science and Intellectual Authority in Mid-Nineteenth-Century Britain," 9.

90. Tyndall, "On Force," 61.

91. Ibid., 62.

92. See Smith, *The Science of Heat*, 180–81.

93. For the relationship between the Darwin's *On the Origin of Species* and the launching of *Macmillan's*, see Gowan Dawson, "The Cornhill Magazine and Shilling Monthlies in mid-Victorian Britain," in *Science in the Nineteenth-Century Periodical: Reading the Magazine of Nature*, ed. Geoffrey N. Cantor et al. (Cambridge: Cambridge University Press, 2004).

94. Thomson, "Age of the Sun's Heat," 391–92.

95. Ibid., 393.

96. Ibid., 388.

97. Ibid. Also, in a sense, the infinite iron bar of Thomson's earlier considerations appears to share a spirit with the potential model of the universe Thomson invokes here.

98. Ibid., 388–89.

99. Smith observes that this passage is consistent with Thomson's "theology of nature." Smith, *Science of Energy*, 173.

100. As Smith has observed, this choice of publication was also significant, its editorial direction under the minister and reformist Norman Mcleod. See ibid., 26–30.

101. William Thomson and P. G. Tait, "Energy," *Good Words* 3 (December 1862): 601. The majority of the article was given to an explanation very like that found in Helmholtz's contemporaneous "On the Conservation of Force," an introduction to a series of lectures given in Karlsruhe beginning at end of 1862. Note that Helmholtz himself sustained the concept of *Kraft* in his lecture, distinguishing between the intensity and the quantity of force. Hermann von Helmholtz, "On the Conservation of Force," *Popular Lectures on Scientific Subjects*, vol. 1 (1895; London: Routledge/Thoemmes, 1996).

102. Thomson and Tait, "Energy," 601.

103. See chapter 9, "North Britain *versus* Metropolis: Territorial Controversy in the History of Energy" in Smith, *The Science of Energy*, esp. 183–84.

104. John Tyndall, "On Force," esp. 64. For an extended analysis of the Tyndall and Tait controversy as "a contest for scientific authority," see Smith, *The Science of Energy*, 170–91. See also Heimann, "Mayer's Concept of 'Force,'" 278. Moreover, as Smith emphasizes, Tyndall's support for Darwin and Spencer and his attempt to align evolution and energy were objectionable to Tate and Thomson.

105. Among the "errors we have undertaken to combat," apart from those of the misunderstanding of conservation principles by the confounding of force and energy, "another refers more to the history than to results of the principle." It is in this that the authors find fault with Tyndall, to whom they allude, but refrain from naming: "We were certainly amazed to find in a recent number of another popular magazine, and in an article specially intended for popular information, that

one great branch of our present subject, which we had been accustomed to associate with the great name of Davy, was in reality discovered so lately as twenty years ago by a German physician." "It especially startles us," they claimed, that Mayer should be given priority above Davy "within the very walls wherein Davy propounded his transcendent discoveries." Thomson and Tait, "Energy," 601, 604.

106. Ibid., 606.

107. Ibid.

108. Ibid., 607.

CHAPTER FOUR

1. In this regard, note John Hedley Brooke, "Natural Theology and the Plurality of Worlds," *Annals of Science* 34 (1977): 281 for "the need for caution when looking for parallels between concepts of astronomical and biological 'evolution.'" Here I differ somewhat from commentators who lay central emphasis on nineteenth-century templates, while nevertheless not disputing the importance of their narrative forms.

2. Hermann von Helmholtz, "The Aim and Progress of Physical Science," in *Selected Writings of Hermann von Helmholtz*, ed. Russell Kahl (Middletown: Wesleyan University Press, 1971), 245. Helmholtz went further to claim that such a nation as Germany has "nothing to fear from the presentation of some hasty and partial theories, even if they should appear to threaten the foundations of morality and society." Ibid. There may be an element here of Helmholtz protesting too much, but to the extent that Helmholtz felt insulated from the "animated controversy" regarding Darwin's work, there was also perhaps a conviction in this cultural difference that might have advanced his work. Among other factors, differences between the British and German contexts in midcentury were partly due to the Chambers controversy waning as it crossed national and German borders. Secord's claim must be kept in mind, that *Vestiges* "had substantial international impact too. It was translated twice into German and twice into Dutch, and went through some twenty editions in the United States, where it sold more copies and had more readers than in Britain." James A. Secord, *Victorian Sensation: The Extraordinary Publication, Reception, and Secret Authorship of* Vestiges of the Natural History of Creation (Chicago: University of Chicago Press, 2000), 38. But in contrast to this, Nicolaas Rupke notes that "on the European mainland *Vestiges* went largely unnoticed." Tracing instead the changing meaning of *Vestiges* in translation, he observes that two translations in German and one in Dutch were the only examples. Nicolaas Rupke, "Translation Studies in the History of Science: The Example of 'Vestiges,'" *British Journal for the History of Science* 33, no. 2 (June 2000): 211. Rupke points out the strong contrasts between the moral drawn from the text, as it were, the very few reviews of the translations, and the fact that, as he observes in passing, "the views of the Vestigiarian failed to engender a major controversy among German academics." Ibid., 220-22.

3. Helmholtz, "The Aim and Progress of Physical Science," 245.

4. Referring to it primarily as the theory of Kant and Laplace, as late as 1871, Helmholtz declared the nebular hypothesis "one of the happiest in science, which at first astounds us, and then connects us in all directions with other discoveries, by which the conclusions are confirmed until we have confidence in them." Hermann von Helmholtz, "On the Origin of the Planetary System," *Popular Lectures on Scientific Subjects*, vol. 2 (1895; London: Routledge/Thoemmes, 1996), 184. But Helmholtz did imagine himself as a beneficiary of a tradition of British science as well, particularly through the more recent figures of Faraday and Thomson. For evidence and some endorsement of this point of view, see Cahan, "Civilizing Power of Science," 588-89 and Heidelberger, "Force, Law, and Experiment: The Evolution of Helmholtz's Philosophy of Science," in David Cahan, ed., *Hermann von Helmholtz and the Foundations of Nineteenth-Century Science* (Berkeley: University of California Press, 1993), particularly 480-81, and passim in the general argument on

the importance of Faraday. For Helmholtz's debt to further Kantian work, see ibid. and, e.g., P. M. Heimann, "Helmholtz and Kant: The Metaphysical Foundations of Über die Erhaltung der Kraft," *Studies in History and Philosophy of Science* 5, no. 3 (1974). Helmholtz and Helmholtz scholars lay particular stress on both Kant and Fichte in relation to Helmholtz's approach to the nature of scientific law (as in his initial considerations of the meaning of physical force laws in his early work on conservation of "force"); in cosmological considerations (the nebular hypothesis, but also folded in this, the theory of tidal breaking); in the indirect debts and his own readings of the programs of his teacher Müller, and through his father's attachment to Fichte. Note as well the treatment of Helmholtz's assault on Lebenskraft, its relation to conservation principles and to Müller's school in Timothy Lenoir, *The Strategy of Life: Teleology and Mechanics in Nineteenth-Century German Biology* (Chicago: University of Chicago Press, 1982), particularly chapter 5, "Worlds in Collision." See also Timothy Lenoir, "The Eye as Mathematician: Clinical Practice, Instrumentation, and Helmholtz's Construction of an Empiricist Theory of Vision," *Hermann von Helmholtz and the Foundations of Nineteenth-Century Science*, esp. 112-13.

5. Helmholtz had a particularly high estimation of Thomson as revealed in his correspondence with his wife and father. See Silvanus P. Thompson, *The Life of William Thomson, Baron Kelvin of Largs*, vol. 1 (London: Macmillan, 1910), 310 and 323-24. Thomson's early biographer Thompson advanced the image of a kind of symbiotic relationship between the scholars, a picture not inconsonant with Smith and Wise's account: Helmholtz articulated the first law, Thomson the second; Helmholtz's memoir on vortex-motion in hydrodynamics led to Thomson's on vortex-atoms and so on. Likewise, in establishing his own conception of solar heat and tying it early on to the nebular hypothesis, Helmholtz found in the Thomson of the early 1850s the resources for extrapolating from terrestrial physics to his cosmological considerations.

6. Anson Rabinbach, *The Human Motor: Energy, Fatigue, and the Origins of Modernity* (New York: Basic, 1990), 55.

7. Hermann von Helmholtz, "On the Interaction of Natural Forces," in *Popular Lectures on Scientific Subjects*, vol. 1 (1895; London: Routledge/Thoemmes, 1996), 138.

8. Ibid., 139.

9. Ibid., 148. In the exposition, the contemporary conception of work/quantity of force, the principle of its conversation, and Carnot's entropic law all emerge from confronting the implications of this impossibility.

10. Ibid., 154.

11. Ibid., 165.

12. Ibid., 153-54. Elsewhere in the essay he asks: "Shall we terrify ourselves by this thought? Men are in the habit of measuring the greatness and the wisdom of the universe by the duration and the profit which it promises to their own race; but the past history of the earth already shows what an insignificant moment the duration of the existence of our race constitutes." Ibid., 170.

13. Ibid., 154.

14. Ibid., 155.

15. Ibid., 160. Helmholtz's father gives a sense of some contemporaneous response to this connection to Mosaic cosmology, singling it out as "the only thing I do not like in your lecture." Leo Koenigsberger, *Hermann von Helmholtz*, trans. Frances A. Welby (New York: Dover, 1960), 121. The mathematician Koenigsberger, a onetime Heidelberg colleague of Helmholtz, stressed the popularity and success of what he referred to as the "most brilliant and the best known of Helmholtz's popular addresses." Ibid., 120.

16. Ibid., 170. It is not *completely* clear how totalizing a linear universal history Helmholtz need have been sketching here; the interest in cosmogony was primarily related "to our system," and others could be forming elsewhere. Some form of cyclicity was possible. Nevertheless, in applying the principles of thermodynamics, Helmholtz's tendency here and elsewhere was to produce a linear and totalizing history.

17. Ibid., 168.

18. Ibid., 171.

19. Helmholtz, "On the Conservation of Force," in *Popular Lectures on Scientific Subjects*, vol. 1, 279.

20. Helmholtz, "The Aim and Progress of Physical Science," 221.

21. Ibid., 226.

22. Ibid., 228–29. For an analysis of the changing meaning of force in Helmholtz's natural philosophy, see Heidelberger, "Force, Law, and Experiment," 480–81. For an analysis of "Helmholtzianism" and its specific theorizations and practices in contrast to Weberian and Maxwellian approaches, see Jed Z. Buchwald, "Electrodynamics in Context: Object States, Laboratory Practice, and Anti-Romanticism," in *Hermann von Helmholtz and the Foundations of Nineteenth-Century Science*. Note Buchwald's discussion of force in relation to conservation laws. Ibid., 368–69.

23. Helmholtz, "The Aim and Progress of Physical Science," 237.

24. Ibid., 244.

25. See for example "On Academic Freedom in German Universities" in Helmholtz, *Popular Lectures on Scientific Subjects*, vol. 2, and "The Endeavour to Popularize Science" in Helmholtz, *Selected Writings*, 335–36.

26. Helmholtz, "The Endeavour to Popularize Science," 337–38. For the exchange between Helmholtz and Zöllner, see Jed Z. Buchwald, "Electrodynamics in Context: Object States, Laboratory Practice, and Anti-Romanticism." Helge Kragh credits Zöllner as "probably the first person who seriously applied the notions of non-Euclidean geometries to astronomy, and indeed to cosmology." See Kragh, *Matter and Spirit in the Universe: Scientific and Religious Preludes to Modern Cosmology* (London: Imperial College Press, 2004), 24.

27. Ibid., 339.

28. For Helmholtz's position among the debating British elites, see David Cahan, "The Awarding of the Copley Medal and the 'Discovery' of the Law of Conservation of Energy: Joule, Mayer and Helmholtz Revisited," *Notes and Records of the Royal Society* 66 (2012), available at http://rsnr.royalsocietypublishing.org/content/early/2011/11/15/rsnr.2011.0045, esp. 127. See also Gregor Schiemann, *Hermann von Helmholtz's Mechanism: the Loss of Certainty: A Study on the Transition from Classical to Modern Philosophy of Nature*, trans. Cynthia Klohr (Dordrecht: Springer, 2009), 164, and Hermann von Helmholtz, "Induction und Deduction," in *Vorträge und Reden* vol. 2 (Braunschweig: Friedrich Vieweg und Sohn, 1903).

29. Helmholtz, "On the Origin of the Planetary System," 157, 187.

30. Ibid., 191–92.

31. Ibid., 192.

32. Ibid., 192–93. Helmholtz later included an addendum to his printed lecture, addressing criticisms by Zöllner on Thomson's theory of panspermia, but also indicating his priority in regard to the idea. See Helmholtz, *Popular Lectures on Scientific Subjects*, vol. 2, 196–97. For a broader context in this regard and for the positioning of Helmholtz's physics in relation to Zöllner's Weberean program, see Buchwald, "Electrodynamics in Context," particularly 370–73.

33. Helmholtz, "On the Origin of the Planetary System," 195.

34. For the response to the address, including Helmholtz's failure to comment and analysis of Haeckel's response, see Gabriel Finkelstein, *Emil du Bois-Reymond: Neuroscience, Self, and Society in Nineteenth-Century Germany* (Cambridge: MIT Press, 2013), chapter 12, "Limits." For the status of this debate in relation to the state of philosophical thought, and for the prominence of Helmholtz's and du Bois Reymond's scientific work and reflections, see Frederick C. Beiser, *After Hegel* (Princeton: Princeton University Press, 2014).

35. T. H. Huxley, "Administrative Nihilism," in *More Criticisms on Darwin and Administrative Nihilism* (New York: D. Appleton & Co., 1872), 58–59. Note however Stanley's observation that such appeals to the politically stabilizing force of education were not for Huxley in the ascendancy.

Matthew Stanley, *Huxley's Church and Maxwell's Demon: From Theistic Science to Naturalistic Science* (Chicago: University of Chicago Press, 2015), 143. For discussion of Huxley's views on education, particularly in reference to his relationship to the Working Men's College, and the consonance of his educational vision to the very differently situated and evangelical Maxwell, see "The Goals of Science Education: The Working Men's College" in ibid. Note however that in contrast to "those who claimed science as a weapon in favor of revolutionary politics or against Christianity," Maxwell "explicitly framed his own science teaching as a remedy for those dangers." Ibid., 137.

36. Huxley, "Administrative Nihilism," 62, 69–71. He found defenses of this position in Alexander von Humboldt's famed brother Wilhelm's and in J. S. Mill's "distrust of legislative interference." Ibid., 70. He countered the further defenses of his X-Club colleague Spencer, who based his position instead on comparing the development of "the body physiological and the body politic." Ibid., 70–71.

37. Ibid., 75–76.

38. From biographical assessments of Thomson's geological work by the geologist J. W. Gregory in 1908, to a treatment closely following Gregory by the biographer S. P. Thompson, through to the more contemporary historical analyses of Joe D. Burchfield, Crosbie Smith, and M. Norton Wise, and others, the extent of Thomson's success in reframing geological and evolutionary arguments has itself been debated. But the large part of the analysis of the period appears to endorse the significance of his role, if particularly perhaps because Darwin himself appeared so susceptible to Thomson's views. For a recent analysis of the debate itself, in reference in particular to shared commitments to the uniformity of natural laws across an emergent divide of naturalistic and theistic science, see Stanley, *Huxley's Church and Maxwell's Demon*, 61–68.

39. William Thomson, "On the Secular Cooling of the Earth," *London, Edinburgh and Dublin Philosophical Magazine and Journal of Science*, 4th series, 25, no. 165 (London: Taylor and Francis, 1863), 1.

40. One of three such, according to Thompson. Thompson, *The Life of William Thomson*,, vol. 1 (London: Macmillan, 1910), 436.

41. William Thomson, "The 'Doctrine of Uniformity,' in Geology Briefly Refuted," *Proceedings of the Royal Society of Edinburgh* 5 (November 1862–April 1866) (Edinburgh: Printed by Neill & Company, 1886), 512.

42. William Thomson, "On Geological Time," in *Popular Lectures and Addresses*, vol. 2, *Geology and General Physics* (London: Macmillan, 1894), 55.

43. Ibid., 10.

44. Ibid., 11–12.

45. Ibid., 33. Thomson adds: "That there is such a tendency has long been known to philosophers of the more abstract kind. It is difficult to say who first promulgated this idea. It has been recently stated that the metaphysician Kant first asserted that the earth's rotation is diminished through the influence of the tides." Ibid. Thomson betrayed little commitment to or interest in the Kantian tradition generally. The idea of tidal breaking was, he emphasized, first presented to him by his brother James. Ibid.

46. Ibid., 44.

47. J. W. Gregory in 1908, in relation to this passage, saw in Thomson's own view less need to invoke catastrophism at either the beginning or the end of his history. Thomson's contradiction of Lyell/Hutton's position, Gregory argues, is "quite inconsistent with the true doctrine of Uniformity. The view that the earth is progressing in an orderly development from its beginning till its end shall come gradually from the exhaustion of its stores of energy, is a grander conception of geological uniformity than belief in unchanging conditions persisting between a sudden beginning, of which Nature retains no impression, and a catastrophic end, of which it gives no hint." Gregory, "Lord Kelvin's Contribution to Geology," *Transactions of the Geological Society of Glasgow* 13, part 2 (1906, 1907, 1908) (Glasgow: Published by the Society, 1908), 176–77. In turn, Gregory himself testifies to the extent to which Thomson's vision of natural laws was understood by the turn of the century

to be as immanent as Hutton's, since the laws themselves appeared to obtain without any transcendental interference. See also Stanley, *Huxley's Church and Maxwell's Demon*, 67.

48. T. Huxley, "Geological Reform," *Quarterly Journal of the Geological Society, London* 25 (1869): xxxviii.

49. Ibid., xlviii.

50. Ibid., xlvii.

51. Ibid., xxxix.

52. Ibid., xli–xlii.

53. Ibid., xlvi.

54. Ibid., xliii–xliv.

55. Ibid., xlvii.

56. Smith and Wise contend in this regard that "Huxley had been anxious to emphasize that a limited time scale—one hundred million years or more—'may serve the needs of geologists perfectly well.' Thus, although eager to point out any possible defects in Thomson's case, he was in fact accepting the probability of a much restricted time scale." Crosbie Smith and M. Norton Wise, *Energy and Empire: A Biographical Study of Lord Kelvin* (Cambridge: Cambridge University Press, 1989), 538.

57. Huxley, "Geological Reform," xlix. For our purposes here, short work can be made of the additional ripostes in the debate. In April of the following year, before the Geological Society of Glasgow, Thomson responded forcefully to Huxley, conceding nothing substantive from his original arguments. He indicted Huxley for missing the central point of his work, that geology could not function as if it were indifferent to physics, and cited Huxley himself in the failure to appreciate this point. Thomson observed that the narrowing of the interval of historical time his calculations precipitated, despite their uncertainty, would still demand significant reform. He saw this belief as shared with prominent geologists. Moreover, in a significant aside, his calculations seemed to him "sufficient to disprove the doctrine that transmutation had taken place through 'descent with modification by natural selection.'" William Thomson, "Of Geological Dynamics," in *Popular Lectures and Addresses*, vol. 2, 89–90.

58. Ibid., 77–78.

59. Thompson in this regard laid particular stress on how Thomson and Huxley came to civil exchange two years later, when Huxley resigned his presidency of the BAAS and introduced Thomson as its new president. Thompson, *Life of Thomson*, vol. 1, 550–51.

60. Thomson discussed panspermia, with life originating from life outside the earth. Huxley, who in correspondence with Hooker characterized Thomson's idea as "creation by cockshy" (i.e., creation by target practice) himself looked to the protoplasm as the possible key to life. He had already singled out the protoplasm in debate with Thomson, and his more extended reflections suggested to an exercised audience (despite his agnostic care) that in his view life itself definitively evolved from nonliving materials. For a discussion of the extensive response to Huxley's views, particularly his "On the Physical Basis of Life," and Huxley's larger role in managing questions of the ultimate origins of life, see James E. Strick, *Sparks of Life: Darwinism and the Victorian Debates over Spontaneous Generation* (Cambridge, MA: Harvard University Press, 2000). Note also Thomson's references to this address in his own the following year. William Thomson, "Presidential Address" in *Report of the 41st Meeting of BAAS held at Edinburgh in 1871* (London: John Murray, 1872), ciii–civ, and Strick's comments in reference to this exchange, Strick, *Sparks of Life*, 21–22. For Tyndall as well universal history was to embrace the origins of life. In correspondence with Pasteur and partly by way of experiments on light, Tyndall plunged into the question of the nature of life, finding that much of the dust in the air was organic and, at a later point, that bacteria could show a heat resistance, impacting the question of sterilization and contaminants that might remain in contemporary experiments attempting to demonstrate spontaneous generation. Once again, it might seem that Thomson was on one side of a question, with outspoken members of the X-Club

Huxley and Tyndall on the other. Strick's extended analysis of Tyndall's and Huxley's roles (and his fuller analysis of the spontaneous generation debate) has shown the extent the which Huxley had initially to remain more circumspect on the question of spontaneous generation and its relationship to evolutionary theory than the still more outspoken Tyndall. Ibid., 79.

61. Ruth Barton, "John Tyndall, Pantheist: A Rereading of the Belfast Address," *Osiris*, 2nd series, 3 (1987): 113. Note that Stanley claims in explanation of the angry response of "theistic scientists" in particular that "it appeared to the theists that Tyndall was trying to use his position as president of the BAAS to enforce his naturalism: precisely the sort of institution-based coercion of belief that both parties [theists and naturalists] had agreed was antithetical to science." Stanley, *Huxley's Church and Maxwell's Demon*, 189.

62. As suggested by historians Barton, Frank M. Turner, and Ursula DeYoung.

63. Stephen Shin Kim, "Fragments of Faith: John Tyndall's Transcendental Materialism and the Victorian Conflict between Religion and Science" (PhD diss., Drew University, 1988).

64. Ursula DeYoung, *A Vision of Modern Science: John Tyndall and the Role of the Scientist in Victorian Culture* (New York: Palgrave Macmillan, 2011), 65.

65. Tyndall cited Bacon in this regard: "For at a time when all human learning had suffered shipwreck these planks of Aristotelian and Platonic philosophy, as being of a lighter and more inflated substance, were preserved and came down to us, while things more solid sank and almost passed into oblivion." John Tyndall, *Address Delivered before the British Association Assembled at Belfast: With Additions* (London: Longmans, Green, 1874), 3. The citation comes from Bacon's "On Principles and Origins according to the Fables of Cupid and Coelum." See Francis Bacon, *The Works of Bacon*, vol. 5, ed. James Spedding et al. (London: Longman and Co., 1858), 466. Note that here Bacon argues that the fact that the elder mythic Cupid is represented as parentless may be "the greatest thing of all": "For nothing has corrupted philosophy so much as this seeking after the parents of Cupid; that is, that philosophers have not taken the principles of things as they are found in nature, and accepted them as a positive doctrine, resting on the faith of experience; but they have rather deduced them from the laws of disputation, the petty conclusions of logic and mathematics, common motions, and such wanderings of the mind beyond the limits of nature. Therefore a philosopher should be continually reminding himself that Cupid has no parents, lest his understanding turn aside to unrealities; because the human mind runs off in these universal conceptions, abuses both itself and the nature of things, and struggling towards that which is far off, falls back on that which is close at hand." Ibid., 463.

66. Turner refers to Tyndall's philosophy of science as sharing with Huxley and Spencer, among others, in a broader "scientific naturalism" that he characterizes as "naturalistic because they referred to no causes not present in empirically observed nature and . . . scientific because they interpreted nature through three major mid-century scientific theories. These were the atomic theory of matter, the conservation of energy, and evolution." Frank M. Turner, "John Tyndall and Victorian Scientific Naturalism," in *John Tyndall, Essays on a Natural Philosopher*, ed. W. H. Brock, N. D. McMillan, and R. C. Mollan (Dublin: Royal Dublin Society, 1981), 174. See also Frank M. Turner, *Between Science and Religion: The Reaction to Scientific Naturalism in Late Victorian England* (New Haven: Yale University Press, 1974). Note Turner's claim that "what Huxley, Spencer, [mathematician W. K.] Clifford and other naturalistic writers said about the nature of truth was not original. The roots of their opinions lay in the empirical philosophy of the English Enlightenment and in the works of Comte and J. S. Mill." Ibid., 18.

67. For connections between Tyndall's previous sentiments as to the limits of science and those of du Bois-Reymond, see Finkelstein, *Emile du Bois-Reymond*, 283. Note that along with Helmholtz and Huxley, Tyndall mentions du Bois-Reymond in the address itself, as one who in his "less technical writings" demonstrates a "breadth of literary culture." Tyndall, *Address Delivered before the British Association Assembled at Belfast*, 62.

68. Barton, "John Tyndall, Pantheist," 121.

69. Tyndall, *Address Delivered before the British Association Assembled at Belfast*, 54.

70. Ibid., 54, 55.

71. Ibid., 55.

72. Ibid.

73. Ibid., 58.

74. Ibid., 59.

75. Ibid., 61.

76. Ibid., 61. Note that this is also the form of the statement appearing in John Tyndall, "Address by the President" in *Report of the Forty-Fourth Meeting of the British Association for the Advancement of Science Held at Belfast in August 1874* (London: John Murray, 1875), xcv.

77. Tyndall, *Address Delivered before the British Association Assembled at Belfast*, xi.

78. Ibid.

79. Ibid., xi-xii.

80. Ibid., xii.

81. Ibid., 36.

82. Ibid., 36-37.

83. With his "On the Physical Basis of Life," Huxley had already faced accusations that his views were marked by uncomplicated materialism, also on the question of life, the protoplasm, and spontaneous generation. See Strick, *Sparks of Life*. The agnosticism marking his views itself made claims for the polite reach of secular reasoning. Ibid., 79.

84. T. H. Huxley, "Science and Culture," in *Science and Culture and Other Essays* (London: Macmillan and Company, 1882): 18.

85. Ibid., 9.

86. Matthew Arnold, "Literature and Science," in *Discourses in America* (London: Macmillan and Company, 1885), 112. Note that the debate between Huxley and Arnold has been seen by some commentators as a forbearer of the two cultures debate of the following century. See for example Stefan Collini's introduction in C. P. Snow, *The Two Cultures: With Introduction by Stefan Collini* (Cambridge: Cambridge University Press, 1998).

87. T. H. Huxley, *Evolution & Ethics and other Essays* (London: Macmillan and Co., 1894), 80.

88. Ibid., 81-82.

89. Huxley and Tyndall were not the only ones to endorse the picture of cosmic evolution: the account of material to organic evolution was finding itself in a number of British practitioner-popularizers, apart from Tyndall, perhaps most notably the astronomer Robert Ball and, if a "liminal figure" in his status as "one who straddled the worlds of the practitioner and the science writer," Richard Proctor. Lightman, *Victorian Popularizers*, 299, and, for the discussion of Ball, 397-421. For Proctor's adherence to "cosmical and biological evolutions" indebted, as he saw it, to Spencer, see ibid., 314, and Proctor's wider activity, see chapter 6, "The Science Periodical: Proctor and the Conduct of 'Knowledge,'" in ibid. Their collective work drew on generic templates circulating earlier in the century, whether from Nichol, Chambers, or also from Helmholtz, partly by way of Tyndall's translations. Though consensus had not been reached with Thomson, conviction over cosmic evolution was emerging in the positions of forceful popularizers such as Huxley, Tyndall, and Ball. With that conviction came the grounds for taking seriously ethical anxieties like Huxley's.

90. Adrian Desmond notes how Huxley at this time and in this lecture struggled to articulate a vision that avoided the individualism of Spencer or the collectivism of Kropotkin or Wallace. Desmond, *Huxley: From Devil's Disciple to Evolution's High Priest* (Reading, MA: Addison-Wesley, 1994), 599. Huxley was also still setting the record straight regarding the points of agreement and disagreement between Thomson and himself. In a preface to an 1894 volume of his collected essays, straining the historical record, he claimed "no one can have asserted more strongly than I have done, the necessity of looking to physics and mathematics for help in regard to the earliest history

of the globe." T. H. Huxley, *Discourses: Biological and Geological, Collected Essays*, vol. 8 (London: Macmillan, 1894), ix. In the same preface, he remarked: "I have not been one of those fortunate persons who are able to regard a popular lecture as a mere *hors d'œuvre*, unworthy of being ranked among the serious efforts of a philosopher; and who keep their fame as scientific hierophants unsullied by attempts—at least of the successful sort—to be understanded of the people." Ibid., v.

91. Hermann von Helmholtz, "The Endeavor to Popularize Science: Preface to the German Translation of John Tyndall's *Fragments of Science*," in *Selected Writings*, 331.

92. Ibid., 332.

93. For Mach, as for Huxley or Arnold, such an opposition was not simply a straightforward rejection of the virtues of the opposing educational program. See John T. Blackmore, *Ernst Mach: His Work, Life and Influence* (Berkeley: University of California Press, 1972), esp. chapter 10, "Educational Theory and Textbooks."

94. For distinctions between Mach's and other forms of positivism, see ibid., 164. See also Ernst Mach, *Contributions to the Analysis of Sensations*, trans. C. M. Williams (1890; Chicago: Open Court, 1897), esp. "Introductory Remarks Antimetaphysical."

95. Blackmore, *Ernst Mach*, 105. For the introduction of the term "Mach number," see ibid., 112. Mach saw his own awakening to causal thought in a visit to a water mill, a more placid technology that Helmholtz, for one, had appealed to among his figures/technologies demonstrating the conservation of energy/force.

96. Blackmore observes: "In a sense, Mach's experimental discovery supported the original accusation, namely, that the French Chassepôt bullets caused illegal 'crater-like wounds.' But since it was not true that they included an explosive charge, the probable explanation lay in their high velocity, and of course, in the shape of the projectile." Blackmore, *Ernst Mach*, 112.

97. See Mach, *Contributions to the Analysis of the Sensations*, 19-21.

98. It should be noted that Newton's own preferred cosmogony changed across his life. For an analysis of this changing cosmogony, see David Kubrin, "Newton and the Cyclical Cosmos: Providence and the Mechanical Philosophy," *Journal of the History of Ideas* 28, no. 3 (July–September 1967).

99. See Stephen G. Brush, "Poincaré and Cosmic Evolution," *Physics Today* 33, no. 3 (1980) and June Barrow-Green, *Poincaré and the Three Body Problem* (Providence: American Mathematical Society, 1997).

100. Henri Poincaré, "On the Stability of the Solar System," *Nature* (June 23, 1898): 183.

101. For the details of this episode, see Peter Galison, *Einstein's Clocks, Poincaré's Maps: Empires of Time* (New York: W. W. Norton, 2003), esp. 62-74, and Barrow-Green, *Poincaré and the Three Body Problem*.

102. The outcomes of Poincare's mathematics would suggest that unstable orbits, in the sense of those that would never return near their current positions, formed (in the language of an emergent mathematics) a set of measure zero: the probability was effectively nil.

103. For quote and analysis, see Brush, "Poincaré and Cosmic Evolution," 45-46. It was, in a way, an option also considered two decades later by Eddington. See the following chapter.

104. Poincaré, "On the Stability of the Solar System," 185.

105. Ernest Rutherford, "Radium—The Cause of the Earth's Heat," *Harper's Magazine*, February 1905, 390.

106. Ibid., 391. Historians give some partial assent to this claim, narrowly understood. For ways in which confidence in Thomson's account was beginning to erode prior to the acceptance of radiation, see Smith and Wise, *Energy and Empire*, 544, for Thomson's debate with his former student John Perry. Note also in this regard, David Bloor: "By the last decade of the century the geologists had plucked up courage to tell Kelvin that he must have made a mistake. This new found courage was not because of any dramatic new discoveries, indeed, there had been no real change in the evidence available. What had happened in the interim was a general consolidation in geology

as a discipline with a mounting quantity of detailed observation of the fossil record." David Bloor, *Knowledge and Social Imagery* (1976; Chicago: University of Chicago Press, 1991), 6.

107. Burchfield quotes Rutherford's description of an address to the Royal Institute in 1904: "I came into the room, which was half dark, and presently spotted Lord Kelvin in the audience and realized that I was in for trouble at the last part of the speech dealing with the age of the earth, where my views conflicted with his. To my relief, Kelvin fell fast asleep, but as I came to the important point, I saw the old bird sit up, open an eye and cock a baleful glance at me! Then a sudden inspiration came, and I said Lord Kelvin had limited the age of the earth, provided no new source of heat was discovered. That prophetic utterance refers to what we are now considering tonight, radium! Behold! the old boy beamed upon me." Burchfield, *Lord Kelvin and the Age of the Earth*, 164.

108. Soddy repeated the same phrase as well, but in reference instead to Tait's treatment of the heat of the sun: "The almost prophetic qualification [of Tait] with regard to 'chemical agencies of a far more powerful order than anything we meet with' will be appreciated at the present time." See Frederick Soddy, *Radio-activity: An Elementary Treatise from the Standpoint of the Disintegration Theory* (England: "Electrician" Printing and Publishing Company Ltd, 1904), 185. Here Soddy subjects to critique the major planks of Thomson's age-of-the-world arguments, internal heat, tidal retardation, and the age of the sun. Ibid., 183–85. He also presents the emergence of the study of radioactivity as the product of a harmonized relationship between chemistry and physics as represented by Marie Curie and Rutherford, respectively. Ibid., iii.

109. William Thomson, "On the Age of the Sun's Heat," *Macmillan's Magazine*, March 1862, 393.

110. Rutherford, "Radium," 393.

111. In the lecture "On Force," Tyndall had stated dramatically: "This pound of coal, which I hold in my hand, produces by its combination with oxygen an amount of heat which, if mechanically applied, would suffice to raise a weight of a 100 lbs. to a height of 20 miles above the earth's surface." Tyndall, "On Force," 60.

112. For an analysis of the cultural reaction to radium, see Lawrence Badash, "Radium, Radioactivity, and the Popularity of Scientific Discovery," *Proceedings of the American Philosophical Society* 122, no. 3 (June 9, 1978) and, with an emphasis on Soddy, see Luis Campos, "The Birth of Living Radium," *Representations* 97, no. 1 (Winter 2007).

113. Smith and Wise, *Energy and Empire*, 551.

114. An example of this touch of magic almost seeming to defy thermodynamic constraints was evident in Rutherford's description of the properties of radium: "In addition to its penetrating radiations, radium also emits heat at a comparatively rapid rate. Curie and Laborde recently observed the striking fact that a piece of radium always keeps itself at a temperature of several degrees above the surrounding air. The amount of heat emitted from radium is sufficient to melt more than its weight of ice per hour. This rate of heat emission is continuous and, so far as observation has gone, does not decrease appreciably with the time. In the course of a year, one pound of radium would emit as much heat as that obtained from the combustion of one hundred pounds of the best coal, but at the end of that time the radium would apparently be unchanged and would itself give out heat at the old rate. It can be calculated with some confidence that, although the actual amount of heat per year to be derived from the radium must slowly decrease with the time, on an average it would emit heat at the above rate for about one thousand years." Rutherford, "Radium," 393. For further analysis, see Smith and Wise, 609, and Burchfield, 169–70.

115. Soddy, *Radio-activity*, v.

116. Ibid., 188–89. Soddy endorsed the first law: "The essential law of the universe may be that the total quantity of energy is constant." But, having already reinvoked Maxwell's demons, Soddy put to use one of the possible lessons from it, "that what we call low-grade and high-grade energy is the expression of the limited means at our disposal for utilising it." Ibid., 189.

117. Ibid., 189. Also, "If we lived in a sub-atomic instead of a molecular world possibly the significance of the terms might be reversed. The universe would then appear as a conservative

system, limited with reference neither to the future nor the past, and demanding neither an initial creative act to start it nor a final state of exhaustion as its necessary termination." Ibid., 189. Note that Campos claims Soddy became "well-known for giving a number of well-attended public lectures in London and Cambridge whose recurring theme was the immense storehouse of energy." Campos, "Living Radium," 12.

118. Rutherford, "Radium," 396.

119. A Darwinian vision that even Kelvin could endorse: Thomson, "Presidential Address," cv.

CHAPTER FIVE

1. See Stephen G. Brush, "A Geologist among Astronomers: The Rise and Fall of the Chamberlin-Moulton Cosmogony," *Journal of the History of Astronomy* 9 (1978), and J. D. North, *The Norton History of Astronomy and Cosmology* (New York: W. W. Norton, 1995), 462–63.

2. And in the Anglo-American context, the natural sciences were increasingly treated as the *only* sciences. For contemporary expressions such as "emergent men of the new time," "these coming men," "the men of the new republic," and "the new mass of capable men" in relation to the future of science, technology, and society, see H. G. Wells, *Anticipations of the Reaction of Mechanical and Scientific Progress upon Human Life and Thought* (New York: Harper & Brothers, 1901). Here, Wells "anticipated" a future world super-state and took positions consonant with those his idolized former instructor Huxley maintained in his debate with Arnold, predicting that the emergent techno-scientific republic would be run by active, critical new men, recalling nineteenth-century images of the vigorous, surgent middle classes. The citizens of his new republic would be educated through new critically minded school systems, producing workers unencumbered by the stale humanist educations preceding them and able to build, adapt, and use new technologies.

3. See Peter J. Bowler, *Evolution: The History of an Idea* (Berkeley: University of California Press, 2003), esp. chapter 7, "The Eclipse of Darwinism: Scientific Evolutionism, 1875–1925." As Bowler observes, the phrase "the eclipse of Darwinism" originates with Julian Huxley.

4. Perhaps the majority of the central histories of modern cosmology and universal expanse begin with Einstein and with the Russian mathematician Friedmann. The literature on the subject is extensive, and I will not revisit it here or below. Instead, I want to attend to only a few moments in their relation to the questions of syntheses in formation, historical and otherwise. The emphasis in this chapter is both on their work as a kind of historiography of the natural world and a historiography initially rejecting a global/universal history. This in turn helps situate the historical syntheses of others, such as like Shapley, Lemaître, and from the 1930s, George Gamow.

5. As seen in the previous chapter, a version of cosmic evolution had already been explicitly articulated by Huxley, and advanced by others such as Helmholtz, but it was not a consensus scientific view in the early part of the twentieth century.

6. And as was the case for Haldane, for Shapley the ethics recommended or contested by the materialistic underpinnings of organic life would become the basis for a central part of his broader activity. See the next chapter.

7. Peter Galison, "Kuhn and the Quantum Controversy," *British Journal for the Philosophy of Science* 32, no. 1 (March 1981): esp. 74.

8. Stephen G. Brush, *The Kind of Motion We Call Heat* (Amsterdam: North Holland, 1976), 672–701.

9. Einstein's first paper in the journal *Annalen der Physik*, facilitated by Planck, who was then its editor, overtly suggested the quantization of energy, the treatment of energy in terms of discrete packets. The paper described this treatment as heuristic, but thereby helped advance the statistical idiom. See Gerald Holton, "Origins of the Special Theory of Relativity," in Holton, *Thematic Origins of Scientific Thought: Kepler to Einstein* (Cambridge, MA: Harvard University Press, 1988),

esp. 192–94. Whether Einstein's work actually followed that of Mach was subject to debate, then and now, though he self-consciously moved increasingly to a Planckian position. Note further that Einstein was famously dismissive of "superfluous" ether concepts. In 1905, in the third of his celebrated *Annalen der Physik* papers of that year, he set himself against the notion of a luminiferous ether as part of his elimination of "absolute stationary space" and postulation of the principle of relativity. Once again the stability of the world picture and the question of the ether were interrelated, if now in less direct terms than in the past. But whatever the role of the critique of the notion of the ether might have been in Einstein's thought, the overall perspective began to suggest that enough was amiss with dominant nineteenth-century worldviews as to merit significant reinterrogation.

10. The proto-hylozoistic matter of that other champion of atomic theory, Tyndall, would however be sifted by other hands. For a recent analysis of Maxwell's demon in relation to his beliefs concerning consciousness and free will, see Matthew Stanley, *Huxley's Church and Maxwell's Demon: From Theistic Science to Naturalistic Science* (Chicago: University of Chicago Press, 2015), 233–40.

11. Even as late as 1930, the astrophysicist and defender of relativity A. S. Eddington noted that there were those who believed that time itself is like a convention, comparing it to the directions "left" and "right." Such a view, which he disputed, would render the notion of universal history a thoroughly subjective one.

12. Reasoning over whether science itself was independent of the time and place in which it was articulated revealed opposing stances. Planck himself, originally an admirer of Mach, found by 1908 Mach's scientific exposition unacceptably anthropomorphic, threatening Planck's own embrace of a timeless unity of worldview promised by the physical sciences. To Planck's mind, human natural sciences should be related to the same truth as that confronting the science of the Martians, regardless of which civilization was more advanced. Max Planck, "The Unity of the Physical Universe," in *A Survey of Physical Theory*, trans. R. Jones and D. H. Williams (New York: Dover, 1993), 18. Whatever the point of view, from Mach through to Planck, the principles for unity, synthesis, and history related through natural scientific knowledge were all being contested together.

13. For an assessment of Einstein's production in this period, see Abraham Pais, *Subtle Is the Lord—The Science and Life of Albert Einstein* (Oxford: Oxford University Press, 2005), 242.

14. And indeed it was partly the force of Hubble's discoveries of the recession of the galaxies that, thirteen years later, finally convinced Einstein to abandon that "quasi-stability." Note that Pais indicates Einstein's particular interest in coming to know astronomers in the move to Berlin. See Pais, *Subtle Is the Lord*, 240.

15. On the importance of symmetry and the continuity of Einstein's work, see Gerald Holton, "The Origins of the Special Theory of Relativity" and "On Trying to Understand Scientific Genius," in *Thematic Origins of Scientific Thought*.

16. Philipp Frank explains this case in reference to Einstein's cosmological reflections. Such a model would, in that endless region of flat space, conflict with "Mach's principle," that inertial motion is determined through its relation to the "fixed stars." Note that Frank here cites Shapley's research and that of his collaborators demonstrating that "matter does *not* form an island in infinite empty space" and that it is "plausible to assume with Einstein that on the average the entire universe is uniformly filled with matter." Philipp Frank, *Einstein: His Life and Times*, trans. George Rosen (New York: De Capo, 1953), 135–36.

17. Albert Einstein, "Cosmological Considerations on the General Theory of Relativity," reprinted in H. A. Lorentz, A. Einstein, H. Minkowski, and H. Weyl, *The Principle of Relativity: A Collection of Original Memoirs on the Special and General Theory of Relativity*, trans. W. Perrett and G. B. Jeffery (New York: Dover, 1923), 188.

18. Ibid., 188.

19. Ibid., 179–80. The language Einstein uses here testifies to his readiness to play with the literary forms of scientific publication, a point that must be balanced against the fact that the paper

was first presented to and published as part of the minutes to a meeting of the Prussian Academy of the Sciences. With regard to these literary technologies, and their role in Einstein's seminal 1905 papers, in particular, see Peter Galison, *Einstein's Clocks, Poincaré's Maps: Empires of Time* (New York: W. W. Norton, 2003). And the idea of exploration, both physical and conceptual, began to take increasingly greater prominence in the literature on the scientific history of the world. See also Matthew Stanley, "So Simple a Thing as a Star: The Eddington-Jeans Debate over Astrophysical Phenomenology," *BJHS* 40, no. 1 (March 2007): 81. See also Matthew Stanley, *Practical Mystic: Religion, Science, and A. S. Eddington* (Chicago: University of Chicago Press, 2007).

20. Ibid., 183–84. The analogy with the surface of the earth itself suggests the possibility of a finite, yet open space. That we find exploratory and topographical metaphors here in the technical solution is in many ways unsurprising. The mathematics in which Einstein develops the theory is itself the mathematical surveying language introduced the century before, the language of charting paths, of computing the nature of surfaces, of exploring the shape of the world.

21. Ibid., 178. Frank observes that this question raised problems in the late nineteenth century. Frank, *Einstein*, 135. See also Philipp Frank, "Cosmological Difficulties with the Newtonian Theory of Gravitation," in J. D. North, *The Measure of the Universe: A History of Modern Cosmology* (New York: Dover, 1990), 16–23.

22. As Helge Kragh states it, "The solution considered by Einstein was static in the sense that the curvature of the universe was independent of time." Helge Kragh, "The Beginning of the World: Georges Lemaître and the Expanding Universe," *Centaurus* 32, no. 2 (July 1987): 115.

23. There does appear to be an affinity between this desire for a stable universe and Einstein's desire for stable, universal political orders. As a biographical matter, Einstein drew affinities between his physical and political views (even if he possibly attested these facetiously). See, for example, Jimena Canales, "Einstein, Bergson, and the Experiment that Failed: Intellectual Cooperation at the League of Nations," *Modern Language Notes* 120, no. 5 (2005).

24. Rallying support for his steady-state cosmology, Fred Hoyle later argues for this understanding of the principle. See Fred Hoyle, *The Nature of the Universe* (New York: Harper and Brothers, 1950), 124.

25. George Gamow comments with regard to the cosmological constant that "true, the new term had a rather strange physical interpretation, representing a repulsive force which increases with the distance between the two objects and depends on the mass of only one of them. But nothing was too much to save the universe!" George Gamow, *My World Line: An Informal Autobiography* (New York: Viking Press, 1970), 44.

26. In his treatment of one of the founding figures of modern geology, James Hutton, Stephen Jay Gould observes that a perfect cyclical pattern of change in the world vacates the possibility of history. Stephen Jay Gould, *"Time's Arrow, Time's Cycle: Myth and Metaphor in the Discovery of Geological Time* (Cambridge: Harvard University Press, 1987). For further discussion, see Ilya Kliger and Nasser Zakariya, "Organic and Mechanistic Time and the Limits of Narrative," *Configurations* 15, no. 3 (Fall 2007).

27. Kragh, "The Beginning of the World," 115.

28. Einstein, "Cosmological Considerations on the General Theory of Relativity," 188.

29. For a critical survey of the various contenders to the title of discoverer of an "expanding universe," including Edwin Hubble (with and without his collaborator Milton Humason), Vesto Slipher, Willem de Sitter, Alexander Friedmann, Georges Lemaître, and to a lesser extent, Howard Percy Roberston and Richard Chase Tolman, see Helge Kragh and Robert W. Smith, "Who Discovered the Expanding Universe?," *History of Science* 41, no. 2 (2003). Despite the critical evaluation of the notion of discovery itself, it is Lemaître that emerges in the article as the largely unsung hero of universal expansion. Kragh and Smith's work in this and in their other texts demonstrates the difficulty of choosing specific historical actors to represent a change in a collective understanding. Here, my attempt is to attend to certain contributors to that change primarily in regarding them as privileged witnesses to shifts in the formation of scientific history and synthesis both. For analysis

of the question of discovery specifically in the context of astronomy, see Steven Dick, *Discovery and Classification in Astronomy: Controversy and Consensus* (Cambridge: Cambridge University Press, 2013).

30. Harlow Shapley, *The Orbits of Eighty-Seven Eclipsing Binaries—A Summary* (Princeton, NJ: Princeton University, 1913). Eclipsing binaries are gravitationally bound stars in orbit around their center of mass that pass in front of each other as seen from Earth.

31. Helen Sawyer Hogg, "Harlow Shapley and Globular Clusters," *Proceedings of the Astronomical Society of the Pacific* 77 (1965): 337.

32. This was true of "Kapteyn's Universe," in particular, as proposed by the Dutch astronomer Jacobus Kapteyn. See A. Blaauw, s.v. "Kapteyn, Jacobus Cornelius," *Dictionary of Scientific Biography*, vol. 7 (New York: Charles Scribner's Sons, 1981). Kapteyn was an energetic and program-building astronomer working mainly on stellar distributions and motions. He also apparently admired Shapley, recommending him first and foremost in a letter to George Ellery Hale in what Owen Gingerich characterized as "the opening salvo in what would become a protracted struggle to fill the Harvard position left vacant by [Edward Charles] Pickering's unexpected death." Owen Gingerich, "How Shapley Came to Harvard, or Snatching the Prize from the Jaws of Debate," *Journal of the History of Astronomy* 19, no. 3 (1988): 201. Hale himself, it should be added, was ultimately somewhat vague in regard to Shapley's appointment, observing that he would be unwilling to see Mount Wilson under Shapley's direction. Ibid., 203.

33. See Heber D. Curtis and Harlow Shapley, "The Scale of the Universe," *Bulletin of the National Research Council* 2 part 3, no. 11 (May 1921): 172. M. A. Hoskin demonstrates that the published version is a heavily revised version of the original debate and includes reproductions of the final ten pages of Shapley's sixteen-page address (according to Shapley's original typescript). See M. A. Hoskin, "The 'Great Debate': What Really Happened," *Journal of the History of Astronomy* 7 (1976): 169, and its appendix, the text, "Shapley's Washington Address," 175–78. This version of the address contains further statements concerning the implications of the scale of the universe for the self-conception of humanity: "If man had reached his present intellectual position in a later geological era, he might not have been led to these vain conceits concerning his position in the physical universe, for the solar system is going rapidly away from the center of the local cluster. . . . Remembering these delusions, relative to his physical status in the universe, may we not appropriately ask if man is also biologically blindfolded? Does he, perhaps, hold his self-assumed and self-defined position at the peak of animal development as a victim of psychological fallacy?" Ibid., 177. See ibid., note 12, for comparison with the (in this instance) close printed version. Owen Gingerich has also pointed out that the published version is a heavily revised and transformed account of the debate. See Owen Gingerich, s.v. "Shapley, Harlow" *Dictionary of Scientific Biography*, vol. 12 (New York: Charles Scribner's Sons, 1975). Shapley also makes sure to underscore his overall point about humanity's vanity and marginality: "If man had reached his present intellectual position in a later geological era, he might not have been led to these vain conceits concerning his position in the physical universe, for the solar system is rapidly receding from the galactic plane, and is moving away from the center of the local cluster." Curtis and Shapley, "The Scale of the Universe," 192.

34. For an extended treatment of the relationship between Eddington's science and his Quakerism, see Stanley, *Practical Mystic*. For specific treatment of his expedition, see Matthew Stanley, "'An Expedition to Heal the Wounds of War': The 1919 Eclipse and Eddington as Quaker Adventurer," *Isis* 94, no. 1 (2003).

35. Arthur S. Eddington, "The Internal Constitution of the Stars," *Nature* 106, no. 2653, (September 2, 1920): 18. In Eddington's celebrated book of the same name, he makes similar points, but the language is toned down. See Arthur S. Eddington, *The Internal Constitution of the Stars* (1926; Cambridge: Cambridge University Press, 1988), 289–90 and 294–95. For a sense of how debatable Eddington's claims in relation to stellar energy were at the time, see Stanley, "So Simple a Thing as a Star," and Karl Hufbauer, *Exploring the Sun: Solar Science since Galileo* (Baltimore: Johns Hopkins University Press, 1991).

36. Eddington, "The Internal Constitution of the Stars," 18.

37. Ibid.

38. Helge Kragh observes that Friedmann "introduced into cosmology two concepts of revolutionary importance, the age of the world and the creation of the world." Helge Kragh, *Cosmology and Controversy: The Historical Development of Two Theories of the Universe* (Princeton: Princeton University Press, 1996), 24. Here Kragh must be taken to be referring to the modern, scientific cosmology, articulated in general relativistic language. Equally, Kragh is perhaps implicitly taking very seriously the notion that a theorist such as Kelvin is not introducing either a cosmogony proper, in his own terms, or a cosmogony as understood by the lights of a later theory. Nevertheless, despite crediting him with introducing both the oscillatory and expanding universe models into relativistic cosmology, and indeed despite Lemaître's own assessment that Friedmann largely anticipated the *mathematical* work of Lemaître's own later papers, Kragh (like Lemaître) does not credit Friedmann with introducing the theory of the Big Bang. (Gamow seems to hold a similar view, despite his celebrating Friedmann's achievements as opening up "a new era in cosmology.") As Kragh argues in relation to Friedmann's papers, "The age of the universe appears as a mathematical curiosity, not a possible physical reality." Ibid., 25. Kragh adds in this regard, "The mathematical character of Friedmann's work is further illustrated by the fact that he indicated a solution with negative density as one of the possible solutions to his equations." Ibid. Note that Alexei Kojevnikov by contrast promotes the priority of Friedmann precisely because of Friedmann's interpretation of mathematical singularity.

39. Alexei Kojevnikov emphasizes how disturbing was Friedmann's early emphasis on the universe exploding out of a mathematical singularity, distinguishing him from Eddington and Lemaître. At the same time, Kojevnikov captures the extent to which Friedmann's mathematical visions were consonant with other Russian revolutionary cultural conceptions reconceiving time, the perpetuity of life, and the possibilities of new worlds. Alexei Kojevnikov, "Space-Time, Death-Resurrection, and the Russian Revolution: The Cultural Field of Big Bang Cosmology" (lecture, History of Science Society Annual Conference, San Diego, CA, November 18, 2012).

40. Eduard Tropp, Viktor Ya. Frenkel, and Artur D. Cherne, *Alexander A. Friedmann: The Man Who Made the Universe Expand*, trans. Alexander Dron and Michael Burov (Cambridge: Cambridge University Press, 1993), 114-15.

41. It is a picture largely endorsed as well by more contemporary historiography, in the works of Kragh and Smith. Kojevnikov further emphasizes the importance of Frederiks, given his proximity to David Hilbert and German mathematics resulting from his internment in the war.

42. Prior to his advisor's untimely death, Gamow had been hoping to work on relativistic cosmology under Friedmann, and his interest in detailing Friedmann's cosmological contributions suggests something of the importance Gamow took Friedmann to have. Gamow, *My World Line*, 41. Gamow goes on to describe how Einstein tried to extend his general relativistic considerations of 1915 to the cosmos as a whole; how this effort, as we've seen, suggested an instability in the universe similar to that predicted by a Newtonian model, where all things must eventually be destabilized under the force of gravity; and how this instability suggested to Einstein a modification to insure against that instability—or, at least, to leave a stable universe as a general relativistic possibility.

43. Gamow observes that "much later, when I was discussing cosmological problems with Einstein, he remarked that the introduction of the cosmological term was the biggest blunder he ever made in his life," a description much repeated thereafter. Gamow, *My World Line*, 41.

44. Ibid., 45. Einstein in fact sent two brief notes to *Zeitschrift für Physik*, the first to the journal to claim that Friedmann had made a mistake, the second to retract his own claim and admit Friedmann's work to have been valid.

45. This translation of the phrase is given in Tropp et al., *Alexander A. Friedmann*, 119.

46. Ibid., 122.

47. Though Friedmann's most direct successor, Gamow, certainly does not share such an attitude toward the need for mathematics in popular treatments and, along with Weinberg in the following generation, relegates the mathematics to the appendices ("for interested readers"), there is a rigor to Gamow's arguments, revealed, as we will see, by their close relationship to his technical publications. It might not be altogether unfair to suggest that there is a consonance between the style of the technical work elaborated by these scientists and that of their popularization: more mathematically rigorous in each case with Friedmann, more physically freewheeling, as we will see, with Gamow.

48. It is, regardless, a trope of universal history, one even made use of by the farcical Gamow, who himself appeals to St. Augustine's conceptions of time.

49. Alexander Friedmann, "Über die Krümmung des Raumes," *Zeitschrift für Physik* 10, no. 1 (1922): 38586. Kojevnikov notes that Friedmann clearly preferred the periodic or oscillatory universal model, of collapse into and then explosion out of a singularity.

50. In the published records of the debate, they refer to themselves and to one another in this way: the "Astronomer" says this, the "Biologist" that, and the "Metaphysician" yet something else. See Harlow Shapley, "Life Throughout the Universe," *Harvard Alumni Bulletin*, February 7, 1924, and Harlow Shapley, Edward C. Jeffrey, and Kirsopp Lake, "The Origin of Life," *Forum* 72 (1924).

51. Shapley, "Life Throughout the Universe," 514.

52. Ibid.

53. The symposium allowed Shapley to observe that "formerly the theologian and philosopher were alone considered competent to study the problem of life's origin. Later the rights of the biologist were recognized in such speculations, and now we see that the origin is a problem for the chemist, and that the geologist and astronomer are concerned." Ibid.

54. For Richard Proctor's championing the possibility of life on Mars, see Bernard Lightman, *Victorian Popularizers of Science: Designing Nature for New Audiences* (Chicago: University of Chicago Press, 2007), 307-8. For discussion of the claims to observe Martian canals, see chapter 10, "The Battle over the Planet of War," in Michael J. Crowe, *The Extraterrestrial Life Debate 1750–1900: The Idea of Plurality of Worlds from Kant to Lowell* (Cambridge: Cambridge University Press, 1986). *The War of the World* by H. G. Wells appeared in 1898, itself invoking a universal history predicated on the nebular hypothesis. Wells, *The War of the Worlds* (Leipzig: B. Tauchnitz, 1898), 12.

55. For example, he observes that "there no longer seems to be a sharp line between the lifeless and the living. The transition stage from inorganic to organic appears, to many scientists, to be merely a field of chemistry as yet unthoroughly explored." Harlow Shapley, "The Universe and Life," *Harper's*, May 1923, 716. For extended treatment of origins of life research in the Soviet Union and the specific relationship between that work and Haldane's treatment, Loren Graham, chapter 3, "Origin of Life," in *Science, Philosophy, and Human Behavior in the Soviet Union* (New York. Columbia University Press, 1987).

56. An example of this is Shapley's midcentury *Of Stars and Men*, as we will see. For a more extended treatment of this point, see Steven J. Dick, *Life on Other Worlds: The 20th-Century Extraterrestrial Life Debate* (Cambridge: Cambridge University Press, 1988), 169-71 and passim.

57. Shapley, "The Universe and Life," 716.

58. "Elmer Davis, Newsman, Is Dead; Broadcaster, 68, Headed O.W.I.: Commentator With Audience of 12 Million on Network: Fought McCarthy Ideas," *New York Times*, May 19, 1958. In a later obituary credited to the editors, *Harper's* noted of Davis that "he will be especially remembered by *Harper's* readers as the man who contributed more articles to the magazine than any other freelance writer—sixty-seven of them, including some of the most memorable we have ever had the privilege of publishing." *Harper's*, July 1958, 67.

59. Elmer Davis, *Harper's*, July 1925, 255-56.

60. Harlow Shapley, "How Big is the Universe: Notes on Recent Measurements of the Milky Way," *Harper's*, May 1925, 648.

61. "One of the major problems of astronomy is to decide whether and in what manner a dwarf star can change into a giant, a red star into a blue star, a rare star into a dense star, a single star into a double, or vice versa in each of these cases. If we can answer these questions, we have begun the solution of stellar evolution; and in solving completely the evolution of stars we also answer the questions of the earth's origin, and the questions of the inorganic development on the earth which was a necessary antecedent of terrestrial life." Harlow Shapley, *Starlight* (New York: George H. Doran, 1926), 64.

62. Ibid., 140.

63. Joann Palmeri, "An Astronomer beyond the Observatory: Harlow Shapley as a Prophet of Science," PhD diss., University of Oklahoma, 2000, 45-54. For Shapley's role in the Institute on Religion in an Age of Science as a "center for cosmological speculation," see James Gilbert, *Redeeming Culture: American Religion in an Age of Science* (Chicago: University of Chicago Press, 1997), 273-95.

64. Gilbert and Palmeri situate Shapley's institutional efforts in the context of enthusiasms in the period for interdisciplinarity. For discussions of the extended interest in interdisciplinarity in the American scientific context and in astronomy in particular, see "Shapley's Early Years: The Astronomer as Interdisciplinary Instigator," in Palmeri, "An Astronomer beyond the Observatory," 27-59, and Joann Palmeri, "Bringing Cosmos to Culture: Harlow Shapley and the Uses of Cosmic Evolution," in *Cosmos & Culture: Cultural Evolution in a Cosmic Context*, ed. Steven J. Dick and Mark L. Lupisella (Washington, DC: National Aeronautics and Space Administration, Office of External Relations, History Division, 2009). For Shapley's efforts in regard to the continuing Conference on Science, Philosophy and Religion from the late 1930s on (the first conference was held in September 1940), relating to interdisciplinarity more in the sense of "integrating science, philosophy and religion," see Gilbert, *Redeeming Culture*, 74-93. For extended arguments on "interdisciplinary" or "borderland" science, see Ronald E. Doel, *Solar System Astronomy in America: Communities, Patronage, and Interdisciplinary Science, 1920-1960* (Cambridge: Cambridge University Press, 1996), and Glenn E. Bugos, "Managing Cooperative Research and Borderland Science in the National Research Council, 1922-1942," *Historical Studies in the Physical and Biological Sciences* 20, no. 1 (1989).

65. Palmeri, "An Astronomer beyond the Observatory," 48. His efforts later in the decade to build his cosmogonic institute were ultimately unsuccessful, but "later" rather than "sooner" origins-themed institutes emerged in numbers the decades after Shapley's death.

66. Charles Willis Thompson, "A History of the World Written in True Perspective," *New York Times*, June 10, 1928, 64.

67. The nebular hypothesis is the departure point of Wells's chronology, a measure perhaps of its continuing persuasiveness. H. G. Wells, *The Outline of History: Being a Plain History of Life and Mankind*, vol. 1 (New York: Macmillan Company, 1920), 5-6.

68. Geoffrey Parsons, "Black Science," *Harper's*, June 1927, 112.

69. Shapley, *Starlight*, x.

70. Parsons, "Black Science," 107 and for comments below on Shakespeare, 112.

71. Emil du Bois-Reymond, "Ueber die Grenzen des Naturerkennens," in *Reden von Emil du Bois-Reymond*, vol. 1 (Leipzig: Verlag von Veit & Comp, 1886), 130.

72. Shapley, Jeffrey, Lake, "The Origin of Life," 517.

73. Harlow Shapley, *Flights from Chaos: A Survey of Material Systems from Atoms to Galaxies* (New York: McGraw-Hill, 1930), and Harlow Shapley, *Star Clusters*, Harvard Observatory Monographs no. 2 (New York: McGraw Hill, 1930). Here, the question of scale is in contrast to, most obviously, the historical. Of the subject of star clusters in general, he observed: "The general study of clusters deals with a wide variety of subjects. It involves, for instance, the problems of supergiant stars, stellar luminosity curves, irregularities in stellar distribution, star streaming, island universes, and the genesis of galactic systems; it considers primarily, however, the composition, structure, distribution, and cosmic position of the easily recognizable galactic and globular clusters, and in the following chapters these groups will receive almost exclusive attention." Shapley, *Star Clusters*,

2-3. As for the impact of *Star Clusters*, Hogg attested its great usefulness some thirty-five years after its publication. See Helen Sawyer Hogg, "Harlow Shapley and Globular Clusters," *Proceedings of the Astronomical Society of the Pacific* 77 (1965): 343.

74. "It should be remembered, however, that systems rather than bodies have been under discussion. A list of the kinds of material bodies would be simpler and shorter; the main items, perhaps, would be quanta, electrons, atoms, molecules, meteorites (satellites), (planets), stars, (nebulosity), the classes in parentheses being of doubtful necessity if meteorites are broadly conceived. If we were developing this classification of bodies, we should also enter into the field of variable star classification, and considerations of spectra of stars." Shapley, *Flights from Chaos*, 162.

75. "In my own studies I have had recent need to classify the types of galaxies, to look over the various kinds of star clusters, and to set up provisional categories among binary stars and star clouds. With the hope of contributing somewhat to the clarification of our view of organizations throughout the cosmos, I propose in the following pages to extend my classifications, and work through the general taxonomic problems in the field of cosmogony. We may fail to attain the desired clarity of view, but at least we shall realize useful by-products—the appraisal, for example, of current researches, the deduction of working hypotheses, exploratory glimpses into fields bordering on various branches of science, and a listing of the unknowns near the inner and outer boundaries of the material world." Ibid., 13.

76. "Very uncommonly, it appears, do the gases have the opportunity of liquefying and freezing into the waters, rocks, and organic phenomena of a planetary surface. Organisms, congealed out of normally gaseous substances, exist exotically in a world that is chiefly composed of hot and hungry atomic nuclei, of the electrons which they forever capture and lose again, and the radiant energy that arises from these violent activities below the surfaces of stars." Ibid., 46.

77. Note in this regard: "To make the survey of material systems useful and comprehensive it will be well to stop here with the superficial view and start at once at the bottom of the series, going upward class by class. Perhaps the series of principal divisions, from the minutest to the greatest, will form the most attractive phase of the panorama. The most useful part of the survey may be the provisional subclassifications that can be proposed and discussed; their practical value lies chiefly in the new interpretations they suggest, the affiliations they bring forth, *and the guidance they give to further research.*" Ibid., 25, emphasis mine.

78. Helge Kragh and Dominique Lambert have extensively traced this history. Lambert notes that the difficulty was over corrections Lemaître made to the ballistics calculations of his instructor. Lambert claims that the episode prevented Lemaître from becoming an officer. See Dominique Lambert, "Georges Lemaître: The Priest Who Invented the Big Bang," in *Georges Lemaître: Life, Science and Legacy*, ed. R. D. Holder and S. Mitton (Heidelberg: Springer, 2013), 9.

79. See Kragh, *Cosmology and Controversy*, 28.

80. For this and other biographical matter, see Dominique Lambert, *Un atome d'univers: La vie et l'œuvre de Georges Lemaître* (Brussels: Éditions Lessius/Racine, 2000) and online materials relating to the Georges Lemaître papers at the Université catholique de Louvain, https://www.uclouvain.be/en-316446.html.

81. Kragh, *Cosmology and Controversy*, 29.

82. See Allie Vibert Douglas, *The Life of Arthur Stanley Eddington* (London: Nelson, 1956), 111.

83. Georges Lemaître, "Un univers homogène de masse constante et de rayon croissant, rendant compte de la vitesse radiale des nébuleuses extra-galactiques," *Annales de la Société Scientifique de Bruxelles* 47 (1927).

84. Documents from Archives Lemaître, Université catholique de Louvain, Centre de recherché sur le Terre et le climat G. Lemaître, Louvain-la-Neuve, Belgium. The article draws two headings from the AAAS meeting: Alva Johnston, "Anti-Vivisection Bill Protested by Scientists: Council at Cleveland Cites Discoveries Gained by Experiments with Dogs; 30,000 Galaxies Mapped: Results Deny Einstein Tenet on Star's Distribution," *New York Herald Tribune*, January 1, 1931.

By contrast, a report from the same day in the *New York Times*, also in the Lemaître archives, emphasizes instead that the thrust of Shapley's observations "differs radically from the conception of the universe reached by Einstein, De Sitter, Father Le Maitre, and other world-famous scientists, including Professor Richard C. Tolman of the California Institute of Technology." William L. Laurence, "Dr. Shapley Depicts Expanding Universe: New Cosmic Conception, Based on Harvard Data, Revealed to Cleveland Convention; Einstein Idea Disputed: New Theory Holds Star Map is Non-Static and Non-Uniform in Distribution," Archives Lemaître.

85. Arthur Eddington, "On the Instability of the Einstein World," *Monthly Notices of the Royal Astronomical Society* 90 (May 1930): 668.

86. Equal to $\pi \div (2 \times \sqrt{\lambda})$. For this point made by Eddington, ibid., 668 and 671.

87. Ibid., 672.

88. Arthur Eddington, "The End of the World: From the Standpoint of Mathematical Physics," *Supplement to Nature* 127, no. 3203 (March 21, 1931): 449.

89. Ibid.

90. Ibid. This is an explicit distinction between what we might call the conscious, felt time and an empty (or "mechanistic") time. The former is the notion of time that persists while the universe is changing and while a consciousness can witness and participate in that change; the latter is the persistence of time as one of the dimensions of an endless space-time that admits no experience, a space and time where there is no world.

91. Ibid., 451.

92. Ibid., 449–50.

93. Ibid., 450. Eddington further clarifies a property of this dilemma: "The dilemma is this:— Surveying our surroundings, we find them to be far from a 'fortuitous concourse of atoms'. The picture of the world, as drawn in existing physical theories, shows arrangement of the individual elements for which the odds are multillions to 1 against an origin by chance. Some people would like to call this non-random feature of the world purpose or design; but I will call it non-committally anti-chance." Design too, what is later termed the "anthropic principle," is also part of the dilemma.

94. As far as I can tell, this metaphor is altogether absent in Lemaître's work, whose sensibilities were more classically minded. Nevertheless, the cinematic would become a nearly ubiquitous element of universal histories.

95. Eddington, "End of the World," 448.

96. Ibid.

97. Like Friedmann, Lemaître published one semipopular monograph on his theory of a primeval atom, *The Primeval Atom: An Essay on Cosmogony*, a collection of previous lectures. It expands on an article he wrote in response to Eddington's, also published in *Nature*. As he puts it: "While reading an interesting paper by Eddington on the origin and the end of the world, the idea which is developed in this book occurred to me." Georges Lemaître, *The Primeval Atom: An Essay on Cosmogony* (Toronto: D. Van Nostrand, 1950), 17.

98. Georges Lemaître, "The Beginning of the World from the Point of View of Quantum Theory," *Nature* 127 (May 9, 1931): 706.

99. Ibid.

100. Though there is an oscillation on this very point in the future narratives of the Big Bang, it is Eddington's that appears to be the understanding more firmly adopted.

101. Lemaître, "The Beginning of the World," 706.

102. Lambert, *Un atome d'univers*, 113.

103. In a 1945 lecture entitled "The Primeval Atom" given to the annual session of Société Helvétique des Sciences Naturelles in Fribourg, Lemaître suggests as much: "We are now going to imagine that the entire universe existed in the form of an atomic nucleus which filled elliptical space of convenient radius in a uniform manner." Reprinted in Lemaître, *The Primeval Atom*, 140.

In the same lecture, Lemaître queries: "Who knows if the evolution of theories of the nucleus will not, some day, permit the consideration of the primeval atom as a single quantum?" Ibid., 142.

104. Ibid., 77–78.

105. By the beginning of 1933, the *Washington Post* was reporting a very favorable reception by Einstein: "Dr. Albert Einstein, enthralled by the report of the Jesuit priest and professor of the University of Louvain, Belgium, warmly congratulated Lemaître. 'It is the most pleasant, beautiful and satisfying interpretation of cosmic radiation,' said the Berlin professor of relativity. 'This picture has less objections and conjures less contradiction than any other theory of the cosmic ray source.'" "Scientist Has New Theory of Origin of Cosmic Rays: Lemaître, Priest, Claims Beams Are Evidence of Superradiation That Existed 10,000,000,000 Years Ago When Expanding Universe Began," *Washington Post*, January 12, 1933, 3. In some sense, then, it could be said that cosmic radiation was long connected to Big Bang theories, even if in ways considered invalid by contemporary lights. Cosmic rays are now believed to come from stellar sources. Moreover, the very different cosmic background radiation is believed to result from the state of the early universe when it was in effect an expanding black body, released at a point when, as a result of the expansion, the radiation was able to pass transparently through matter.

106. "The natural tendency of matter to break up into smaller and smaller particles, which is evidenced so strikingly in radioactive transformations, can also be observed in the grains of light or photons which can make up the various forms of radiation: X-rays, light, radiant heat." Lemaître, *The Primeval Atom*, 73. And: "Thus it seems that, in thermal exchanges as well as radioactive transformations, evolution takes place with the increase in the number of particles: photons or atoms. . . . Evolution takes place, therefore, from the simple to the composite, and not from the diffuse to the condensed. The origin of the world is not really a primeval nebula but, rather, a sort of primeval atom whose products of disintegration form the present world." Ibid., 74. Lemaître argues for this by considering the number of photons per cubic centimeter and the energy per cubic centimeter, the former proportional to the cube of the absolute temperature, the latter to its fourth power. This consideration allows him to conclude that combinations of "two thermal radiations in equilibrium at different temperatures" constructed by connecting two different volumes that enclose them produces an increase in the number of photons. Ibid., 73. From this he further concludes that the number of photons is always increasing, regardless of the thermal exchange—that, derived alternatively, "the increase in entropy comes back, therefore, to an increase in the number of photons." Ibid.

107. A measure of how much that consensus had yet to be established on this matter in the period is the treatment of alternative possibilities given by Sir James Jeans in his popular text *The Mysterious Universe*. In this regard, still in 1933, Jeans asserted that "yet other things than speed are capable of reddening light; for instance, sunlight is reddened by the mere weight of the sun, it is reddened still more by the pressure of the sun's atmosphere; it is further reddened, although in a different way, in its passage through the earth's atmosphere, as we see at sunrise or sunset. The light emitted by certain stars of a different kind [the later quasars?] is reddened in a mysterious way we do not yet understand. Furthermore, on de Sitter's theory of the universe, distance alone produces a reddening of light. . . . None of these causes seems capable of explaining the observed reddening of nebular light, but quite recently Dr. Zwicky of the California Institute of Technology has suggested that still another cause of reddening may be found in the gravitational pull of stars and nebulae on light passing near them." James Jeans, *The Mysterious Universe* (New York: Macmillan Company, 1933), 78–79. When so many different causes follow the point that "the only reason for thinking that the distant nebulae are receding from us is that the light we receive from them appears redder than it ought normally to be," Jeans's interest in sustaining doubt as to the expansion of the universe is clear. Ibid., 77.

108. In reference to the fireworks universe described above, Gamow states with regard to Lemaître, "The author of these spectacular views did not choose to follow up on the details of the

fragmentation process by means of strict mathematical analysis. This chore fell on the shoulders of Maria Meyer [*sic*] and Edward Teller of Chicago, who arrived, quite independently, at similar views concerning the origin of atomic species." Gamow, *My World Line*, 53. However, as we will see, much of this latter work appealed to research in nuclear reaction rates that took its cue from data and theory emerging in the context of the Manhattan Project and the later hydrogen bomb efforts. For such theorists, the cosmological was an offshoot of the nuclear work and not their major theoretical interest. The fact is that, as Gamow asserts, their work nevertheless supplied substance to Lemaître's fireworks theory in attempting to articulate the mechanisms for the emergence of all the elements. Though the later successful theory allowed for the production of the heavier elements in the stars, lighter and heavier, from some first, generally contracted state of the universe, that theory itself suggests how critical a nuclear-research background was to the work Gamow charges Lemaître with not having carried out.

109. Lemaître, *The Primeval Atom*, 72-73.

110. Ibid., 75.

111. Ibid., 75. Hardening features of the contemporary natural history are present here, some forty years before it emerges as the contemporary myth: speaking of the relative youth of the universe, the reintroduction of the idea of natural historical "monuments of history," the appeal to the telescope as a machine of time travel, and counterfactual considerations as to alternative evolutions of life.

112. Sir Arthur Eddington, *The Expanding Universe* (Cambridge: Cambridge University Press, 1933), 56.

113. Ibid., 60.

114. In this regard, see Stanley, "'An Expedition to Heal the Wounds of War,'" 57-89.

115. *New York Times*, February 22, 1933, 18.

116. To take just one example: "What is the object of making these risky extrapolations to regions of space and time remote from our practical experience? It might be a sufficient answer to say that we are explorers." Eddington, *The Expanding Universe*, 27. This is not the most urgent of reasons, however, Eddington goes on to explain. Also, as Stanley has observed, the Quaker language of "adventurer," as well as the Quaker tradition of dangerous adventuring, resonates throughout Eddington's work.

117. Here, the emphasis of the message is that we (all scientists) need to work together to understand our world in order to help to prevent future wars and to reconcile people in the aftermath of war. Decades later, as we will see, Carl Sagan's message was by contrast that the history of the universe allows us (all people) to understand ourselves as already united, as the questing and voiced consciousness of the universe, a unity that has to be recognized if we hope to avoid destroying ourselves.

118. "The Exploding Universe," *New York Times*, May 24, 1931.

119. Ransome Sutton, "Abbe Expounds Cosmos Theory," *Los Angeles Times*, December 4, 1932, A1.

120. For example, Helge Kragh: "The beginning of the world was a radical concept, which was difficult to accept, and to which scientists had to accustom themselves." Kragh, *Cosmology and Controversy*, 57.

121. It is interesting to note that in his 1945 lecture to the Royal Belgian Society of Engineers and Industrialists on January 10, as published in *Ciel et Terre* in the March–April issue of the same year, Lemaître traces past attempts at natural history, a lecture beginning with a methodological counterpoint to Poincaré, and along with Buffon and Kant, appeals to Laplace. Indeed Lemaître's own work establishes the following genealogy of his own "Primeval Atom" theory: Buffon's *Natural History*, the first volume in 1749 considering "The Theory of the Earth"; Kant's cosmogony composed in the 1755 *General Natural History and Theory of the Heavens*; and Laplace's 1796 *L'exposition du systeme du monde*, after which, among the mathematical outgrowths of this work, "it suffices to mention" Roche, Poincaré, and Jeans.

122. This can be judged by a related, partly parenthetical point Kragh makes: "As far as the physical aspects of the Lemaître model were concerned, including the exploding primeval atom, they were ignored in the scientific literature of the 1930s (but were sometimes mentioned in the popular literature)." Kragh, *Cosmology and Controversy*, 56.

123. Thus for example, in the *Los Angeles Times* article "Abbe Expounds Cosmos Theory," which has the subtitle "Explanation 'Simple,' Aver Scientific Listeners," the author Sutton goes on to report: "At the 1931 meeting of the British Association for the Advancement of Science, after listening to Lemaître, Prof. W. de Sitter declared: 'There cannot be the slightest doubt that Lemaître's theory is essentially true and must be accepted as a very real and important step toward a better understanding of nature. It not only gives a complete solution of the difficulties it was intended to solve, a solution of such simplicity as to make it appear self-evident—like Columbus's famous solution of the problem of standing an egg on end—but it also incidentally contains answers to some questions of long standing." Sutton, "Abbe Expounds Cosmos Theory," A1.

124. Originary hypotheses that emphasized a falling inward of sorts had an extended pedigree. Steven Dick has pointed out to me that Alfred Russell Wallace argued for such a falling inward at the beginning of the century. For Wallace's views, see the final chapter of Alfred Russell Wallace, *Man's Place in the Universe: A Study of the Results of Scientific Research in Relation to the Unity or Plurality of Worlds* (New York: McLure, Philipps & Co., 1904), 302–4. In reference to a longer history, Galileo, and Newton in analyzing Galileo's arguments, had offered speculations on the creation of the current organization of the solar system in terms of a vision of things falling into place, planets dropped from some distant point to fall into an orbit around the sun, when for them the sun was the center of the universe and when space and time were somehow a given within that cosmogony. See for example I. Bernard Cohen, "Galileo, Newton, and the Divine Order of the Solar System," in *Galileo, Man of Science*, ed. Ernan McMullin (New York: Basic, 1967); Alexandre Koyré, "Newton, Galileo, and Plato," in Koyré, *Newtonian Studies* (Cambridge, MA: Harvard University Press, 1965); and Eric Meyer, "Galileo's Cosmogonical Calculations," *Isis* 80, no. 3 (September 1989): 456.

125. Such could be argued was the case in the context of the scientific community until the late 1960s, when the Big Bang had finally begun to be of widespread interest and acceptance, over against the theory of a steady-state universe.

126. Parsons, "Black Science," 107. It should be noted that Parsons did not find "the remotest hope that [the newer sciences] can reach solid conclusions or yield much sound advice for years to come." Ibid.

127. There were those who argued that the universe is something researchers are not entitled to imagine can be made reconcilable with our data: "At the British Association meeting in 1931 the problem was emphasized by de Sitter. He could see no way out of the dilemma except suggesting that perhaps the expansion of the universe and the evolution of stellar systems were unconnected processes which only happened to take place side by side." Helge Kragh, *Cosmology and Controversy*, 74. More radically, de Sitter suggested that "after all the 'universe' is an hypothesis, like the atom, and must be allowed the freedom to have properties and to do things which would be contradictory and impossible for a finite material structure." Ibid., 75. Such a view would preclude anything like the notion of the history of the world.

CHAPTER SIX

1. John Milton, *Paradise Lost* (Oxford: Oxford University Press, 2005), 153. In an undated letter in the Seth MacFarlane Collection of the Carl Sagan and Ann Druyan Archive, Manuscript Division Library of Congress, Washington, DC, J. B. S. Haldane sends a close recollection of this passage to Carl Sagan as an example of evolutionary thought.

2. Correspondence Eric Collins, Recorder Naval Department Loyalty Hearing Board to G. Gamow, May 31, 1950, George Gamow and Barbara Gamow Papers, Manuscript Division, Library

of Congress, Washington, DC. From the late 1940s, the FBI investigated allegations that Gamow's escape was "planned by the soviets," that Gamow "might be a Russian spy." Inquiries into his background had been provoked at least in part by Gamow's own applications for security clearance needed for military consultancy, but partly, too—at least in one case—through the initiative of an acquaintance writing to a government official, to Senator John Bricker of Ohio, who thereafter contacted the FBI (in the case of Bricker, by letter to Herbert Hoover). Gamow was also of interest to the FBI as a function of his connections with others under investigation. Two named in the files are the Brazilian physicist Mário Schenberg and the plastic surgeon Johannes Fredericus Samuel Esser, the latter of whom had been cleared of charges of espionage by the time of the investigation into Gamow. The strong majority of those the FBI interviewed who felt they knew Gamow well enough to express an opinion asserted their confidence in Gamow's loyalty (if not in his ability to maintain rigid security decorum) and in his account of his defection. George Gamow FBI file (116-HQ-12246 and 12-HQ-18180), via a Freedom of Information Act Request (1227772-0). My thanks to Alex Wellerstein for sharing this file with me.

3. Correspondence G. Gamow to E. Collins, October 1950, *A Gamowian Miscellany* Red Book 3, Box 30, George Gamow and Barbara Gamow Papers, Manuscript Division, Library of Congress.

4. "Interrogatories," May 31, 1950, and Correspondence G. Gamow to Mr. Mitchell, May 21, 1967, Box 5, George Gamow and Barbara Gamow Papers, Manuscript Division, Library of Congress.

5. Bukharin, occupying the position of "Coordinator of Scientific and Technical Research," was by then in the political decline that would end with his execution.

6. Enclosure to Interrogatory Reply, October 1950, Box 5, p. 3, George Gamow and Barbara Gamow Papers, Manuscript Division, Library of Congress.

7. Correspondence G. Gamow to Mr. Mitchell, May 21, 1967, Box 5, George Gamow and Barbara Gamow Papers, Manuscript Division, Library of Congress.

8. See Philipp Frank, *Einstein, His Life and Times*, trans. George Rosen, ed. Suichi Kusaka (New York: Knopf, 1953), 226. The more exact statement is as follows: "When Einstein set out with his wife for California in the fall of 1932, and as they left the beautiful villa in idyllic Caputh [the Einsteins's country home], Einstein said to her: 'Before you leave our villa this time, take a good look at it.' 'Why?' she asked. 'You will never see it again,' Einstein replied quietly." Ibid.

9. Ibid., 239-40.

10. In the United Kingdom, J. B. S. Haldane, who had long since been given to political commentary weighing science and ethics together, worked to find scholars new institutional homes, at his own University College London and beyond. See J. B. S. Haldane, "Science and Ethics," *Harper's*, June 1928 and, in relation to repatriation efforts, Ronald Clark, *J. B. S.: The Life and Work of J. B. S. Haldane* (New York: Coward-McCann, 1968), 105-6. Harlow Shapley, from early in 1930s, also invested his energies in initially fitful efforts to find new homes for those in need. Frank, who had succeeded Einstein in the latter's position in Prague, became one among those whom Shapley sought to help. See Bessie Zaban Jones, "To the Rescue of the Learned: The Asylum Fellowship Plan at Harvard, 1938-1940," *Harvard Library Bulletin*, vol. 32 (1984), Gerald Holton, "Ernst Mach and the Fortunes of Positivism in America," *Isis* 83, no. 1 (1992), and, for discussion more specifically of the relationship between Frank and Shapley, see Nasser Zakariya, "Making Knowledge Whole: Genres of Synthesis and Grammars of Ignorance," *Historical Studies in the Natural Sciences* 42, no. 5, esp. 454-58. In the course of that effort, Frank's own philosophy of science and efforts at intellectual unification came into the orbit of Shapley's synthesizing and interdisciplinary initiatives. In turn, in both its early European and later American phases, the Unity of Science movement advanced in the United States by Frank played a role in the very different synthesis ultimately emerging through the contrasting genres of universal history offered by Shapley and by Gamow.

11. Though Gamow and Shapley appear to have been aware of each other in this period, they do not seem to have met until the early 1950s, by which point both of them were writing fuller narratives that their work had been promising for a longer period.

12. There is an extensive literature on the evolutionary synthesis. Those accounts are important to the emergence of the account given here, though I will not seek to summarize them or to engage debates within them explicitly.

13. See Ernst Mayr and William B. Provine, "The Evolutionary Synthesis," *Bulletin of the American Academy of Arts and Sciences* 34, no. 8 (May 1981).

14. Karl Philip Hall, "Purely Practical Revolutionaries: A History of Stalinist Theoretical Physics," PhD diss., Harvard University, 1999, 55.

15. Gamow had met Houtermans in 1928, while attending the summer session of Göttingen University, and the work of the paper was itself partly based on work Gamow had previously done on nuclear reactions. The collaboration between Atkinson and Houtermans was aided by direct consultation of Gamow, an impact evident in the extended reference to Gamow's work within the Atkinson-Houtermans paper itself. Another sign of the impact of Gamow's work was that Houtermans and his first wife, Charlotte, later translated into German Gamow's monograph on the *Constitution of the Atomic Nucleus and Radioactivity*.

16. Edward Teller, "The Work of Many People," *Science*, new series 121, no. 3139 (February 25, 1955): 267.

17. This account appears in George Gamow, *My World Line: An Informal Autobiography* (New York: Viking, 1970).

18. By contrast, there's a touch more emphasis on how much Gamow may have weighed the proposal—how it might have stopped him short: "I decided to decline that proposal, and I am glad that I did because it certainly would not have worked." Ibid., 121.

19. It is a lecture that Teller, himself recounting Gamow's tale, places in a direct genealogical line with the concepts and innovations that produced the hydrogen bomb—a point returned to below.

20. Enclosure to Interrogatory Reply, October 1950, Box 5, and Letter to Mr. Mitchell, May 21, 1967, Box 5, George Gamow and Barbara Gamow Papers, Manuscript Division, Library of Congress.

21. As early as 1931, in his slim text *Constitution of Atomic Nuclei and Radioactivity*, Gamow considered the relative abundance of different types of nuclei (as opposed to elements as a whole), but also stated that "the abundance of different elements depends not only on the special properties of their nuclei but also on the cosmological factors governing the origin of the elements. The chemical analysis of the universe (in practice of the earth's crust and of meteorites) gives us only the abundance of different chemical elements." George Gamow, *Constitution of Atomic Nuclei and Radioactivity* (Oxford: Clarendon Press of Oxford University Press, 1931), 21. But Gamow doesn't pursue this question further at this point: "Considerations based on the relative abundance of the elements are interesting, but not of very great importance for the theory of nuclear constitution, as the observed abundance depends also on astronomical and geological factors." Ibid., 21-23. The book, it might be added, is dedicated to "the Cavendish Laboratory Cambridge"—a sign perhaps of Rutherford's importance to Gamow.

22. For an extended treatment of Brownian motion, see Stephen G. Brush, "Brownian Movement," in *The Kind of Motion We Call Heat: A History of the Kinetic Theory of Gases in the 19th Century, Book 2: Statistical Physics and Irreversible Processes* (Amsterdam: North Holland, 1976), 655701. For Max Born's assessment of Einstein's particular importance, see ibid., 673. For Mach's views, see "Mach," in Brush, in *The Kind of Motion We Call Heat: A History of the Kinetic Theory of Gases in the 19th Century, Book 1: Physicists and Atomists* (Amsterdam: North Holland, 1976).

23. Indeed, the titles of Gamow's series of papers almost seem to speak to V. M. Goldschmidt's earlier series "Geochemische Verteilungsgesetze der Elemente," or the "Geochemical Distribution Laws of the Elements." For an exposition of these latter, see C. E. Tilley, "Victor Moritz Goldschmidt 1888-1947," *Obituary Notices of Fellows of the Royal Society* 6, no. 17 (November 1948). In 1917, Goldschmidt became chairman of the Norwegian Commission for Raw Materials and Director of the Raw Materials Laboratory. According to Tilley, "The origins of modern geochemistry

may be said to date from this period," in particular, to the effort, in Goldschmidt's own words, "to find the general laws and principles which underlie the frequency and distribution of the various chemical elements in nature—the basic problem of geochemistry. I proposed to attack the problem from the viewpoint of atomic physics and atomic chemistry." Ibid., 55. Here, applying "both old and new methods of chemical and physical analysis," including spectrographic and X-ray methods, Goldschmidt in Tilley's view achieved his scientific objective.

24. George Gamow, "Nuclear Transformations and the Origins of the Chemical Elements," *Ohio Journal of Science* 35, no. 5 (1935): 406. The observation that such nuclear transformations could occur at tremendous heats also allowed Gamow to exploit a cooking metaphor he used repeatedly throughout his work: that the earliest elements of the world were cooked at the right temperatures and in the right conditions to produce the fundamental elements of matter in the ratios, distributions, and forms in which they are found in the world today. On this view, the worlds and galaxies were the baked goods of primeval ovens. Here, the historian Helge Kragh attributes much to Gamow's training: "First and foremost, Gamow was a nuclear physicist, neither a mathematician nor an astronomer. Influenced by his own and others' work in nuclear physics, he conceived the early universe as a nuclear oven in which the elements constituting our present universe were once cooked. This was a persistent theme in Gamow's cosmology. Although the theme was not entirely new, it was only with Gamow that it was developed systematically. He thereby provided cosmology with a new perspective, linking the science of the universe intimately to nuclear physics." Helge Kragh, *Cosmology and Controversy: The Historical Development of Two Theories of the Universe* (Princeton: Princeton University Press, 1996), 80.

25. The answers to what continue to be considered fundamental cosmological questions thus turned on the recent history of the physical sciences. But they also must have turned on a broader historicizing spirit from which scientists were by no means insulated, then or now. The historicizing drive of the nineteenth century persisted in the sciences in the interwar and postwar periods.

26. Gamow, "Nuclear Transformations and the Origins of the Chemical Elements," 406.

27. This question of element creation in this early paper is a rather ironic one, contrasted with later events, since this is a point of disputation when argued by Fred Hoyle thereafter in the development of Hoyle's steady-state theory. As Gamow concludes the 1935 paper: "We see thus that there are many different processes inside of a star giving rise to formation and transformation of different elements and the hope may be justified that further investigation will clarify the relative importance of various processes and lead to a complete explanation of the relative abundance of different elements in the universe." Ibid., 413. Hoyle over a decade later looked to the stars as sites of the production of heavy elements, a tack scorned by Gamow before it was embraced by cosmologists more generally. But, as we see here, Gamow was exploring this possibility years before, if in a different theoretical configuration.

28. Here, I'm building on the insights of Mikhail Bakhtin. See "Forms of Time and the Chronotope in the Novel: Notes toward a Historical Poetics," in *The Dialogic Imagination: Four Essays*, ed. Michael Holquist, trans. Caryl Emerson and Michael Holquist (1981; Austin: University of Texas Press, 2004) and in "The Bildungsroman and Its Significance in the History of Realism (Toward a Historical Typology of the Novel)" in *Speech Genres and Other Late Essays*, ed. Caryl Emerson and Michael Holquist, trans. Vern W. McGee (Austin: University of Texas Press, 1986).

29. Despite the successes of the *Tompkins* series, Gamow in correspondence was somewhat dismissive of the relations between fiction and science. In response to an inquiry from a representative of the Romanian newspaper *Romania Libera* the year before his death, Gamow stated: "When I write 'fiction', as for example the Mr Tompkins stories, I present them in the form of dreams (anybody can dream!) with a strictly scientific background. Most, or rather all, science fiction stories I see now have a lot of fiction and no science." Letter, July 19, 1967, Romania Libera Folder, Box 5, George Gamow and Barbara Gamow Papers, Manuscript Division, Library of Congress. No science fiction he knew of, continued Gamow, had helped advance science. *Tompkins* was instead a picaresque fiction meant to relate scientific truths. But regardless of the contrast between

Tompkins and generic science fiction, Gamow indicated something of the difficulty in regarding any such fabulous narrative as potentially offering orientation to future research. It should also be added that the picaresque could inform scientific biography more generally, as might be argued in the case of Richard Feynman. Though her conception and characterization are not the same as that pursued here, Sharon Traweek makes this point in her usage of the picaresque: "In their careers, physicists journey from romantic readings of others' lives, through handing on mimetic tales of heroic action and quests for survival, to becoming skilled practitioners of gossip and rhetoric. They complete the circle by telling erotic tales about physics, tales transformed into romance for the next generation of neophytes. These stories reflect how physicists come to care passionately about who they are and what they are doing. Together they form a picaresque cycle, which chronicles a journey that begins necessarily with innocence and reports its loss; it depicts the growth of strength and pain of betrayal, hails the achievement of success, recognizes the signs of grace in eminent discoveries, looks back in erotic nostalgia, and lastly eulogizes the heroic dead. In Western culture the picaresque genre is usually reserved for stories about men, not women: women are not seen as gaining strength and wisdom through the rambunctious loss of their 'innocence.'" Sharon Traweek, *Beamtimes and Lifetimes: The World of High Energy Physicists* (Cambridge, MA: Harvard University Press, 1988), 103–4.

30. There are different resonances possible here, and a more traditional notion of physical exploration is at times at stake, but these accounts tended more to the notion of exploration in the mind's eye, of surveying the world through observational astronomy rather than doing what was not yet possible, literally traveling to other worlds (though Gamow briefly explores this possibility).

31. George Gamow, *Mr. Tompkins in Paperback* (Cambridge: Cambridge University Press, 1965), vii.

32. *Discovery: A Monthly Popular Journal of Knowledge*, ed. A. D. Russell, no. 1. January 1920, 3. Scanning several collected volumes of the magazine suggests that it did indeed make a good-faith effort, if with an increasing emphasis on science.

33. C. P. Snow, "Collected Volume 1938," *Discovery: A Monthly Popular Journal of Knowledge*, new series 1, nos. 1–9, April–December 1938, 1.

34. Gamow, *My World Line*, 156. In the first sentence of the preface of the first collected edition of the stories, Gamow recalls that "in the winter of 1938 I wrote a short, scientifically fantastic story (not a science fiction story) in which I tried to explain to the laymen the basic ideas of the theory of curvature of space and the expanding universe."

35. All but the last were boldly advertised on the front cover of the magazine, while the last in the May 1939 edition mentioned their coming publication in book form by Cambridge University Press.

36. George Gamow, in Snow, "Collected Volume 1938," 431. The "primitive savage" trope appears elsewhere in Gamow, at times in a way disturbing to today's sensibilities. See, for example, Gamow's repeated invocation of Hottentots and, more facetiously, Hungarians, in George Gamow, *One Two Three . . . Infinity: Facts and Speculations of Science* (New York: Viking, 1947).

37. "The author hopes that the experiences of Mr. Tompkins will help the reader to build a clear picture of the world in which we are living." Gamow, in Snow, "Collected Volume 1938," 431.

38. The manipulation of fundamental constants could be construed as a scalar transformation of its own, effecting a different type of transformation than that linked to the cartographical. In Tompkins it is nevertheless linked to the cartographical in at least two ways, either by situating Tompkins in the spatiotemporal mapping of an alternative world or, what can come to the same, by providing a fictive device for rescaling Tompkins in this world, in order to give him the experience of scales he cannot access. At the same time, such a notion of scales can be linked to a limited foundationalist vision predicated on fundamental science as being what uncovers the underpinnings of the phenomenological world—Tompkins' initials encoding that idea. Through these accounts, Gamow suggested that there are fundamental facts about the world that constrain the basic features of all lives, upon which they depend. But through Tompkins he does not build up

an orienting synthesis of knowledge connecting the disciplines through foundationalist expository and phenomenological links or by systematically delineating disciplines as a matter of the comparative scales which they take as their object.

39. George Gamow, *Mr. Tompkins in Wonderland* (1940; Cambridge: Cambridge University Press, 1964), 1-2.

40. For these stories, see ibid., "First Dream: Toy Universe," and "Third Dream: City Speed Limit." The exposition of these dreams is altered somewhat in later collections. Compare Gamow, *Mr. Tompkins in Paperback*.

41. Gamow, *Mr. Tompkins in Paperback*, 5-6.

42. George Gamow, "Nuclear Energy Sources and Stellar Evolution," *Physical Review* 53 (April 1, 1938): 598.

43. Edward Teller and George Gamow, "On the Origins of Great Nebulae," *Physical Review* 55 (April 1, 1939): 654.

44. Ibid., 657.

45. In the preface, Gamow takes the further opportunity to connect the sun directly to cosmogony: "As our Sun represents only one member of the numerous families of stars scattered through the vast spaces of the universe, the answer to this solar problem necessarily involves also the question of the evolutionary history of stars, and this brings us back to the fundamental puzzle concerning the creation of the stellar universe." George Gamow, preface, *The Birth and Death of the Sun* (New York: Viking, 1940), vii.

46. Ibid., 227.

47. Ibid.

48. Though he certainly tested definitions of life, looking at what he regarded as life's major features in order to show that such features could also be predicated of "dead" matter.

49. What allowed this unapologetic flexibility on Oparin's part is perhaps worthy of examination in itself.

50. This is language Oparin uses with "dead" material, as he refers to it in the first edition, approvingly quoting Carus Sterne on the subject of "the struggle for existence in the realm of [sulphur] crystals!" A. I. Oparin, "The Origin of Life," reprinted in J. D. Bernal, *The Origin of Life* (Cleveland: World, 1967), 215.

51. Loren Graham, *Science, Philosophy, and Human Behavior in the Soviet Union* (New York: Columbia University Press, 1987), 72. For example, in a passage Graham points to, consider Oparin in 1926 in relation to the question of metabolism: "the more closely and accurately we get to know the essential features of the processes which are carried out in the living cell, the more strongly we become convinced that there is nothing peculiar or mysterious about them, nothing that cannot be explained in terms of the general laws of physics and chemistry." Oparin, "The Origin of Life," 214.

52. And indeed in the published notes to his *Dialectics of Nature*, Engels observed: "How little Comte can have been the author of his encyclopaedic arrangement of the natural sciences, which he copied from Saint-Simon, is already evident from the fact that it only serves him for the purpose of *arranging the means of instruction* and *course of instruction*, and so leads to the crazy *enseignement intégral*, where one science is always exhausted before another is even broached, pushing a basically correct idea to a mathematical absurdity." Frederick Engels, *Dialectics of Nature*, trans. Clemens Dutt (New York: International, 1940), 262.

53. Note that, according to Graham, it was only in the late 1930s that Oparin achieved "international stature" with Oparin's 1936 edition and its first English translation in 1938. Graham, *Science, Philosophy, and Human Behavior in the Soviet Union*, 82.

54. As to their very real differences, Graham observes that "a number of writers who actually differed considerably with Oparin on details considered themselves in agreement. Haldane in his 1929 article, for example, hypothesized—in contrast to Oparin—a primitive earth atmosphere rich in carbon dioxide, described the first 'living or half-living things' as 'probably large molecules,' and did

not mention coacervates, coagula, or gels. These are important points of difference. Yet the hypothesis became known as the Haldane-Oparin (or Oparin-Haldane) one and even now is frequently referred to in that way." Graham, *Science, Philosophy, and Human Behavior in the Soviet Union*, 82. I have nevertheless found it necessary to treat their work together, since they were treated as putting forward a shared or mutually reinforcing theory by synthesizers such as Shapley.

55. Graham has argued that the links between materialism (dialectical or otherwise), on the one hand, and the question of the origins of life, on the other, need further research before anything compelling might be argued, even as he favors the view that in the case of Oparin and others, dialectical materialism did impact their scientific work. Peter J. Bowler gives a brief gloss of John Farley's argument for the importance of dialectical materialism in Oparin's thought. See Peter J. Bower, *Evolution: The History of an Idea* (Berkeley: University of California Press, 2003), 339-40. Likewise, we might suggest that Oparin's research offers not so much the hypothesis that dialectic materialism was a necessary spur or fertile ground for the present-day conception, but rather that the question of life, itself susceptible to materialistic explanation, confirmed these thinkers in a kind of dialectics or at least proved so little a hindrance that they could refine their concepts of each together. The materialistic conquest of the question of the origin of life (as opposed to its essence for an ambivalent Erwin Schrödinger in the physicist's speculations on life in relation to physical laws) might very well suggest the materialist conceptual conquest of the political sphere, as well. It certainly held promise for any who hoped to provide a synthesis not only of scientific disciplines, but of science and culture more widely still.

56. Haldane, "On Being the Right Size," *Harper's*, March 1926, 427. Note that his collected edition of essays of 1928 includes this essay and begins with an article, "On Scales."

57. Ibid., 427.

58. J. B. S. Haldane, "The Origin of Life," reprinted in Bernal, *The Origin of Life*, 246.

59. Bernal, *The Origin of Life*, 250.

60. The latter of these Oparin did not have in his 1924 treatment, implicitly suggesting something of what Bernal thought to be the case concerning the question of credit.

61. The previous year, in a 1928 lecture inveighing against the "scientific ethics" of the British theologian and Dean of St. Paul's, William Ralph Inge, the so-called Gloomy Dean, Haldane claimed: "I am not a materialist, but I do not think that the influence of materialism on ethics is wholly bad." J. B. S. Haldane, "Science and Ethics," 3. Explicitly promoting a latter-day Comtean collective being, in the published version of the lecture, Haldane fully aligned science and universalism. See J. B. S. Haldane, *Science and Ethics: Conway Memorial Lecture delivered at Essex Hall on April 18, 1928* (London: Watts & Co., 1928), esp. 15 and 34-41. The *Harper's* article is a version of this address; likewise, see Haldane, "Science and Ethics," esp. 2, and 8-10. Note that Peter Bowler has examined the work of Inge in the broader context of Anglican Modernism, Dean's views exemplary of the pro-scientific dimensions of that movement. Peter J. Bowler, *Reconciling Science and Religion: The Debate in Early-Twentieth-Century Britain* (Chicago: University of Chicago Press, 2001), esp. chapter 8, "Anglican Modernism."

62. Haldane, "The Origin of Life," 249. The question of the relationship of scales, of the microscopic to the macroscopic, also affected the understanding of the origins of life. However, Haldane and Oparin emphasized very different sizes than did Shapley. In their treatments, contrasting with Tyndall's speculation on proto-organic molecules, drilling down in scales while moving back in time tracked down the shift from the nonliving to the living. Haldane once more offered a redolent image: "from the point of view of size, the bacillus is the flea's flea, the bacteriophage the bacillus' flea; but the bacteriophage's flea would be the dimensions of an atom, and atoms do not behave like fleas. In other words, there are only about as many atoms in a cell as cells in a man. The link between living and dead matter is therefore somewhere between a cell and an atom." Ibid., 244. If it isn't clear from the context, Haldane is referring here to the following famed lines that he cites explicitly: "Big fleas have little fleas/Upon their backs to bite 'em;/the little ones have lesser ones/And so ad infinitum." Ibid.

63. Note that J. D. Bernal in 1945 had claimed by contrast that "it is no use any longer attempting to present science as a series of pictures of the beauties or the mysteries of the universe and of nature. People have had enough of that already; it belongs to a time when individual and social security and the general running of society could be taken for granted." J. D. Bernal, "Transformation in Science," *Scientific Monthly* 61, no. 6 (December 1945): 474.

64. See, for example, J. R. Baker, *Julian Huxley: Scientist and World Citizen 1887-1975* (Paris: United Nations Educational, Scientific and Cultural Organization, 1978), 43-44, who dates Huxley's conception of the "human or psycho-social phase of evolution" to 1913, when Huxley was living in Texas.

65. For an enlightening exchange concerning Huxley's views on eugenics, see Garland E. Allen, "Julian Huxley and the Eugenical View of Human Evolution," along with the responses by Diane B. Paul, "The Value of Diversity in Huxley's Eugenics," and Elazar Barkan, "The Dynamics of Huxley's views on Race and Eugenics," in *Julian Huxley: Biologist and Statesman of Science: Proceedings of a Conference held at Rice University 25-27 September,* ed. C. Kenneth Waters and Albert Van Helden (Houston: Rice University Press, 1992).

66. He had been president and vice president of the Association of Scientific Workers, helped found the Political Economic Planning Committee, and was a part of the Next Five Year Group before becoming the founding director of UNESCO.

67. Julian Huxley, *If I Were Dictator* (New York: Harper and Brothers, 1934), 84.

68. Ibid., 58.

69. Julian Huxley, *Evolution: The Modern Synthesis* (New York: Harper and Brothers, 1943), 28.

70. For an overview, see, for example, Vassiliki Betty Smocovitis, *Unifying Biology: The Evolutionary Synthesis and Evolutionary Biology* (Princeton: Princeton University Press, 1996).

71. He credited all of these scientists in his 1943 text, *Evolution: The Modern Synthesis.* Huxley also gave special thanks to Lancelot Hogben, who so emphasized the differences between the "Particulate Inheritance" of Haldane, Wright, and Fischer and the "blending inheritance" of Darwin that he rejected the term "Darwinism" and preferred an alternative to "Natural Selection." See Huxley, *Evolution: The Modern Synthesis,* 27. Hogben also advanced a decidedly populist vision of mathematics in his highly popular *Mathematics for the Million.*

72. The questions of the relationship of science to the larger cultural concerns of the mid-twentieth century and of the unity of science itself interested scientific practitioners and commentators across widely different historical circumstances with no less diverse viewpoints. Both prior to and after World War II, European scientists and philosophers working on these issues attempted to find refuge in the Anglophone world, including survivors of the Vienna Circle (otherwise termed the Ernst Mach Society), the Frankfurt School, and Ernst Cassirer.

Both Haldane and Shapley attempted to help resituate European scholars displaced by World War II, with Shapley's efforts on the part of threatened European scholars beginning as early as 1933. See chapter 6, n. 10. Among those with whom Shapley came into increasing contact were members of the Vienna Circle or their intellectual heirs and hence with the Unity of Science movement (UoS), which took shape as an outgrowth of the Vienna Circle. In the 1930s, the movement had brought together scientists and philosophers, among them Moritz Schlick, Rudolf Carnap, Niels Bohr, Bertrand Russell, and Philipp Frank. Before the migration of their major strength to the English-speaking world, they found common cause in the shared belief in and active promotion of the unity of science.

However, they had varying conceptions of the nature of that unity. In the 1938 volume of the *International Encyclopedia of Unified Science,* Russell contended that "the unity of science . . . is essentially a unity of method, and the method is one upon which modern logic throws much new light." The philosopher John Dewey emphasized social unity instead, and the linguist Charles Morris addressed scientific unity through semiotics as a unity of language/syntactics, a unity via objects of analysis/semantics and a unity of practices/pragmatics. Otto Neurath, ed. in chief, *International Encyclopedia of Unified Science* (Chicago: University of Chicago Press, 1938), 41, 33–34, 69–70. Neurath for his part introduced the volume depicting science as collectively producing a "mosaic of knowledge," its diverse elements coming together into a shared whole, an encyclopedia: "An encyclopedia and not a system is the genuine model of science as a whole. An encyclopedic integration of scientific statements, with all the discrepancies and difficulties which appear, is the maximum of integration which we can achieve." Ibid., 3, 20.

From the perspective of the UoS, the discipline of philosophy considered the traditional site of the integration of knowledge—a point of view consonant in this instance with Cassirer's—needed reformation in light of scientific truths and of scientific unity, with a particular emphasis on the questions of shared scientific method and language. Despite the diversity of views it embraced, the UoS did not generally treat this synthesis as if it would happen by itself, and it emphasized above all the construction of a unified scientific language in order to advance further integration. Though this did not require logical foundationalism, members such as Carnap invoked such a picture. See Rainer Hegselmann, "Unified Science: The Positive Pole of Logical Empiricism," in *Unified Science: The Vienna Circle Monograph Series originally edited by Otto Neurath*, trans. Hans Kaal (Dordrecht: D. Reidel, 1987), esp. xiii. For a more contemporary account, see Philipp Frank, "The Institute for the Unity of Science," *Synthese* 6, nos. 3/4 (July-August 1947). The breadth of these views raises the question of whether Neurath's "mosaic" or encyclopedia as a synthetic form functioned primarily as a collection of dogmas of unity, largely consonant in broad terms, or further provided a compelling topos central to emerging genres of synthesis.

73. Huxley, *Evolution: The Modern Synthesis*, 483.

74. Ibid., 485.

75. "True human progress consists in increases of aesthetic, intellectual, and spiritual experience and satisfaction." Ibid., 576.

76. Some commentators have seen in Julian Huxley a partial throwback to an early Victorian age. See in particular Colin Divall, "From a Victorian to a Modern: Julian Huxley and the English Intellectual Climate," in Waters and Van Helden, *Julian Huxley*.

77. Though doing so threatens to ossify such periodizations, historians have recognized this to a degree in Huxley's case, in trying to diagnose him as more Victorian than Edwardian.

78. See C. H. Waddington, *Science and Ethics, an Essay by C. H. Waddington, together with a Discussion between the Author and E. W. Barnes, et al.* (London: George Allen and Unwin, 1942).

79. Haldane in ibid., 43. Waddington held himself to be in agreement with Haldane's remarks. Note by contrast that in his earlier foray into future, his famed 1923 *Daedalus; or, Science and the Future*, the younger Haldane had remarked of the elder Huxley: "The time has gone by when a Huxley could believe that while science might indeed remould traditional mythology, traditional morals were impregnable and sacrosanct to it." J. B. S. Haldane, *Daedalus; or, Science and the Future; a paper read to the Heretics, Cambridge, on February 4th, 1923* (New York: E. P. Dutton, 1924), 90.

80. Bernal in Waddington, *Science and Ethics*, 115. The varied understandings of scientific myths is returned to in the next chapter.

81. See Ralph Desmarais, "Jacob Bronowski: A Humanist Intellectual for an Atomic Age, 1946–1956," *British Journal for the History of Science* 45, no. 4 (December 2012), esp. 6.

82. In World War I, he took part in trench warfare, organized and taught bombing, and investigated the effects of gas in the combat zone. In World War II, he advised air intelligence and the navy while sustaining his complicated engagement with the British Communist Party.

83. He published his own book-length critique of any too naïve application of genetics to politics as *Heredity and Politics* (New York: W. W. Norton, 1938). For further background on Haldane's eugenic views, see Diane B. Paul, "A War on Two Fronts: J. B. S. Haldane and the Response to Lysenkoism in Britain," *Journal of the History of Biology* 16, no. 1 (Spring 1983): esp. 2–8.

84. With regard to the latter, Bernal ultimately linked his theories to geochemistry and to speculations on life elsewhere through the study of carbonaceous materials from meteorites. J. D. Bernal, "The Physical Basis of Life," *Proceedings of the Physical Society*, section B, 62, no. 10 (1949). See also Andrew Brown, *J. D. Bernal: The Sage of Science* (Oxford: Oxford University Press, 2006), esp. chapter 18, "The History and Origin of Life." For the debts of researchers on the structure of DNA and in molecular biology more broadly, see ibid., chapter 17, "The Physical Basis of Life."

85. See Robert K. Merton, "Science and the Social Order," in Merton, *The Sociology of Science*, ed. Norman W. Storer (1938; Chicago: University of Chicago Press, 1973), esp. 262n24 and Merton, "The Normative Structure of Science" (1942), esp. 274n14.

86. Jessica Wang, *American Science in an Age of Anxiety: Scientists, Anticommunism, and the Cold War* (Chapel Hill: University of North Carolina Press, 1999), esp. 55-57.

87. However difficult to characterize, the US discussion in which so many elements were in play and in which Shapley and Urey played their parts in turn generated interest and impetus in the United Kingdom. For an emphasis on Vannevar Bush and his *Science, the Endless Frontier*, see William McGucken, *Science, Society and the State: The Social Relations of Science Movement in Great Britain, 1939-1947* (Columbus: Ohio State University Press, 1984), 318-19. See also Merton, "Science and the Social Order" and "The Normative Structure of Science." For Shapley's and Urey's involvement, see Jessica Wang, "Liberals, the Progressive Left, and the Political Economy of Postwar American Science: The National Science Foundation Debate Revisited," *Historical Studies in the Physical and Biological Sciences* 26, no. 1 (1995). It should be kept in mind that Bush backed expert control of federal funding for science, as opposed to the more egalitarian and New Deal vision of Senator Harley M. Kilgore. See Ibid., 149. Advocacy for expert planning did not map neatly onto any political stance and was also linked to the demands for broad-based scientific literacy.

88. Urey himself pursued origins of life research, his student Stanley Miller carrying out experiments simulating primitive atmospheres in order to demonstrate that the building blocks of organic matter could emerge from nonorganic materials.

89. Gamow, *My World Line*, 149.

90. It was work that was apparently productive for Gamow and his then masters and future doctoral student Robert Alpher, but with its own frustrations: "Gamow and Alpher, with a third researcher named James Pickering, developed a model for the ramjet missile using radioactive waste material as a heat source. When the work was submitted for publication, the navy and the FBI appeared on the scene, told us our work was classified beyond our classification status, and confiscated our notes." Ralph A. Alpher and Robert Herman, *Genesis of the Big Bang* (Oxford: Oxford University Press, 2001), 71. During the war, Alpher worked in the US Naval Ordnance Laboratory and the Navy Bureau of Ordnance, continuing his applied work throughout his doctoral research. The third of the cosmogonic trinity yet to be, Robert Herman, worked in a section of the Office of Scientific Research and Development on quality control of proximity fuses for antiaircraft rockets and shells. Ibid., 72. Several of the central papers of the Gamow set in the late 1940s acknowledge support by the Bureau of Ordnance US Navy.

91. Gamow, *My World Line*, 150-51.

92. George Gamow, *Biography of the Earth: Its Past, Present, and Future* (New York: Viking, 1941), vii-viii.

93. Ibid., 234.

94. Ibid., 235.

95. Ibid.

96. George Gamow, "The Origin of Chemical Elements," *Journal of the Washington Academy of Sciences* (December 15, 1942): 353.

97. Ibid., 354.

98. Ibid., 355.

99. Ibid.

100. Of this work, Kragh comments that in late 1942, Gamow was speculating on the beginning of the universe as a vast fission. Gamow "must soon have realized that it was, in fact, excluded, or at least implausible, for he never returned to comparing the big bang with an enormous atomic bomb of the fission type. In spite of its speculative nature and brief lifetime, the hypothesis deserves to be remembered because it was the first attempt to give a physical picture of the big bang." It is clear however that, by Kragh's analysis, Gamow takes several years before dismissing the fission model entirely. Thus, as Kragh notes, "In the fall of 1945 Gamow thus had the essential idea of a revised big bang model building on a combination of nuclear physics and the Friedmann-Lemaître equations. Apparently he still believed that the big bang started with fission processes." Kragh, *Cosmology and Controversy*, 106, 107.

101. It should be added that the preface to the work is dated September 1942. Moreover, the copyright date is given as 1944, but the publication date on the title page is 1945. George Gamow, *Mr Tompkins Explores the Atom* (New York: Macmillan, 1945).

102. Ibid., 24.

103. Ibid., 42.

104. Ibid., 94.

105. Ibid., 95.

106. George Gamow, *Atomic Energy in Cosmic and Human Life: Fifty Years of Radioactivity* (Cambridge: Cambridge University Press, 1946), ix.

107. Ibid., 159

108. For a treatment of this history through an analysis of changing representations of place, see Jeffrey Sasha Davis, "Representing Place: 'Deserted Isles' and the Reproduction of Bikini Atoll," *Annals of the Association of American Geographers* 95, no. 3 (September 2005).

109. Correspondence G. Gamow to USN Captain McKeehan, 1946, in Gamow Miscellany, Binder, Box 30, George Gamow and Barbara Gamow Papers, Manuscript Division, Library of Congress. Gamow adds here: "Admiral Nimitz originally planned to attend all 10 lectures but was too busy and asked me to tell him everything in just one afternoon in his office."

110. George Gamow, "Expanding Universe and the Origin of Elements," *Physical Review*, "Letters to the Editor," September 13, 1946, 572.

111. "The transition from the slowed down to the free expansion took place at the epoch when the two terms [the linear scale of the universe and the average density of matter] were comparable, i.e. when l [the linear dimension] was about one thousandth of its present value. At this epoch the gravitational clustering of matter into stars, stellar clusters, and galaxies, probably must have taken place." Ibid., 573.

112. Ibid.

113. Ibid. Many of these details and developments were produced by Alpher and by his colleague Herman.

114. Teller, "The Work of Many People." As more of in the following chapter, Gamow's theory was repeatedly linked to an explosive and indeed thermonuclear beginning to the universe. Gamow at times made this connection himself, but also, in relation to potential misunderstandings, was annoyed by it. Thus, in a 1968 interview conducted in his home by Charles Wiener, Gamow observes, "Well, I don't like the word 'big bang;' I never call it 'big bang,' because it is kind of cliché. This was invented, I think, by steady-state cosmologists—'big bang' and also the 'fire ball' they call it, which has nothing to do with it—it's not fire ball at all. Nothing to do with the fire ball of atomic bomb. I call it radiation regime and meta regime. And there is no ball because you never see seams in it, so there is no ball if it means the surface of a ball." American Institute of Physics, "Oral Histories: George Gamow," https://www.aip.org/history-programs/niels-bohr-library/oral-histories/4325.

115. George Gamow, *One Two Three . . . Infinity*, 333–34.

116. Ibid., v.

CHAPTER SEVEN

1. See Ritchie Calder, "Bernal at War" in Brenda Swann and Francis Aprahamian, *J. D. Bernal: A Life in Science and in Politics* (London: Verso, 1999): esp. 165.

2. There has been recent research on the role of Bronowski during the war and thereafter as a public intellectual in the United Kingdom. See in particular Ralph Desmarais, "Jacob Bronowski: A Humanist Intellectual for an Atomic Age, 1946–1956," *British Journal for the History of Science* 45, no. 4 (2012). Desmarais emphasizes some of the dissonances between Bronowski's war work, his evasiveness with regard to it, and the scientific humanism he promulgated. See also Lisa Jardine, "Dad's Slide Rule Armageddon," *London Times*, November 28, 2010.

3. Jacob Bronowski, "Science and Human Values 1: The Creative Mind," *Universities Quarterly* 10, no. 3 (May 1956): 247–48.

4. Ibid., 247.

5. For an analysis of common associations and images in the nuclear age, see Spencer R. Weart, *Nuclear Fears: A History of Images* (Cambridge, MA: Harvard University Press, 1988).

6. Snow credits Bronowski in "The Two Cultures, a Second Look." See *C. P. Snow, The Two Cultures* (Cambridge: Cambridge University Press, 1998), esp. 54. Both Bronowski and Snow were members of the Gaitskell Group, a group that worked to align Labour Party politics with what its members regarded as the common interests of science and society. See Guy Ortolano, *The Two Cultures Controversy: Science, Literature and Cultural Politics in Postwar Britain* (Cambridge: Cambridge University Press, 2009), esp. 173–78.

7. For a discussion of the political dimensions of even the "purest" general relativity concerns, see Benjamin Wilson and David Kaiser, "Calculating Times: Radar, Ballistic Missiles, and Einstein's Relativity," in Naomi Oreskes and John Krige, ed., *Science and Technology in the Global Cold War* (Cambridge, MA: MIT Press, 2014).

8. Jacob Bronowski, "Science and Human Values 2: The Habit of Truth," *Universities Quarterly* 10, no. 4 (August 1956): 334.

9. Jacob Bronowski, "Science and Human Values 3: The Sense of Human Dignity," *Universities Quarterly* 11, no. 1 (November 1956): 26. To Bronowski's mind, given the origins of European ethics, its non-Western counterparts depended less on a commitment to truth. He went so far as to report on a cultural inability of Japanese people—the "bland, kindly civilizations of the East"—to answer with basic facts during his 1945 trip. While noting Japanese physicist's Hideki Yukawa's prediction of the meson, Bronowski at the same time saw Eastern societies as not typified by the kind of institutions that would allow work like Hideki's to thrive. Ibid., 35.

10. His colleagues more persuaded by Marxist thought largely held consonant views—ethics was seen as part of the historical, dialectical materialistic process.

11. In attempting to resituate the two-cultures debates in terms of an ideological battle over contemporaneous British society, rather than in terms of disciplinary divisions, Ortolano analyzes the "era of meritocracy." From this point of view, what the two antagonists, C. P. Snow and F. R. Leavis, had in common is at least as important, if not more so, than what separated them. Ortolano places them in the context of a "meritocratic moment" in the decades after the war, in which "specialized professionals" directed English institutions. See Ortolano, *The Two Cultures Controversy*, 1 and 16–22.

12. Julian Huxley, *UNESCO: Its Purpose and Its Philosophy* (Washington, DC: Public Affairs Press, 1948), 6. The document is struck through with such language, for example, of the "evolutionary progress" of "world evolutionary humanism." The divisions of knowledge, along with the purpose of the humanities, are accordingly viewed through the lens of evolutionary humanism. Ibid., 44–49.

13. Categorizations based on attitudes toward planning and expertise were not always hard and fast. Huxley was at times considered a planner and a progressive, but his own views shifted over time, and as many commentators observe, he tended to hold less radical views than those of

Haldane and Joseph Needham. At the same, Haldane appeared to express less faith in contemporary practical eugenics than did Huxley, when eugenics was so much a part of the imagination of planned societies.

14. William McGucken, *Scientists, Society, and State: The Social Relations of Science Movement in Great Britain 1931-1947* (Columbus: Ohio State University Press, 1984), 284.

15. Patrick Petitjean, "The Joint Establishment of the World Federation of Scientific Workers and of UNESCO after World War II," *Minerva* 46 no. 2 (2008): 256, 258. For Needham's relation to Marxism, see Gregory Blue, "Joseph Needham, Heterodox Marxism and the Social Background to Chinese Science," *Science and Society* 62, no. 2 (Summer 1998).

16. Harlow Shapley, *Through Rugged Ways to the Stars* (New York: Charles Scribner's Sons, 1969), 146-47.

17. Ibid.

18. Ibid., 148-49.

19. As Peter Galison has emphasized, this second phase of the Unity of Science movement cannot be easily assimilated to the first. The activity of the American UoS was based more on a non-semiotic/-logical interdisciplinarity, concentrating on the circulation of shared techniques and concepts. Galison has referred to this as a "science unified in pieces." Holton and Galison have documented the efforts of the participants, including Shapley. See Gerald Holton, "Ernst Mach and the Fortunes of Positivism in America," *Isis* 83, no. 1 (1992), and "Philipp Frank at Harvard University: His Work and His Influence," *Synthese* 153, no. 2 (2006); and Peter Galison, "The Americanization of Unity," *Daedalus* 127, no. 1, *Science in Culture* (1998): esp. 64-65.

20. For Shapley's efforts in regard to the conference, see James Gilbert, *Redeeming Culture: American Religion in an Age of Science* (Chicago: University of Chicago Press, 1997), 74-93. For his further role in the Institute on Religion in an Age of Science as a "center for cosmological speculation," see ibid., 273-95.

21. Philipp Frank, *Relativity: A Richer Truth* (Boston: Beacon Press, 1950), xiii.

22. Ibid., xv.

23. See Harlow Shapley, "The Fourth Adjustment," *American Scholar* 25 (1956).

24. Shapley was partly working through the theories of his senior colleague, Professor of Chemistry and of Biochemistry Lawrence Joseph Henderson. See Lawrence J. Henderson, *The Fitness of the Environment; an Inquiry into the Biological Significance of the Properties of Matter* (New York: Macmillan, 1913).

25. "Pope Pius XI Prize to Astronomer Shapley," *Harvard Crimson*, December 1, 1941. Five years earlier, in 1936, Lemaître had joined the Pontifical Academy of Sciences, reestablished that year by Pope Pius XI.

26. Frank, *Relativity*, 76-77.

27. See Owen Gingerich, s.v. "Shapley, Harlow," *Dictionary of Scientific Biography*, ed. C. C. Gillespie (New York: Charles Scribner and Sons, 1975). See also Ronald Edmund Doel, *Solar System Astronomy in America: Communities, Patronage and Interdisciplinary Science, 1920-1960* (Cambridge: Cambridge University Press, 1996). For an enthusiastic assessment of Shapley's establishment of the graduate program, see Zdeněk Kopal, *Of Stars and Men: Reminiscences of an Astronomer* (Bristol: A. Hilger, 1986), 162. For this evaluation of his collaborative research on the Magellanic Clouds following his appointment to the observatory, see Bart J. Bok, "Shapley's Researches on the Magellanic Clouds," *Publications of the Astronomical Society of the Pacific* 77, no. 459.

28. See Jessica Wang, *American Science in an Age of Anxiety: Scientists, Anticommunism, and the Cold War* (Chapel Hill: University of North Carolina Press, 1999), esp. 128, and 118-30. For a more personal recollection, see Kopal, *Of Stars and Men*, esp. 164.

29. J. D. Bernal, "The Physical Basis of Life," *Proceedings of the Physical Society* 62, no. 9 (1949): 537.

30. Ibid., 538.

31. Ibid., 544.

32. Ibid., 557.

33. Ibid.

34. Ibid., 558.

35. Ibid., 541.

36. Ibid.

37. Ibid.

38. Alpher recalled in this regard that "the main obstacle initially was our lack of detailed knowledge of the probability of various kinds of nuclear reactions among the species likely to be present at the onset of nucleosynthesis. We had just emerged from World War II, and data for such reactions—reactions driven by high temperatures and densities—were still classified or just beginning to be declassified from the Manhattan Project." Ralph A. Alpher and Robert Herman, *Genesis of the Big Bang* (Oxford: Oxford University Press, 2001), 75. Alpher's point, one corroborated by Gamow, is evinced by even a casual survey of the papers.

39. As Alpher and Herman later saw the matter, "What [Hughes] measured was the probability that a nucleus would absorb a neutron and become a new nucleus.... On the basis of Hughes's suggestive work, Alpher and Gamow calculated in a very approximate way the growth of abundances by neutron-capture reactions." Ibid., 76.

40. The story of the authorship is perhaps more famous than the contents of 1948 paper itself. Referring to the computations for the atom building described by this theory, Gamow remarks in a publication four years later that "the results of these calculations were first announced in a letter to *The Physical Review*, April 1, 1948. This was signed Alpher, Bethe, and Gamow, and is often referred to as the 'alphabetical article.' It seemed unfair to the Greek alphabet to have the article signed by Alpher and Gamow only, and so the name of Dr. Hans A. Bethe (*in absentia*) was inserted in preparing the manuscript for print. Dr. Bethe, who received a copy of the manuscript, did not object, and, as a matter of fact, was quite helpful in subsequent discussions. There was, however, a rumor that later, when the α, β, δ theory went temporarily on the rocks, Dr. Bethe seriously considered changing his name to Zacharias." George Gamow, *The Creation of the Universe* (New York: Viking, 1952), 65.

41. George Gamow, "The Origin of Chemical Elements," *Physical Review* 73, no. 7 (April 1, 1948): 803.

42. Ralph Alpher, "A Neutron-Capture Theory of the Formation and Relative Abundance of the Elements," *Physical Review* 74, no. 11 (December 1, 1948): 1581.

43. With such a picture in mind, Alpher and Gamow formulated a basic equation on the basis of this process, where the rate of building up of atoms of a given weight is proportional to the number of the atoms of the previous weight, less the current number of atoms of that weight, where the terms are weighted by a given atom's likelihood of capture—the capture cross-section. The idea behind the basic building-up equation is that atoms of previous weight are in a position to capture neutrons to form the next heaviest elements. At the same time, the number of atoms of that weight decrease when they capture neutrons and form elements of the next higher weight still. The likelihood of that capture in each case is taken into account by the capture cross-sections. It is in relation to this basic equation that the authors appealed to Hughes's data and, on this empirical ground, attempted to reproduce some of the element abundance data.

44. In the follow-up to the paper based on his thesis, Alpher conceded many open problems with the theory, particularly because it could not yet *actually* explain the emergence of heavier elements. Since there are no stable nuclei of atomic weight 5 and 8 (and 11), the building-up process from lighter elements could not cross the "mass-5 and mass-8 crevasses," as Gamow referred to them. Alpher, Gamow, and Herman were nevertheless optimistic about its future. Alpher and Herman published a succession of articles working out the details of the theory, the first of these papers appearing in the pages of the *Physical Review* two weeks after Alpher's Ylem paper.

45. Maria G. Mayer and Edward Teller, "On the Origin of Elements," *Physical Review* 76, no. 8 (October 15, 1949): 1226.

46. Ibid.

47. Ibid., 1229.

48. Thus, three years later, in the first edition of *Creation of the Universe*, Gamow states that Mayer and Teller "begin with the stage at which the individual fragments were reduced to a size of several miles in diameter and to a mass comparable to that of an average star (Lemaître's 'atom stars')." Gamow, *The Creation of the Universe*, 54.

49. William L. Laurence, "Physicists Trace Creation Miracle: Scientific Version of Genesis Provided by 3 Men's Studies, Disclosed at Meeting Here," *New York Times*, January 30, 1949. I haven't been able to obtain a transcript of the lecture; it's unclear how explicit the dramatic structure was in the lecture itself.

50. Gamow, *Creation of the Universe*, 137. In their later history, Alpher and Herman could likewise still draw a comparison: "It now appears that these elements [hydrogen, helium, lithium and their isotopes], perhaps along with small amounts of beryllium and boron (collectively called the light elements), must have been synthesized during the early minutes after the Big Bang, when conditions prevailed not unlike those associated with the temperature and pressure within the explosion of a hydrogen bomb." Alpher and Herman, *Genesis of the Big Bang*, 19.

51. Robert Sachs, "Maria Goeppert-Mayer, 1906-1972, a Biographical Memoir" (Washington, DC: National Academy of Sciences, 1979), 318-19, and interview of Edward Teller by Karen Fleckenstein on September 9, 1983, Niels Bohr Library & Archives, American Institute of Physics, College Park, MD, www.aip.org/history-programs/niels-bohr-library/oral-histories/4372.

52. Edward Teller, "The Work of Many People," *Science*, new series 121, no. 3139 (February 25, 1955): 272. It was approximately during this period of his consulting on the hydrogen bomb that Gamow was investigated by the Navy Department Loyalty Hearing Board. The first letter on this account dates in May 1950. Recalling their previous acquaintance on the Bikini tests, Gamow apparently applied directly to Admiral Nimitz, who had recently been appointed chairman of "The President's Commission on Internal Security and Individual Rights." The letter of acknowledgment is dated February 7, 1951. Correspondence Harry E. Benson to G. Gamow, February 7, 1951, *A Gamowian Miscellany* Red Book 3, Box 30, George Gamow and Barbara Gamow Papers, Manuscript Division, Library of Congress. The draft of Gamow's original letter, dated the previous month, reveals his irritation and the apparent confusion in the minds of the Board as to Gamow's knowledge and role. Correspondence G. Gamow to Admiral Nimitz, January 24, 1951, *A Gamowian Miscellany* Red Book 3, Box 30, Gamow Papers. As Gamow notes on his copy of the acknowledgement, "This commission never started working, and was soon dissolved." Gamow's papers were returned to him. Of Gamow's response to the interrogatory, Gamow adds to his copy in his own hand, "Few months after this was sent in, I was informed by Dr. Urner Liddel (from ONR) that my explanations were found unsatisfactory by the Board, and didn't do anything about it as far as Navy Dep. is concerned. G.G." Correspondence Eric Collins, Recorder Naval Department Loyalty Hearing Board to G. Gamow, May 31, 1950, Gamow Papers.

53. George Gamow, "On Relativistic Cosmogony," *Reviews of Modern Physics* 21, no. 3 (July 1949): 369. Here he extends the jest of the Alpher-Bethe-Gamow paper, noting that "the neutron-capture theory of the origin of atomic species developed by Alpher, Bethe, Gamow, and Delter [Robert Herman] suggests that different atomic nuclei were formed by the successive aggregation of neutrons and protons which formed the original hot ylem during the early highly compressed stages in the history of the universe." Ibid.

54. Ralph. A. Alpher and Robert C. Herman, "Theory of the Origin and Relative Abundance Distribution of the Elements," *Reviews of Modern Physics* 22, no. 2 (April-June 1950): 203.

55. Ibid., 207.

56. Ibid., 208. Gamow's argument against this theory in the more popular literature is that it does not explain the formation of the lighter elements without appeal to something like the frozen equilibrium theory, which he cannot see as easily reconcilable with the primeval atom. Gamow, *The Creation of the Universe*, 55. Also see the second, 1961 edition, 54.

57. Helge Kragh is largely dismissive of this account, noting that Bondi and Gold did not recall a connection between the film and the theory. See Helge Kragh, *Cosmology and Controversy: The Historical Development of Two Theories of the Universe* (Princeton: Princeton University Press, 1996), 173–74. What is more relevant for considerations here is the readiness to appeal to these temporal constructions through film, even if as historical reconstruction. Moving images, more specifically, were already a resource drawn on to domesticate the different possible histories. We've seen an example of this in the generation before Gamow, in the work of Eddington. Curiously enough, Gamow soon characterized the steady-state theory through conceptions of motion pictures: "If we were to make a motion picture representing the views of Bondi, Gold, and Hoyle, and run it backward, it would seem at first that all the galaxies on the screen were going to pile up as soon as we reached the date 1.7 billion years ago. But, as the film continues to run backward, we would notice that the nearby galaxies, which were approaching our Milky Way system from all sides, threatening to squeeze it into a pulp, would fade out into thin space long before they became a real danger. And, before the second-nearest number could converge on us . . . our own galaxy would fade too." Gamow, *The Creation of the Universe*, 32.

58. Fred Hoyle, *The Nature of the Universe* (New York: Harper and Brothers, 1950), 124.

59. George Gamow, *The Creation of the Universe*, vii.

60. Ibid., 139.

61. And both texts of 1952, the generalist and the brief technical letter, move toward the "hypothesis of the gaseous protogalaxies." Computation and reasoning were shared across the writings as well, suggesting that not only is the generalist text a representation and expansion of the technical, but the technical is itself a condensation and working into brevity of more generalist considerations. George Gamow, "The Role of Turbulence in the Evolution of the Universe," *Physical Review* 86, no. 2 (February 1952): 251.

62. Ibid.

63. Gamow, *The Creation of the Universe*, 30.

64. Ibid., 30, emphasis in the original.

65. This in a certain sense again recalls the Huttonian view of the history of the world—a succession of scenes, with only the present one readily visible. As opposed to Hutton's earth, there are no previous traces of past universes in Gamow's account. But as with Hutton, who balanced mechanistic and organic principles in order to produce his natural history, a similar balancing is at stake in Gamow's account, where aspects of the endless, organic whole are more radically unrepresentable.

66. Gamow, *The Creation of the Universe*, 29–30.

67. Ibid., 30.

68. One can always pose the question of what happened before the time of the narrative as established by any historical account, examining how the account crafts the beginning as the beginning, but usually on the presumption that there is at least time itself preceding that historical beginning.

69. Gamow, *Creation of the Universe*, 137.

70. As Gamow phrases it, "in the dim pregalactic past we perceive a glimpse of a metaphysical 'St. Augustine's Era' when the universe, whatever it was made of, was involved in a gigantic collapse." Ibid. Even aside from the wider context of his assertions, the perception of a glimpse of a metaphysical speculation suggests intellectual insight rather than detection by the senses or physical instruments of the senses.

71. As here: "Thus the table of relative abundances of atomic species may be thought of as the oldest document pertaining to the history of our universe." Ibid., 44.

72. Erwin Schrödinger, *Nature and the Greeks* and *Science and Humanism* (1954; Cambridge: Cambridge University Press, 1996), 108.

73. Alexander von Humboldt appealed to the term "Kosmographie" a century before to describe his own vision of a universalist discipline, which Shapley may have known. The term well predates this usage. See Maria M. Portuiondi, *Secret Science: Spanish Cosmography and the New World* (Chicago: University of Chicago Press, 2009).

74. Cosmography was a course notable for its reading list, reflecting Shapley's scholarly connections, interest in language, and attention to the syntheses at play in the works of others. The texts are generalist, interdisciplinary, and international, beginning with required readings of Jacob Bronowski's *Common Sense of Science* and A. I. Oparin's *The Origin of Life*, and including Rachel Carson's *The Sea around Us*, J. B. S. Haldane's *Science and Human Life*, Schrödinger's *What Is Life?*, Georges Lemaître's *The Primeval Atom*, and Philipp Frank's *Relativity: A Richer Truth*, among potential essay readings. Harlow Shapley, Required Readings, Fall Term 1954-55, Natural Sciences 115 and Short Essay Assignment, Fall Term 1954-55, Box 19, HUG 4773.10, Folder Natural Sciences 115, Harlow Shapley Papers, Harvard University Archives, Cambridge, MA.

75. See Joann Palmeri, "Astronomer beyond the Observatory: Harlow Shapley as Prophet of Science" (PhD diss., University of Oklahoma, 2000), 154-55, for a discussion of the choice of "Cosmography" over "Cosmogony" and esp. 169-71 for the reception of the course.

76. This in turn helped to shape a network of scholars actively interested in the promulgation of larger questions of universal history and set up one of the prominent narratives of a history of everything. For the archival trace of the course evolution, see Courses of Instruction, series HUC 8500.16, including boxes 7-10, Harvard University Archives. See also Official Registrar of Harvard University XLV-10—Summer School, series HU 75.25, Harvard University Archives for the Cosmogony course taught in the summer school program. Field, along with Gerrit L. Verschuur and Cyril Ponnamperuma, published *Cosmic Evolution: An Introduction to Astronomy* in the late 1970s, an introductory textbook taking "cosmic evolution" as its unifying theme. Field and Chaisson themselves collaborated on another text, the 1985 *The Invisible Universe: Probing the Frontiers of Astrophysics*. And Chaisson extended these efforts through an extensive list of publications and educational initiatives—efforts that, as with Shapley, have been both more narrowly research-based and academic, as well as consultative, generalist, and popular.

77. Shapley to Gamow, October 20, 1952, Box 89, HUG 4773.10, Folder George Gamow, Harlow Shapley Papers, Harvard University Archives, Cambridge, MA.

78. Note that Hoyle published a book in 1964, the title of which seems a play on Shapley's: Fred Hoyle, *Of Men and Galaxies* (Seattle: University of Washington Press, 1964). The book was based on Hoyle's lectures in the Danz series at the University of Washington, in which Francis Crick soon followed with *Of Molecules and Men*: Francis Crick, *Of Molecules and Men* (Seattle: University of Washington Press, 1966). Neither text appears to make overt mention of Shapley.

79. Harlow Shapley, *Of Stars and Men: The Human Response to an Expanding Universe* (Boston: Beacon Press, 1958), 51.

80. Shapley saw it in this light himself. See Palmeri, "Astronomer beyond the Observatory," 223-34.

81. The revised 1964 edition of *Of Stars and Men* included images from the animated film. Gamow's *Tompkins* series also inspired a film in which Gamow appeared, Stan Brakhage's *Mr. Tompkins inside Himself*. See next chapter.

82. Pat McGilligan and Faith Hubley, "An Interview," *Film Quarterly* 42, no. 2 (1988-1989): 14.

83. Shapley, *Of Stars and Men*, 21.

84. Ibid., 22-23.

85. Ibid., 37. Despite Gamow having lectured in his Cosmography course, this text suggests that Shapley didn't strongly distinguish between Lemaître's models and Gamow's.

86. Nevertheless, this alphabet does link to wider and later work, sharing a spirit for example with the presentation of Field and Chaisson's *The Invisible Universe*. There, the authors emphasize

that different wavelengths of radiation offer different surveys of the universe, and illustrate those wavelengths by associating them with a spectrum of objects of different sizes, from mountains down to the atomic nucleus. George B. Field and Eric J. Chaisson, *The Invisible Universe: Probing the Frontiers of Astrophysics* (Boston: Birkhäuser, 1985), esp. 4 and 7.

87. Michael Gordin suggested to me the link between the use of alphabets by these scientists and the broader interest in language. See in this regard Michael Gordin, *Scientific Babel: How Science Was Done Before and After Global English* (Chicago: University of Chicago Press, 2015). Lily Kay discusses the specific relationship between the biological discourse Gamow helped define and linguistics and research into ancient languages. See Lily Kay, *Who Wrote the Book of Life: A History of the Genetic Code* (Stanford, CA: Stanford University Press, 2000), esp. 128. Note also that in Bernal's 1947 lecture, he compared the logic upon which there appeared to be a common chemical basis to life, a conclusion he found in T. H. Huxley, to etymological work: "This is exactly the same logic that shows, for example, that all so-called Indo-European languages have a common set of root words, however much they have deviated afterwards." Bernal, "Physical Basis of Life," 548.

88. Reprinted in James D. Watson, *Genes, Girls, and Gamow* (Oxford: Oxford University Press, 2001).

89. This claim is made in the "first official club circular" of July 4, 1955. The official RNA-Tie Club letterhead lists the officers as follows: "Geo Gamow—Synthesiser; Jim Watson—Optimist; Francis Crick—Pessimist; Martynas Ycas [*sic*]—Archivist; Alek Rick—Lord Privy Seal." When Gamow writes this first circular to Cys (Martynas), he signs off as "Ala" above the typed words "The Synthesizer." Correspondence G. Gamow to Martynas Yčas, July 4, 1955, Folder Yčas, Martynas Re: RNA Coding, 1955, Box 7, George Gamow and Barbara Gamow Papers, Manuscript Division, Library of Congress.

90. E.g., "Synthesiser" in the spelling of the RNA-Tie Club letter-head. George Gamow, "first official club circular," July 4, 1955, Box 30, Folder RNATIE Club George Gamow and Barbara Gamow Papers, Manuscript Division, Library of Congress.

91. See George Gamow, *My World Line: An Informal Autobiography* (New York: Viking, 1970), 145, and Watson, *Genes, Girls, and Gamow*, 59-61.

92. See for example the conclusion of George Gamow, "Information Transfer in the Living Cell," *Scientific American* 193, no. 4 (1955): 78. Lily Kay also brings out the scalar structure at play in this piece, and provides an example of a generalist text preceding a specialist one: "That article, which raised its gaze from the microsanctum of nucleic acids-proteins transactions to the panoramic vistas of cells and organismic life, appeared a year earlier than its specialized counterpart." Kay, *Book of Life*, 154.

93. George Gamow, *Matter, Earth, and Sky* (1958; Englewood Cliffs: Prentice Hall, 1965), v-vi. In correspondence of 1955, Gamow asked to meet Shapley to discuss aspects of his textbook as it was being composed. Gamow to Shapley, September 11, 1955, Box 89, HUG 4773.10, Folder George Gamow, Harlow Shapley Papers, Harvard University Archives, Cambridge, MA. In the text, Gamow treats the senior astronomer's galactic discoveries, particularly Shapley's use of the period-luminosity relationship of cepheids, to help institute the standard candle method for determining distances to globular clusters. See Gamow to Shapley, March 28 [1955?], Box 89, HUG 4773.10, Folder George Gamow, Harlow Shapley Papers.

94. Gamow, *Matter, Earth, and Sky*, vi.

95. Bronowski, "The Sense of Human Dignity," 42.

96. His address was given in September 1955. An article based on the address, the source of my remarks, was published in the same year and republished as Jacob Bronowski, "The Educated Man in 1984," *Science*, new series 123, no. 3200 (April 27, 1956).

97. Ibid., 711. Bronowski referred to a 1945 British White Paper on the development of atomic energy during the Second World War. Churchill's letter is addressed to "General Ismay for Chiefs

of Staff Committee" and, according to the White Paper, dates from August 30, 1941. See *Manhattan Project: Official history and documents*. Microform. Washington, DC: University Publications of America, 1977, roll 1, book 1, volume 4, chapter 8, part 2, page 3. In the same sentence Bronowski quotes, Churchill goes on to note that "I feel we must not stand in the path of improvement, and I therefore think that action should be taken in the sense proposed by Lord Cherwell." Ibid., 3.

98. Bronowski, "The Educated Man in 1984," 711.

99. "H. G. Wells used to write stories in which tall, elegant engineers administered with perfect justice a society in which other people had nothing to do except to be happy: the Houyhnhnms administering the Yahoos. Wells used to think this a very fine world: but it was only 1984 or Aldous Huxley's *Brave New World*. A world run by specialists for the ignorant is, and will be, a slave world." Ibid., 712.

100. Like the Romanes lecture, this series also had its famed history: Arnold's retort to Huxley had been a Rede lecture.

101. Recent scholarship has testified to how much the debate that emerged—setting science against literature, Snow against Leavis—was itself an ideological argument. See Ortolano notes above. Nevertheless, its resonance in the United States was taken up, and often explicitly, by Americans who explicitly sought out a literary and increasingly mythic register, some who found in science as much or more aesthetic force than those arts the humanistic disciplines celebrated (such as E. O. Wilson) or those whose arguments tended to run counter to this (such as Eiseley).

CHAPTER EIGHT

1. More so apparently than the Huxleys living in the same period, at least in the view of American commentators. The comparison is drawn by a number of authors. See, for example, Andrew J. Angyal, *Loren Eiseley* (Boston: Twayne, 1983). These specific words come from Rudolph Umland, a writer who had known Eiseley since his college years. See Gale E. Christianson, *Fox at the Wood's Edge: A Biography of Loren Eiseley* (New York: Henry Holt, 1990), 108. The impact of Huxley's writing on Eiseley—particularly Huxley's *On a Piece of Chalk*, a piece to which Steven Weinberg also responded—is recounted by a number of authors. See "On a Piece of Chalk," in Steven Weinberg, *Dreams of a Final Theory* (New York: Vintage, 1993). Weinberg cites the volume of Huxley's lecture edited by Eiseley. Ibid., 284.

2. Thomas Huxley, *On a Piece of Chalk*, ed. Loren Eiseley (New York: Scribner, 1967), 10.

3. Stanley L. Miller, "A Production of Amino Acids under Possible Primitive Earth Conditions," *Science*, new series 117, no. 3046 (May 15, 1953). In the more prosaic prose of the scientific paper, Miller refers to the more general category of "electrical discharge," which would simulate lightning among other natural electric phenomena. Ibid., 528.

4. Eiseley in Huxley, *On a Piece of Chalk*, 16–18.

5. Six months earlier, in the April 25, 1953, issue of *Nature*, Watson and Crick's article, "A Structure for Deoxyribose Nucleic Acid" was published. Several foundational articles on the molecular structure of DNA also appeared that year, as did the results of Miller's advisor Harold Urey, who, along with Gerard Kuiper, Carl Friedrich von Wiezsäcker, and others, had in recent years become actively interested in the origins of the solar system, in a revival of "monistic" theories—theories positing an endogenous development of the solar system and, in the twentieth century, regarded at times as revised nebular hypotheses. See for example, Stephen G. Brush, *A History of Modern Planetary Physics* (Cambridge: Cambridge University Press, 1996), esp. 134–37. For a more extended treatment, see Brush, *Fruitful Encounters: The Origin of the Solar System and of the Moon from Chamberlin to Apollo* (Cambridge: Cambridge University Press, 2008). To this Urey added theorizations of the origins of life on earth. Apart from that very recent work, Miller's results cited only the prior work of Oparin and Bernal on the same subject.

6. Loren Eiseley, "The Secret of Life," *Harper's*, October 1953, 65.

7. Loren Eiseley, *The Invisible Pyramid* (London: Rupert Hart-Davis, 1971), 106.

8. However, science and its technological products were generally for Eiseley not one and the same. To this extent, he was not singular. To take just one example, in *Science and Humanism*, Erwin Schrödinger had observed that in looking to understand the value of the sciences, "a great many people ... are inclined to answer this question by pointing to the practical consequences of scientific achievements in transforming technology, industry, engineering, etc. . . . Few scientists will agree with this utilitarian appraisal of their endeavor." Schrödinger, *Nature and the Greeks* and *Science and Humanism* (1954; Cambridge: Cambridge University Press, 1996), 105-6. He further shed doubt on whether those practical consequences were generally conducive to happiness. Ibid., 107.

9. For recent work examining such a relationship of the sciences to larger cultural movements in the period, see Michael Gordin's analysis of Immanuel Velikovsky, which reveals the extent to which newer cultural contexts allowed for a wider hearing of Velikovsky's work (understood as science) in the last decade of Velikovsky's life, a reception Velikovsky first embraced, but from which ultimately he felt the need to withdraw. See Michael D. Gordin, *The Pseudoscience Wars: Immanuel Velikovsky and the Birth of the Modern Fringe* (Chicago: University of Chicago Press, 2012), in particular 163-93. Also see David Kaiser's analysis of the importance of cultural interests such as parapsychology to the physical sciences from the mid-1970s: David Kaiser, *How the Hippies Saved Physics: Science, Counterculture, and the Quantum Revival* (New York: W. W. Norton, 2011).

10. Eiseley, *The Invisible Pyramid*, 105.

11. Ibid., 106.

12. J. B. S. Haldane, *Daedalus or Science and the Future* (New York: E. P. Dutton, 1924), 29.

13. J. B. S. Haldane, "The Future of Man," *Harper's*, March 1932, 449.

14. Recall Bernal's critique of Waddington that "he has thrown away old myths and sanctions, but has felt the need to introduce new ones for justification." Bernal in C. H. Waddington, *Science and Ethics, an Essay by C. H. Waddington, together with a Discussion between the Author and E. W. Barnes, et al.* (London: George Allen and Unwin, 1942), 115.

15. J. D. Bernal, *The World, The Flesh and The Devil: An Enquiry into the Future of the Three Enemies of the Rational Soul* (London: Kegan Paul, Trench, Trubner & Co., 1929), 7.

16. So, for example, in *Essay on Man*, Cassirer saw mythic thought as correlated to synthetic worldviews predicated on sentiment, while modern scientific thought was correlated to analytic methods, to "classification and systematization." Ernst Cassirer, *An Essay on Man: An Introduction to a Philosophy of Human Culture* (Garden City, NY: Doubleday, 1953), 108. At the same time, Cassirer saw it as the task of a "philosophical synthesis" to discern a unity across the fields of cultural endeavors, predicated on actions, processes, and ultimately ends. Ibid., 95.

17. A number of commentators have discussed Eiseley as an essayist. Perhaps the most extended analysis is given in Leslie E. Gerber and Margaret McFadden, *Loren Eiseley* (New York: Frederick Ungar, 1983): "Preeminently and fundamentally, Loren Eiseley is an essayist." Ibid., 21. They argue that the essay form, through to its roots in Montaigne, suited Eiseley's subjects and personal tone: "Eiseley shared the discomfort that modern essayists have felt with urban, mechanized mass society." Ibid., 23. Peter Heidtmann emphasizes the synechdocal identifications Eiseley was able to construct "between his own condition and past, and the condition and past of the planet" or, identically, "the analogical connections between the microcosm of himself and the macrocosm of the earth." By these identifications, made possible through Eiseley's fieldwork and poetic ability, Eiseley "established himself as the brooding hero of his personal myth." Peter Heidtmann, *Loren Eiseley: A Modern Ishmael* (Hamden, CT: Archon, 1991), 41. It is a myth Heidtmann finds seductive: "And it is this figure—the wanderer or solitary seeker who, with Sisyphean perseverance, refuses to give up his struggle despite the threat of ultimate extinction—that captures our imagination." Ibid., 41.

18. Bernal argued that the older myths were internal to the science and expressed through it: "It was not, as we so often think, that Science had to fight an external enemy, the Church; it

was that the Church itself—its dogmas, its whole way of conceiving the Universe—was within the scientists themselves." Bernal in Waddington, *Science and Ethics*, 115. And still, Bernal had implied, within Waddington.

19. At one point, Barbara Gamow authored a light-hearted "New Genesis" on the basis of the cosmogonic theories of George Gamow's network, teasing Hoyle with the same words. Gamow, *My World Line*, 127.

20. Lily Kay, *Who Wrote the Book of Life? A History of the Genetic Code* (Stanford: Stanford University Press, 2000), 128.

21. Across the sources I've seen, it was not atypical to find Gamow read through Lemaître's earlier work.

22. He had quoted Lemaître's poetic figure of humanity as living on a cinder witnessing "the slow fading of suns." This poetic turn he combined with the storehouse motif, in words very close to Kelvin's: "Their stores of radioactive energy, though great, are not inexhaustible, and unless some unknown source of renewal awaits them, they are destined to go dark." Loren Eiseley, "Is Man Alone in Space?," *Scientific American* 189, no. 1 (July 1953): 81. This is a telling sentence, not only as a silent paraphrase, but in the neat replacement of the gravitational energies involved in Kelvin's picture with the contemporary emphasis on radioactive energies.

23. Helge Kragh, *Cosmology and Controversy: The Historical Development of Two Theories of the Universe* (Princeton: Princeton University Press, 1996), esp. 191.

24. Gamow is not entirely systematic with his updating. In both the 1952 and 1961 editions, he includes the sentence: "Since we know from observational evidence that the average distances between neighboring galaxies are about a hundred times greater than their mean diameters, we must place the separation era at about one-hundredth of the present age of the universe, at about the date when the universe was some 30 million years old." George Gamow, *The Creation of the Universe* (New York: Viking, 1952) and, Gamow, *The Creation of the Universe*, rev. ed. (New York: Viking, 1961), 78 in both editions. "If we could get H. G. Wells' 'time machine' and go back to the year 30,0000,000 A. C. (After Creation), we would find ourselves floating in an almost complete vacuum, comparable to that which exists today in the space between the stars inside our galaxy." Ibid., 79. The ease in the addition of two billion years to the antiquity of Gamow's Big Squeeze, and the slips in updating it, obliquely summarized how nearly a decade after the original publication, the Big Bang theory, as the Big Squeeze was increasingly being called, despite Gamow's preferences, had weathered the preceding years. Hoyle, for his part, did little to revise his characterization of Gamow's theory in the 1960 edition of his own *Nature of the Universe*.

25. Some element of the resistance to the change in the account may have to do with a kind of textual inertia working against extensive revisions. But in other contexts, Gamow was more than willing to make broad changes to his casting of scientific theory in and through popular texts, particularly when he regarded the story he had told as no longer valid. Over a larger time span, Gamow was eager to repudiate an earlier text altogether. Thus, in the December 1963 preface to his *A Star Called the Sun*, Gamow observed that "twenty-four years is much too long a period of time for any book on a rapidly developing subject to stand its ground and today *The Birth and Death of the Sun* is hopelessly obsolete. Thus I promised my publishers to write an entirely new book on the same subject, on the condition that they subject to auto-da-fé all the remaining copies of the old book except for those needed for bibliographic purposes. It is only natural that I dedicate my new book to the memory of the old one." George Gamow, *A Star Called the Sun* (New York: Viking, 1964), viii.

26. Gamow, *The Creation of the Universe*, 1st and rev. eds., 20.

27. Harlow Shapley, *The View from a Distant Star* (New York: Basic, 1964), 178. For Hubley quote, see Pat McGilligan and Faith Hubley, "An Interview," *Film Quarterly* 42, no. 2 (1988-1989): 14.

28. Shapley, *The View from a Distant Star*, 179.

29. Ibid., 176.

30. Ibid., 176-81.

31. Ibid., 179.

32. See Loren Eiseley, "Wallace and the Brain," in *Darwin's Century: Evolution and the Men Who Discovered It* (New York: Doubleday, 1958), esp. 304–14. See also Eiseley, *Invisible Pyramid*, 19.

33. For discussion of Verdansky's biosphere, as well as connections to Seuss, Teilhard, mathematician and Bergsonian Édouard Le Roy, and ecologist George Evelyn Hutchinson, see Alexej M. Ghilarov, "Vernadsky's Biosphere Concept: An Historical Perspective," *Quarterly Review of Biology* 70, no. 2 (1995) and Jonathan D. Oldfield and Denis J. B. Shaw, "V. I. Vernadskii and the Development of Biogeochemical Understandings of the Biosphere, c. 1880s–1968," *British Journal for the History of Science* 46, no. 2 (2013).

34. See Amy Mandelker, "Semiotizing the Sphere: Organicist Theory in Lotman, Bakhtin, and Vernadsky," *PMLA* 109, no. 3 (May 1994). In "The Physical Basis of Life," Bernal had emphasized Goldschmidt's usage of "biosphere," understood as "consisting materially of the group of complex organic compounds found almost exclusively in the watery layer on the surface of the earth, the hydrosphere, or in the adjoining regions of the atmosphere or parts of the lithosphere clearly derived from it." Bernal, "Physical Basis of Life," *Proceedings of the Physical Society* 62, no. 9 (1949): 539. For Goldsmith's importance in Vernadsky's reception, see Oldfield and Shaw, "Vernadskii and the Development of Biogeochemical Understandings of the Biosphere," 297.

35. W. I. Vernadsky, "The Biosphere and the Noösphere," *American Scientist* 33, no. 1 (January 1945): 7.

36. The term was associated with Teilhard. It was claimed at various points either to have been coined by him in the 1920s or to have been coined in discussion with him. Vernadsky noted that it was Le Roy who introduced the concept in 1927, in dialogue with Teilhard. Ibid., 9.

37. See Jacques Grinevald's "Introduction: The Invisibility of the Vernadskian Revolution," in Vladimir I. Vernadsky, *The Biosphere*, trans. David B. Langmuir (New York: Copernicus, 1998), esp. 22–23 and 24–25.

38. Julian Huxley noted an ambiguity in "noosphere" between "the total pattern of thinking organisms (i.e. human beings) and their activity," and the "special environment of man," which also included all humanity's intellectualized constructions, whether material or not. Huxley tentatively suggested "noosystem" for the latter, and "noosphere" exclusively for the former. See Pierre Teilhard de Chardin, *The Phenomenon of Man*, trans. Bernard Wall (New York: Harper, 1959), 13–14.

39. He refers to a passage of Huxley's "Science and Ethics" essay as "some masterly lines recently written," a text he finds in the 1932 collection *The Inequality of Man and Other Essays*.

40. Teilhard de Chardin, *The Phenomenon of Man*, 29.

41. Ibid., 54.

42. Bernal had asserted that "the events out of which so complicated a thing as the general state of the universe is built, form neither one indivisible whole nor a set of equally independent units, but consist of complexes . . . of which the components are themselves complex parts. This hierarchy of complexes is not imagined to have any objective validity." Bernal, *The World, The Flesh and The Devil*, 11.

43. Teilhard de Chardin, *The Phenomenon of Man*, 244. In footnotes, he drew a parallel between his arguments and Haldane's: Haldane had postulated the existence of "rudimentary forms" of mind in matter, at least as seen from the point of view of future observers. Teilhard also tied Haldane's more tentative invocation of the Comtean Great Being in his "Science and Ethics" essay to the conception of the Omega Point. Ibid., 57, 268. Whatever the thickness of the intersection between Teilhard and Haldane, the politics they saw as embracing and furthering such a metasynthesis were not so easily reconciled. Teilhard listed Communism, National Socialism, and the factory systems as examples of the "perversion of the rules of nöogenesis"—organizations of humanity that he saw as conducive to a dangerous mechanical simulacrum of holism. Ibid., 257. To Haldane, in contrast to his own earlier doubts as to the proper size of socialism, Communism had come closest to solving the question of the right political and scientific totality, a political animal whose right size might be global.

44. And apparently Haldane. However, I have not found a direct record of Haldane's opinions, thought they are alluded to in a number of sources. We return to Dobzhansky below.

45. See George Gaylord Simpson, "On the Remarkable Testament of the Jesuit Paleontologist Pierre Teilhard de Chardin," *Scientific American* 202 (April 1960): 204. Ernst Mayr invoked Teilhard in reference to the wrong kind of teleology in biologically inflected thought: there was the legitimate programmatic (teleonomic) purpose on the individual level, but no overall purpose to evolution as whole. Ernst Mayr, "Cause and Effect in Biology," *Science*, new series 134, no. 3489 (November 10, 1961): 1503.

46. He notes in the preface to the published form of the lectures that the subject of man's future was proposed to him by the BBC. Peter Medawar, *The Future of Man* (New York: Mentor, 1959), x. The first printing was in September 1961. For the transcripts of the lectures, see http://www.bbc.co.uk/radio4/features/the-reith-lectures/transcripts/1948/.

47. Vassiliki Betty Smocovitis underscores the centennial's synthetic program with organizer Sol Tax's attention to anthropology, as well arguing for the impetus the centennial gave to creationism, particularly through the negative response to Huxley's evolutionary humanistic convocational address or "secular sermon." Smocovitis, "The 1959 Darwin Centennial Celebration in America," *Osiris* 14 (1999): 305, 315–316, 320. Note that Muller recommended the participation of a young Carl Sagan. Ibid., 295n105.

48. Sol Tax and Charles Callender, ed., *Evolution after Darwin: The University of Chicago Centennial*, Issues in Evolution, vol. 3 (Chicago: University of Chicago Press, 1964), 58–59.

49. Ibid., 49.

50. Ibid., 64.

51. Julian Huxley, "The Evolutionary Vision: The Convocation Address," in *Evolution after Darwin*, ed. Tax and Callender.

52. Medawar, *The Future of Man*, 60. The words are identical to the transcript.

53. Medawar might have had directly in mind the Reith Lectures of the year before given by the radio astronomer Bernard Lovell, entitled "The Individual and the Universe." For the transcripts, see http://www.bbc.co.uk/radio4/features/the-reith-lectures/transcripts/1948/.

54. Medawar, *The Future of Man*, 94–95.

55. P. B. Medawar, "Critical Notice: *The Phenomenon of Man* by Pierre Teilhard de Chardin," *Mind*, new series 70, no. 277 (January 1961): 101.

56. Ibid., 105.

57. Ibid.

58. Ibid., 106.

59. For example, in his second Reith Lecture "The Meaning of Fitness," Medawar claimed that "it is of the physical nature of nucleic acids that they can offer up for selection ever more complex sets of genetical instructions, can propose ever more complex solutions of the problem of remaining alive and reproducing. Every now and again one of these more complex solutions will be accepted, and so there is always a certain pre-existing inducement or authority for evolution to have what, in retrospect, we call an 'upward trend.'" He immediately qualified: "This is extremely lame and halting but, as I said in my first lecture, it is a more useful way of trying to explain the phenomenon than to talk about a 'vital force' of some kind which inspires organisms to advance in evolutionary history."

60. Shapley, *The View From a Distant Star*, 35.

61. Ibid., 84.

62. Ibid., 84–85.

63. Haldane aimed at something similar to Medawar's initial intent with his Reith Lectures, in his paper, "Biological Possibilities in the Next Ten Thousand Years," not a new subject to the author of *Daedalus*. At the start of his remarks, Haldane outlined several possible future cataclysms, but the central questions of Haldane's remarks in the conference turn on his final, preferred alternative:

a collective world organization, allowing him at the same time to examine the underpinnings of views such as those of Huxley. Gordon Wolstenholme, ed., *Man and His Future: A Ciba Foundation Volume* (Boston: Little, Brown, 1963).

64. He made this point in the context of the observation about "what may be the most important ethical fact about applied science, namely, that it magnifies pre-existing evils until they are seen to be intolerable. . . . When an evil is sufficiently magnified everyone recognizes it as an evil, and that is one of the things science does." Ibid., 381.

65. Ibid., 284-88. Note Bronowski's claim in the conference: "The deeper effect of science over the past three hundred years has been, not in the accumulation of true facts, but in making people aware that the very search for what is factually true is itself an ethical activity." Ibid., 370.

66. For a later statement of appreciation of Ciba, see Peter Medawar, *Memoir of a Thinking Radish: An Autobiography* (Oxford: Oxford University Press, 1986), 116.

67. Julian Huxley, in *Man and His Future*, ed. Wolstenholme, 2. Huxley also claimed that an integrated curriculum "would go a long way, not only towards bridging the gap between C. P. Snow's two cultures, but also towards making the process of learning much more interesting and enjoyable." Ibid., 188. Many of Huxley's points intersect with other texts, such as Julian Huxley, "Education and Humanism," 1962, in *Evolutionary Humanism* (1964; New York: Prometheus Books, 1992), and his Darwin Centennial address.

68. Huxley, in *Man and His Future*, ed. Wolstenholme, 6.

69. Ibid., 21-22.

70. Loren Eiseley, *Darwin's Century: Evolution and the Men Who Discovered It* (New York: Doubleday, 1958), 306.

71. Ibid., 310-12, 321.

72. Ibid., 296.

73. Loren Eiseley, *The Mind as Nature* (New York: Harper Row, 1962), 44-46.

74. The course was described as discussing "man's relationship to the universe, the answers of the major religions to this problem, the origins of the earth, and the development of the moral concepts which help to distinguish man from other creatures." "Dr Eiseley Launches New McHarg TV Program," *University of Pennsylvania Almanac* 7, no. 6 (February 1961): 2. See also, "Huxley, Julian" and "Television" Files, Box 2, Loren Corey Eiseley Papers, University of Pennsylvania Archives, Philadelphia, PA.

75. It is not clear Eiseley ever saw the film. See Correspondence to Gloria Glissmeyer from Lester Novros, February 7, 1975, Loren Corey Eiseley Papers, University of Pennsylvania Archives.

76. His papers include various inquiries as to potential projects, as well as requests for his work. See, for example, "The Unexpected Universe by Loren Eiseley: Reprint Permissions," Box 16. These requests were not always satisfied: for example, a request on the part of the English Department of the Laboratory School of the University of Illinois to use an article the materials of which were to be employed in *The Unexpected Universe* was turned down. Correspondence from Loren Eiseley to Edward R. Levy, November 29, 1966, Loren Corey Eiseley Papers, University of Pennsylvania Archives.

77. Loren Eiseley et al., *The Shape of Likelihood: Relevance and the University* (Tuscaloosa: University of Alabama Press, 1971), 6. The quote is from Joseph Campbell, *The Masks of God: Primitive Mythology* (New York: Viking, 1969), 3. Note that the year before, Bronowski and Eiseley had spent time together and in May 1970, Bronowski wrote Eiseley to declare that "I greatly profited by our several discussions on the American Indian and his technologies." On letterhead declaring his affiliation with the Council for Biology in Human Affairs, Bronowski asked a further set of questions relating to human-centered natural history. Correspondence, to Loren Eiseley from Jacob Bronowski, May 7, 1970, Box 2, "Bronowski, J," Loren Corey Eiseley Papers, University of Pennsylvania Archives. For background on Bronowski's role in the council and at the Salk Institute, see Suzanne Bourgeois, *Genesis of the Salk Institute: The Epic of Its Founders* (Berkeley: University of California Press, 2013), esp. chapter 11, "Biology in Human Affairs." These points of natural history might also have been of interest to Bronowski as background research for his "Ascent of Man" project, a television series he had recently begun working on with the BBC and to which we return.

78. Ibid., 48.

79. Loren Eiseley, "Little Men and Flying Saucers," *Harper's*, March 1, 1953, 86.

80. Ibid., 91.

81. Steven Dick, *Life on Other Worlds: The 20th-Century Extraterrestrial Life Debate* (Cambridge: Cambridge University Press, 1998), 212; and, see further, chapter 7, "SETI: The Search for Extraterrestrial Intelligence," for a history of SETI into the 1990s. See also Stephen J. Garber, "Searching for Good Science: The Cancellation of NASA's SETI Program," *Journal of the British Interplanetary Society* 52 (1999) and Steven J. Dick and James E. Strick, *The Living Universe: NASA and the Development of Astrobiology* (New Brunswick, NJ: Rutgers University Press, 2005), esp. chapter 6, "The Search for Extraterrestrial Intelligence."

82. Eiseley, "Is Man Alone in Space?," 86.

83. "Astronautical and Aeronautical Events of 1962," Report of the National Aeronautics and Space Administration to the Committee on Science and Astronautics, US House of Representatives, Eighty-Eighth Congress, First Session (June 12, 1963), 280.

84. Eiseley, *The Invisible Pyramid*, 132.

85. Edward O. Wilson, *Naturalist* (Washington, DC: Island, 1994). See in particular 218-37. In relation to the microcosm of Harvard, Wilson reflected on this in anthropological terms, the molecular and cellular biologists dominating and eradicating other subcultures/tribes. Ibid., 220-21.

86. By the time of the address, Dobzhansky and Eiseley were already acquainted, at least through correspondence. In a letter from Dobzhansky to Eiseley dating to 1959, Dobzhansky makes clear that they have yet to meet in person. Correspondence, Th. Dobzhansky to Loren Eiseley, June 30, 1959, Dobzhansky, Theodosius File, Box 2, Loren Corey Eiseley Papers, University of Pennsylvania Archives.

87. See Theodosius Dobzhansky, *Evolution, Genetics, and Man* (New York: John Wiley and Sons, 1955), 1-2, 21.

88. Theodosius Dobzhansky, "Biology, Molecular and Organismic," *American Zoologist* 4, no. 4 (November 1964): 443. As with Mayr's division between functional and evolutionary biology, Dobzhansky split biology into molecular and organismic. And as Mayr had clarified his categorization of biological cause in dialogue with and in partial objection to the views of the philosopher Ernest Nagel, Dobzhansky used Nagel's philosophical formulation of reduction to demonstrate the limits of any reductionist spirit. He believed such a vision was at play in Comte's nineteenth-century positivism and found that in the 1960s, Comte was practically unknown or no longer a subject of discussion. Despite that fact, "the Comtian hierarchy of sciences, and the faith in reduction as the intent of scientific inquiry, nevertheless persist and are seldom questioned among scientists, especially among biologists." Ibid., 446.

89. Ibid., 444.

90. Ibid., 449.

91. Ibid.

92. Ibid., 451.

93. Theodosius Dobzhansky, "Mendelism, Darwinism, and Evolutionism," *Proceedings of the American Philosophical Society* 109, no. 4, Commemoration of the Publication of Gregor Mendel's Pioneer Experiments in Genetics (August 18, 1965): 214.

94. Ibid., 214-15. Teilhard explicitly defends his use of "orthogenesis" in a footnote: "On the pretext of its being used in various questionable or restricted senses, or of its having a metaphysical flavour, some biologists would like to suppress the word 'orthogenesis.' But my considered opinion is that the word is essential and indispensible for singling out and affirming the manifest property of living matter to form a system in which 'terms *succeed each other* experimentally, following the constantly increasing values of centro-complexity.'" Teilhard de Chardin, *Phenomenon of Man*, 108.

95. Teilhard de Chardin, *Phenomenon of Man*, 244, and Dobzhansky, "Mendelism, Darwinism, and Evolutionism," 215.

96. In a later interview, he confessed that by the late 1960s he had "developed a head of steam against particle physics." He also says that he "built up quite a head of resentment against

particle physics." Interview of Philip Anderson by Alexei Kojevnikov on June 29, 2000, Niels Bohr Library & Archives, American Institute of Physics,College Park, MD, http://www.aip.org/history /ohilist/23362_4.html. Established physics departments at Yale, Columbia, and Princeton had not, he noted, invested in condensed matter physics. And, since his own election to it in 1967, he had witnessed scientists being overlooked or blocked from induction to the National Academy of Science on the basis of their subdiscipline.

97. P. W. Anderson, "More Is Different," *Science*, new series 177, no. 4047 (August 4, 1972): 396.

98. Ibid., 393. For background on and analysis of Anderson's views among an array of scientists weighing in on the question of foundationalism and unity, see in particular Jordi Cat, "The Physicists' Debate on Unification at the End of the 20th Century," *Historical Studies in the Physical and Biological Sciences* 28, no. 2 (1998), and Joseph D. Martin, "Fundamental Disputations: The Philosophical Debates that Governed American Physics, 1939-1993," *Historical Studies in the Natural Sciences* 45, no. 5 (November 2015).

99. He began with a simplified example of symmetry breaking, the theorization of which he had been central in developing. Recalling a university lesson that in a stationary state, no physical system will have an electric dipole moment, Anderson took up another case familiar for him from graduate school, the ammonia molecule. Its pyramid structure would, he observed, invert, so that an ammonia particle at the top of a pyramid made up of itself and a floor of three hydrogen atoms would flip. A truly stationary state had to be understood as the superposition of two states: nitrogen at the peak of the pyramid and the inverted upside down pyramid, standing on its nitrogen head. Such a superposition would therefore abide by those general considerations that disallowed the possibility of such a stationary dipole. However, as the molecules under examination grew larger, reaching the size of sugars associated with organic processes, inversion effects such as with ammonia became impossible. "At this point we must forget about the possibility of inversion and ignore the parity symmetry [of left-handed and right-handed structures]: the symmetry laws have been, not repealed, but broken." Ibid., 394.

100. This was most evident in Anderson's claim that "in some sense, structure—functional structure in a teleological sense, as opposed to mere crystalline shape—must also be considered a stage, possibly intermediate between crystallinity and information strings, in the hierarchy of broken symmetries." The hierarchy of such broken symmetries, like the breaking of left-right symmetries in the movement up to large molecules, traced the formation of new basic entities with new laws at new scales. "To pile speculation on speculation, I would say that the next stage could be hierarchy or specialization of function, or both. At some point we have to stop talking about decreasing symmetry and start calling it increasing complication. Thus, with increasing complication at each stage, we go on up the hierarchy of the sciences." Ibid., 396.

101. Ibid.

102. Ibid.

103. Ibid., 393. Anderson thus shared with Whewell an emphasis on syllogistic descent, but so emphatically as to deny, contra Herschel, the possibility of any meaningful ascent.

104. He drew near his conclusion with a sentiment not from Engels but a related one looking to Marx, that "quantitative differences become qualitative ones." Ibid., 396.

105. In a 1961 *Scientific American* article on gravity Gamow quoted from the laboratory diaries of Faraday, invoking the famous case of the unification of electricity and magnetism. Pointing to Faraday's failure to find a further connection to gravity, Gamow noted that most scientists in his day regarded this underlying possibility of a "unified field theory" as "futile." George Gamow, "Gravity," *Scientific American* 204, no. 3 (March 1961): 102.

106. Henri Lefebvre, *The Urban Revolution*, trans. Robert Bononno (1970; Minneapolis: University of Minnesota Press, 2003), 74-75. These elements function as dogmas of unity to the extent that they do not form concrete and concretized possibilities. All these concepts Lefebvre found unsatisfactory and saw instead the possibility of launching an idea of "urban strategy" in which urban

phenomena/actuality serve as model instead. Ibid., 76. The urban is "pure form: a place of encounter, assembly, simultaneity," characterized by notions of "centrality" (as for example in the institutions of the state) and "poly-centrality" (as with trends to "dispersion and segregation"). Ibid., 119-21.

107. The logic of the foundational and the picaresque syntheses were not in this period as clearly pervasive as the historical or indeed the scalar synthesis underpinning narratives of scalar traverse. More work is necessary to tease them out of some of the different branches of scientific production, and they did not as obviously motivate or justify collective scientific engagements. They require further analysis to see their roles in other synthetic schemes, but they offer some clue as to why prominent scientists in the period often found them less compelling as far the formation and conceptualization of professional networks and popular representations were concerned.

108. Correspondence G. Gamow to Martynas Yčas, December 6, 1966, Folder Yčas, Martynas Re: Book, 1964-67, Box 7, George Gamow and Barbara Gamow Papers, Manuscript Division, Library of Congress.

109. Michael J. Golec, "Optical Constancy, Discontinuity, and Nondiscontinuity in the Eameses' *Rough Sketch*," in *The Educated Eye: Visual Culture and Pedagogy in the Life Sciences*, ed. Nancy Anderson and Michael R. Dietrich (Hanover: Dartmouth College Press, 2012), 164. Golec discusses Charles Eames's notion of "found education" (see below), which bears an interesting relationship to the pedagogical dimensions of the synthetic genres considered here. See esp. 163.

110. Arthur H. Compton, introduction to Kees Boeke, *Cosmic View: The Universe in 40 Jumps* (New York: John Day, 1957), 3.

111. Ibid., 4.

112. Ibid. The invocation of *Alice in Wonderland* yet again suggests how much the fabulaic and the scalar are woven together in synthesizing patterns.

113. Ibid., 21. But also note: "We advise you, reader, to concentrate your thought on this return journey. Try to picture how what you see in front of you would extend and extend as you came down . . ." (ellipsis in the original). Ibid., 32. The language of such journeying suggests a falling through frames, but as a discursive matter, rather than the time characteristic of physical possibility.

114. Golec, "Optical Constancy, Discontinuity, and Nondiscontinuity," 168.

115. Golec finds in the use of the man in *Rough Sketch* the gendered construction of science. Ibid., 171.

116. Golec ties Bronowski's filmic efforts to Eames's notion of found knowledge, a notion that examines discontinuous conceptions of knowledge making. Ibid., 164-65.

CHAPTER NINE

1. See, for example, "Multiculturalism and Universalism: A History and Critique," ed. John Higham, special issue, *American Quarterly* 45, no. 2 (June 1993) for contributions by John Higham, Gerald Early, Gary Gerstle, Nancy A. Hewitt, and Vicki L. Ruiz, structured on the basis of responses to Higham's reflections in defense of a variety of what he termed "American Universalism." For the recognition of past provincially minded universalisms, see David Hollinger, *Postethnic America: Beyond Multiculturalism* (New York: Basic, 2006), 52-54.

2. David A. Hollinger, "How Wide the Circle of the 'We'? American Intellectuals and the Problem of the Ethnos since World War II," *American Historical Review* 98, no. 2 (April 1993): 319.

3. Ibid. Here I must emphasize that despite academic multicultural critique, science popularization did posit and sustain a culturally resonant species-wide voice. I observe and attend to a slightly different social pattern than does Hollinger.

4. Harold R. Isaacs, *Idols of the Tribe: Group Identity and Political Change* (Cambridge, MA: Harvard University Press, 1975), 11.

5. Hollinger, "How Wide the Circle of the 'We'?," 319. In his book *Postethnic America: Beyond Multiculturalism*, Hollinger promotes this point to the body of the text, and clarifies the notion of masculinity here adduced: "*The Ascent of Man* might have done very well if it had appeared fifteen or twenty years earlier, in the era of Steichen's *Family of Man*. Bronowski's vigorously species-centered account of the progressive steps taken by human beings, most often through the activities of male inventors and scientists, was soon all but forgotten." Hollinger, *Postethnic America*, 57.

6. For example, Steven Woolgar and Bruno Latour claimed that "scientists in our laboratory constitute a tribe whose daily manipulation and production of objects is in danger of being misunderstood, if accorded the high status with which its outputs are sometimes greeted by the outside world." Steve Woolgar and Bruno Latour, *Laboratory Life: The Construction of Scientific Facts* (Princeton: Princeton University Press, 1986), 29. Or in the course of usage: "Our anthropological observer is thus confronted with a strange tribe who spend the greatest part of their day coding, marking, altering, correcting, reading, and writing." Ibid., 49. Meanwhile, anthropologists such as Jack Goody could be read as upsetting dichotomies used to separate anthropological studies of East versus West, past versus present, further extending the possibility of painting the modern emergence of science using a more particularizing palette of anthropological concepts that could in turn reflect back on the activity of science in the West. See Jack Goody, *The Domestication of the Savage Mind* (Cambridge: Cambridge University Press, 1977), 9, 36-37. For later reflections on the possibilities of these and other exercises, see Bruno Latour, "Postmodern? No, Simply Amodern! Steps towards an Anthropology of Science," *Studies in the History and Philosophy of Science* 21, no. 1 (1990).

7. Hollinger, *Postethnic America*, 57.

8. Here, I have in mind Werner Sollors's analysis of consent versus descent relations in the context of American culture. See below and in particular, Werner Sollors, *Beyond Ethnicity: Consent and Descent in American Culture* (New York: Oxford University Press, 1986).

9. Attenborough interview included in materials with the DVD release of the series.

10. Jacob Bronowski, *The Ascent of Man* (Boston: Little, Brown, 1973), 15. Much of the accompanying text is faithful to the series. When the words are identical, I use the punctuation chosen in the text. When the words differ, I use the formulations given in the broadcast.

11. For the transcript of the interview see Marjorie E. Hoachlander, *The Ascent of Man: A Multiple of Uses? A Case Study*, vol. 2 (New York: Center for Advanced Study in Education, Graduate School and University Center of the City University of New York, 1977), 72. Rita Bronowski recalled that her husband claimed, "'If I do anything at all, it will be what I am working on now,' which was about 'human specificity'—what makes humans different from animals." Ibid.

12. Bronowski makes this point in the preface to the book accompanying the series. He also makes a contrast that no longer as clearly obtains: "The printed book . . . is not remorselessly bound to the forward direction of time, as any spoken discourse is. The reader can do what the viewer and the listener cannot, which is to pause and reflect, turn the pages back and the argument over, compare one fact with another and, in general, appreciate the detail of evidence without being distracted by it." Bronowski, *The Ascent of Man*, 14.

13. Also in Bronowski, *The Ascent of Man*, 19-20.

14. Ibid., 20.

15. UNESCO World Heritage List, "Wieliczka and Bochnia Royal Salt Mines," http://whc.unesco.org/en/list/32.

16. The views he presents here also predate the series, so late in his life. See Jacob Bronowski, *The Common Sense of Science* (London: Heinemann, 1951), 10.

17. Bronowski, *The Ascent of Man*, 279, and television series, "The Drive for Power."

18. Margaret and Geoffrey Burbridge, William Fowler, and Fred Hoyle, "Synthesis of the Elements in Stars," *Reviews of Modern Physics* 29, no. 4 (October 1957).

19. Bronowski, *The Ascent of Man*, 340.

20. These were not idiosyncratic positions. In a not altogether distant example, Fred Hoyle airs some consonant views in the first lecture in *Of Men and Galaxies*, "Motives and Aims of the Scientist." Hoyle, *Of Men and Galaxies* (Seattle: University of Washington Press, 1964), 4.

21. The distinctive punctuation is as in the text.

22. Geoffrey Wansell, "Bronowski, Messiah of Culture," London *Times*, April 25, 1973, 18.

23. Among the reviews of books and text, see Robert Kirsch, "'Ascent' of a Unique Animal," *Los Angeles Times*, August 15, 1974, who emphasizes the breadth and depth of Bronowski's popularization.

24. Les Brown, "Colleges Offer 'Ascent of Man': 25,000 Students Will Earn Credits With TV Series," *New York Times*, January 7, 1975.

25. Tom Zito, "The Paradox of Man, and the Poetry Of His 'Ascent': 'The Ascent of Man,'" *Washington Post*, January 7, 1975.

26. See, for example, Jane Oppenheimer, "The Ascent of Man by J. Bronowski," *Quarterly Review of Biology* 50, no. 3 (September 1975) and Bruce S. Eastwood, "Concepts of Science: A Personal View: The Ascent of Man by Jacob Bronowski," *Isis* 66, no. 3 (September 1975).

27. Brown, "Colleges Offer 'Ascent of Man.'"

28. Hoachlander, *The Ascent of Man*, 1:4 and 2:5.

29. Ibid., 1:7 and 2:17.

30. "A third of the schools offered *The Ascent of Man* as an interdisciplinary course, making it available through extension studies, Continuing Education or an equivalent framework for nontraditional courses. In 19 percent of the schools, it was offered within the Humanities Department; in 16 percent, within History; and in 17 percent, within Science. Placement in Anthropology was reported by 11 percent of the schools, most of these being four-year private institutions." Ibid., 2:18.

31. Student responses to survey data were ultimately not robust enough to give more than an "indicative" rather than a representative sense of its use. Ibid., 1:9. The sample had a large number above the age of twenty-six (roughly half) and more women than men (57 to 43 percent), reflecting, it was believed, the advertisement of the course as a stay-at-home offering, with a large percentage (72 percent) working at the same time. Despite such constraints, the limited interactions with instructors, and critiques of exams within the courses in the sample, the rate of attrition among respondents was low (92 percent apparently completed the courses) and even of those who dropped the course, the majority reported watching the remaining series. Ibid., 1:10.

32. See Gerard P. Kuiper, "On the Origin of the Lunar Surface," *Proceedings of the National Academy of Sciences of the United States of America* 40, no. 12 (December 15, 1954): and Urey's response, Harold C. Urey, "Some Criticisms of 'On the Origin of the Lunar Surface Features,' *Proceedings of the National Academy of Sciences of the United States of America* 41, no. 7 (July 15, 1955). See also Dale P. Cruikshank, *Gerard Peter Kuiper, 1905–1973, A Biographical Memoir* (Washington, DC: National Academy of Sciences, 1993).

33. For this and the early history of the life sciences and exobiology at NASA, see Steven J. Dick and James E. Strick, *The Living Universe: NASA and the Development of Astrobiology* (New Brunswick: Rutgers University Press, 2004), esp. chapter 2, "Organizing Exobiology: NASA Enters Life Science." Also see Audra J. Wolfe, "Germs in Space: Joshua Lederberg, Exobiology, and the Public Imagination, 1958–1964," *Isis* 93, no. 2 (June 2002).

34. Carl Edward Sagan, "Physical Studies of Planets," PhD diss., University of Chicago, 1960, 12.

35. Container 13, Seth MacFarlane Collection of the Carl Sagan and Ann Druyan Archive, Manuscript Division, Library of Congress, Washington, DC. Haldane and Sagan were engaged in correspondence over the possibility of communication with intelligences elsewhere, appealing it seems to the terms of the newly minted Drake Equation (Frank Drake having first published on it in 1961). Haldane found Sagan's estimates of some of the individual factors of the equation too optimistic.

36. As Strick and Dick observe, "with NASA as the matchmaker, the two fields" concerned with the origins of life and extraterrestrial life "had been wed, merged together to create a new discipline, exobiology. So exhilarating was the wedding that by the early 1970s hardly anyone could imagine that working on the origin of life problem had not always been part and parcel of the search for life on other planets." Dick and Strick, *The Living Universe*, 24.

37. For example, the eighth such conference, "The Problems of Stellar Evolution and Cosmology" (1942) and the "The Physics of Living Matter" (1946-47). Washington Conferences on Theoretical Physics, 1st-10th Conferences, 1933-1947, Folder: 8th Wash. Conf—1942, George Gamow Papers, Niels Bohr Library, Center for the History of Physics. The existence of Sagan's conference was brought to my attention by Ann Druyan.

38. In a 1976 interview, Sagan recollected a conversation with a high school science teacher who cited Harlow Shapley as an example of astronomer who likely earned a salary through the profession, suggesting to the young Sagan that it was a practical pursuit. See Tom Head, ed., *Conversations with Carl Sagan* (Jackson: University Press of Mississippi, 2006), 28. In her collaborative work with Sagan, Druyan credits Eiseley in particular as another model.

39. Robert Walgate, "Worrying Wins Out," *New Scientist* 17 (February 1977), 404.

40. Druyan recalls an episode in which Sagan, having successfully invited Gamow to speak at the University of Chicago in his graduate student lecture series, was the object of the sarcasm of faculty members, the lecture having been published in the local press. Ann Druyan, interview with the author, July 21, 2009. On misleading reports, see William Poundstone, *Carl Sagan: A Life in the Cosmos* (New York: Henry Holt, 1999), 43-44.

41. Among these, he later underscored "high energy behavior of Feynman graphs, second-class weak interaction currents, broken symmetries, scattering theory, muon physics." See Steven Weinberg, "Steven Weinberg—Biographical," Nobelprize.org, http://www.nobelprize.org/nobel _prizes/physics/laureates/1979/weinberg-bio.html.

42. See Edward O. Wilson, *Naturalist* (Washington, DC: Island, 1994), esp. 312-14.

43. Ibid., esp. 44-45, and 110-11.

44. Wilson recalls interactions with Weinberg in which Weinberg expressed admiration for Wilson's published ideas and their reception before the publication of Weinberg's *The First Three Minutes*. E. O. Wilson, correspondence with the author, June 19, 2010.

45. Wilson, *Naturalist*, 374.

46. Edward O. Wilson, *Sociobiology: The New Synthesis* (Cambridge, MA: Harvard University Press, 1975). In November 1975, the *New York Review of Books* had published a sharp-edged attack on *Sociobiology* following a largely positive review given by C. H. Waddington. Elizabeth Allen et al., "Against 'Sociobiology,'" *New York Review of Books*, November 13, 1975. The letter depicted Wilson's work as empirically baseless and ethically compromised, linked to views that had helped nurture repression and atrocity and conspiring to establish as natural the social prerogatives of those such as Wilson. With fellow geneticist Jon Beckwith, one of the signees of the *New York Review of Books* letter, Richard Lewontin established the Sociobiology Study Group (SSG), which survived several years thereafter, with Lewontin and Stephen Jay Gould eventually playing limited or no roles. Neil Jumonville, "The Cultural Politics of the Sociobiology Debate," *Journal of the History of Biology* 35, no. 3 (Autumn 2002): 573 and 585. Placing the formation of this group in the mid-1970s in the context of American cultural and intellectual history, Jumonville emphasizes that the formation of the SSG took place when "student activists of the previous decade moved into faculty positions and began to use their academic positions to address the unfinished reform agenda. . . . Into this charged atmosphere the new discipline of sociobiology emerged." Ibid., 569. Jumonville argues multiculturalist criticism was due to Wilson's liberal universalism, to the fact that "he denied that there were significant multicultural differences to be preserved and honored between races and ethnicities." Ibid., 587.

47. Carl Sagan, *The Cosmic Connection: An Extraterrestrial Perspective* (New York: Anchor, 1973), 194.

48. Ibid., 195.

49. Ibid., 6, and 1-6 for the sweep of natural and human history.

50. Ibid., 7.

51. Ibid., 26-27.

52. Ibid., 30.

53. Ibid., 37, 39.

54. One of Eiseley's biographers, Gale E. Christian, puts the matter this way: "In his own way, Eiseley was no less a protester than the youth whose confrontational tactics caused him to move furtively about campus, slipping in and out of back doors. Ostensibly a work of social criticism, *The Invisible Pyramid* is more an aging poet's fugue, the eternal question of human existence set in counterpoint against the flight of *Apollo 11*." Gale E. Christian, *Fox at the Wood's Edge: A Biography of Loren Eiseley* (New York: Henry Holt, 1990), 408.

55. Loren Eiseley, *The Invisible Pyramid* (London: Rupert Hart-Davis, 1971), 131-32, 45.

56. Hannah Arendt, *The Human Condition* (Chicago: University of Chicago Press, 1958), 1-2. For analysis of Arendt's claims in relation to those of Martin Heidegger and Hans Blumenberg in the context of "Earthrise era" and what he terms after Arendt the "post-Earthrise condition," see Benjamin Lazier, "Earthrise; or, The Globalization of the World Picture," *American Historical Review* 116, no. 3 (June 2011): 602.

57. Hannah Arendt, "The Conquest of Space and the Stature of Man," *American Scholar* 32 (1963): 536.

58. Ibid., 540.

59. As Howard McCurdy has observed, NASA saw an immense expansion in the early 1960s, with its budget and employee base increasing by more than factors of two and eight respectively. Moreover, "by the 1965 fiscal year, NASA officials were spending over 8% of the entire gross national product or 4.4 percent of all federal outlays that year." Howard E. McCurdy, *Inside NASA: High Technology and Organizational Change in the U.S. Space Program* (Baltimore: Johns Hopkins University Press, 1993), 101. In *The Invisible Pyramid*, Eiseley expressed outrage at the monies involved: "The estimated cost of placing the unmanned Surveyor 3 upon the moon mounted to more than eighty million dollars. Just one unmanned space probe, in other words, equaled or exceeded the entire endowment of many a good college or university; the manned flight of Apollo 12 cost two hundred and fifty millions. The total space program is inconceivably costly, yet the taxpayer, up until recently, accepted it with little question. By contrast, his elected officials frequently boggle over the trifling sums necessary to save a redwood forest or to clear a river of pollution." Eiseley, *Invisible Pyramid*, 89-90.

60. Loren Eiseley, "Is Man Alone in Space?," *Scientific American* 189, no. 1 (July 1953): 86.

61. Sagan, *The Cosmic Connection*, 67.

62. Ibid., 249.

63. Ibid., 255 and Friedrich Schlegel, *Philosophical Fragments*, trans. Peter Firchow (Minneapolis: University of Minnesota Press, 1991), 96.

64. Sagan, *The Cosmic Connection*, 257.

65. Ibid.

66. See "Textbook list for Astronomy 102/104—Carl Sagan," Folder 14 "Astronomy 102/104—1974-1976" Container 254, and "Textbook list for Astronomy 102/104—Spring, 1978—MWF 12:20 pm" Folder "Spring 1978 (1 of 3)," Container 255, Carl Sagan and Ann Druyan Archive, Manuscript Division, Library of Congress. Astronomy 102 was a three-hour-per-week lecture course with one hour of laboratory work; Astronomy 104 was the same course without the laboratory hour.

67. Steven Weinberg, "The Early Universe," video of lecture, Cabot Library Collections, Science Center, Harvard University, 1973.

68. Weinberg, "The Early Universe," 1973.

69. George Gamow, *The Creation of the Universe*, 1st ed. (New York: Viking, 1951), 44.

70. This sensibility is sustained, thematized more explicitly in the later considerations of Steven Weinberg, *Dreams of Final Theory* (New York: Pantheon, 1992) into an extended discussion on the relationship between history and accident.

71. Ralph A. Alpher and Robert Herman, *The Genesis of the Big Bang* (New York: Oxford University Press, 2001), vi. Jürgen Habermas had made the point in the late 1960s that "the more specialized research becomes, the greater the distances important information must traverse in order to enter the work of another expert. Physicists may even use *Time Magazine* to inform themselves about new developments in technology and chemistry." Jürgen Habermas, *Toward a Rational Society: Student Protest, Science and Politics* (Boston: Beacon, 1970), 77.

72. Thus, in *The First Three Minutes*, "A cinematic treatment seems appropriate: frame by frame, we will watch the universe expand and cool and cook. We will also try to look a little way into an era that is still clothed in mystery—the first hundredth of a second, and what went before." Steven Weinberg, *The First Three Minutes: A Modern View of the Origin of the Universe* (Boston: Beacon, 1977), 9.

73. Ibid., 4.

74. Ibid., 154. The word "farce" here should be read against an earlier statement: "It is almost irresistible for humans to believe that we have some special relation to the universe, that human life is not just a more-or-less farcical outcome of a chain of accidents reaching back to the first three minutes, but that we were somehow built in from the beginning." Ibid.

75. Karl Marx, *The Eighteenth Brumaire of Louis Bonaparte* (London: Electric Book Company, 1852), 7.

76. Astronomy 102/104, Spring 1978, Container 255, Carl Sagan and Ann Druyan Archive, Manuscript Division, Library of Congress.

77. Wilson regarded *On Human Nature* as the third of a trilogy, one "that unfolded without my being consciously aware of any logical sequence until it was nearly finished." The first two texts were his 1971 *The Insect Societies*, and his 1975 *Sociobiology: The New Synthesis*. Edward O. Wilson, *On Human Nature* (Cambridge, MA: Harvard University Press, 1978), ix; Edward O. Wilson, *The Insect Societies* (Cambridge, MA: Harvard University Press, 1971).

78. Wilson, *On Human Nature*, x.

79. Ibid., 2. Wilson here cites the 1976 publication of a lecture, "The Forces of Nature," that Weinberg delivered to the American Academy of Sciences in October of the previous year. This first chapter of Wilson's book is itself strongly based on his own February 1976 lecture (and October publication) "The Social Instinct," also delivered to the academy.

80. Steven Weinberg, "The Forces of Nature," *Bulletin of the American Academy of Arts and Sciences* 24, no. 4 (January 1976): 29, and J. B. S. Haldane, *Possible Worlds and Other Papers* (New York: Harper and Brother Publishers, 1928), 298.

81. Wilson, *On Human Nature*, 192.

82. Ibid., 201.

83. A set of beliefs he likely saw at play in the attack of the SSG. For retrospective evidence of this, see E. O. Wilson, "Science and Ideology," *Academic Questions* 8, no. 3 (Summer 1995): 77–78.

84. Wilson, *On Human Nature*, 196.

85. Ibid., 196. Wilson articulates an extended complement to the foundationalist view, partly through his notion of an "antidiscipline," the "special adversary relation that often exists when fields of study at adjacent levels of organization first begin to interact. For chemistry there is the antidiscipline of many-body physics; for molecular biology, chemistry; for physiology, molecular biology; and so on upward through the paired levels of increasing specification and complexity." Ibid., 7.

86. Ibid., 201.

87. Ibid., 208.

88. In his "Possible Worlds" essay, Haldane argued that the universe is differently constituted from the point of view of different creatures—that the "idea of a thing" as "a portion of experience conceived as public and ethically neutral" was "perhaps man's greatest intellectual achievement." See J. B. S. Haldane, "Possible Worlds," 277–78.

89. Wilson, *On Human Nature*, 209.

90. Carl Sagan, *The Dragons of Eden: Speculations on the Evolution of Human Intelligence* (New York: Random House, 1977), 7 and 8.

91. Ibid., 7. Wilson refers to *Dragons of Eden* in the endnotes to *On Human Nature*, not on the question of myth, but on the possibility of primate language.

92. Such an account, with the world on a platform upon an elephant upon a turtle and thereon turtles, appears earlier in Clifford Geertz's 1973 "Thick Description: Toward an Interpretive Theory of Culture," in *The Interpretation of Cultures: Selected Essays* (New York: Basic, 1973), 28-29. My thanks to Bud Bynack for recalling this. Stephen Hawking also put a version of this story to use in his 1988 *A Brief History of Time*, in another case of the invocation of myth and contemporary universal history.

93. Carl Sagan, *Broca's Brain: Reflections on the Romance of Science* (New York: Random House, 1979), 312. "I do not insist that these connections between religion and perinatal experience are correct or original," he noted, announcing at least some of the more contemporaneous thinkers on mythology upon whom he actively drew. "Many of them are at least implicit in the ideas of Stanislav Grof and the psychoanalytic school of psychiatry, particularly Otto Rank, Sandor Freneczi and Sigmund Freud. But they are worth thinking about." Ibid., 311.

94. Ibid., 336-37.

95. Wilson's formulations could be expressed in provocative terms, appearing almost calculated to offend critics such as those in SSG: "That transition [to the acceptance of the scientific epic over religion] will proceed at an accelerating rate. Man's destiny is to know, if only because societies with knowledge culturally dominate societies that lack it. Luddites and anti-intellectuals do not master the differential equations of thermodynamics or the biochemical cures of illness. They stay in thatched huts and die young." Wilson, *On Human Nature*, 207.

96. Marx, *The Eighteenth Brumaire of Louis Bonaparte*, 7.

CHAPTER TEN

1. As suggested throughout, there are moreover strong consonances and connections between these authors and texts: Wilson cited both Weinberg (his 1976 American Academy of Arts and Science lecture) and Sagan (*The Dragons of Eden*) in *On Human Nature*. Weinberg and Wilson each gave lectures to the American Academy of Arts and Sciences on subjects that formed part of their narratives (and for Wilson, a chapter of his book). *The Dragons of Eden* has a section entitled "On Human Nature" that could perhaps have been one of Wilson's cues to the composition and/or titling of his book on the same topic. All three were Harvard scientists for extended periods and at least briefly for overlapping periods, giving prominence to Harvard research, in particular, in the sciences, social sciences, and humanities. The history of science and historians of science figured in all their work: Gerald Holton, Owen Gingerich, and I. Bernard Cohen were acknowledged or cited in at least one of each of their texts. Wilson and Sagan won the Pulitzer Prize in general nonfiction in successive years (1977 and 1978) for *The Dragons of Eden* and *On Human Nature*, respectively. Weinberg, by contrast, won the Steel Foundation Science Writing Award, awarded by the American Institute of Physics. These authors thus were central to a scientific literary canon that they constructed and perpetuated through their work, work that spoke to several audiences at once.

2. Steven Weinberg, *The First Three Minutes: A Modern View of the Origin of the Universe* (Boston: Beacon, 1977), viii.

3. Edward O. Wilson, *On Human Nature* (Cambridge, MA: Harvard University Press, 1978), 189.

4. Ibid., 201.

5. According to Chaisson, not long after his initial appointment at Harvard, he met Shapley unexpectedly, providing the opportunity to discuss Chaisson's idea of reviving Shapley's own earlier interdisciplinary, synthetic efforts. They began to plan the course, but Shapley died within days of the exchange. Nevertheless, such a course did emerge at Harvard, first under the direction of Field and thereafter Chaisson. Eric Chaisson, correspondence with the author, July 22, 2014, and Eric Chaisson, "The Old Man in the Corner," unpublished article.

6. Field had also taken courses in thermodynamics and nuclear physics with Gamow in the early fifties. George Field, interview by Richard Hirsch, July 14, 1980, Niels Bohr Library and Archives, American Institute of Physics, College Park, MD, https://www.aip.org/history-programs/niels-bohr-library/oral-histories/4602-1.

7. "The story of cosmic evolution itself is a cultural myth hereby told in the form of this book—a myth, because it's admittedly a simplification of an extremely elaborate approximation of reality." Eric Chaisson, *Cosmic Evolution: the Rise of Complexity in Nature* (Cambridge, MA: Harvard University Press, 2001), 199, and *Epic of Evolution: Seven Ages of the Cosmos* (New York: Columbia University Press, 2005), 426.

8. Keay Davidson, *Carl Sagan: A Life* (New York: John Wiley and Sons, 1999), 319.

9. Ibid., 320.

10. Ann Druyan, interview with the author, July 21, 2009, and correspondence with the author, December 3, 2009.

11. Container 254, Seth MacFarlane Collection of the Carl Sagan and Ann Druyan Archive, Manuscript Division, Library of Congress, Washington, DC.

12. The overlap is evident in part from the course's original proposed description and the list of lectures also suggests as much. Memorandum to Carl Sagan from P. D. Drake, January 15, 1973, Container 254, and Container 255 Carl Sagan and Ann Druyan Archive, Manuscript Division, Library of Congress.

13. Ann Druyan, interview, July 21, 2009 and correspondence, December 3, 2009, with the author. I nevertheless generally refer to Sagan alone below, partly because he functioned as the narrator for the series. However, this threatens to overstate his role.

14. "Cosmos Voyage to Far Reaches of Universe on 'Spaceship of Imagination' Took Years of Scientific, Artistic Work to be Accurate, Visually Exciting," Background Feature, Special Effects, Ben Kubasik Inc, Folder Carl E. Sagan, Clips 1980s #4/3/15, Carl Sagan Papers, Cornell University Archives.

15. The sequence in question is near the start of the eleventh episode, "The Persistence of Memory."

16. Correspondence Gentry Lee to Stanley Mailman, April 12, 1977, and Adrian Malone to Carl Sagan, September 12, 1977, Container 511 Carl Sagan and Ann Druyan Archive, Manuscript Division, Library of Congress.

17. "Carl Sagan's 'Cosmos' Will Mean Worlds to You," *Chicago Tribune*, September 26, 1980, A16.

18. Karl Guthke, "Nightmare and Utopia: Extraterrestrial Worlds from Galileo to Goethe," *Early Science and Medicine* 8, no. 3 (2003): 180.

19. For storyboard description of this sequence, see Container 578 Carl Sagan and Ann Druyan Archive, Manuscript Division, Library of Congress.

20. Carl Sagan, *Cosmos* (New York: Random House, 1980), 200.

21. Ibid., 201.

22. So such observers may well have to confront the experience of the paradox that "although many physicists had found . . . difficult to swallow" experiments have verified. As a result, "the only alternative to it would be the assumption that earthborn life under all circumstances remains bound to a time concept that demonstrably does not belong among 'true realities,' but among mere appearances. We have reached the stage where the Cartesian radical doubt of reality as such, the first philosophical answer to the discoveries of science in the modern age, may become subject to physical experiments that would make short shrift of Descartes's famous consolation, I doubt therefore I am, and of his conviction that, whatever the state of reality and of truth as they are given to the senses and to reason, you cannot 'doubt of your doubt and remain uncertain whether you doubt or not'." Hannah Arendt, "The Conquest of Space and the Stature of Man," *American Scholar* 32 (1963): 535–36.

23. Correspondence Carl Sagan to Adrian Malone, August 26, 1979. In a letter dated April 18 of the same year, Paul C. Huang—who among his achievements, founded the board game Dynasty—petitioned

Sagan to include scientific developments in Asia. Paul C. Huang to Carl Sagan, April 18, 1979. On the same day that Sagan wrote to Malone, he wrote to Huang to explain the difficulties but also the ongoing efforts to overcome a Westernizing frame. Carl Sagan to Paul C. Huang, August 26, 1979. Container 511, Carl Sagan and Ann Druyan Archive, Manuscript Division, Library of Congress.

24. These words are identical in the text. Sagan, *Cosmos*, 258.

25. This was a few years prior to Sagan's nuclear activism, associated in particular with his championing the theory of nuclear winter, though *Cosmos* itself functioned as an attempted political intervention. Matthew Stanley has provided an analysis of this work with an eye toward the context of the American religious prophetic traditions. Matthew Stanley, "The Top of the Error Bars: Carl Sagan and the Prophetic Precautionary Principle," unpublished talk.

26. For the use of these terms, see discussions below in relation to Northrop Frye.

27. If we were to attempt to cast the more specific genre in Frye's terms, we would necessarily have a kind of hybrid: in terms of heroic structure, this is a romance. In terms of its narrative voice/ambition, it is an "epic myth," just as Sagan suggests in the final moments of the broadcast. But whether as a romance it is comic or tragic is not decided. The account hovers between the two, mirroring more popular hopes and anxieties at the time of its production and broadcast. Moreover, the possibility of the comic end is linked to that of continued authorship—the international community of scientists prosecuting their science together, demonstrating both the universality at stake here and the manner in which it renders mutual annihilation absurd.

28. Correspondence Adrian Malone and Maggie Morris to Carl Sagan, November 10, 1977, Container 511, Carl Sagan and Ann Druyan Archive, Manuscript Division, Library of Congress.

29. Ibid.

30. See for example Vivian Sobchack, "Toward a Phenomenology of Nonfictional Film Experience," in *Collecting Visible Evidence*, ed. Jane M. Gaines and Michael Renov (Minneapolis: University of Minnesota Press, 1999).

31. As Sobchack observes of another case, "as we watch images of Fala or John F. Kennedy or the Gulf War, aspects of what we see are taken up by us as unknown in their existential specificity, yet because have some general cultural knowledge of them, their past and or present existence is posited by us nonetheless—if, however, always in a way qualified by our lack of personal knowledge." Ibid., 243. The third-person "us" appealed to here is itself the marker of the assumption of shared, uniform responses to the films analyzed.

32. "In terms of audience ratings," as the *New York Times* reported near the conclusion of the screenings, "it is among the most popular of domestic series produced for the Public Broadcasting Service. While it dropped in the first few weeks from a rating of 10 in major markets to a level of about 5, it has been bouncing between 4 and 6 since then." The *Times* added that "one must realize that commercial network programs that are considered certain failures usually get audience ratings only somewhere around the 13 level. But this, remember, is public television, which was caught up in the habit of 'narrowcasting' long before cable carved out a marketplace presence in that preserve." John J. O'Connor, "TV View: Putting 'Cosmos' into Perspective," *New York Times*, December 14, 1980, D36.

33. John J. O'Conner, who referred to *Ascent of Man* as "superb," expressed ambivalence about Sagan's enthusiasm, but also about the visual focus on him: "Viewers may have wondered, justifiably, what was supposed to impress them most: the stunning galaxies or Dr. Sagan's profile." Ibid. See also Richard A. Baer Jr., "TV: Carl Sagan's Narrow View of the Cosmos," *Wall Street Journal*, October 24, 1980, 35.

34. Beverly Beyette, "Carl Sagan Is a Busy Man in the Universe," *Los Angeles Times*, April 3, 1985, H1.

35. Tom Shales, "'Cosmos'—Public TV's Big Bang," *Washington Post*, September 28, 1980, K1.

36. John J. O'Connor, "TV View: 'Cosmos'—A Trip Into Outer Space," September 28, 1980, D39. There seemed to be some consensus that the first episode was among the weaker of the

series. For evidence of this, see, for example, Tom Zito, "Carl Sagan, Cosmonaut," *Washington Post*, September 19, 1980, reporting a very mixed reception before the premiere screening at the National Academy of Sciences.

37. Cecil Smith, "A Window Opens on the 'Cosmos,'" *Los Angeles Times*, September 25, 1980, G1.

38. Baer, "TV: Carl Sagan's Narrow View of the Cosmos," 35. In a similar comparison, Cecil Smith in the *Los Angeles Times* noted "we [the viewers] join Dr. Sagan in a space vehicle with a curious vaulted ceiling, rather like a church ('Everything I do' says Malone, 'is faintly liturgical')." Smith, "A Window Opens on the 'Cosmos.'"

39. Sixteen years later, Robert C. Cowen in the *Christian Science Monitor* declared that "the second half of the 20th century has been a golden age for cosmic exploration, and Carl Sagan had been its prophet." Robert C. Cowen, "Carl Sagan: Science Poet, Science Prophet," *Christian Science Monitor*, December 24, 1996.

40. "Sagan's television series is subtitled 'a personal voyage.' The book isn't, though, and probably should have been." Peter Gormer, "Tempo: Carl Sagan's 'Cosmos'—A Valuable Work despite the Excesses and the Skeptics," *Chicago Tribune*, October 8, 1980, A3.

41. Cecil Smith, "The Best TV Programs of 1980," *Los Angeles Times*, January 1, 1981, G1.

42. O'Connor, "TV View: Putting 'Cosmos' into Perspective."

43. Marilynn Preston, "Carl Sagan's 'Cosmos' Will Mean Worlds to You," *Chicago Tribune*, September 26, 1980, A16.

44. See, for example, Lee Margulies, "Adult Education: Telecourses Set for PBS Stations," *Los Angeles Times*, June 16, 1981, G6, and Zerita S. Walther, "PBS Offering Credits For College on TV," *New York Times*, November 15, 1981, EDUC62.

45. "Public Broadcasting Service Takes TV Classes to Millions," *Christian Science Monitor*, January 8, 1984, ED13.

46. Quoted in Sherry Walton, Ronald E. Kutz, and Lowell Thompson, "The Integrated Day Comes to College," *Social Studies* (March–April 1986): 83.

47. Ibid., 83–84.

48. Ibid., 85.

49. Sherry Walton, telephone interview with the author, June 17, 2014. Walton noted having been strongly drawn to Sagan's work prior to the experiment.

50. Walton, Kutz, and Thompson, "The Integrated Day Comes to College," 87.

51. It needs continually to be kept in mind that—as in the cases of Frye, or Sagan, or Wilson for that matter—the word "myth" and related terms such as "narrative" and "romance" do not convey any one specific attitude to the question of truth that is being put forward. They are neither lies nor simply verifiable empirical truths. Traditional myths and romances can themselves put forward truths or worldviews, claims around which a society constructs itself. There may be a temptation regardless to see myth primarily as literary fiction. But even if these accounts are seen as literature, it would be a mistake to identify fiction and literature as such—or to imagine that the application of narrative terms to science somehow equates science with fiction. Narrative in any case is not in the first instance always and perhaps even ever free from claims to truth, just as claims to truth often (at the very least) involve/rely on narrative forms.

52. Joseph Campbell, *The Masks of God: Primitive Mythology* (New York: Viking, 1959), 5.

53. Ibid.

54. Joseph Campbell, *The Inner Reaches of Outer Space: Metaphor as Myth and as Religion* (Novato, CA: New World Library, 2002), xix.

55. Ibid., xx–xxi.

56. Roland Barthes, *Mythologies*, trans. Richard Howard [and] Annette Lavers (1957; New York: Hill and Wang, 2012).

57. "For the philosopher, physics has taken over the distancing function of myth: It neutralizes everything, without exception. But above all it lets us comprehend, for the first time, what had been at issue—with the inadequate means of myth too—all along. Only work on myth—even if it is

the work of finally reducing it—makes the work of myth manifest." Hans Blumenberg, *The Work on Myth*, trans. Robert M. Wallace (Cambridge, MA: MIT Press, 1990), 118.

58. "Myths are stories that are distinguished by a high degree of constancy in their narrative core and by an equally pronounced capacity for marginal variation." Ibid., 34. Also, in reference to "unit myths," Blumenberg notes: "The fundamental patterns of myth are so sharply defined [*prägnant*], so valid, so binding, so gripping in every sense, that they convince us again and again and still present themselves as the most useful material for any search for how matters stand, on a basic level, with human existence." Ibid., 150-51.

59. Ibid., 162.

60. Note that Frye reviewed Jacob Bronowski's 1943 Blake study, *A Man without a Mask*, observing that the book "perhaps reflects the uneasy social conscience of a slightly later period, expressed in the dictum that poetry must be either tendentious or unreal." Northrop Frye, *Northrop Frye on Milton and Blake*, ed. Angela Esterhammer (Toronto: University of Toronto Press, 2005), 185. A decade later, contributing a chapter on Blake to an edited volume on the English Romantic poets and essayists, Frye included Bronowski's volume on Blake among other criticism, judging it "a crisp and incisive study." His brief mention concluded that whereas "it had always been obvious that Blake's poetry was in part a poetry of social protest," it was Bronowski's book that "was the first to show in detail how wide open Blake's eyes were and how much of the life around him he absorbed and recorded." Ibid., 284. By the 1980s, Frye's stature in the study of literature was such as also to lead to proposals for Frye to anchor a television series on literature based on the model of Kenneth Clarke's *Civilization* and Bronowski's *The Ascent of Man*. What ultimately resulted was a series produced and conceived by Robert Sandler, consisting of edited, videotaped lectures, themselves also transcribed into a book. As of this writing, the edited lectures are available online. See, e.g., http://heritage.utoronto.ca/content/bible-and-english-literature-northrop-frye-full-lecture-1. The models of Clarke, Bronowski, and perhaps Sagan, were echoing out and in subtle ways affecting the message and content of narrative exposition.

61. Clearly, I do not pursue here any straightforward application of Frye—neither a general archetypical critical exercise nor through it a transhistorical characterization of myth.

62. "The greatness of Frye, and the radical difference between his work and that of the great bulk of garden-variety myth criticism, lies in his willingness to raise the issue of community and to draw basic, essentially social, interpretive consequences from the nature of religion as collective representation." Fredric Jameson, *The Political Unconscious: Narrative as a Socially Symbolic Act* (Ithaca: Cornell University Press, 1982), 68 and see also 69.

63. Northrop Frye, *Creation and Recreation* (Toronto: University of Toronto Press, 1980), 5.

64. Ibid., 5-6. Recall, as well, Roland Barthes's earlier appeal to a similar image in explicating myth as a semiological system, drawing a distinction between mythic signifier as pointing to meaning, on the plane of language (where it itself is a sign), and mythic signifier as form, on the plane of myth: "The meaning is always there to *present* the form; the form is always there to *outdistance* the meaning. And there never is any contradiction, conflict, or split between the meaning and the form: they are never at the same place. In the same way, if I am in a car and I look at the scenery through the window, I can at will focus on the scenery or on the windowpane. At one moment I grasp the presence of the glass and the distance of the landscape; at another, on the contrary, the transparence of the glass and the depth of the landscape; but the result of this alternation is constant: the glass is at once present and empty to me, and the landscape unreal and full. The same thing occurs in the mythical signifier: its form is empty but present, its meaning absent but full." Barthes, *Mythologies*, 233. In the analogy between Frye's train scene at night and Barthes's drive in the day (so it would seem), the train window glass with its varying reflective power (operating regardless of my will) compares to the car window's invariably transparent glass (at or beyond the boundaries of which I can will myself to focus). Whereas reflection in the glass aligns to meaningful representation for Frye, for Barthes it is the transparency of the glass that reveals the absent yet fuller meaning the window enframes. And whereas for Frye, myth is aligned with the mirror

alone, for Barthes the signification of myth (its sign, the third term of its own semiological system) is constituted through the oscillation between the material presence of the glass and its transparency.

65. Frye, *Creation and Recreation*, 8.

66. I will concentrate here on what Frye terms his historical criticism (his theory of modes) to characterize the shape of the universal histories in these materials and the use of his ethical criticism in addressing questions of audience. Frye's overall project, developed in his *Anatomy of Criticism*, is to expand on the categories of Aristotle's *Poetics* and the categories of medieval scriptural reading in order to try in effect to classify all of narrative. Though Frye ultimately presents this work as a historical progression of an ahistorical set of possibilities, his terms do not demand or depend on this assumption, and I do not ascribe to it here. Thus, though I take Seymour Chatman and various poststructuralist criticisms seriously on the matter of the need for historicization, I cannot share the at times implicit conclusion that this in turn means dismissing Frye's work.

67. His is now an often-deprecated line of literary-historical theory, and also an often partially defended one. Part of the reason for the resistance to Frye is that his effort was involved in a kind of deductive genre analysis threatening to treat generic literary forms as stable and ahistorical, even if they emerged in particular historical moments. One might see in this a similar criticism in line with the question of species: to try to provide a system that captured all the species in the world suggested that the framework for discussing species difference was somehow decided upon prior to all the contingencies that determined what a given species is. We somehow can know which way species are going to evolve before the fact of their evolution. Likewise the genres, however clearly they appear to be cultural-historical products. But, it would be a philosophical mistake to imagine that such apparently ahistorical systems are less insightful and a historical mistake not to assess their specific connections to the historical periods in which they emerge.

68. The sequence functions as a documentary within a documentary, a simulated history that brings to attention another strand of documentary film theorization, that of simulation. A number of theorists have expanded on the idea of the "subjunctive" or "conditional" documentary, in which historical or potential future events are elaborated upon in a documentary setting or in a manner meant to provoke the documentary consciousness. See Carl Plantinga, *Rhetoric and Representation in Nonfiction Film* (Cambridge: Cambridge University Press, 1997); Mark J. P. Wolf, "Subjunctive Documentary: Computer Imaging and Simulation," in *Collecting Visible Evidence*, ed. Jane M. Gaines and Michael Renov (Minneapolis: University of Minnesota, 1999); and Paul Ward, "The Future of Documentary? 'Conditional Tense' Documentary and the Historical Record," in *Docufictions: Essays on the Intersections of Documentary and Fictional Filmmaking*, ed. Gary D. Rhodes and John Parris Springer (Jefferson, NC: McFarland, 2006). Much of this literature is centered on the attempt to distinguish fiction from nonfiction film or to argue that such a distinction is impossible. The relevant matter here, however, is that film in the conditional mode is intrinsically structured as an argument, insofar as it places viewers in a position (often an awkward or even logically impossible one) of witnesses to events that they could not have seen, but the verisimilitude of the (re)creation of which they are asked to judge. In *Phenomenon*, Teilhard de Chardin had appealed to a filmic metaphor to avoid "any misunderstanding about the degree of reality which I accord to the different parts of the film I am projecting. When I try to picture the world before the dawn of life, or life in the Palaeozoic era, I do not forget that there would be a cosmic contradiction in imagining a man as spectator of those phases which ran their course before the appearance of thought on earth." Instead his argument is not a matter of a witnessing but an argumentative deduction: "I do not pretend to describe them as they really were, but rather as we must picture them to ourselves so that the world may be true for us at this moment." Pierre Teilhard de Chardin, *The Phenomenon of Man*, trans. Bernard Wall (New York: Harper, 1959), 35. *Cosmos* in turn connects the aesthetic principles of the simulation and the logic of sequence to its epistemological argument—its claims about the principles of the world it enacts. In this regard, the historical simulation need not be principally filmic to be structured in this way.

In *The First Three Minutes*, Weinberg claims to see his reader as "a smart old attorney" expecting to "hear some convincing arguments before he makes up his mind." Weinberg, *First Three Minutes*, viii.

69. It concludes with the beginning of another, in images of India, a reminder the extent to which the promotion of universality was linked here, as with other universal histories, to a projection of a cultural other.

70. These elements go by various names, with various different emphases. In response to three of Aristotle's poetical aspects, *mythos*, *ethos*, and *dianoia*, Frye distinguishes between the "internal fiction of the hero and his society" and the "external fiction which is a relation between the writer and the writer's society." This in turn produces four literary elements to consider, "the hero, the hero's society, the poet and the poet's readers." Frye, *Anatomy of Criticism*, 52–53. The internal hero-centered fiction involves the plot or *mythos* as well as character and setting as included in *ethos*. The external poet-centered fiction engages the themes or thought, the dianoia of the text. As with the dimension of power, alienation/individuation track both the external and internal fictions of any literary text, deciding not only internally whether the hero is a tragic or comic one, but also externally whether the poet is constructed more as spokesperson for their society as with the "'epic' tendency" or more as an individual, as with the "lyric" counterpart.

71. For example, folk heroes or legendary ones—in the American context, one might think of a Paul Bunyan or John Henry—are romantic heroes. The representation of scientists considered here is not, to be sure, in the same cultural register as Bunyan or Henry, but scientists are nevertheless "romantic" in Frye's sense in relation to the power and status they have over the social and natural world. Other possibilities in Frye's typology include what he calls High or Low Mimetic, where the hero is more powerful than or as powerful as those around him or her, respectively, or the ironic, the hero being at the mercy of the social and natural world. These need not be regarded as hard and fast distinctions.

72. Wilson, *On Human Nature*, 203–4. It is hard to resist the suspicion that the image of the astronomer that Wilson had in mind was Sagan, an association all the easier to make for Wilson's readers after 1980.

73. Here, as with the hero, the gradations of power work down, from the mythic narrator to the ironic narrator who undercuts him- or herself at every turn.

74. The question of different formulations of the distinction between myth and epic will be discussed subsequently.

75. By the light of their conclusions, all narratives according to Frye fall into either the comic or tragic modes. He argues that the hero's alienation from society is the chief characteristic of the tragic. So we might think of King Lear as tragic as a result of Lear's increasing isolation. In contrast to the tragic possibility, the comic—a term with no derogatory sense—is for Frye marked by reconciliation. It is used in the same spirit as in Dante's *Divine Comedy* (or as in Jane Austen's novels, where all the characters will arrive at their proper places in the world). Dante details a cosmology that allows for the possibility of the reconciliation of his heroes—the literary Dante himself—with all the world (or all the world that matters). It also might be noted here in passing that Sagan's account attempts to speak to the future in proposing the care that must be taken with regard to the technology it has created.

76. A reproduction of the image is available from NASA at http://www.nasa.gov/multimedia /imagegallery/image_feature_623.html.

77. "Only a God can save Us: Der Spiegel's Interview with Heidegger," *Philosophy Today* 20, no. 4 (Winter 1976): 277.

78. Ibid., 288.

79. Campbell, *The Inner Reaches of Outer Space*, 18. For historical context and argument as to the importance of the fact of this photograph, see Benjamin Lazier, "Earthrise; or, The Globalization of the World Picture," *American Historical Review* 116, no. 3 (June 2011).

80. Ibid., 625–26.

CHAPTER ELEVEN

1. See, for example, Diane Vaughan, *The Challenger Launch Decision: Risky Technology, Culture, and Deviance at NASA* (Chicago: University of Chicago Press, 1996), and Howard E. McCurdy, *Inside NASA: High Technology and Organizational Change in the U.S. Space Program* (Baltimore: Johns Hopkins University Press, 1993).

2. According to one science writer, "the loss of Challenger was significant, delaying its [the HST's] launch an additional four years." Robert Zimmerman, *The Universe in a Mirror* (Princeton: Princeton University Press, 2008), 144.

3. The euphemism "credibility problem" is Zimmerman's. Ibid. With regard to its history of reinvention, the central mission of a lunar landing of the NASA of the 1960s, even while it was being accomplished, was superseded by the objective of space transit, the promise of easy travel to stations in orbit. Historians have seen this transformation as linked to public intolerance to vast expenditure entailed by the costs of the Vietnam War and social unrest, factors among others that undercut the central place of NASA in the battles of the Cold War. By the lights of such scholarship, the growing caution of the institution over the course of the 1980s and into the early 1990s, the necessity to continue to justify a budget against this increasingly intolerant public and Congress, a more regulative and intrusive civil service, along with the headaches and heartaches of the space shuttle and Hubble space telescope missions made for an organization less effective in defining its own raison d'être and with significantly less of its own former luster. See Vaughan, *The Challenger Launch Decision*, and McCurdy, *Inside NASA*.

4. Hallam Stevens has to a large degree blamed the failure of the SSC in comparison with the success with Fermilab on the "fragmentations of the physics community" overawing a compelling narrative: "Fragmentations over strings, SDI, and condensed matter resulted in a cacophony of competing voices laying claim to 'fundamental' science and drowning out the persuasive force of the symmetry narrative." This symmetry narrative was the "narrative of elegance, simplicity, and fundamental symmetries of the universe that had been told by high-energy physicists in the 1960s and the 1970s," a narrative itself fragmenting with the fragmentation Stevens sees in the community itself. Hallam Stevens, "Fundamental Physics and Its Justifications," *Historical Studies in the Physical and Biological Sciences* 34, no. 1 (2003): esp. 196-97. By contrast, partly in response to such views, Joseph Martin looks to different philosophical commitments in solid state and particle physics, arguing overall that "philosophical outlooks, as elements of scientific discourse, shaped the American physics community's internal politics and external interaction," and conversely— outlooks that in turn shaped and responded to the political and disciplinary climate in debates over the SSC. Joseph D. Martin, "Fundamental Disputations: The Philosophical Debates that Governed American Physics, 1939-1993," *Historical Studies in the Natural Sciences* 45, no. 5 (2015): 56.

5. For a contemporaneous assessment of the state of the physics community and physical theory, with particular attention to Anderson's philosophical critique, see Silvan S. Schweber, "Physics, Community and Crisis in Physical Theory," *Physics Today* 46, no. 11 (November 1993).

6. National Aeronautics and Space Administration, *Origins 2003: Roadmap for the Office of Space Science Origins Theme*, 2002, 68, available at http://solarsystem.nasa.gov/multimedia/downloads /Origins_Roadmap_03.pdf.

7. Among participating institutions were the California and Massachusetts Institutes of Technology, the University of California at Berkeley and at Santa Cruz, and the Universities of Arizona and Michigan.

8. The NASA Origins Project has since been redefined as the NASA Cosmic Origins Project, which covers later plans and roadmaps: see http://cor.gsfc.nasa.gov/docs.

9. Language in terms of, for example, the mathematical formulation of a scientific theory versus the non-mathematically conveyed presentation of a scientific quest or saga, of *logos* versus *mythos*. For the use of these terms in the context of the study of myth, in which logos is largely

aligned with explanatory reason, but cannot be separated from myth, see Blumenberg, *Work on Myth*, e.g., 11-12. Note that Blumenberg argues that "the boundary line between myth and logos is imaginary and does not obviate the need to inquire about the logos of myth in the process of working free of the absolutism of reality. Myth itself is a piece of high-carat 'work of logos.'" Ibid., 12.

10. For an extended discussion of the HST's specifications, see "Appendix 5: The Hubble Space Telescope," in Robert W. Smith, *The Space Telescope: A Study of NASA, Science, Technology and Politics* (Cambridge: Cambridge University Press, 1989).

11. For these and other facts relating to the Space Telescope Science Institute see, for example, ibid.

12. True for Lyman Spitzer perhaps above all, the astrophysicist who conceived of a space telescope decades earlier and engaged in a enduring effort to build it.

13. As Steven J. Dick and James E. Strick have shown, there were also at least two other NASA space science efforts that affected the emergence of the Origins program, the Space Interferometry Mission and the Exploration of Neighboring Planetary Systems. Each group organized reports that were ultimately embraced by the NASA Origins project. See Steven J. Dick and James E. Strick, *The Living Universe: NASA and the Development of Astrobiology* (New Brunswick: Rutgers University Press, 2005), 173-76. Though I will not give an extended analysis of the connection between these reports here, their language shares in the sensibility and form of the narrative voiced most prominently through Sagan. In several contexts, Sagan's popular voice was explicitly cited by various members of the NASA astrobiology community, in particular. This is not to suggest that Sagan's narrative coercively produced these other stories or that there are not important consonances with other voices—Eric Chaisson's is a prominent example. The limited point here is that there are strong consonances with Saganesque characterizations. Thus, in regard to a series of such early astrobiology workshops at NASA's Ames Research Center, Dick and Strick draw attention to "the 'Pale Blue Dot' workshop, referring to planet Earth as described in Carl Sagan's 1994 book with the same title." Dick and Strick make a further claim difficult both to prove and to deny, that "although Sagan was not directly involved in the development of astrobiology at Ames, he was in many ways a guiding spirit, even after his early death in 1996." Ibid., 211.

14. "HST & Beyond" Committee, "Exploration and the Search for Origins: A Vision for Ultraviolet Optical Infrared Astronomy" (Washington, DC: Association of Universities for Research in Astronomy, 1996), 10.

15. Ibid., 14.

16. Alan Dressler, correspondence with the author, October 10, 2006.

17. "HST & Beyond" Committee, "Exploration and the Search for Origins," 7.

18. Ibid., 6.

19. Ibid., 5.

20. Ibid., 6.

21. Ibid., 4.

22. Dick and Strick, *The Living Universe*, 29.

23. See Stephen J. Garber, "A Political History of NASA's SETI Program," in *Archaeology, Anthropology, and Interstellar Communication*, ed. Douglas A. Vakoch (Washington, DC: The NASA History Series, 2014), and "Searching for Good Science: The Cancellation of NASA's SETI Program," *Journal of the British Interplanetary Society* 52 (1999). See also Steven J. Dick, "NASA High Resolution Microwave Survey (HRMS): Historical Perspectives," *Space Science Reviews* 64 (1993).

24. George Gaylord Simpson, "The Nonprevalence of Humanoids," *Science*, new series 143, no. 3608 (February 21, 1964): 770. For Sagan and Pollack's research, see Carl Sagan and James B. Pollack, "Windblown Dust on Mars," *Nature* 223 (August 23, 1969). Evidence in favor of their hypothesis was soon found in the images produced by Mariner 9. See "Recent Mariner Results," *Nature* 237 (May 12, 1972).

25. Bartholomew Nagy, Warren G. Meinschein, and Douglas J. Hennessy, "Mass Spectroscopic Analysis of the Orgueil Meteorite: Evidence for Biogenic Hydrocarbons," *Annals of the New York Academy of Sciences* 93, no. 25 (1961): 34.

26. Simpson, "The Nonprevalence of Humanoids," 770-71.

27. Simpson did not endorse the findings of Nagy, Meinschein, and Hennesy, directing readers to the doubts cast over their analysis. See Edward Anders and Frank W. Fitch, "Search for Organized Elements in Carbonaceous Chondrites," *Science*, new series 138, no. 3548 (December 28, 1962): 1392-99.

28. Simpson, "The Nonprevalence of Humanoids," 774-75.

29. Ernst Mayr, "The Probability of Extraterrestrial Life," in *Extraterrestrials: Science and Alien Intelligence* (Cambridge: Cambridge University Press, 1985), 25. Mayr also critiques Weinberg's foundationalist position. See Jordi Cat, "The Physicists' Debate on Unification," *Historical Studies in the Physical and Biological Sciences* 28, no. 2 (1998): esp. 266n39, 270-71, and 297.

30. Mayr, "The Probability of Extraterrestrial Life," 29. For the resistance of the long-serving senator from Wisconsin, William Proxmire, and Sagan's role in overcoming it, see Garber, "A Political History of NASA's SETI Program," 26-27. As far back as 1978, Proxmire had awarded to SETI the "Golden Fleece Award" targeting programs perceived as a waste of public funds.

31. Donald L. Savage, James Hartsfield, and David Salisbury, "Meteorite Yields Evidence of Primitive Life on Early Mars," press release 96-160, August 7, 1996, available at http://www2.jpl.nasa.gov/snc/nasa1.html.

32. David S. McKay et al., "Search for Past Life on Mars: Possible Relic Biogenic Activity in Martian Meteorite ALH84001," *Science*, new series 273, no. 5277 (August 1996).

33. Ibid., 924.

34. Among these observations were potential fossilized bacterialike "ovoid and elongate forms," as well as the abundance of polycyclic aromatic hydrocarbons, organic compounds that could result from "biogenic processes."

35. Ibid., 929.

36. Also, the European Space Agency's Philae Lander has announced the detection of organic molecules on Comet 67P/Churyumov-Gerasimenko. Paul Rincon, "Comet Landing: Organic Molecules Detected by Philae," November 19, 2014, BBC News, http://www.bbc.com/news/science-environment-30097648.

37. David Ballingrud, "Mars Mission: A Search for Life," *St. Petersburg Times*, November 3, 1996, South Pinellas edition, 1A.

38. United States Congress, "Life on Mars?," Hearing before the Subcommittee on Space and Aeronautics of the Committee on Science, US House of Representatives, 104th Congress, 2nd sess. (Washington, DC: Government Printing Office, September 12, 1996), 1.

39. Ibid., 4.

40. Ibid.

41. Ibid., 5.

42. Ibid., 23.

43. Carl Sagan to Wesley Huntress and John H. Gibbons, October 26, 1996, Personal Papers of Claude Canizares.

44. Ibid.

45. Space Science Workshop, *The Search for Origins: Findings of a Space Science Workshop*, 1996 Canizares Papers.

46. Statement by the Vice President Following the Space Symposium, December 11, 1996, Canizares Papers. See also David L. Chandler, "Scientists Map Research Strategy; Briefing Gore, They Emphasize Origins of Life, Universe," *Boston Globe*, December 12, 1996, city edition, A29.

47. The historian of science Steven J. Dick was also present at the meeting.

48. Claude Canizares, *Transcript of Opening Remarks*, Vice President's Symposium on Space Science, December 11, 1996, Canizares Papers.

49. Weinberg voiced such critique, but so did opponents of the SSC such as Anderson, who critiqued the Space Station Freedom as "scientifically unsound" and "badly mismanaged," in explicit contrast to the SSC, which he argued was neither. See First Session on the Department of Energy's Superconducting Super Collider Project, Joint Hearing before the Committee on Energy and Natural Resources and the Subcommittee on Energy and Water Development of the Committee on Appropriations, United States Senate, One Hundred Third Congress, August 4, 1993 (Washington, DC: Government Printing Office, 1993), 60.

50. Chandler, "Scientists Map Research Strategy," A29.

51. NASA, *Origins 2003*, 1.

52. Ibid., 9.

53. Ibid., 72. This text appeared both in the *Roadmap* and individually on the NASA Origins Website, no longer active.

54. Ibid., 74.

55. It also reveals other strains on Frye's characterizations of the comic and the tragic. Other than a kind of hybrid terminology that must be adopted if it were to apply Frye directly (epic or romantic heroes, mythic authorship), the limits of the power of the narrative also demand attention, the limits of the heroes within it and the breadth of the audience outside it. These limits are a part of Frye's analysis, insofar as they help establish the genre of a given text. But they are deeply embedded in the poetics here, just as, Blumenberg would argue, they are in older myths, as much to require an eye both to demonstrable power and to its obscured conceptual complement, positive powerlessness. Even the character of ancient myths was determined as much by the horizons and constraints on the power of the Gods—their struggles with themselves, their subjection to fate—as they were by any show of their direct strength. Here science and technology oscillate between dangerous and indefinite power, what forever might transcend them, and what might in fact crush them, however far science should progress.

56. For example, "By looking far out into space, and thus back in cosmic time, the HST has demonstrated with the immediacy of a simple picture how different the universe appeared when it was half its present age. The HST has made it possible for anyone to see and understand the concept of an evolving universe." "HST and Beyond" Committee, "Exploration and the Search for Origins," 2.

57. As Lillian Hoddeson and Adrienne W. Kolb have characterized it, the fate of the SSC cannot be reduced to any one factor, because it "is an epic tale with multiple sub-plots involving a myriad of actors, including physicists, engineers, contractors, officials in federal agencies, military-industrial managers, politicians and journalists"—a field of interaction subject to "subtle processes" changing the varied relationships connecting these actors. Lillian Hoddeson and Adrienne W. Kolb, "The Superconducting Super Collider's Frontier Outpost, 1983–1988," *Minerva* 38, no. 3 (2000): 271–72. Their analysis draws on the "frontier" as a culturally resonant framework in the United States, "conceptualizing the tensions between physicists and their Washington patrons in the late 1980s" and helping to understand the fateful shifts in control the authors find, from physicists to Washington managers. Ibid., 273 and 308. Two points are immediate from this argument: keeping in mind the envelope of causes and processes affecting the SSC helps analysis to avoid casting any individual one as singularly coercive and, in a point that needs more examination, the frontier and epic functioned and function as categories that are themselves resonant in multiple ways, for both scientists and historians. In diagnosing the failure of the SSC, Michael Riordan posits a "two cultures" thesis within the technosciences, laying particular emphasis on conflicts between the individual cultures of science and engineering, in particular, between high-energy physicists and that of engineers "hailing from the military industrial complex." See Michael Riordan, "A Tale of Two Cultures: Building the Superconducting Super Collider, 1988–1993," *Historical Studies in the Physical and Biological Sciences* 32, no. 1 (2001).

58. One electron volt measures the energy gained by one electron crossing a one-volt potential difference.

59. US Department of Energy, Office of Energy Research, Division of High-Energy Physics, "Appendix A: The Superconducting Supercollider Project: A Summary," *High Energy Physics Advisory Panel's Subpanel on Vision for the Future of High-Energy Physics*, May 1994, DOE/ER-0614P. For more background on the initiation of the SSC, see Adrienne Kolb and Lillian Hoddeson, "The Mirage of the 'World Accelerator for World Peace' and the Origins of the SSC, 1953-1983," *Historical Studies in the Physical and Biological Sciences* 24, no. 1 (1993): esp. 112-24.

60. The lecture was adapted into a *Science* article entitled "Origins." The citations here are to that adaptation. See Steven Weinberg, "Origins," *Science*, new series 230, no. 4721 (October 1985).

61. Ibid., 17.

62. Ibid.

63. Ibid., 18.

64. The work of Kolb and Hoddeson suggests how characteristic Weinberg's views were here, drawing perhaps on the longstanding idea of "a cooperative world-wide particle accelerator"—the "Very Big Accelerator" (VBA)—a "utopian concept, which stood in relief against the backdrop of the Cold War" involving "American, Asian, Western, and Eastern European physicists" and which "was a primary concern of leading particle physicists between 1976 and 1983." Kolb and Hoddeson, "The Mirage of the 'World Accelerator for World Peace,'" 101-2. For Kolb and Hoddeson, the SSC was the death of the VBA, because the "balance between [international] cooperation and competition in planning the 20 TeV on 20 TeV collider was upset when prominent American physicists shifted their support from the VBA to the SSC." Ibid., 124.

65. Weinberg, "Origins," 18.

66. Steven Weinberg, "Elementary Particles and The Laws of Physics: The 1986 Dirac Memorial Lectures," videotaped lecture at the Niels Bohr Library, American Institute of Physics. This line was slightly rewritten, as was most of the lecture, and published in the text Richard P. Feynman and Steven Weinberg, *Elementary Particles and the Laws of Physics: The 1986 Dirac Memorial Lectures* (Cambridge: Cambridge University Press, 1987).

67. See Kolb and Hoddeson, "World Accelerator for World Peace" and Daniel Kevles, "Preface 1995: The Death of the Superconducting Super Collider in the Life of American Physics," in Kevles, *The Physicists: The History of a Scientific Community in Modern America*, rev. ed. (New York: Knopf, 1995), esp. xix-xx.

68. Sheldon L. Glashow and Leon M. Lederman, "The SSC: A Machine for the Nineties," *Physics Today* 38, no. 3 (March 1985): 30-31. For the original text, see George Gamow, *One Two Three . . . Infinity: Facts and Speculations of Science* (1947; New York: Dover, 1974), 163.

69. Gamow, *One Two Three . . . Infinity*, 162. These are the words just before the fuller passage cited by Glashow and Lederman.

70. Glashow and Lederman, "The SSC," 32-33.

71. Martin observes the following shift: "The position Anderson articulated in 1972 still grounded his opposition to the SSC, but his arguments in the 1990s were blunter. He claimed that laws governing complex systems were independent rather than not derivable. He leaned heavily on social and technological relevance in a way he had not in 1972. Anderson's views evolved in response to developments in the organization of science." Martin, "Fundamental Disputations," 747.

72. Superconducting Super Collider: Hearing before the Subcommittee on Energy Research and Development of the Committee on Energy and Natural Resources, United States Senate, One Hundredth Congress, first session on the Department of Energy's funding request for the Superconducting Super Collider, April 7, 1987 (Washington, DC: Government Printing Office: 1988), 63.

73. P. W. Anderson, "More Is Different," *Science*, new series 177, no. 4047 (August 4, 1972): 395.

74. Superconducting Super Collider Hearing, April 7, 1987, 63-64.

75. Ibid., 68-69.

76. Alan Lightman and Roberta Brawer, *Origins: The Lives and Worlds of Modern Cosmologists* (Cambridge, MA: Harvard University Press, 1990), 466.

77. As promised, he did return to it in 1992 in *Dreams of a Final Theory: The Scientist's Search for the Ultimate Laws of Nature*: "In my 1977 book, *The First Three Minutes*, I was rash enough to remark that 'the more the universe seems comprehensible, the more it seems pointless.' I did not mean that science teaches us that the universe is pointless, but rather that the universe itself suggests no point." This was open to interpretation, but it did not entail any resignation: "I hastened to add that there were ways that we ourselves could invent a point for our lives, including trying to understand the universe. But the damage has been done: that phrase has dogged me ever since." The astronomer Gérard de Vaucouleurs "said that he thought my remark was 'nostalgic.' Indeed it was—nostalgic for a world in which the heavens declared the glory of God." Steven Weinberg, *Dreams of a Final Theory: The Scientist's Search for the Ultimate Laws of Nature* (New York: Vintage, 1993), 255-56.

78. Peter Galison, "Metaphysics and Texas," *New Republic*, September 1993, 40.

79. Ibid., 51.

80. Weinberg, *Dreams of a Final Theory*, 55. This is not to say that Weinberg promotes a naïve view that all questions can simply be reduced to these foundational ones. Indeed, he at times goes out of his way to observe that this may never be possible. See ibid., 28.

81. "It is simply a logical fallacy to go from the observation that science is a social process to the conclusion that the final product, our scientific theories, is what it is because of the social and historical forces acting in this process. A party of mountain climbers may argue over the best path to the peak, and these arguments may be conditioned by the history and social structure of the expedition, but in the end either they find a good path to the peak or they do not, and when they get there they know it. (No one would give a book about mountain climbing the title *Constructing Everest*.)" Ibid., Weinberg, *Dreams*, 188.

82. "Popper and many others who believe in an infinite chain of more and more fundamental principles might turn out to be right. But I do not think that this proposition can be argued on the grounds that no one has yet found a final theory. That would be like a nineteenth-century explorer arguing that, because all previous arctic explorations over hundreds of years had always found that however far north they penetrated there was still more sea and ice left unexplored to the north, either there was no North Pole or in any case no one would ever reach it. Some searches do come to an end." Ibid., 230-31.

83. "The intransigent metaphysics comes to the surface especially in discussions of the origin of the universe. . . . Stephen Hawking has offered what may be a better analogy; it makes sense to ask what is north of Austin or Cambridge or any other city, but it makes no sense to ask what is north of the North Pole." Ibid., 172-73.

84. "If history is any guide at all, it seems to suggest that there *is* a final theory. In this century we have seen a convergence of the arrows of explanation, like the convergence of meridians toward the North Pole. Our deepest principles, although not yet final, have become steadily more simple and economical." Ibid., 231-32.

85. Ibid., 240. Weinberg's sentiments recall the "pessimism" of the "next decimal" trope in prior reflections on the state of the physical sciences. See Lawrence Badash, "The Completeness of Nineteenth-Century Science," *Isis* 63, no. 1 (March 1972).

86. Ibid., 245. Emphasis added. More explicitly, Weinberg observes: "The lessons of religious experience can be deeply satisfying, in contrast to the abstract and impersonal worldview gained from scientific investigation. Unlike science, religious experience can suggest a meaning for our lives, a part for us to play in a great cosmic drama of sin and redemption, and it holds out to us a promise of some continuation after death. For just these reasons, the lessons of religious experience seem to me indelibly marked with the stamp of wishful thinking." Ibid., 255.

87. See Steven Weinberg, "Anthropic Bound on the Cosmological Constant," *Physical Review Letters* 59, no. 22 (1987) and Weinberg, "The Cosmological Constant Problem," *Reviews of Modern Physics* 61, no. 1 (January 1989).

88. Note that Sherrilyn Roush has defended the weak anthropic principle (WAP), which she characterizes as follows: "Our evidence about the universe (what we observe) may be restricted

by the conditions necessary for our presence as observers." She notes that "The WAP is confused with, but is not the same as, what I call the Trivial Anthropic Principle (TAP): Since human beings are actual, they must be possible. Unlike WAP, that is a tautology." Sherrilyn Roush, "Copernicus, Kant, and the Anthropic Cosmological Principles," *Studies in History and Philosophy of Modern Physics* 34 (2003): 21. Roush further quotes Weinberg as claiming in correspondence that "the WAP is common sense, while the SAP [Strong Anthropic Principle] is nonsense." Ibid., 22. For a recent discussion and assessment of the anthropic principle in the context of the history of science, see Matthew Stanley, "From Ought to Is: Physics and the Naturalistic Fallacy," *Isis* 105, no. 3 (September 2014).

89. Weinberg, "The Cosmological Constant Problem," 7.

90. Weinberg, *Dreams of a Final Theory*, 167.

91. Other connotations of the term evoke the thicker history of these scientific lives. In focusing on the Jason Division of the Institute of Defense Analysis, Kevles notes that "as the war in Vietnam escalated, some [of the Jasons], like the Harvard physicist Steven Weinberg, declined to contribute to that effort." Kevles, *The Physicists*, 402. Stevens emphasizes that Weinberg's choice is in the context of the call for "social awareness within physics" on the part of an increasing number of physicists responding to "the anti-science movement." Stevens, "Fundamental Physics and Its Justifications," 61 and 169.

92. The press gave prominence to such characterizations of the SSC, for example, "In today's floor debate, Senator Jim Sasser . . . called the 54-mile oval tunnel built under the plains of Waxahachie a bottomless money pit." Clifford Krauss, "Senate Gives the Supercollider Another Chance at Survival" *New York Times*, October 1, 1993, A21. Herman Wouk entitled his SSC-related novel *A Hole in Texas*.

93. See, for example, Weinberg, *Dreams of a Final Theory*, 272. The comparison also appears in congressional testimony, where at times both projects, the SSC and the space station, are taken as examples of "Big Science," for example, United States Senate, 139 Congressional Record (September 29, 1993), 22933. Note Arendt observed a similar judgment decades earlier: "Even today, when billions of dollars are spent year in and year out for highly 'useful' projects that are the immediate results of the development of pure, theoretical science, and when the actual power of countries and governments depends upon the performance of many thousands of researchers, the physicist is still likely to look down upon all these space scientists as mere 'plumbers.'" To this, she commented: "The sad truth of the matter, however, is that the lost contact between the world of the senses and appearances and the physical world view had been reestablished not by the pure scientist but by the 'plumber.'" Hannah Arendt, "The Conquest of Space and the Stature of Man," *American Scholar* 32 (1963): 534.

94. Kevles, *The Physicists*, xxxv. As Kevles further argues, "The vote against the SSC was thus not a vote against science or for an end to the longstanding partnership of science and government; rather it signified a redirection of the partnership's aims in line with the felt needs of post–Cold War circumstances." Ibid., xli.

95. T. H. Geballe and J. M. Rowell, "Funding the SSC," *Science*, new series 259, no. 5099 (February 26, 1993): 1237.

96. G. H. Trilling, "Support for the SSC," *Science*, new series 260, no. 5109 (May 7, 1993): 737.

97. Geballe and Rowell, "Funding the SSC," 1237.

98. If again, as Joseph Martin has observed, "dulling the subtle argument of 'More Is Different' and embroidering his stance with elements of the Weinberg Criterion," now referring to *Alvin* Weinberg's characterization of the "scientific merit" of a field in terms of its benefits to other disciplines. Martin, "Fundamental Disputations," 747. Martin emphasizes that the earlier arguments of "More Is Different" did not avail themselves of and remained independent from the "fecundity" principle (as Martin terms it) suggested by the Weinberg Criterion. Martin sees the shift in position toward fecundity and the invocation of a blunter characterization of fundamentality as not only a philosophical but pragmatic response to the widening difference of status and funding between solid state and particle physics: "Anderson, by accepting the innate view of fundamentality

more typical of the reductionist account, denied particle physics an exclusive claim to the privilege physical knowledge enjoyed, as a matter of course, over chemical, biological, or social scientific knowledge." Martin, "Fundamental Disputations," 736.

99. Superconducting Super Collider Hearing, August 4, 1993, 78-79.

100. Deborah Shapley, "The Last Behemoth," *Washington Post*, August 4, 1993, p. A17.

101. Here he explicitly invoked the "[Alvin] Weinberg Criterion": "In science policy we judge the importance of a field or discovery by quantifying the impact it has first on neighboring fields of science." By such a measure, "the SSC even if successful would rank very, very low." Superconducting Super Collider Hearing, August 4, 1993, 9-10.

102. Ibid., 10.

103. Ibid., 8.

104. Ibid., 50. The language in his prepared statement is a shade more confident: "Many of us think that if we keep tracing our explanations to deeper and deeper levels, if we keep pushing back the frontier of scientific knowledge, then sooner or later we will come to a final theory, a set of simple principles that govern everything in the universe. Certainly our theories have become steadily simpler, explaining more and more in terms of fewer and fewer fundamental principles. We can not promise that the clues that will be provided by the super collider will lead immediately to a final theory. But with the super collider we can start to move again toward our goal." Ibid., 55.

105. Ibid., 58.

106. Ibid., 71-72 for this exchange.

107. Ibid., 68-69.

108. United States Senate, 139 Congressional Record, (September 29, 1993), 22923.

109. Ibid., 22937.

110. Ibid., 22937.

111. Weinberg, *Dreams of a Final Theory*, 277.

112. Edward O. Wilson, *On Human Nature* (Cambridge, MA: Harvard University Press, 1978), 201. It was a plea with its own clear precursors, perhaps most clearly in the writings of Haldane.

113. The next chapter returns to this subject, to the novel, popular, and specific forms of the epic, and the newer problematics addressed through its poetics.

CHAPTER TWELVE

1. As preceding chapters have in part documented, public television in particular functions as a central site for understanding the public conception of contemporary scientific cosmology, both because it is formative of this opinion and because it is a collaborative platform for its articulation. There are strong continuities between public television documentaries, university lectures and lecture circuits, research programs and work, and other textual sources, historical, philosophical, and so on. The documentaries function as a synthesis of these elements. In turn, they are employed in varied intellectual, pedagogical, and political contexts. This connection is apparent in the series themselves: *Evolution* spotlighted many of the popularizers of the previous decades, the leading figures of the summits and debates of the preceding chapters, E. O. Wilson and Stephen Jay Gould among them, with the famed primatologist and conservationist Jane Goodall acting as "Overall Spokesperson" for the project and one of the series advisors. The different series added to their numbers astrophysicist and Hayden Planetarium director Neil deGrasse Tyson, as well as the experimental psychologist and cognitive scientist Stephen Pinker and the ethologist Richard Dawkins. Further evidence of the reach of television as a whole is suggested by the private, not-for-profit National Opinion Research Center (NORC). Through its General Social Survey, sponsored primarily by the National Science Foundation, NORC has surveyed scientific education in the United States since the early 1970s. For most of this period, television is cast as the public's major source of information about science. Over the cumulative period of 1972-2006, it constitutes 40 percent of what

surveyed adults name as their central source of information about science. See NORC at the University of Chicago, Dataset: General Social Surveys, 1972-2006 [Cumulative File], http://www.norc
.org/GSS+Website/Data+Analysis. Over the first decade of the millennium, as film formats shifted and opinion was increasingly crafted by and expressed through Internet sources, these numbers have begun to shift, as well, particularly among people with more extensive science educational background. But the presence of these documentaries on the Internet has also been felt. New media have not meant the end of the documentary influence. Instead, older documentaries have been revived—as with the continuing life of *Cosmos* on the Internet and in new media. According to Ann Druyan, on the day of its debut in 2008 DVD format, *Cosmos* without advertisement became the number one bestselling, nonfiction television series available on iTunes from the day of its release and for the following thirty days. Ann Druyan, interview with the author, Ithaca, NY, July 21, 2009. This popularity also extends to related materials, given the sustained attention to Sagan on the Internet.

2. See, for example, Jane Gaines on amplification through images in the radical political work of documentary: Jane M. Gaines, "The Production of Outrage: The Iraq War and the Radical Documentary Tradition," *Framework* 48, no. 2 (Fall 2007): 36-44.

3. The public television documentary is worthy of particular attention in another sense, because through its simulations, it established a tradition of visual representation for universal history. The outlook that public television produced is in many ways consonant with what is and was found in other popular films and books. As Julie Benyo, former WGBH director of pedagogical outreach, observes, television and the Web in this respect can "help you see the unseen." Julie Benyo, interview with the author, Brighton, MA, October 12, 2007. This is true also in relation to what, by the lights of its own account, could not be seen in principle: the birth of the universe, an event that could have no witnesses (in regard to which, see chapter 10, n. 68, and n. 75 below.) There has been throughout the history charted here an emphasis and desire for the visual, on simulation if not actual witnessing, specifically through the idea of film, from Eddington's turn-of-the-century articles in *Nature* to Weinberg's *The First Three Minutes*. The reconstruction of these events partly satisfies then a longing to *see*, to film, to produce visual documentation of a universal history, to actualize the metaphor of film that was used throughout the history of modern cosmology, relating the concepts and events at play within it. All these points have wider implications for the documentaries as knowledge-producing efforts. These series enact a cosmological vision animating other media, all of which collectively establish what it means for a wider community to know the universe, to adopt a new myth.

4. Thomas Levenson, interview with the author, October 18, 2006.

5. For the connection to the song, see Donald C. Johanson and Maitland Armstrong Edey, *Lucy: The Beginnings of Humankind* (New York: Simon & Schuster, 1990), 18.

6. For transcripts, see http://www.pbs.org/wgbh/nova/transcripts/2106hum1.html, http://www.pbs.org/wgbh/nova/transcripts/2107hum2.html, and http://www.pbs.org/wgbh/nova/tran scripts/2108hum3.html.

7. Julie Benyo, interview with the author. Hutton himself presented the matter as one of personal fascination and circumstance. Patricia Brennan, "'EVOLUTION,' Exploring Darwin's Theory," *Washington Post*, September 23, 2001, Y06. Paula Apsell, executive producer of NOVA, shared an interest in a film on evolution, but the controversial subject matter gave the producers pause. When the series was broadcast, Hutton explained that "because of the controversy" between creationism and Darwinian evolution, "everybody knew that [*Evolution*] would be hard to fund ... It was one of those ideas that everybody knew should be done, but nobody quite knew how to do it." Ibid.

8. Ronald Numbers, *The Creationists: From Scientific Creationism to Intelligent Design* (Cambridge, MA: Harvard University Press, 2006). Note that Matthew Stanley traces the conceptual grounds for the debate over ID to the construction of an opposition between what he terms "theistic science" (the "close embrace" of science and Christianity) and scientific naturalism, the rise

of the latter eclipsing the former, producing the conditions according to which disputants over ID generally agree that science is inherently naturalistic. Matthew Stanley, *Huxley's Church and Maxwell's Demon: From Theistic Science to Naturalistic Science* (Chicago: University of Chicago Press, 2015), esp. Introduction, chapter 7, "How the Naturalists 'Won,'" and Conclusion.

9. Numbers, *The Creationists*, 380.

10. Ibid., 381.

11. Michael Behe, *Darwin's Black Box: The Biochemical Challenge to Evolution* (New York: Free Press, 1996), 39.

12. Julie Benyo, interview with the author.

13. Hutton was explicit on this point: "'We think what we're doing is helping in a sense to defuse the issue,' said Hutton, who started working on 'Evolution' less than five months before the Kansas decision. 'Not only do we define evolution as to what it is, but we also say what it isn't. It isn't about religion. It doesn't mean that you can't believe in God. It doesn't have any impact on those sorts of beliefs at all because that's not what it's about.'" Brennan, "'EVOLUTION,' Exploring Darwin's Theory," Y06.

14. Larry Witham, "'PBS's' Evolution Generates a Debate," *Washington Times*, September 27, 2001, D1. The use of Zogby by the Discovery Institute itself prompted response and criticism. See Chris Mooney, "Polling for ID," originally posted on the Web site of the Committee for Skeptical Inquiry, September 11, 2003, and now archived here: https://web.archive.org/web/20030919021746 /http://www.csicop.org/doubtandabout/polling/.

15. Jonathan Wells, "Evolution for the Masses," *Washington Times*, September 23, 2001, B3.

16. The Discovery Institute, *Getting the Facts Straight: A Viewer's Guide to PBS's* Evolution (Seattle: Discovery Institute Press, 2001), 9.

17. Ibid., 10.

18. Julie Benyo, interview with the author.

19. Ibid.

20. Public Broadcasting Service, *Evolution*, available at http://www.pbs.org/wgbh/evolution.

21. Of this effort, Hutton claimed that "it's the first interactive, comprehensive Web site of its kind. But you can also go deeper through the Web site portals. And the teacher-training materials are most important. We've created an eight-week on-line continuing education seminar." Brennan, "'EVOLUTION,' Exploring Darwin's Theory," Y06. As these materials suggest, the documentaries were no longer produced primarily for broadcast. In a format itself at least as old as Kenneth Clarke's BBC *Civilization* series, public television documentary series were produced with accompanying books. But for these new documentaries, the accompanying Web sites, with their extensive pedagogic materials produced in tandem with the series (referred to in the episodes themselves), extended the active use of the documentary materials well beyond the broadcast, a continuing life that promoted their scientific accounts.

22. Figures supplied by Julie Benyo.

23. Richard Hutton, forward to Carl Zimmer, *Evolution: The Triumph of An Idea* (New York: Harper Perennial, 2002), xv.

24. Ibid., xvii.

25. As with the varied uses of "myth" (e.g., chapter 8), the language of story in this context merits being taken as seriously as possible, a language invoked directly and repeatedly in the episodes, as well as to recommend the series and to critique it. It is used to recommend and construct the contemporary account that the series reflected, a story that is an argument for the body of knowledge it forms. It is also used by the detractors, as seen above, as a way to dismiss it, as a "mere" story, much in the way that the word "theory" was understood differently by the different parties to the evolution-creationism debate. To the film producers, "story" was either politically and epistemologically neutral or it was affirming—science produces true stories, true histories. To ID critics and young-earth proponents, these terms rendered the knowledge claims of the documentary suspect, as mere theory, mere story.

26. Hutton, foreword to Zimmer, *Evolution*, xvii.

27. Julie Benyo, interview with the author.

28. A review in *BioScience*, the journal of the American Institute of Biological Sciences, criticized the series on precisely this point: "What about God?" "is also the segment that I found the most troubling, troubling not because of what it presents, but because of what it does not present. . . . The real challenge to teaching evolution today comes from members of the 'intelligent design' community, who argue that their brand of creationism is really 'creation science' and who have been fighting to get this very, very bad understanding of science into the curriculum." Wayne W. Carley, "One for the Faithful: A Review of the Television Series 'Evolution,'" *BioScience* 52, no. 4 (April 2002): 385.

29. "Equally important, NOVA shows the human story behind the science story. Whether exploring a galaxy or an atom, the series delves into the personalities responsible for the discoveries, and the social consequences of events in the lab." "Nova's Approach" in "About Nova," formerly available at http://www.pbs.org/wgbh/nova/about/appr.html, and archived here: https://web.archive.org/web/20021020035559/http://www.pbs.org/wgbh/nova/about/appr.html.

30. Discovery Institute, *Getting the Facts Straight*, 13.

31. Ibid., 14.

32. The distinctions here between "pivotal" and "lesser" are simpler, more straightforward recastings of Seymour Chatman's "kernel" and "satellite" events. See Seymour Chatman, *Story and Discourse: Narrative Structure in Fiction and Film* (Ithaca: Cornell University Press, 1978), 53-54.

33. "K-T" refers to the "Cretaceous-Tertiary" event.

34. Hutton, foreword to Zimmer, *Evolution*, xvi.

35. Hutton comments that the second episode, "Great Transformations," was a particular challenge "because it covers 3.8 billion years of life in 55 minutes. I was always worried as to whether that show would work. But it does. It's wonderful, it's brilliant. I love it." Brennan, "'EVOLUTION,' Exploring Darwin's Theory," Y06.

36. "Will we ever truly comprehend such immensity of time? Probably not. But to develop a better understanding of evolutionary change in its proper historical context, we must try. This timeline, and the events portrayed along it, provided a framework for doing so." Public Broadcasting Service, *Evolution*, http://www.pbs.org/wgbh/evolution/change/deeptime/index.html.

37. This distinction is again a slight modification and simplification of Chatman's terms of "story time" and "discourse time." Chatman, *Story and Discourse*, 62-63.

38. Public Broadcasting Service, *Evolution*, "What About God?," http://www.pbs.org/wgbh/evolution/religion/index.html.

39. "Bipartite" is how Karl Guthke terms the structure of this narrative. Karl S. Guthke, *The Last Frontier: Imagining Other Worlds from the Copernican Revolution to Modern Science Fiction* (Ithaca: Cornell University Press, 1990).

40. Thomas Levenson, interview with the author, Cambridge, MA, October 18, 2006.

41. Ibid.

42. Ann Druyan dates Tyson's connection with Sagan from a letter Tyson sent Sagan while Tyson was still a teenager. Sagan invited Tyson to Ithaca, and the two met to discuss Tyson's future, though Tyson ultimately chose to attend Harvard as an undergraduate, rather than Cornell. Ann Druyan, interview with the author, July 22, 2009. See also Tyson's reminiscences, which also date his meeting Sagan to his adolescent years, if with slight differences in the account, formerly available at the Hayden Planetarium Web site (at http://www.haydenplanetarium.org/tyson/read/speeches/carlsaganeulogy), now archived at https://web.archive.org/web/20090625175413/http://www.haydenplanetarium.org/tyson/read/speeches/carlsaganeulogy. This connection is also drawn upon in the new *Cosmos* of 2014.

43. Neil deGrasse Tyson, quoted in Leslie Mullen, "The Origins Umbrella," *Astrobiology Magazine*, January 26, 2005, available at http://www.astrobio.net/interview/1414/sitemap.php.

44. Ibid.

45. Thomas Levenson, interview with the author, February 1, 2008.

46. Ibid.

47. Hence, "Well, it turns out, Earth became a habitable planet only after a series of devastating disasters in its early years," among other similar formulations. Public Broadcasting service, NOVA, *Origins*, "Origins: Earth is Born," http://www.pbs.org/wgbh/nova/space/origins-series-overview .html#origins-earth-born.

48. A similar observation is made in a slightly different way in the third episode, in reference to the K-T event. Tyson asks: "But what if you turned back the clock? What if that asteroid had taken a slightly different course and missed Earth completely? Little mammals may never have gotten their chance because the dinosaurs could still be in charge today. And instead of me, one of them would be hosting this show!"

49. Again, Tyson: "Some scientists say that the key to our evolution was Earth's long and relatively peaceful history."

50. The episode presents the theory as Bob Dicke's, and, in Wilkinson's words, a theory "so original" as to undercut any threat that their own work to find the background radiation would be scooped.

51. Thomas Levenson, interview with the author, Cambridge, MA, October 18, 2006.

52. Ibid.

53. Ibid.

54. Thomas Levenson, interview with the author, Cambridge, MA, February 1, 2008.

55. Thomas Levenson, interview with the author, Cambridge, MA, October 18, 2006.

56. The *New York Times* review, in particular, makes the comparison more a matter of function than presence, making somewhat awkward comparisons in its own turn: "A quarter-century after Carl Sagan's 'Cosmos' series, television's astro-impresario is Neil deGrasse Tyson, director of the Hayden Planetarium. 'Origins' is meant to send the young astrophysicist into orbit as a space-savvy celebrity, with a companion book that Dr. Tyson wrote with Donald Goldsmith on sale this week. Dr. Tyson is like a higher-pitched James Earl Jones, with that actor's full-size ego and his own rubbery kid-friendly body language." Ned Martel, "Mysteries of Life, Time and Space (and Green Slime)," *New York Times*, E5. Tyson himself comments on the connections in reception between Sagan and *Origins* in Mullen, "The Origins Umbrella." Note that the series as a whole was a finalist for the International Documentary Association Distinguished Documentary Achievement Award in the Limited Series category, and the final hour of the series won the National Academies Communication Award.

57. I do not have independent confirmation of audience size.

58. Jason Lisle, "Preliminary Comments on the PBS-TV Series 'Origins,'" available at http:// www.answersingenesis.org/docs2004/0929PBSOrigins.asp.

59. See for example Lisle's video introduction to the Creation Museum's Stargazer's Planetarium: "A lot of folks have gone to planetariums and you get some good science but you also get s ome evolutionary storytelling. You hear about the Big Bang, and the billions of years and how the universe popped into existence all by itself and the stars formed all by themselves." See n. 67 below.

60. Jason Lisle, "Preliminary Comments."

61. Ibid.

62. NOVA *Origins*, http://www.pbs.org/wgbh/nova/origins.

63. NOVA Teachers Help Center, http://www.pbs.org/wgbh/nova/teachers/faq.html#q05. Apart from *Evolution*'s extensive program to contact and distribute *Evolution*-related materials to every high school in the United States, Benyo also related the effort to prepare an "executive briefing packet" for the NOVA film *Judgment Day: Intelligent Design on Trial* to "every high school principal, to every school board president and to every superintendent in the country." Julie Benyo, interview with the author. NOVA vigorously promotes its materials to schools across the country.

64. NOVA "Origins: Earth Is Born," NOVA Teacher's Guide, http://www.pbs.org/wgbh/nova /education/programs/3111_origins.html.

65. Ibid.

66. What Erich Auerbach says of scripture might also be said then of the final story put forward in these documentaries, as matter not of literary register but of content: "The world of the Scripture stories is not satisfied with claiming to be a historically true reality—it insists that it is the only real world, is destined for autocracy. All other scenes, issues, and ordinances have no right to appear independently of it, and it is promised that all of them, the history of all mankind, will be given their due place within its frame, will be subordinated to it. The Scripture stories do not, like Homer's, court our favor, they do not flatter us that they may please us—they seek to subject us, and if we refuse to be subjected, we are rebels." Erich Auerbach, *Mimesis: The Presentation of Reality in Western Literature* (1946; Princeton: Princeton University Press, 2003), 14–15.

67. The video is archived at https://web.archive.org/web/20090828075417/http://creationmuseum.org/whats-here/exhibits/planetarium.

68. They were also not unique to the period. Relating Darwin's reaction to James Croll's work particularly in light of Thomson's arguments, Joe Burchfield points to the following analogy, particularly impressing Darwin with the meaning of time spans on the order of millions of years: "Croll suggested that an actual experiment be done by hanging a narrow strip of paper 83 ft 4 in. long around a large room. The full length of the tape would then represent 1,000,000 years, while 100 years—the limit of time that he felt could actually be conceived of in human experience— would be represented by a mark only 1/10in. from the end. In this way he believed the inconceivable vastness of 1,000,000 years, let alone Kelvin's 100,000,000 years, could be appreciated if not fully grasped." Joe D. Burchfield, "Darwin and the Dilemma of Geological Time," *Isis* 65, no. 3 (September 1974): 310. Even before these exercises became textual and filmic tropes, Croll's model in Burchfield's reading already suggests the difficulty in using them to illustrate the limits of understanding without at the same time undercutting that illustrative use in appealing to the models as an imaginative mode for grasping those time spans.

69. This metaphorical structure is perhaps beneath the surface in the series in the treatment of the seeding of life on earth through comets, which makes of the earth an egg to be fertilized. And of course, the more crass connotations of the "Big Bang" itself as the source of all creation connect back to Sagan's speculations in *The Dragons of Eden* as to the links between imagining human birth and universal birth.

70. Here Frye comes closer together with Auerbach in their characterizations of myth: "Myths stick together because of the cultural forces impelling them to do so: these forces are not primarily literary, and mythologies are mainly accepted as structures of belief or social concern rather than imagination," in Northrop Frye, *The Secular Scripture: A Study of the Structure of Romance* (Cambridge, MA: Harvard University Press, 1976), 18. Note that Frye uses the more heroic characterization in the *Anatomy of Criticism* and does not seek to relate them directly.

71. The fourth of the *Origins* series, for example, shows two different teams competing in the attempt to get refined images of the microwave background radiation (which itself would demonstrate something of the early inhomogeneity of the universe). But here the Holy Grail remains the same for both teams: better images. As far as the episode is concerned, there is complete agreement as to what constitutes a better image and what a better image means for understanding of the history of the universe.

72. For Tyson's views on these points, see NOVA, "A Conversation with Neil Tyson," September 28, 2004, http://www.pbs.org/wgbh/nova/origins/tyson.html, and Mullen, "The Origins Umbrella."

73. Claude Canizares, interview with the author, Cambridge, MA, November 17, 2008.

74. This recalls Mikhail Bakhtin's formulations of epic, in particular, that so far as the conventional body of literature is concerned, the epic is "completed," is in fact "already antiquated." Indeed what Bakhtin characterizes as the three constitutive features of the epic—that its subject is a national epic past, that national tradition serves as its source, and that an absolute epic distance separates it from the quotidian—all have a strong resonance with the scientific rather than the

literary epic at stake in the materials here. See M. M. Bakhtin, "Epic and Novel," in *The Dialogic Imagination: Four Essays by M. M. Bakhtin*, ed. Michael Holquist, trans. Caryl Emerson and Michael Holquist (Austin: University of Texas Press, 1981). The documentary epiphany also recalls Teilhard de Chardin's disclaimer regarding his "film." See Pierre Teilhard de Chardin, *The Phenomenon of Man*, trans. Bernard Wall (New York: Harper, 1959), 35, discussed here in chapter 10, n. 68.

75. Despite the extensive differences between Genesis and the modern scientific cosmology as represented in the series, there is also a particular canniness in the explicit comparison that the critics of *Origins* and *Evolution* draw between biblical authors and modern cosmologists. The idea of Moses inscribing his own chronicle and death into his writing of the first five biblical books—a sometime theological, if not historical view—places him in a similar position of timeless epiphany. This position can be clarified by comparing and contrasting it to the notion of "auto-historiography," the attempt to locate oneself in a universal history in order to provide life with meaning. The primary objective of the scientists as myth writers is the crafting of a final story that only *implicitly* includes their own lives. For a discussion of the concept of auto-historiography, see Ilya Kliger, "Auto-historiography: Genre, Trope and Modes of Emplotment in Alexander and Natalie Herzen's Narratives of the Family Drama," *Russian Literature* 61 (Winter 2007).

76. Jonathan Wells, "Public Schools Still Using PBS's Evolution," *Evolution News and Views*, June 16, 2007, http://www.evolutionnews.org/2007/06/public_schools_still_using_pbs003770.html.

77. Julie Benyo, interview with the author.

78. Carley, "One for the Faithful: A Review of the Television Series 'Evolution,'" 384.

79. The series itself depicts the first meeting between Sagan and Tyson, a passing of a mantle.

80. See chapter 10.

81. By consonance and contradistinction, the uneasy difference between description and explanation has at times emphasized the comparative brevity of explanation.

82. See chapter 10, n. 68 for relevant literature.

83. Adelaide Mena, "Science Historians Critique New 'Cosmos' Series," *Catholic News Agency*, http://www.catholicnewsagency.com/news/historians-of-science-critique-new-cosmos-series/.

84. Animation, it is perhaps worth emphasizing, has a long history of varied use and the attendant theorization of it, and its contemporary use in film and media need not function in ways consistent with mainstream or televisual prime-time connotations.

85. Max Horkheimer had accused earlier synthetic visions of science as reflecting bourgeois truths and reinforcing uncritical modes of scientific research. Max Horkheimer, "The Latest Attack against Metaphysics," in *Critical Theory: Selected Essays* (New York: Continuum, 2002). He might have regarded such universal history as a kind of pervasive bourgeois myth that Roland Barthes found permeating his society. The new series might court this criticism by way of fashion: the mirroring, sleek, silver pin that is Tyson's new "Ship of the Imagination" threads through the cosmos, Tyson standing at its spherical eye as a kind of executive officer navigating from a moving glass-framed office. Views are panoramic and in motion; music, intonation, and image are often in crescendo, punctuated by the transition to the next advertisement.

86. Ursula Goodenough, *The Sacred Depths of Nature* (New York: Oxford University Press, 1998). Goodenough was herself in conversation with the Institute on Religion in an Age of Science (ibid., xi–xii), and also reflects on and attempts to address Weinberg's "poignant nihilism" (ibid., 10). She situates her project in relation to Snow's language of "two cultures" (ibid., xx).

CODA

1. Georg Lukács, *The Theory of the Novel: A Historico-philosophical Essay on the Forms of Great Epic Literature*, trans. Anna Bostock (Cambridge, MA: MIT Press, 1971), 49. Lukács notes that he

first drafted the book at the outbreak of World War I and that it was published in journal form in 1916, before being published as a book in 1920.

2. Ibid., 38.

3. Edward O. Wilson, *On Human Nature* (Cambridge, MA: Harvard University Press, 1978), 202.

4. Ibid.

5. Steven Weinberg, *Dreams of a Final Theory: The Scientist's Search for the Ultimate Laws of Nature* (New York: Vintage, 1993), 242.

6. Weinberg, *Dreams of a Final Theory*, 32.

7. Ibid.

8. Ibid., 37. To this, he added the important further qualification: "The separation of law and history is a delicate business, one we are continually learning how to do as we go along." Ibid., 38. So it remains possible both that what appears to be universal will in fact be historical and what appears to be historical could ultimately be deduced from universals.

9. Recently however, Weinberg has noted that "I'm not as sure as I once was about the future of quantum mechanics. It is a bad sign that those physicists today who are most comfortable with quantum mechanics do not agree with one another about what it all means." For Weinberg's discussion and the exchange following it, see Steven Weinberg, "The Trouble with Quantum Mechanics," *New York Review of Books* (January 19, 2017) and N. David Mermin, Jeremy Bernstein, Michael Nauenberg, Jean Bricmont, and Sheldon Goldstein, et al., "Steven Weinberg and the Puzzle of Quantum Mechanics," *New York Review of Books* (April 6, 2017).

10. John F. W. Herschel, *A Preliminary Discourse on the Study of Natural Philosophy* (London: Printed for Longman, Rees, Orme, Brown, and Green and John Taylor, 1831), 18.

11. Lukács, *The Theory of the Novel*, 49.

12. See Dipesh Chakrabarty, "The Climate of History: Four Theses," *Critical Inquiry* 35, no. 2 (Winter 2009): 221. Chakrabarty has elaborated on the challenge he sees climate change posing to postcolonial studies and the contemporary disciplinary cartography dividing broad approaches within history: "Postcolonial Studies and the Challenge of Climate Change," *New Literary History* 43, no. 1 (2012) and "Climate and Capital: On Conjoined Histories," *Critical Inquiry* 41, no.1 (2014).

13. Chakrabarty, "The Climate of History," 213.

14. Ibid., 220-22.

15. For these terms, see for example David Christian, "The Case for 'Big History,'" *Journal of World History* 2, no. 2 (Fall 1991); Edmund Russell, "Evolutionary History: Prospectus for a New Field," *Environmental History* 8, no. 2 (April 2003); and Daniel Lord Smail, "In the Grip of Sacred History," *American Historical Review* 110, no. 5 (December 2005). I have reviewed this historiography elsewhere, in Nasser Zakariya, "Is History Still a Fraud?" *Historical Studies in the Natural Sciences* 43, no. 5 (2013).

16. David Christian, "The Return of Universal History," *History and Theory* 49, no. 4 (2010). See also Kerwin Lee Klein, "In Search of Narrative Mastery: Postmodernism and the People without History?" *History and Theory* 34, no. 4 (1994).

17. For different visions of nation in relation to space exploration and historiography, see Asif A. Sidiqqi, "Competing Technologies, National(ist) Narratives, and Universal Claims: Toward a Global History of Space Exploration," *Technology and Culture* 51, no. 2 (April 2010).

18. If any field at all was given pride of place, it was the new field of astrobiology.

Index